Functional Inequalities: New Perspectives and New Applications

Mathematical
Surveys
and
Monographs

Volume 187

Functional Inequalities: New Perspectives and New Applications

Nassif Ghoussoub
Amir Moradifam

American Mathematical Society
Providence, Rhode Island

EDITORIAL COMMITTEE

Ralph L. Cohen, Chair Benjamin Sudakov
Michael A. Singer Michael I. Weinstein

2010 *Mathematics Subject Classification.* Primary 42B25, 35A23, 26D10, 35A15, 46E35.

For additional information and updates on this book, visit
www.ams.org/bookpages/surv-187

Library of Congress Cataloging-in-Publication Data

Library of Congress Cataloging-in-Publication Data has been applied for by the AMS. See www.loc.gov/publish/cip.

Copying and reprinting. Individual readers of this publication, and nonprofit libraries acting for them, are permitted to make fair use of the material, such as to copy a chapter for use in teaching or research. Permission is granted to quote brief passages from this publication in reviews, provided the customary acknowledgment of the source is given.

Republication, systematic copying, or multiple reproduction of any material in this publication is permitted only under license from the American Mathematical Society. Requests for such permission should be addressed to the Acquisitions Department, American Mathematical Society, 201 Charles Street, Providence, Rhode Island 02904-2294 USA. Requests can also be made by e-mail to reprint-permission@ams.org.

© 2013 by the American Mathematical Society. All rights reserved.
The American Mathematical Society retains all rights
except those granted to the United States Government.
Printed in the United States of America.

∞ The paper used in this book is acid-free and falls within the guidelines
established to ensure permanence and durability.
Visit the AMS home page at http://www.ams.org/

10 9 8 7 6 5 4 3 2 1   18 17 16 15 14 13

To my son Joseph Ghoussoub
To my parents Saed and Iran Moradifam

Contents

Preface	xi
Introduction	xiii
Part 1. Hardy Type Inequalities	**1**
Chapter 1. Bessel Pairs and Sturm's Oscillation Theory	3
1.1. The class of Hardy improving potentials	3
1.2. Sturm theory and integral criteria for HI-potentials	9
1.3. The class of Bessel pairs	14
1.4. Further comments	17
Chapter 2. The Classical Hardy Inequality and Its Improvements	19
2.1. One dimensional Poincaré inequalities	19
2.2. HI-potentials and improved Hardy inequalities on balls	21
2.3. Improved Hardy inequalities on domains with 0 in their interior	24
2.4. Attainability of the best Hardy constant on domains with 0 in their interior	26
2.5. Further comments	28
Chapter 3. Improved Hardy Inequality with Boundary Singularity	31
3.1. Improved Hardy inequalities on conical domains with vertex at 0	31
3.2. Attainability of the Hardy constants on domains having 0 on the boundary	34
3.3. Best Hardy constant for domains contained in a half-space	38
3.4. The Poisson equation on the punctured disc	41
3.5. Further comments	42
Chapter 4. Weighted Hardy Inequalities	45
4.1. Bessel pairs and weighted Hardy inequalities	45
4.2. Improved weighted Hardy-type inequalities on bounded domains	49
4.3. Weighted Hardy-type inequalities on $\mathbb{R}^n$	52
4.4. Hardy inequalities for functions in $H^1(\Omega)$	54
4.5. Further comments	57
Chapter 5. The Hardy Inequality and Second Order Nonlinear Eigenvalue Problems	59
5.1. Second order nonlinear eigenvalue problems	59
5.2. The role of dimensions in the regularity of extremal solutions	61
5.3. Asymptotic behavior of stable solutions near the extremals	62
5.4. The bifurcation diagram for small parameters	65

5.5. Further comments 67

Part 2. Hardy-Rellich Type Inequalities 69

Chapter 6. Improved Hardy-Rellich Inequalities on $H_0^2(\Omega)$ 71
  6.1. General Hardy-Rellich inequalities for radial functions 71
  6.2. General Hardy-Rellich inequalities for non-radial functions 74
  6.3. Optimal Hardy-Rellich inequalities with power weights $|x|^m$ 78
  6.4. Higher order Rellich inequalities 83
  6.5. Calculations of best constants 85
  6.6. Further comments 90

Chapter 7. Weighted Hardy-Rellich Inequalities on $H^2(\Omega) \cap H_0^1(\Omega)$ 93
  7.1. Inequalities between Hessian and Dirichlet energies on $H^2(\Omega) \cap H_0^1(\Omega)$ 93
  7.2. Hardy-Rellich inequalities on $H^2(\Omega) \cap H_0^1(\Omega)$ 101
  7.3. Further comments 107

Chapter 8. Critical Dimensions for 4th Order Nonlinear Eigenvalue Problems 109
  8.1. Fourth order nonlinear eigenvalue problems 109
  8.2. A Dirichlet boundary value problem with an exponential nonlinearity 110
  8.3. A Dirichlet boundary value problem with a MEMS nonlinearity 113
  8.4. A Navier boundary value problem with a MEMS nonlinearity 118
  8.5. Further comments 121

Part 3. Hardy Inequalities for General Elliptic Operators 123

Chapter 9. General Hardy Inequalities 125
  9.1. A general inequality involving interior and boundary weights 125
  9.2. Best pair of constants and eigenvalue estimates 132
  9.3. Weighted Hardy inequalities for general elliptic operators 134
  9.4. Non-quadratic general Hardy inequalities for elliptic operators 137
  9.5. Further comments 141

Chapter 10. Improved Hardy Inequalities For General Elliptic Operators 143
  10.1. General Hardy inequalities with improvements 143
  10.2. Characterization of improving potentials via ODE methods 147
  10.3. Hardy inequalities on $H^1(\Omega)$ 151
  10.4. Hardy inequalities for exterior and annular domains 154
  10.5. Further comments 156

Chapter 11. Regularity and Stability of Solutions in Non-Self-Adjoint Problems 157
  11.1. Variational formulation of stability for non-self-adjoint eigenvalue problems 157
  11.2. Regularity of semi-stable solutions in non-self-adjoint boundary value problems 159
  11.3. Liouville type theorems for general equations in divergence form 161
  11.4. Further remarks 167

Part 4. Mass Transport and Optimal Geometric Inequalities 169

Chapter 12.	A General Comparison Principle for Interacting Gases	171
12.1.	Mass transport with quadratic cost	171
12.2.	A comparison principle between configurations of interacting gases	173
12.3.	Further comments	179
Chapter 13.	Optimal Euclidean Sobolev Inequalities	181
13.1.	A general Sobolev inequality	181
13.2.	Sobolev and Gagliardo-Nirenberg inequalities	182
13.3.	Euclidean Log-Sobolev inequalities	183
13.4.	A remarkable duality	185
13.5.	Further remarks and comments	189
Chapter 14.	Geometric Inequalities	191
14.1.	Quadratic case of the comparison principle and the HWBI inequality	191
14.2.	Gaussian inequalities	193
14.3.	Trends to equilibrium in Fokker-Planck equations	196
14.4.	Further comments	197

Part 5. Hardy-Rellich-Sobolev Inequalities — 199

Chapter 15.	The Hardy-Sobolev Inequalities	201
15.1.	Interpolating between Hardy's and Sobolev inequalities	201
15.2.	Best constants and extremals when 0 is in the interior of the domain	203
15.3.	Symmetry of the extremals on half-space	206
15.4.	The Sobolev-Hardy-Rellich inequalities	208
15.5.	Further comments and remarks	211
Chapter 16.	Domain Curvature and Best Constants in the Hardy-Sobolev Inequalities	213
16.1.	From the subcritical to the critical case in the Hardy-Sobolev inequalities	213
16.2.	Preliminary blow-up analysis	219
16.3.	Refined blow-up analysis and strong pointwise estimates	227
16.4.	Pohozaev identity and proof of attainability	236
16.5.	Appendix: Regularity of weak solutions	240
16.6.	Further comments	243

Part 6. Aubin-Moser-Onofri Inequalities — 245

Chapter 17.	Log-Sobolev Inequalities on the Real Line	247
17.1.	One-dimensional version of the Moser-Aubin inequality	247
17.2.	The Euler-Lagrange equation and the case $\alpha \geq \frac{2}{3}$	250
17.3.	The optimal bound in the one-dimensional Aubin-Moser-Onofri inequality	252
17.4.	Ghigi's inequality for convex bounded functions on the line	258
17.5.	Further comments	262
Chapter 18.	Trudinger-Moser-Onofri Inequality on $\mathbb{S}^2$	263
18.1.	The Trudinger-Moser inequality on $\mathbb{S}^2$	263
18.2.	The optimal Moser-Onofri inequality	267

18.3.	Conformal invariance of J_1 and its applications	270
18.4.	Further comments	272

Chapter 19. Optimal Aubin-Moser-Onofri Inequality on $\mathbb{S}^2$ 275
- 19.1. The Aubin inequality   275
- 19.2. Towards an optimal Aubin-Moser-Onofri inequality on $\mathbb{S}^2$ 277
- 19.3. Bol's isoperimetric inequality   283
- 19.4. Further comments   287

Bibliography   289

Preface

This book is not meant to be another compendium of select inequalities, nor does it claim to contain the latest or the slickest ways of proving them. It is rather an attempt at describing how most functional inequalities are not merely the byproduct of ingenious guess work by a few wizards among us, but are often manifestations of natural mathematical structures and physical phenomena. Our main goal here is to show how this point of view leads to "systematic" approaches for proving the most basic functional inequalities, but also for understanding and improving them, and for devising new ones - sometimes at will, and often on demand.

Our aim is therefore to describe how a few general principles are behind the validity of large classes – and often "equivalence classes" – of functional inequalities, old and new. As such, Hardy and Hardy-Rellich type inequalities involving radially symmetric weights are variational manifestations of Sturm's theory on the oscillatory behavior of certain ordinary differential equations. Similarly, allowable non-radial weights in Hardy-type inequalities for more general uniformly elliptic operators are closely related to the resolution of certain linear PDEs in divergence form with either a prescribed boundary condition or with prescribed singularities in the interior of the domain.

On the other hand, most geometric inequalities including those of Sobolev and Log-Sobolev type, are simply expressions of the convexity of certain free energy functionals along the geodesics of the space of probability measures equipped with the optimal mass transport (Wasserstein) metric. Hardy-Sobolev and Hardy-Rellich-Sobolev type inequalities are then obtained by interpolating the above two classes of inequalities via the classical ones of Hölder.

Besides leading to new and improved inequalities, these general principles offer novel ways for estimating their best constants, and for deciding whether they are attained or not in the appropriate function space. In the improved versions of Hardy-type inequalities, the best constants are related to the largest parameters for which certain linear ODEs have non-oscillatory solutions. Duality methods, which naturally appear in the new "geodesic convexity" approach to geometric inequalities, allow for the evaluation of the best constants from first order equations via the limiting case of Legendre-Fenchel duality, as opposed to the standard method of solving second order Euler-Lagrange equations.

Whether a "best constant" on specific domains is attained or not, is often dependent on how it compares to related best constants on limiting domains, such as the whole space or on half-space. These results are based on delicate blow-up analysis, and are reminiscent of the prescribed curvature problems initiated by Yamabe and Nirenberg. The exceptional case of the Sobolev inequalities in two dimensions initiated by Trudinger and Moser can also be linked to mass transport methods, and some of their recent improvements by Onofri, Aubin and others are both interesting and still challenging. They will be described in the last part of the monograph.

The parts dealing with Hardy and Hardy-type inequalities represent a compendium of an approach mostly developed by –and sometimes with– my (now former) students Amir Moradifam and Craig Cowan. The weighted second order inequalities have some overlap with the books of B. Opic and A. Kufner and the one by A. Kufner and L-E Persson, which have been the standard references on the subject. The part on fourth order inequalities reflect more recent developments.

The "mass transport" approach to geometric inequalities follows closely work with my former student X. Kang and postdoctoral fellow Martial Agueh. It is largely based on the pioneering work of Cedric Villani, Felix Otto, Robert McCann, Wilfrid Gangbo, Dario Cordero-Erausquin, Bruno Nazareth, Christian Houdré and many others. Unfortunately, we do not include here another related "entropy-energy" approach to functional inequalities that is closely intertwined with the study of large time asymptotics of evolution equations, typically diffusive or hypocoercive kinetic equations. This approach has been developed and used extensively by A. Arnold, J. Carillo, L. Desvillettes, J. Dolbeault, A. Jungel, P. Markowich, G. Toscani, A. Unterreiter and C. Villani, to name a few.

The chapters dealing with Hardy-Sobolev type inequalities follow work done with my students Chaogui Yuan, and Xiaosong Kang, as well as my collaborator Frederic Robert. Finally, much of the progress on the –still unresolved– best constant in Moser-Onofri-Aubin inequalities on the 2-dimensional sphere was done with my friends and collaborators, Joel Feldman, Richard Froese, Changfeng Gui, and Chang-Shou Lin. I owe all these people a great deal of gratitude.

Nassif Ghoussoub, Vancouver, August 2012

Introduction

"Functional inequalities" are often manifestations of natural physical phenomena as they often express very general laws of nature formulated in physics, biology, economics and various aspects of engineering. They also form the basis of fundamental mathematical structures such as the calculus of variations, which has over and over again proved itself to be one of the most powerful and far-reaching tools available for advancing our understanding of mathematics and much of its applications.

One can then try to venture a statement to the effect that functional inequalities essentially reflect the fact that there are all kinds of *"energies"* out there that are somewhat controlled by *"entropy"*, the latter being royally represented by the Dirichlet and Hessian integrals and their variations. The challenge becomes then to identify which energies they are and how to mathematically substantiate such claims. This book is an attempt to describe how a select few general but basic principles can naturally point to the validity of large classes of functional inequalities, and often lead to systematic ways of proving them. It consists of six parts, which –though interrelated– are meant to reflect the unified mathematical structure behind various collections of inequalities.

We will only be presenting in this book inequalities involving entropies associated with the Laplacian $-\Delta$ and the bi-Laplacian Δ^2. We shall not cover the equally interesting "relativistic theory" corresponding to $\sqrt{-\Delta}$ and other fractional powers of the Laplacian. For an outline of that theory and its applications to the stability of matter problem, we refer the interested reader to the recent lecture notes of R. Seiringer [**246**].

In Part I, we deal with Hardy-type inequalities involving radially symmetric weights and their improvements. The classical Hardy inequality asserts that for $n \geq 3$,

$$(0.1) \qquad \int_{\mathbb{R}^n} |\nabla u|^2 dx \geq (\tfrac{n-2}{2})^2 \int_{\mathbb{R}^n} \tfrac{u^2}{|x|^2} dx \quad \text{for } u \in \mathcal{D}^{1,2}(\mathbb{R}^n),$$

which imply that for any potential V on $\mathbb{R}^n$ such that $V(x) \geq -\tfrac{(n-2)^2}{4|x|^2}$, the ground state energy (i.e., the bottom of the spectrum) of the corresponding Schrödinger operator $-\Delta + V$ is finite. This translates into the stability of various potentials whose singularities are slightly "worse" than those in the natural $L^{n/2} + L^\infty$ class such as the Coulomb potential $V(x) = -\tfrac{1}{|x|}$. More recently, it was observed that the inequality can be improved once restricted to compactly supported functions on bounded domains. In particular, one can show that on the unit ball B in $\mathbb{R}^n$,

$$(0.2) \qquad \int_B |\nabla u|^2 dx - (\tfrac{n-2}{2})^2 \int_B \tfrac{u^2}{|x|^2} dx \geq z_0 \int_B u^2 dx \quad \text{for } u \in H_0^1(B),$$

where $z_0 = 2.4048...$ is the first zero of the Bessel function z_0. This yields –for example– that the Schrödinger operator $-\Delta - \frac{(n-2)^2}{4|x|^2}$ on bounded domains of $\mathbb{R}^n$ is positive definite. The story here is the newly discovered link between various improvements of this inequality confined to bounded domains and Sturm's theory regarding the oscillatory behavior of certain linear ordinary equations, which we review in Chapter 1.

In Chapter 2, we first identify suitable conditions on a non-negative C^1-function P defined on an interval $(0, R)$ that will allow for the following improved Hardy inequality to hold on every domain Ω contained in a ball of radius R:

$$(0.3) \qquad \int_\Omega |\nabla u|^2 dx - \left(\tfrac{n-2}{2}\right)^2 \int_\Omega \tfrac{u^2}{|x|^2} dx \geq \int_\Omega P(|x|) u^2 dx \quad \text{for } u \in H_0^1(\Omega).$$

It turned out that a necessary and sufficient condition for P to be a *Hardy Improving Potential* (abbreviated as *HI-potential*) on a ball B_R, is for the following ordinary differential equation

$$(0.4) \qquad y'' + \frac{1}{r} y' + P(r) y = 0,$$

to have a positive solution on the interval $(0, R)$. Elementary examples of HI-potentials are $P \equiv 0$ on any interval $(0, R)$, $P \equiv 1$ on $(0, z_0)$, where $z_0 = 2.4048...$ is the first zero of the Bessel function J_0, and more generally $P(r) = r^{-a}$ with $0 \leq a < 2$ on $(0, z_a)$, where z_a is the first root of the largest solution of the equation $y'' + \frac{1}{r} y' + r^{-a} y = 0$. Other examples are $P_\rho(r) = \frac{1}{4r^2 (\log \frac{\rho}{r})^2}$ on $(0, \frac{\rho}{e})$, but also $P_{k,\rho}(r) = \frac{1}{r^2} \sum_{j=1}^{k} \left(\prod_{i=1}^{j} \log^{(i)} \frac{\rho}{r} \right)^{-2}$ on $(0, \frac{\rho}{e^{e^{e^{\cdot^{\cdot^{\cdot^{e(k-\text{times})}}}}}}})$.

Besides leading to a large supply of explicit Hardy improving potentials, this connection to the oscillatory theory of ODEs, gives a new way of characterizing and computing best possible constants such as

$$(0.5) \qquad \beta(P, R) := \inf_{\substack{u \in H_0^1(\Omega) \\ u \neq 0}} \frac{\int_\Omega |\nabla u|^2 \, dx - \frac{(n-2)^2}{4} \int_\Omega |x|^{-2} |u|^2 \, dx}{\int_\Omega P(|x|) u^2 \, dx}.$$

On the other hand, we shall see in Chapter 3 that the value of the following best constant

$$(0.6) \qquad \mu_\lambda(P, \Omega) := \inf_{\substack{u \in H_0^1(\Omega) \\ u \neq 0}} \frac{\int_\Omega |\nabla u|^2 \, dx - \lambda \int_\Omega P(|x|) u^2 \, dx}{\int_\Omega |x|^{-2} |u|^2 \, dx}$$

and whether it is attained, depend closely on the position of the singularity point 0 vis-a-vis Ω. It is actually equal to $\frac{(n-2)^2}{4}$, and is never attained in $H_0^1(\Omega)$, whenever Ω contains 0 in its interior, but the story is quite different for domains Ω having 0 on their boundary. In this case, $\mu_\lambda(P, \Omega)$ is attained in $H_0^1(\Omega)$ whenever $\mu_\lambda(P, \Omega) < \frac{n^2}{4}$, which may hold or not. For example, $\mu_\lambda(P, \Omega)$ is equal to $\frac{n^2}{4}$ for domains that lie on one side of a half-space.

In Chapter 4, we consider conditions on a couple of positive functions V and W on $(0, \infty)$, which ensure that on some ball B_R of radius R in $\mathbb{R}^n$, $n \geq 1$, the following inequality holds:

$$(0.7) \qquad \int_B V(|x|) |\nabla u|^2 dx \geq \int_B W(|x|) u^2 dx \quad \text{for } u \in C_0^\infty(B_R).$$

A necessary and sufficient condition is that the couple (V, W) forms a *n-dimensional Bessel pair* on the interval $(0, R)$, meaning that the equation

$$(0.8) \quad y''(r) + \left(\frac{n-1}{r} + \frac{V_r(r)}{V(r)}\right) y'(r) + \frac{W(r)}{V(r)} y(r) = 0,$$

has a positive solution on $(0, R)$. This characterization allows us to improve, extend, and unify many results about weighted Hardy-type inequalities and their corresponding best constants. The connection with Chapter 2 stems from the fact that P is a HI-potential if and only if the couple $(1, \frac{(n-2)^2}{4} r^{-2} + P(r))$ is a Bessel pair. More generally, the pair

$$(0.9) \quad \left(r^{-\lambda}, \frac{(n-\lambda-2)^2}{4} r^{-\lambda-2} + r^{-\lambda} P(r)\right)$$

is also a n-dimensional Bessel pair on $(0, R)$ provided $0 \leq \lambda \leq n - 2$. Again, the link to Sturm theory provides many more examples of Bessel pairs.

Hardy's inequality and its various improvements have been used in many contexts such as in the study of the stability of solutions of semi-linear elliptic and parabolic equations, of the asymptotic behavior of the heat equation with singular potentials, as well as in the stability of eigenvalues for Schrödinger operators. In Chapter 5, we focus on applications to second order nonlinear elliptic eigenvalue problems such as

$$(0.10) \quad \begin{cases} -\Delta u = \lambda f(u) & \text{in } \Omega \\ u = 0 & \text{on } \partial \Omega, \end{cases}$$

where $\lambda \geq 0$ is a parameter, Ω is a bounded domain in $\mathbb{R}^n$, $n \geq 2$, and f is a superlinear convex nonlinearity. The bifurcation diagram generally depends on the regularity of the extremal solution, i.e., the one corresponding to the largest parameter for which the equation is solvable. Whether, for a given nonlinearity f, this solution is regular or singular depends on the dimension, and Hardy-type inequalities are crucial for the identification of the critical dimension, as well as for proving uniqueness of solutions for small values of λ.

Part II deals with the Hardy-Rellich inequalities, which are the fourth order counterpart of Hardy's. In Chapter 6, we show that the same condition on the couple (V, W) (i.e, being an n-dimensional Bessel pair) is also key to improved Hardy-Rellich inequalities of the following type: For any radial function $u \in C_0^\infty(B_R)$ where B_R is a ball of radius R in $\mathbb{R}^n$, $n \geq 1$, we have

$$(0.11) \quad \int_B V(|x|) |\Delta u|^2 dx \geq \int_B W(|x|) |\nabla u|^2 dx$$
$$+ (n-1) \int_B \left(\frac{V(|x|)}{|x|^2} - \frac{V_r(|x|)}{|x|}\right) |\nabla u|^2 dx.$$

Moreover, if

$$(0.12) \quad W(r) - \frac{2V(r)}{r^2} + \frac{2V_r(r)}{r} - V_{rr}(r) \geq 0 \quad \text{on } [0, R),$$

then the above inequality holds true for all $u \in C_0^\infty(B_R)$ and not only for the radial functions among them. By combining this with the inequalities involving the Dirichlet integrals of Chapter 4, one obtains various improvements of the Hardy-Rellich inequality for $H_0^2(\Omega)$. In particular, for any bounded domain Ω containing

0 with $\Omega \subset B_R$, we have the following inequality for all $u \in H_0^2(\Omega)$,

$$(0.13) \quad \int_\Omega |\Delta u|^2 dx \geq \frac{n^2(n-4)^2}{16} \int_\Omega \frac{u^2}{|x|^4} dx$$
$$+ \frac{\beta(P;R)(n^2 + (n-\lambda-2)^2)}{4} \int_\Omega \frac{P(|x|)}{|x|^2} u^2 dx,$$

where $n \geq 4$, $\lambda < n-2$, and where P is a HI-potential on $(0,R)$ such that $\frac{P_r(r)}{P(r)} = \frac{\lambda}{r} + f(r)$, $f(r) \geq 0$ and $\lim_{r \to 0} rf(r) = 0$.

In Chapter 7, we explore Hardy-type inequalities for $H^1(\Omega)$-functions, i.e., for functions which do not necessarily have compact support in Ω. In this case, a penalizing term appears in order to account for the boundary contribution. If a pair of positive radial functions (V, W) is a n-dimensional Bessel pair on an interval $(0, R)$, and if B_R is a ball of radius R in $\mathbb{R}^n$, $n \geq 1$, then there exists $\theta > 0$ such that the following inequalities hold:

$$(0.14) \quad \int_{B_R} V(x)|\nabla u|^2 dx \geq \int_{B_R} W(x)u^2 dx - \theta \int_{\partial B_R} u^2 ds \quad \text{for } u \in H^1(B_R),$$

while for radial functions $u \in H^2(B_R)$,

$$(0.15)$$
$$\int_{B_R} V(|x|)|\Delta u|^2 dx \geq \int_{B_R} W(|x|)|\nabla u|^2 dx + (n-1) \int_{B_R} \left(\frac{V(|x|)}{|x|^2} - \frac{V_r(|x|)}{|x|}\right)|\nabla u|^2 dx$$
$$+ [(n-1) - \theta)V(R)] \int_{\partial B_R} |\nabla u|^2 dx.$$

The latter inequality is also satisfied by all functions in $H^2(B)$ provided condition (0.12) holds. The combination of these two inequalities lead to various weighted Hardy-Rellich inequalities on $H^2 \cap H_0^1$.

In Chapter 8, we include some applications of the improved Hardy-Rellich inequalities to fourth order nonlinear elliptic eigenvalue problems of the form

$$(0.16) \quad \begin{cases} \Delta^2 u = \lambda f(u) & \text{in } \Omega \\ u = \Delta u = 0 & \text{on } \partial\Omega, \end{cases}$$

as well as their counterpart with Dirichlet boundary conditions. In particular, they are again crucial for the identification of "critical dimensions" for such equations involving either an exponential or a singular supercritical nonlinearity.

Part III addresses Hardy-type inequalities for more general uniformly elliptic operators. The issue of allowable non-radial weights (to replace $\frac{1}{|x|^2}$) is closely related to the resolution of certain linear PDEs in divergence form with either prescribed conditions on the boundary or with prescribed singularity in the interior. We also include L^p-analogs of various Hardy-type inequalities.

In Chapter 9, the following general Hardy inequality is associated to any given symmetric, uniformly positive definite $n \times n$ matrix $A(x)$ defined in Ω with the notation $|\xi|_A^2 := \langle A(x)\xi, \xi\rangle$ for $\xi \in \mathbb{R}^n$:

$$(0.17) \quad \int_\Omega |\nabla u|_A^2 dx \geq \frac{1}{4} \int_\Omega \frac{|\nabla E|_A^2}{E^2} u^2 dx \quad \text{for all } u \in H_0^1(\Omega)$$

The basic assumption here is that E is a positive solution to $-\text{div}(A\nabla E)\, dx = \mu$ on Ω, where μ is any nonnegative nonzero finite measure on Ω. The above inequality is then optimal in either one of the following two cases:

- E is an interior weight, that is $E = +\infty$ on the support of μ, or
- E is a boundary weight, meaning that $E = 0$ on $\partial\Omega$.

The case of an interior weight extends the classical Hardy inequality in many directions. One application is the following *multipolar Hardy inequality*, which states that if $x_1, ..., x_k$ are preassigned singularities in $\mathbb{R}^n$, then there exists a potential V, which behaves like $\frac{1}{|x-x_i|^2}$ near each singularity x_i such that

$$(0.18) \qquad \int_{\mathbb{R}^n} |\nabla u|^2 dx \geq \frac{1}{(n-2)^2} \int_{\mathbb{R}^n} V(x) u^2 dx \quad \text{for } u \in H_0^1(\Omega).$$

In particular, there exists $C > 0$ such that

$$(0.19) \qquad \int_{\mathbb{R}^n} |\nabla u|^2 \, dx \geq C \sum_{i=1}^k \int_{\mathbb{R}^n} \frac{u^2}{|x-x_i|^2} \, dx \quad \text{for } u \in H_0^1(\Omega).$$

The case of a boundary weight extends the following so-called *Hardy's boundary inequality*, which holds for any bounded convex domain $\Omega \subset \mathbb{R}^n$ with smooth boundary:

$$(0.20) \qquad \int_\Omega |\nabla u|^2 dx \geq \tfrac{1}{4} \int_\Omega \frac{u^2}{\text{dist}(x,\partial\Omega)^2} dx \quad \text{for } u \in H_0^1(\Omega).$$

Moreover the constant $\frac{1}{4}$ is optimal and not attained. One also obtains other Hardy inequalities involving more general distance functions. For example, if Ω is a domain in $\mathbb{R}^n$ and M a piecewise smooth surface of co-dimension k ($k = 1, ..., n$). Setting $d(x) := \text{dist}(x, M)$ and assuming $k \neq 2$ and $-\Delta d^{2-k} \geq 0$ in $\Omega \backslash M$, then

$$(0.21) \qquad \int_\Omega |\nabla u|^2 dx \geq \tfrac{(k-2)^2}{4} \int_\Omega \frac{u^2}{d(x)^2} dx \quad \text{for } u \in H_0^1(\Omega \backslash M).$$

The inequality is not attained in either case, and one can therefore get the following improvement for the case of a boundary weight:

$$(0.22) \qquad \int_\Omega |\nabla u|_A^2 dx \geq \frac{1}{4} \int_\Omega \frac{|\nabla E|_A^2}{E^2} u^2 dx + \frac{1}{2} \int_\Omega \frac{u^2}{E} d\mu, \qquad u \in H_0^1(\Omega)$$

which is optimal and still not attained. Optimal weighted versions of these inequalities are also established, as well as their L_p-counterparts when $p \neq 2$. Many of the Hardy inequalities obtained in the previous chapters can be recovered via the above approach, by using suitable choices for E and $A(x)$.

In Chapter 10, we investigate the possibility of improving (0.17) in the spirit of Chapters 4 and 5, namely whether one can find conditions on non-negative potentials V so that the following improved inequality holds:

$$(0.23) \qquad \int_\Omega |\nabla u|_A^2 dx - \tfrac{1}{4} \int_\Omega \frac{|\nabla E|_A^2}{E^2} u^2 dx \geq \int_\Omega V(x) u^2 dx \text{ for } u \in H_0^1(\Omega).$$

Necessary and sufficient conditions on V are given for (0.23) to hold, in terms of the solvability of a corresponding linear PDE. Analogous results involving improvements are obtained for the weighted versions. Optimal inequalities are also obtained for $H^1(\Omega)$.

We conclude Part III by considering in Chapter 11, applications of the Hardy inequality for general uniformly elliptic operators to study the regularity of stable solutions of certain nonlinear eigenvalue problems involving advection such as

$$(0.24) \qquad \begin{cases} -\Delta u + c(x) \cdot \nabla u = \frac{\lambda}{(1-u)^2} & \text{in } \Omega, \\ u = 0 & \text{on } \partial\Omega, \end{cases}$$

where $c(x)$ is a smooth bounded vector field on $\bar{\Omega}$.

In Part IV, we describe how the Monge-Kantorovich theory of mass transport provides a framework that encompasses most geometric inequalities. Of importance is the concept of *relative energy of ρ_0 with respect to ρ_1* defined as:

$$(0.25) \qquad H_V^{F,W}(\rho_0|\rho_1) := H_V^{F,W}(\rho_0) - H_V^{F,W}(\rho_1),$$

where ρ_0 and ρ_1 are two probability densities, and where the *Free Energy Functional* $H_V^{F,W}$ is defined on the set $\mathcal{P}_a(\Omega)$ of probability densities on a domain Ω as:

$$(0.26) \qquad H_V^{F,W}(\rho) := \int_\Omega \left[F(\rho) + \rho V + \frac{1}{2}(W \star \rho)\rho \right] dx.$$

In other words, $H_V^{F,W}$ is the sum of the internal energy $H^F(\rho) := \int_\Omega F(\rho) dx$, the potential energy $H_V(\rho) := \int_\Omega \rho V dx$ and the interaction energy $H^W(\rho) := \frac{1}{2} \int_\Omega \rho(W \star \rho) \, dx$. Here F is a differentiable function on $(0, \infty)$, while the confinement (resp., interactive) potential V (resp., W) are C^2-functions on $\mathbb{R}^n$ satisfying $D^2 V \geq \mu I$ (resp., $D^2 W \geq \nu I$) for some $\mu, \nu \in \mathbb{R}$.

In Chapter 12, we describe Brenier's solution of the Monge problem with quadratic cost, which yields that the Wasserstein distance $W(\rho_0, \rho_1)$ between two probability densities ρ_0, ρ_1 supported on domains X (resp., Y) of $\mathbb{R}^n$, i.e.,

$$(0.27) \qquad W(\rho_0, \rho_1)^2 = \inf \left\{ \int_X |x - s(x)|^2 dx; \, s \in S(\rho_0, \rho_1) \right\}$$

is achieved by the gradient $\nabla \varphi$ of a convex function φ. Here $S(\rho_0, \rho_1)$ is the class of all Borel measurable maps $s : X \to Y$ that "push" ρ_0 into ρ_1, i.e., those which satisfy the change of variables formula,

$$(0.28) \qquad \int_Y h(y)\rho_1(y)dy = \int_X h(s(x))\rho_0(x)dx \quad \text{for every} \ h \in C(Y).$$

This fundamental result allows one to show that for certain natural candidates F, V and W, the corresponding free energy functionals $H_V^{F,W}$ are convex on the geodesics of optimal mass transport joining two probability densities in $\mathcal{P}_a(\Omega)$. This convexity property translates into a very general inequality relating the relative total energy between the initial and final configurations ρ_0 and ρ_1, to their entropy production $I_{c^*}(\rho|\rho_V)$, their Wasserstein distance $W_2^2(\rho_0, \rho_1)$, as well as the Euclidean distance between their barycenters $|b(\rho_0) - b(\rho_1)|$,

$$(0.29)$$
$$H_{V+c}^{F,W}(\rho_0|\rho_1) + \frac{\lambda + \nu}{2} W_2^2(\rho_0, \rho_1) - \frac{\nu}{2} |b(\rho_0) - b(\rho_1)|^2 \leq H_{c + \nabla V \cdot x}^{-nP_F, 2x \cdot \nabla W}(\rho_0) + I_{c^*}(\rho|\rho_V).$$

Here $P_F(x) := xF'(x) - F(x)$ is the *pressure function* associated to F, while c is a Young function (such as $c(x) = \frac{1}{p}|x|^p$), c^* is its Legendre transform, while $I_{c^*}(\rho|\rho_V)$ is the *relative entropy production-type function of ρ measured against c^** defined as

$$(0.30) \qquad I_{c^*}(\rho|\rho_V) := \int_\Omega \rho c^* \left(-\nabla \left(F'(\rho) + V + W \star \rho \right) \right) dx.$$

Once this general comparison principle is established, various – new and old – inequalities follow by simply considering different examples of internal, potential and interactive energies, such as $F(\rho) = \rho \log \rho$ or $F(\rho) = \rho^\gamma$, and V and W are convex functions (e.g., $V(x) = \frac{1}{2}|x|^2$), while W is required to be even.

The framework is remarkably encompassing even when $V = W \equiv 0$, as it is shown in Chapter 13 that the following inequality, which relates the internal energy of a probability density ρ on $\mathbb{R}^n$ to the corresponding entropy production contains almost all known Euclidean Sobolev and log-Sobolev inequalities:

$$\text{(0.31)} \qquad \int_\Omega [F(\rho) + n P_F(\rho)] \, dx \leq \int_\Omega \rho c^\star \left(-\nabla(F' \circ \rho) \right) \, dx + K_c.$$

The latter constant K_c can always be computed in terms of F and the Young function c. For example, one can easily deduce the optimal Euclidean Log-Sobolev inequality, which holds for all $p \geq 1$, that is for all $f \in W^{1,p}(\mathbb{R}^n)$ such that $\|f\|_p = 1$, we have

$$\text{(0.32)} \qquad \int_{\mathbb{R}^n} |f|^p \log(|f|^p) \, dx \leq \tfrac{n}{p} \log \left(C_p \int_{\mathbb{R}^n} |\nabla f|^p \, dx \right),$$

where

$$\text{(0.33)} \qquad C_p := \begin{cases} \left(\tfrac{p}{n}\right) \left(\tfrac{p-1}{e}\right)^{p-1} \pi^{-\tfrac{p}{2}} \left[\dfrac{\Gamma(\tfrac{n}{2}+1)}{\Gamma(\tfrac{n}{q}+1)} \right]^{\tfrac{p}{n}} & \text{if } p > 1, \\[2mm] \dfrac{1}{n\sqrt{\pi}} \left[\Gamma(\tfrac{n}{2}+1) \right]^{\tfrac{1}{n}} & \text{if } p = 1, \end{cases}$$

and q is the conjugate of p ($\tfrac{1}{p} + \tfrac{1}{q} = 1$).

The approach also allows for a direct and unified way for computing best constants and extremals. It also leads to remarkable duality formulae, such as the following associated to the standard Sobolev inequality for $n \geq 3$ and where $2^* := \tfrac{2n}{n-2}$:

$$\text{(0.34)} \quad \begin{aligned} &\sup \left\{ \frac{n(n-2)}{n-1} \int_{\mathbb{R}^n} \rho(x)^{\tfrac{n-1}{n}} dx - \int_{\mathbb{R}^n} |x|^2 \rho(x) dx ; \int_{\mathbb{R}^n} \rho(x) \, dx = 1 \right\} \\ &= \inf \left\{ \int_{\mathbb{R}^n} |\nabla f|^2 dx ; \, f \in C_0^\infty(\mathbb{R}^n), \int_{\mathbb{R}^n} |f|^{2^*} dx = 1 \right\}. \end{aligned}$$

This type of duality also yields a correspondence between ground state solutions of certain quasilinear (or semi-linear) equations, such as "Yamabe's",

$$-\Delta f = |f|^{2^*-2} f \text{ on } \mathbb{R}^n,$$

and stationary solutions of the (non-linear) Fokker-Planck equations $\tfrac{\partial u}{\partial t} = \Delta u^{1-\tfrac{1}{n}} + \text{div}(x.u)$, which –after appropriate scaling– reduces to the fast diffusion equation

$$\frac{\partial u}{\partial t} = \Delta u^{1-\tfrac{1}{n}} \text{ on } \mathbb{R}^+ \times \mathbb{R}^n.$$

Chapter 14 deals with applications to Gaussian geometric inequalities. We first establish the so-called HWBI inequality, which relates the relative total energy H of two probability densities, to their Wasserstein distance W, the Fisher information I, as well as to the distance between their barycenters B. This fundamental inequality – first established and dubbed HWI by Otto-Villani for the classical Tsallis entropy $F(x) = x \log x$ and in the absence of a barycentric term – follows immediately from a direct application of (0.29) with parametrized quadratic Young functions $c_\sigma(x) = \tfrac{1}{2\sigma}|x|^2$ for $\sigma > 0$, coupled with a simple scaling argument:

$$\text{(0.35)}$$
$$H_V^{F,W}(\rho_0 | \rho_1) \leq W_2(\rho_0, \rho_1) \sqrt{I_2(\rho_0 | \rho_V)} - \frac{\mu + \nu}{2} W_2^2(\rho_0, \rho_1) + \frac{\nu}{2} |\mathrm{b}(\rho_0) - \mathrm{b}(\rho_1)|^2.$$

This gives a unified approach for –extensions of– various powerful inequalities by Gross, Bakry-Emery, Talagrand, Otto-Villani, Cordero-Erausquin, and others. As expected, such inequalities also lead to exponential rates of convergence to equilibria for solutions of Fokker-Planck and McKean-Vlasov type equations.

Part V deals with Caffarelli-Kohn-Nirenberg and Hardy-Rellich-Sobolev type inequalities. All these can be obtained by simply interpolating –via Hölder's inequalities– many of the previously obtained inequalities. This is done in Chapter 15, where it is also shown that the best constant in the Hardy-Sobolev inequality, i.e.,

$$(0.36) \quad \mu_s(\Omega) := \inf \left\{ \int_\Omega |\nabla u|^2 dx;\ u \in H_0^1(\Omega) \text{ and } \int_\Omega \frac{|u|^{2^*(s)}}{|x|^s} dx = 1 \right\},$$

where $0 < s < 2$ and $2^*(s) = \frac{2(n-s)}{n-2}$, is never attained when 0 is in the interior of the domain Ω, unless the latter is the whole space $\mathbb{R}^n$, in which case explicit extremals are given. This is not the case when Ω is half-space $\mathbb{R}^n_-$, where only the symmetry of the extremals is shown. Much less is known about the extremals in the Hardy-Rellich-Sobolev inequality (i.e., when $s > 0$) even when $\Omega = \mathbb{R}^n$.

The problem whether $\mu_s(\Omega)$ is attained becomes more interesting when 0 is on the boundary $\partial\Omega$ of the domain Ω. The attainability is then closely related to the geometry of $\partial\Omega$, as we show in Chapter 16, that the negativity of the mean curvature of $\partial\Omega$ at 0 is sufficient to ensure the attainability of $\mu_s(\Omega)$.

In Chapter 17, we consider log-Sobolev inequalities on the line, such as those involving the functional

$$(0.37) \quad I_\alpha(g) = \frac{\alpha}{2} \int_{-1}^{1} (1-x^2)|g'(x)|^2\ dx + \int_{-1}^{1} g(x)\ dx - \log \frac{1}{2} \int_{-1}^{1} e^{2g(x)}\ dx$$

on the space $H^1(-1,1)$ of L^2-functions on $(-1,1)$ such that $\int_{-1}^{1}(1-x^2)|g'(x)|^2 dx < \infty$. There are two key results concerning the functional I_α. The first states that

$$(0.38) \quad \inf_{g \in H^1} I_\alpha(g) = 0 \text{ if } \alpha \geq 1 \quad \text{while} \quad \inf_{g \in H^1} I_\alpha(g) = -\infty \text{ if } \alpha < 1.$$

The other, which is shown in Chapter 17, is that once J_α is restricted to the manifold

$$\mathcal{G} = \left\{ g \in H^1(-1,1);\ \int_{-1}^{1} e^{2g(x)} x dx = 0 \right\},$$

then the following improvement of (0.38) holds:

$$(0.39) \quad \inf_{g \in \mathcal{G}} I_\alpha(g) = 0 \text{ if } \alpha \geq \tfrac{1}{2} \quad \text{while} \quad \inf_{g \in \mathcal{G}} I_\alpha(g) = -\infty \text{ if } \alpha < \tfrac{1}{2}.$$

We also give a recent result of Ghigi, which says that the functional

$$(0.40) \quad \Phi(u) = \int_{-1}^{1} u(x)\ dx - \log\left(\frac{1}{2} \int_{-\infty}^{+\infty} e^{-2u^*(x)}\ dx \right)$$

is convex on the cone $\mathcal{W}$ of all bounded convex functions u on $(-1,1)$, where here u^* denotes the Legendre transform of u, and that $\inf_{u \in \mathcal{W}} \Phi(u) = \log(\frac{4}{\pi})$. Both inequalities play a key role in the next two chapters, which address inequalities on the two-dimensional sphere $\mathbb{S}^2$. It is worth noting that Ghigi's inequality relies on the Prékopa-Leindler principle, which itself is another manifestation of a mass transport context. One can therefore infer that the approach of Part IV should be made more readily applicable to critical Moser-type inequalities.

In Chapter 18, we establish the Moser-Trudinger inequality, which states that for $\alpha \geq 1$, the functional

$$(0.41) \qquad J_\alpha(u) := \alpha \int_{\mathbb{S}^2} |\nabla u|^2 \, d\omega + 2 \int_{\mathbb{S}^2} u \, d\omega - \log \int_{\mathbb{S}^2} e^{2u} \, d\omega$$

is bounded below on the Sobolev space $H^1(\mathbb{S}^2)$, where here $d\omega := \frac{1}{4\pi} \sin\theta \, d\theta \wedge d\varphi$ denotes Lebesgue measure on the unit sphere, normalized so that $\int_{\mathbb{S}^2} d\omega = 1$. We also give a proof of Onofri's inequality which states that the infimum of J_α on $H^1(\mathbb{S}^2)$ is actually equal to zero for all $\alpha \geq 1$, and that

$$(0.42) \qquad \inf\{J_1(u); \, u \in H^1(\mathbb{S}^2)\} = \inf\{J_1(u); \, u \in \mathcal{M}\} = 0,$$

where $\mathcal{M}$ is the submanifold $\mathcal{M} = \{u \in H^1(\mathbb{S}^2); \int_{\mathbb{S}^2} e^{2u} \mathbf{x} \, d\omega = 0\}$. Note that this inequality, once applied to axially symmetric functions, leads to the following counterpart of (0.39)

$$(0.43) \qquad \inf_{g \in H^1(-1,1)} I_\alpha(g) = \inf_{g \in \mathcal{G}} I_\alpha(g) = 0 \quad \text{if } \alpha \geq 1.$$

In Chapter 19, we include results of T. Aubin asserting that once restricted to the submanifold $\mathcal{M}$, the functional J_α then remains bounded below (and coercive) for smaller values of α, which was later conjectured by A. Chang and P. Yang to be equal to $\frac{1}{2}$. We conclude the latest developments on this conjecture, including a proof that

$$(0.44) \qquad \inf\{J_\alpha(u); \, u \in \mathcal{M}\} = 0 \quad \text{if } \alpha \geq \tfrac{2}{3}.$$

The conjecture remains open for $1/2 < \alpha < 2/3$.

We have tried to make this monograph as self-contained as possible. That was not realistic though, when dealing with the applications such as in Chapters 5, 8 and 11. We do however give enough references for the missing proofs.

The rapid development of this area and the variety of applications forced us to be quite selective. We mostly concentrate on certain recent advances not covered in the classical books such as the one by R. A. Adams [4] and V. G. Maz'ya [**224**]. Our choices reflect our taste and what we know –of course– but also our perceptions of what are the most fundamental functional inequalities, the novel methods and ideas, those that are minimally ad-hoc, as well as the ones we found useful in our own research. It is however evident that this compendium is far from being an exhaustive account of this continuously and rapidly evolving line of research. One example that comes to mind are inequalities obtained by interpolating between the Hardy and the Trudinger-Moser inequalities. One then gets the singular Moser-type inequalities, which states that for some $C_0 = C_0(n, |\Omega|) > 0$, one has for any $u \in W_0^{1,n}(\Omega)$ with $\int_\Omega |\nabla u|^n \, dx \leq 1$,

$$(0.45) \qquad \int_\Omega \frac{\exp\left(\beta |u|^{\frac{n}{n-1}}\right)}{|x|^\alpha} \, dx \leq C_0,$$

for any $\alpha \in [0, n)$, $0 \leq \beta \leq \left(1 - \frac{\alpha}{n}\right) n \omega_{n-1}^{\frac{1}{n-1}}$, where $\omega_{n-1} = \frac{2\pi^{\frac{n}{2}}}{\Gamma(\frac{n}{2})}$ is the area of the surface of the unit n-dimensional ball. See for instance [**10, 11**]).

Another direction is the connection between the Hardy inequality and the optimal *logarithmic Sobolev inequality* in $\mathbb{R}^n$ (0.32) for $p = 2$. Del Pino et al. [**114**] recently showed that for $n \geq 3$, there exists a constant $\mathsf{C}_{\text{LH}} > 0$ such that for $u \in \mathcal{D}^{1,2}(\mathbb{R}^d)$ with $\int_{\mathbb{R}^n} \frac{|u|^2}{|x|^2} = 1$, we have

$$(0.46) \qquad \int_{\mathbb{R}^n} \frac{|u|^2}{|x|^2} \log(|x|^{n-2}|u|^2) \, dx \leq \frac{n}{2} \log \left[\mathsf{C}_{\text{LH}} \int_{\mathbb{R}^n} |\nabla u|^2 \, dx \right].$$

Last but not least of what is missing is the Hardy-Leray inequality, which states that the Hardy inequality (with the same constant) holds for vector fields in $\mathbb{R}^n$ [201], as well as its recent improvement by O. Costin and V. Maz'ya [94], who show that for divergence-free fields, the inequality becomes

$$(0.47) \qquad \int_{\mathbb{R}^n} \frac{|\mathbf{u}|^2}{|x|^2} dx \leq \frac{4}{(n-2)^2} \left(1 - \frac{8}{(n+2)^2} \right) \int_{\mathbb{R}^n} |\nabla \mathbf{u}|^2 dx.$$

These recent developments and their variations could have easily added several chapters to this book, but we had to stop somewhere and this is where we stopped, though not before including at least one open problem at the end of each of the 19 chapters.

Glossary of notation. The following list of notation and abbreviations will be used throughout this book. Let Ω be a smooth domain of $\mathbb{R}^n$.

(1) $C^0(\Omega, \mathbb{R}^n)$ (resp., $C_c^0(\Omega, \mathbb{R}^n)$) will denote the space of continuous functions (resp., the space of continuous functions with compact support) on Ω.

(2) For $k \geq 1$, $C^k(\Omega, \mathbb{R}^n)$ (resp., $C_c^k(\Omega, \mathbb{R}^n)$ and sometimes $C_0^k(\Omega, \mathbb{R}^n)$) will denote the space of k-differentiable functions (resp., the space of k-differentiable functions with compact support) on Ω.
If $0 \in \Omega \subset \mathbb{R}^n$, we then denote
$$C_{c,r}^k(\Omega) = \{v \in C_c^k(\Omega) : v \text{ is radial and supp } v \subset \Omega\}.$$

(3) $C^\infty(\Omega, \mathbb{R}^n)$ (resp., $C_c^\infty(\Omega, \mathbb{R}^n)$) will denote the space of infinitely differentiable functions (resp., the space of infinitely differentiable functions with compact support) on Ω. The space of distributions on Ω will be denoted by $\mathcal{D}'(G)$.

(4) If $0 < \alpha \leq 1$, $C^{k,\alpha}(\Omega, \mathbb{R}^n)$ will then denote the subspace of those functions in $C^k(\Omega, \mathbb{R}^n)$ such that the k-th differential is α-Hölder continuous.

(5) For $1 \leq p < +\infty$, $L^p(\Omega)$ will be the space of all integrable functions $u : \Omega \to \mathbb{R}$ equipped with norm
$$\|u\|_{L^p} = \Big(\int_\Omega |u(x)|^p dx\Big)^{\frac{1}{p}}.$$
For $p = +\infty$, $L^\infty(\Omega)$ will be the space of all measurable functions $u : \Omega \to \mathbb{R}$ such that
$$\|u\|_{L^\infty} = \operatorname{ess\,sup}_{x\in\Omega} |u(x)| < +\infty.$$

(6) For $1 \leq p < +\infty$, the space $W^{1,p}(\Omega)$ (resp., $W_0^{1,p}(\Omega)$) is the completion of $C^\infty(\Omega, \mathbb{R}^n)$ (resp., $C_c^\infty(\Omega, \mathbb{R}^n)$ for the norm
$$\|u\|_{W^{1,p}(\Omega)} = \|u\|_p + \|\nabla u\|_p \text{ (resp., } \|u\|_{W_0^{1,p}(\Omega)} = \|\nabla u\|_p).$$
We denote the dual of $W^{1,p}(\Omega)$ by $W^{1,-p}(\Omega)$.

(7) $W^{1,2}(\Omega)$ (resp., $W_0^{1,2}(\Omega)$ will sometimes be denoted by $H^1(\Omega)$ (resp., $H_0^1(\Omega)$) and the dual of $H_0^1(\Omega)$ will be denoted by $H^{-1}(\Omega)$.

(8) More generally, we consider for any $m \in \mathbb{N}$ and $1 \leq p \leq +\infty$, the Banach space $W^{m,p}(\Omega)$ of (classes of) measurable functions $u : \Omega \to \mathbb{R}$ such that $D^\alpha u \in L^p(\Omega)$ in the sense of distributions, for every multi-index α with $|\alpha| \leq m$. The space $W^{m,p}(\Omega)$ will be equipped with the norm
$$\|u\|_{W^{m,p}(\Omega)} = \sum_{|\alpha| \leq m} \|D^\alpha u\|_{L^p}.$$
$W_0^{m,p}(\Omega)$ will be the closure of $C_c^\infty(\Omega, \mathbb{R})$ in $W^{m,p}(\Omega)$, and $W^{-m,q}(\Omega)$ will denote the Banach space dual of $W^{m,p}(\Omega)$, where $\frac{1}{p} + \frac{1}{q} = 1$.

(9) $W^{m,2}(\Omega)$ (resp., $W_0^{m,2}(\Omega)$) will be denoted by $H^m(\Omega)$ (resp., $H_0^m(\Omega)$). They will be Hilbert spaces once equipped with the scalar product
$$\langle u, v \rangle = \int_\Omega u(x)v(x)\,dx.$$
The dual of $H_0^m(\Omega)$ will be denoted by $H^{-m}(\Omega)$.
If $0 \in \Omega$, then
$$H_{0,r}^m(\Omega) = \{u \in H_0^m(\Omega) : u \text{ is radial}\}.$$

(10) For $a \in \mathbb{R}$, we denote by $\mathcal{D}_a^{1,2}(\Omega)$ the completion of $C_0^\infty(\Omega)$ with respect to the norm
$$\|u\|_a^2 = \int_\Omega |x|^{-2a} |\nabla u|^2 dx.$$

(11) If $\Omega \subset \mathbb{R}^n$ contains 0, then $L^2(\Omega; |x|^{-2}dx)$ will denote the space of measurable functions in Ω such that $\int_\Omega \frac{|u|^2}{|x|^2} dx < \infty$. We then set
$$\widehat{H}^1(\Omega) := H^1(\Omega) \cap L^2(\Omega; |x|^{-2}dx).$$

Part 1

Hardy Type Inequalities

CHAPTER 1

Bessel Pairs and Sturm's Oscillation Theory

The ordinary differential equation associated to a non-negative real valued C^1-function P on $(0, +\infty)$,
$$y'' + \frac{1}{r}y' + P(r)y = 0,$$
as well as the equation
$$(V(r)y')' + W(r)y = 0,$$
associated to a pair (V, W) of non-negative functions on $(0, +\infty)$, are central to all results revolving around the inequalities of Hardy and Hardy-Rellich type. We summarize in this chapter the properties of these equations that will be used throughout this book. In particular, we give conditions on P (resp., V and W), which guarantee that the above equations have a positive solution on a non-trivial interval $(0, R)$.

1.1. The class of Hardy improving potentials

DEFINITION 1.1.1. We say that a non-negative real valued C^1-function P is a *Hardy improving potential* –abbreviated as "HI-potentials"– on $(0, R)$, if there exists $c > 0$ such that the equation

$$(\mathcal{B}_{cP}) \qquad y''(r) + \frac{1}{r}y'(r) + cP(r)y(r) = 0,$$

has a positive solution on $(0, R)$.

The class of HI-potentials on $(0, R)$ will be denoted by $\mathcal{B}(0, R)$. Here are a few immediate examples of such functions:

- $P \equiv 0$ is a HI-potential on $(0, R)$ for any $R > 0$. Indeed, It is clear that $\varphi(r) = -\log(\frac{e}{R}r)$ is a positive solution of $(\mathcal{B}_0)$ on $(0, R)$.
- $P \equiv 1$ is a HI-potential on $(0, z_0)$, where $z_0 = 2.4048...$ is the first zero of the Bessel function J_0. Indeed, the latter is a positive solution of $y'' + \frac{1}{r}y' + y = 0$ until it reaches its first zero at z_0.
- For any $\rho \geq Re$, the function $P(r) = \frac{1}{4r^2(\log\frac{\rho}{r})^2}$ is a HI-potential on $(0, R)$, with $(\mathcal{B}_P)$ having the explicit solution $\varphi_\rho(r) = (\log\frac{\rho}{r})^{\frac{1}{2}}$.

We shall later need the following easy result regarding the behaviour of positive solutions of equation $(\mathcal{B}_P)$.

LEMMA 1.1.2. *Assume P is non-negative on $(0, R)$, $a \geq 1$ and that the equation*

(1.1) $$y'' + \frac{a}{r}y' + P(r)y = 0,$$

has a positive solution φ on $(0, R)$. Then,

(1) *φ is decreasing on $(0, R)$.*

3

(2) φ has the following limiting behavior on the boundary:

(1.2) $$\lim_{r \to 0} r \frac{\varphi'(r)}{\varphi(r)} = 0 \quad \text{and} \quad \limsup_{r \to R} \frac{\varphi'(r)}{\varphi(r)} \leq 0.$$

(3) If $P(r) > 0$ on $(0, R)$, then φ is strictly decreasing on $(0, R)$.
(4) If $P(r) > 0$ on $(0, +\infty)$, then (1.1) has no positive solution on $(0, +\infty)$.

Proof: Observe that the function $x(r) = r \frac{\varphi'(r)}{\varphi(r)}$ satisfies the ODE:

(1.3) $$rx'(r) + x^2(r) = -F(r), \quad \text{for } 0 < r \leq \delta,$$

where $F(r) = r^2 P(r) \geq 0$. It follows that $\varphi'(t) \leq 0$ on $(0, R)$ and is therefore decreasing on that interval.

To prove 2), divide equation (1.3) by r and integrate once to obtain

(1.4) $$x(r) \geq \int_r^R \frac{|x(s)|^2}{s} ds + x(R) + \int_r^R \frac{F(s)}{s} ds.$$

It follows that $\lim_{r \downarrow 0} x(r)$ exists. In order to prove that this limit is zero, it therefore suffices to prove that

(1.5) $$G(r) := \int_r^R \frac{x^2(s)}{s} ds < +\infty.$$

Suppose not, that is $G(r) \to \infty$ as $r \to 0$. From (1.3) we have

$$(-rG'(r))^{\frac{1}{2}} \geq G(r) + x(1) + \int_r^R \frac{F(s)}{s} ds.$$

Since $F \geq 0$, and G goes to infinity as r goes to zero, then for r sufficiently small we have $-rG'(r) \geq \frac{1}{2} G^2(r)$, and hence, $(\frac{1}{G(r)})' \geq \frac{1}{2}(\ln(r))'$, which contradicts the fact that $G(r)$ goes to infinity as r tends to zero. It follows that indeed, $\lim_{r \downarrow 0} r \frac{\varphi'(r)}{\varphi(r)} = \lim_{r \downarrow 0} x(t) = 0$.

For 3) it suffices to note that if $P(r) > 0$ on $(0, R)$, then φ cannot have a local minimum in $(0, R)$. Indeed, if $\varphi'(x_0) = 0$ for some $R > x_0 > 0$, and $\varphi''(x_0) \geq 0$, then necessarily $\varphi''(x_0) = 0$ which contradicts the fact that φ is a positive solution of the above ODE. it follows that φ is strictly decreasing on the whole interval $(0, R)$.

4) Suppose that for any $R > 0$, the equation $y''(r) + \frac{a}{r} y' + P(r) y = 0$ has a positive solution φ_R, which is then necessarily strictly decreasing on $(0, R)$. It follows that $\frac{\varphi_R''(r)}{\varphi_R'(r)} \geq -\frac{a}{r}$ on $(0, R)$ which yields that for some $c > 0$, we have $\varphi_R'(r) \leq \frac{-c}{r}$ on $(0, R)$. We can also clearly assume that if $R < R'$, then $\varphi_R \geq \varphi_{R'}$ on $(0, R]$. It then follows that $\varphi_R(R) \leq \varphi_R(1) - c \ln R \leq \varphi_1(1) - c \ln R$ for any $R > 1$. This means that $\varphi_R(R) < 0$ for R large enough, which is clearly a contradiction. $\square$

It is clear that if $P \in \mathcal{B}(0, R)$, then $P \in \mathcal{B}(0, R')$ for any $0 < R' < R$. We shall be interested by the largest such interval, i.e.,

(1.6) $$\delta(P) := \sup \{R; (\mathcal{B}_P) \text{ has a positive solution on } (0, R)\}.$$

In view of the above lemma, $\delta(P)$ can be seen as the first time a particular positive solution of $(\mathcal{B}_P)$ reaches zero.

On the other hand, we shall consider for a given $R > 0$, the largest *HIP-constant* associated to a HI-potential $P \in \mathcal{B}(0,R)$, defined as:

(1.7) $\quad \beta(P;R) = \sup\{c > 0; (\mathcal{B}_{cP}) \text{ has a positive solution on } (0,R)\}.$

The following proposition will be frequently used in the sequel.

PROPOSITION 1.1.1. *Let P be a non-negative C^1-function on an interval $(0,R)$. Then,*

(1) $(\mathcal{B}_P)$ *has a positive solution on $(0,R)$ if and only if it has a positive supersolution φ on $(0,R)$, i.e., if*

(1.8) $\quad \varphi'' + \frac{1}{r}\varphi' + P(r)\varphi \leq 0 \text{ on } (0,R).$

(2) *Consequently, if $\mathcal{B}_P$ has a positive solution on an interval $(0,R)$ for some non-negative C^1-potential $P \geq 0$, then for any C^1-function Q such that $0 \leq Q \leq P$, the equation $(\mathcal{B}_Q)$ has also a positive solution on $(0,R)$.*

(3) *The class $\mathcal{B}(0,R)$ of HI-potentials on $(0,R)$ is a closed convex and solid subset of $C^1(0,R)$.*

(4) *Moreover, for every $P \in \mathcal{B}(0,R)$, the equation (B_{cP}) has a positive solution on $(0,R)$, for all $c \leq \beta(P;R)$.*

Proof: Statement 1) is a direct consequence of proposition 1.2.1. The proofs of 2), 3), and 4) are straightforward and are left to the interested reader. $\square$

For a given HI-potential on an interval $(0,R)$, we shall often be interested in computing its *HIP-constant* $\beta(P;R)$. This is often closely related to finding $\delta(P)$. Indeed, if (B_P) has a positive solution φ on $(0,\delta)$ for some $\delta > 0$, then $\psi(r) = \varphi(\frac{\delta r}{R})$ is a solution for $y''(r) + \frac{1}{r}y' + \frac{\delta^2}{R^2}P(\frac{\delta}{R}r)y = 0$ on $(0,R)$. In other words, the scaled potential $V(x) = \frac{\delta^2}{R^2}P(\frac{\delta}{R}x)$ is then a HI-potential on $(0,R)$. We therefore have the following relations.

PROPOSITION 1.1.2. *If P is a C^1-function such that $(\mathcal{B}_P)$ has a positive solution on $(0,\delta)$ for some $\delta > 0$, then for any $R > 0$, the function Q defined by $Q(x) := P(\frac{\delta}{R}x)$ belongs to $\mathcal{B}(0,R)$, and*

(1.9) $\quad \beta(Q;R) = \frac{\delta(P)^2}{R^2}\beta(P;\delta(P)).$

In particular, if P is also α-homogeneous (i.e. $P(\lambda x) = \lambda^{-\alpha}P(x)$ for some $\alpha > 0$), then

(1.10) $\quad \beta(P;R) = \frac{\delta(P)^{2-\alpha}}{R^{2-\alpha}}.$

$\square$

We now exhibit a few explicit HI-potentials and compute their HIP-constants. We use the following notation.

(1.11) $\quad log^{(1)}(.) = log(.) \quad \text{and} \quad log^{(k)}(.) = log(log^{(k-1)}(.)) \quad \text{for } k \geq 2.$

and

(1.12) $\quad X_1(t) = (1 - log(t))^{-1}, \quad X_k(t) = X_1(X_{k-1}(t)) \quad k = 2, 3, ...,$

THEOREM 1.1.3. *Explicit HI-potentials*

(1) $P \equiv 0$ is a HI-potential on $(0, R)$ for any $R > 0$.
(2) $P \equiv 1$ is a HI-potential on $(0, R)$ for any $R > 0$. Moreover,

(1.13) $$\delta(1) = z_0, \quad \text{and} \quad \beta(1; R) = \frac{z_0^2}{R^2} \text{ for every } R > 0,$$

(3) If $0 \leq a < 2$, then $P(r) = r^{-a}$ is a HI-potential on $(0, R)$ for any $R > 0$. Moreover, there is $z_a > 0$ such that

(1.14) $$\delta(r^{-a}) = z_a, \quad \text{and} \quad \beta(r^{-a}; R) = \frac{z_a^{2-a}}{R^{2-a}} \text{ for every } R > 0,$$

where z_a is the first root of the largest solution of the equation $y'' + \frac{1}{r}y' + r^{-a}y = 0$.

(4) For each $k \geq 1$ and $R > 0$, the function $P_{k,\rho}(r) = \frac{1}{r^2}\sum_{j=1}^{k}\left(\prod_{i=1}^{j} log^{(i)} \frac{\rho}{r}\right)^{-2}$, where $\rho > R(e^{e^{e^{\cdot^{\cdot^{e}}}}})$ $e(k-times)$, is a HI-potential on $(0, R)$ and

(1.15) $$\beta(P_{k;\rho}, R) = \tfrac{1}{4}.$$

(5) For $k \geq 1$ and $R > 0$, the function

$$\tilde{P}_{k,R}(r) = \frac{1}{r^2}\sum_{j=1}^{k} X_1^2(\tfrac{r}{R})X_2^2(\tfrac{r}{R})\ldots X_{j-1}^2(\tfrac{r}{R})X_j^2(\tfrac{r}{R})$$

is a HI-potential on $(0, R)$ and

(1.16) $$\beta(\tilde{P}_{k,R}; R) = \tfrac{1}{4}.$$

Proof: 1) As noted above, for any $R > 0$ the function $\varphi(r) = -log(\frac{e}{R}r)$ is a positive solution of $(\mathcal{B}_0)$ on $(0, R)$.

2) The Bessel function J_0 is a positive solution for equation $\mathcal{B}_P$ with $P \equiv 1$, on $(0, z_0)$, where $z_0 = 2.4048...$ is the first zero of J_0. Moreover, z_0 is larger than the first root of any other solution for $(\mathcal{B}_1)$. Indeed if α is the first root of the an arbitrary solution of the Bessel equation $y'' + \frac{y'}{r} + y(r) = 0$, then we have $\alpha \leq z_0$. To see this let $x(t) = aJ_0(t) + bY_0(t)$, where J_0 and Y_0 are the two standard linearly independent solutions of Bessel equation, and a and b are constants. Assume the first zero of $x(t)$ is larger than z_0. Since the first zero of Y_0 is smaller than z_0, we have $a \geq 0$. Also $b \leq 0$, because $Y_0(t) \to -\infty$ as $t \to 0$. Finally note that $Y_0(z_0) > 0$, so if $b < 0$, then $x(z_0 + \epsilon) < 0$ for ϵ sufficiently small. Therefore, $b = 0$ which is a contradiction. The rest follows from Proposition 1.1.2.

3) will follow from the integral criteria below –as applied in Corollary 1.3.2– while formula (1.14) also follows from Proposition 1.1.2.

4) Note first that $\varphi_{k,\rho}(r) = (\prod_{i=1}^{k} log^{(i)} \frac{\rho}{r})^{\frac{1}{2}}$ is an explicit solution of the equation $(\mathcal{B}_{\frac{1}{4}P_k})$ on $(0, R)$ provided $\rho > R(e^{e^{e^{\cdot^{\cdot^{e}}}}})$ $e(k-times)$. This readily implies that $\beta(P_{k;\rho}, R) \geq \tfrac{1}{4}$. To establish equality, we need to show that equation $(\mathcal{B}_{(\frac{1}{4}+\lambda)P_{k,\rho}})$ has no positive solution for any $\lambda > 0$. Since $P_{k,\rho}(r) = \Sigma_{j=1}^{k} U_j$ where $U_j(r) = \frac{1}{r^2}(\prod_{i=1}^{j} log^{(i)} \frac{\rho}{r})^{-2}$, it suffices to show that equation $(\mathcal{B}_{\frac{1}{4}P_{k-1,\rho}+\lambda U_k})$ which corresponds to the smaller weight $\tfrac{1}{4}P_{k-1,\rho} + \lambda U_k$ has no positive solution for any $\lambda > 0$.

1.1. THE CLASS OF HARDY IMPROVING POTENTIALS

To do that, we assume that there exists a positive function φ on $(0, R)$ such that

$$-\frac{\varphi'(r) + r\varphi''(r)}{\varphi(r)} = \frac{1}{4}\sum_{j=1}^{k-1}\frac{1}{r}\Big(\prod_{i=1}^{j}\log^{(i)}\frac{\rho}{r}\Big)^{-2} + \Big(\frac{1}{4} + \lambda\Big)\frac{1}{r}\Big(\prod_{i=1}^{k}\log^{(i)}\frac{\rho}{r}\Big)^{-2},$$

and work towards a contradiction.

Set $f(r) = \frac{\varphi(r)}{\varphi_k(r)} > 0$, where $\varphi_k = \varphi_{k,\rho}$ defined above, and calculate,

$$\frac{\varphi'(r) + r\varphi''(r)}{\varphi(r)} = \frac{\varphi_k'(r) + r\varphi_k''(r)}{\varphi_k(r)} + \frac{f'(r) + rf''(r)}{f(r)} - \frac{f'(r)}{f(r)}\sum_{i=1}^{k}\frac{1}{\prod_{j=1}^{i}\log^{j}(\frac{\rho}{r})}.$$

Thus,

(1.17) $$\frac{f'(r) + rf''(r)}{f(r)} - \frac{f'(r)}{f(r)}\sum_{i=1}^{k}\frac{1}{\prod_{j=1}^{i}\log^{j}(\frac{\rho}{r})} = -\lambda\frac{1}{r}\Big(\prod_{i=1}^{k}\log^{(i)}\frac{\rho}{r}\Big)^{-2}.$$

If now $f'(\alpha_n) = 0$ for some sequence $\{\alpha_n\}_{n=1}^{\infty}$ that converges to zero, then there exists a sequence $\{\beta_n\}_{n=1}^{\infty}$ that also converges to zero, such that $f''(\beta_n) = 0$, and $f'(\beta_n) > 0$. But this contradicts (1.17), which means that f is eventually monotone for r small enough. We consider the two cases according to whether f is increasing or decreasing:

Case I: Assume $f'(r) > 0$ for $r > 0$ sufficiently small. Then we will have

$$\frac{(rf'(r))'}{rf'(r)} \leq \sum_{i=1}^{k}\frac{1}{r\prod_{j=1}^{i}\log^{j}(\frac{\rho}{r})}.$$

Integrating once we get $f'(r) \geq \frac{c}{r\prod_{j=1}^{k}\log^{j}(\frac{\rho}{r})}$, for some $c > 0$. Hence, $\lim_{r\to 0} f(r) = -\infty$ which is a contradiction.

Case II: Assume $f'(r) < 0$ for $r > 0$ sufficiently small. Then

$$\frac{(rf'(r))'}{rf'(r)} \geq \sum_{i=1}^{k}\frac{1}{r\prod_{j=1}^{i}\log^{j}(\frac{\rho}{r})}.$$

Thus,

(1.18) $$f'(r) \geq -\frac{c}{r\prod_{j=1}^{k}\log^{j}(\frac{\rho}{r})},$$

for some $c > 0$ and $r > 0$ sufficiently small. On the other hand

$$\frac{f'(r) + rf''(r)}{f(r)} \leq -\lambda\sum_{j=1}^{k}\frac{1}{r}\Big(\prod_{i=1}^{j}\log^{(i)}\frac{R}{r}\Big)^{-2} \leq -\lambda\Big(\frac{1}{\prod_{j=1}^{k}\log^{j}(\frac{\rho}{r})}\Big)'.$$

Since $f'(r) < 0$, there exists l such that $f(r) > l > 0$ for $r > 0$ sufficiently small. From the above inequality we then have

$$bf'(b) - af'(a) < -\lambda l\Big(\frac{1}{\prod_{j=1}^{k}\log^{j}(\frac{\rho}{b})} - \frac{1}{\prod_{j=1}^{k}\log^{j}(\frac{\rho}{a})}\Big).$$

From (1.18) we have $\lim_{a\to 0} af'(a) = 0$. Hence, $bf'(b) < -\frac{\lambda l}{\prod_{j=1}^{k}\log^{j}(\frac{\rho}{b})}$, for every $b > 0$, and $f'(r) < -\frac{\lambda l}{r\prod_{j=1}^{k}\log^{j}(\frac{\rho}{r})}$, for $r > 0$ sufficiently small. Therefore,

$\lim_{r \to 0} f(r) = +\infty$, and by choosing l large enouph (e.g., $l > \frac{c}{\lambda}$) we get to contradict (1.18).

The proof of 5) is similar. Indeed, let $D \geq \sup_{x \in \Omega} |x|$, and define
$$\varphi_k(r) = (X_1(\frac{r}{D})X_2(\frac{r}{D})\ldots X_{i-1}(\frac{r}{D})X_i(\frac{r}{D}))^{-\frac{1}{2}}, \quad i = 1, 2, \ldots.$$
Using the fact that $X'_k(r) = \frac{1}{r}X_1(r)X_2(r)\ldots X_{k-1}(r)X_k^2(r)$ for $k = 1, 2, \ldots$, we get
$$-\frac{\varphi'_k(r) + r\varphi''_k(r)}{\varphi_k(r)} = \frac{1}{4r}\Sigma_{j=1}^k X_1^2(\frac{r}{D})X_2^2(\frac{r}{D})\ldots X_{j-1}^2(\frac{r}{D})X_j^2(\frac{r}{D}).$$
This means that $\beta(\tilde{P}_{k;R}, R) \geq \frac{1}{4}$.

One can again show that $\frac{1}{4}$ is the best constant by assuming in contradiction that for some $\lambda > 0$, there exists a positive function φ such that
$$-\frac{\varphi'(r) + r\varphi''(r)}{\varphi(r)} = \frac{1}{4}\sum_{j=1}^{m-1} \frac{1}{r}X_1^2(\frac{r}{D})X_2^2(\frac{r}{D})\ldots X_{j-1}^2(\frac{r}{D})X_j^2(\frac{r}{D})$$
$$+ (\frac{1}{4} + \lambda)\frac{1}{r}X_1^2(\frac{r}{D})X_2^2(\frac{r}{D})\ldots X_{m-1}^2(\frac{r}{D})X_m^2(\frac{r}{D}).$$
Setting $f(r) = \frac{\varphi(r)}{\varphi_m(r)} > 0$, we have
$$\frac{\varphi'(r) + r\varphi''(r)}{\varphi(r)} = \frac{\varphi'_m(r) + r\varphi''_m(r)}{\varphi_m(r)} + \frac{f'(r) + rf''(r)}{f(r)} - \frac{f'(r)}{f(r)}\sum_{i=1}^m \prod_{j=1}^i X_j(\frac{r}{D}).$$
Thus,

(1.19) $$\frac{f'(r) + rf''(r)}{f(r)} - \frac{f'(r)}{f(r)}\sum_{i=1}^m \prod_{j=1}^i X_j(\frac{r}{D}) = -\lambda\frac{1}{r}\prod_{j=1}^m X_j^2(\frac{r}{D}).$$

Arguing as before, we deduce that f is eventually monotone for r small enough, and we consider two cases:

Case I: If $f'(r) > 0$ for $r > 0$ sufficiently small, then we will have
$$\frac{(rf'(r))'}{rf'(r)} \leq \sum_{i=1}^m \frac{1}{r}\prod_{j=1}^i X_j(\frac{r}{D}).$$
Integrating once we get $f'(r) \geq \frac{c}{r}\prod_{j=1}^m X_j(\frac{r}{D})$, for some $c > 0$, and therefore $\lim_{r \to 0} f(r) = -\infty$ which is a contradiction.

Case II: Assume $f'(r) < 0$ for $r > 0$ sufficiently small. Then
$$\frac{(rf'(r))'}{rf'(r)} \geq \sum_{i=1}^m \frac{1}{r}\prod_{j=1}^i X_j(\frac{r}{D})$$
Thus,

(1.20) $$f'(r) \geq -\frac{c}{r}\prod_{j=1}^m X_j(\frac{r}{D}),$$

for some $c > 0$ and $r > 0$ sufficiently small. On the other hand

$$\frac{f'(r) + rf''(r)}{f(r)} \leq -\lambda \sum_{j=1}^{m} \frac{1}{r} \prod_{i=1}^{j} X_j^2 \leq -\lambda (\prod_{j=1}^{m} X_j(\frac{r}{D}))'.$$

Since $f'(r) < 0$, we may assume $f(r) > l > 0$ for $r > 0$ sufficiently small, and from the above inequality we have

$$bf'(b) - af'(a) < -\lambda l (\prod_{j=1}^{m} X_j(\frac{b}{D}) - \prod_{j=1}^{m} X_j(\frac{a}{D})).$$

From (1.20) we have $\lim_{a \to 0} af'(a) = 0$. Hence, $f'(r)) < -\frac{\lambda l}{r} \prod_{j=1}^{m} X_j(\frac{r}{D})$, for $r > 0$ sufficiently small. Therefore, $\lim_{r \to 0} f(r) = +\infty$, and by choosing l large enouph (i.e. $l > \frac{c}{\lambda}$) we contradict (1.20) and the proof of Theorem 1.1.3 is complete. $\square$

1.2. Sturm theory and integral criteria for HI-potentials

The existence of zeros for the solutions of linear ordinary differential equations of the following type

(1.21) $$(a(x)y')' + b(x)y = 0,$$

is of central importance for the identification of the class of HI-potentials – as well as the class of Bessel pairs that will be studied in the next section. There is fortunately a well developed theory to deal with that, starting with Sturm's first comparison principle, whose proof can be found in [**178**]. The rest of the chapter is self-contained.

THEOREM 1.2.1. (**Sturm's First Comparison Theorem**) *Suppose a, b_1, and b_2 are continuous functions on $I := [x_0, x^0]$ and $b_2(x) \geq b_1(x)$ on I. Let $y_1 \not\equiv 0$ be a solution of*

$$(a(x)y')' + b_1(x)y = 0 \quad on \quad I,$$

and assume $y_1(x)$ has exactly n (≥ 1) zeroes on I. Let $y_2(x)$ be a solution of

$$(a(x)y')' + b_2(x)y = 0 \quad on \quad I,$$

satisfying

(1.22) $$\frac{a(x)y_1'(x)}{y_1(x)} \leq \frac{a(x)y_2'(x)}{y_2(x)} \quad \left(resp., \frac{a(x)y_1'(x)}{y_1(x)} \geq \frac{a(x)y_2'(x)}{y_2(x)}\right)$$

at $x = x^0$ (resp., at $x = x_0$). Then $y_2(x)$ has at least n zeroes on I.

Note that the expression of the right [or left] of (1.22) at $x = x^0$ is considered to be ∞ if $y_2(x^0) = 0$ [or $y_1(x^0) = 0$]. In particular, (1.22) holds if $y_2(x^0) = 0$.

PROPOSITION 1.2.1. *Assume $a(x)$ and $b(x)$ are continuous functions on $I := [x_0, x^0]$. If the ordinary differential equation*

(1.23) $$(a(x)y')' + b(x)y = 0,$$

has a positive supersolution φ on I, then it has a positive solution ψ on I with

(1.24) $$\frac{a(x)\varphi'(x^0)}{\varphi(x^0)} = \frac{a(x)\psi'(x^0)}{\psi(x^0)} \quad if \quad \varphi(x^0) \neq 0,$$

and $\psi(x^0) \geq \varphi(x^0) = 0$, otherwise.

Proof: Let φ be a supersolution of the equation (1.23), i.e.
$$(a(x)\varphi')' + b(x)\varphi \leq 0,$$
and assume first that $\varphi(x^0) > 0$. Define
$$q(x) := -\frac{(a(x)\varphi')' + b(x)\varphi}{\varphi(x)} \geq 0.$$
Then φ is a positive solution of the equation
$$(a(x)\varphi')' + (b(x) + q(x))\varphi = 0, \quad x \in (x_0, x^0).$$
Let ψ be the unique solution of the following ordinary differential equation on I
$$(a(x)\psi')' + b(x)\psi = 0, \quad \psi(x^0) = \varphi(x^0), \quad \psi'(x^0) = \varphi'(x^0).$$
Since $q(x) + b(x) \geq b(x)$ on I and since φ does not have any zeroe on I, it follows from Proposition 1.2.1, that $\psi > 0$ on I.

Now assume $\varphi(x^0) = 0$ and let ψ be the unique solution of the following ordinary differential equation on (x_0, x^0)
$$(a(x)\psi')' + b(x)\psi = 0, \quad \psi(\bar{x}) = \varphi(\bar{x}), \quad \psi'(\bar{x}) = \varphi'(\bar{x}),$$
for some $\bar{x} \in (x_0, x^0)$. It follows from Proposition 1.2.1 that $\psi > 0$ on the interval (x_0, x^0) and consequently that $\psi(x^0) \geq 0$. $\square$

DEFINITION 1.2.2. A non-trivial solution of (1.21) is said to be *oscillatory* if there is a sequence $\{t_n\}$ tending to ∞ such that $x(t_n) = 0$. Otherwise, it is said to be *non-oscillatory*.

The first important result in the study of the oscillatory behaviour of solutions of ODEs is the celebrated comparison theorem of Sturm, which deals with second order self-adjoint equations of the form:

(1.25) $$lu \equiv \frac{d}{dx}[a(x)\frac{du}{dx}] + c(x)u = 0$$

(1.26) $$Lv \equiv \frac{d}{dx}[a(x)\frac{dv}{dx}] + C(x)v = 0$$

on a bounded open interval $\alpha < x < \beta$, where $a, c,$ and C are real-valued continuous functions and $a(x) > 0$ on $[\alpha, \beta]$.

THEOREM 1.2.3. **(Sturm's Comparison Theorem)** *Suppose $c(x) < C(x)$ in the bounded interval $\alpha < x < \beta$. If there exists a nontrivial real solution u of $lu = 0$ such that $u(\alpha) = u(\beta) = 0$, then every real solution of $Lv = 0$ has at least one zero in (α, β).*

Proof: Suppose to the contrary that v does not vanish in (α, β). It may be supposed without loss of generality that $v(x) > 0$ and also $u(x) > 0$ in (α, β). Multiplying the above equations with v and u, subtracting the resulting equations, and integrating over (α, β) yields

(1.27) $$\int_\alpha^\beta [(au')'v - (av')'u]dx = \int_\alpha^\beta \beta(C - c)uv dx.$$

Since the integrand on the left hand side is the derivative of $a(u'v - uv')$ and since $C(x) - c(x) > 0$, it follows that

(1.28) $$[a(x)(u'(x)v(x) - u(x)v'(x))]_\alpha^\beta > 0.$$

However, $u(\alpha) = u(\beta) = 0$ by hypothesis, and $u(x) > 0$ in (α, β), while $u'(\alpha) > 0$ and $u'(\beta) < 0$. This contradicts (1.28). □

The following criterion is a consequence of Sturm's Comparison Theorem.

THEOREM 1.2.4. **(Hille-Kneser)** *Set*
$$\omega^* = \limsup_{x \to \infty} x^2 c(x) \quad \text{and} \quad \omega_* = \liminf_{x \to \infty} x^2 c(x).$$

Then the equation

(1.29) $$u'' + c(x)u = 0,$$

is oscillatory if $\omega_* > \frac{1}{4}$ and nonoscillatory if $\omega^* < \frac{1}{4}$. On the other hand, the equation can be oscillatory or nonoscillatory if either ω_* or ω^* equals $\frac{1}{4}$.

Proof: If $\omega_* > \frac{1}{4}$, there exists a $\gamma > \frac{1}{4}$ and a positive number x_0 such that $c(x) - \gamma x^{-2} > 0$ for $x \geq x_0$. Since the Euler equation $v'' + \gamma x^{-2} v = 0$ is oscillatory for $\gamma > \frac{1}{4}$, the Sturm's comparison theorem shows that every solution of (1.29) has arbitrary large zeros.

If $\omega^* < \frac{1}{4}$, there exists $\gamma < \frac{1}{4}$ and a number $x_0 > 0$ such that $c(x) - \gamma x^{-2} < 0$ for $x \geq x_0$. If a solution of (1.29) had arbitrary large zeros, then every solution of $v'' + \gamma x^{-2} v = 0$ would have arbitrary large zeros by Sturm's comparison theorem, which is a contradiction. □

Now we are ready to establish oscillation criteria for equation (1.21).

THEOREM 1.2.5. *Assume a satisfies the condition*

(1.30) $$\int_\alpha^\infty \frac{1}{a(\tau)} d\tau = \infty.$$

(1) *If $a(t)$ and $b(t)$ satisfy for t sufficiently large,*

(1.31) $$a(t)b(t)\left(\int_\alpha^t \frac{1}{a(\tau)} d\tau\right)^2 > \frac{1}{4},$$

then equation (1.21) is oscillatory.

(2) *On the other hand, if $a(t)$ and $b(t)$ satisfy for t sufficiently large,*

(1.32) $$a(t)b(t)\left(\int_\alpha^t \frac{1}{a(\tau)} d\tau\right)^2 < \frac{1}{4},$$

then equation (1.21) is non-oscillatory.

Proof: Set
$$s(t) = \int_\alpha^t \frac{1}{a(\tau)} d\tau \quad \text{and} \quad u(s) = x(t(s)),$$
where $t(s)$ is the inverse function of $s(t)$. Then
$$x'(t) = \frac{ds}{dt} \frac{du}{ds} = \frac{1}{a(t)} \frac{du}{ds},$$
and equation (1.21) is transformed into the equation

(1.33) $$u'' + a(t(s))b(t(s))u = 0,$$

which has the form of (1.29). Since $a(t)$ is positive and satisfies for $t > \alpha$ and satisfies (1.30) the functions $s(t)$ and $t(s)$ are increasing and $s(t) \to \infty$ as $t \to \infty$. Hence equation (1.21) is oscillatory (non-oscillatory) if and only if equation (1.29) is oscillatory (non-oscillatory). Let $c(s) = a(t(s))b(t(s))$. Then conditions (1.31) and (1.32) coincide with those of the Hille-Kneser theorem. The proof is complete. □

We now state another integral criteria for the oscillatory behavior of equation (1.29) which is also due to E. Hille.

THEOREM 1.2.6. *Let c be a continuous function on* $\mathbb{R}$.
(1) *If* $\limsup\limits_{t \to \infty} t \int_t^\infty c(s)ds < \frac{1}{4}$, *then Eq. (1.29) is non-oscillatory,*
(2) *If* $\liminf\limits_{t \to \infty} t \int_t^\infty c(s)ds > \frac{1}{4}$, *then Eq. (1.29) is oscillatory.*

In order to prove Theorem 1.2.6 we first need to establish some preliminary results. Here the key idea is to study the existence of positive solutions of the associated Riccati equation

(1.34) $$v' + v^2 + c(x) = 0.$$

We start with the following simple lemma.

LEMMA 1.2.7. *Suppose that equation (1.29) is non-oscillatory and let u be a solution of (1.29) that is positive for $x \geq \alpha$, then u is monotone increasing and concave downwards for $x > \alpha$. Furthermore, u' is positive and monotone decreasing towards a non-negative limit.*

Proof: From (1.29), we have for $x_2 > x_1 > \alpha$ that

$$u'(x_2) - u'(x_1) = -\int_{x_1}^{x_2} c(t)u(t)dt \leq 0.$$

It follows that u' is non-increasing and that u is concave downwards for $x > \alpha$. Since the graph of u does not intersect the x-axis for $x > \alpha$, we must have that $u'(x) > 0$ for $x > \alpha$. □

The following lemma shows the connection between the oscillatory behavior of solutions of (1.29) and the existence of solutions of the associated Riccati equation (1.34).

LEMMA 1.2.8. *Equation (1.29) is non-oscillatory if and only if the integral equation*

(1.35) $$v(x) = \int_x^\infty [v^2(t) + c(t)]dt$$

has a solution for sufficiently large x.

Proof: If there exists a solution $u(x)$ of (1.29) such that $u(x) \neq 0$ for sufficiently large x, then $v(x) := \frac{u'(x)}{u(x)}$ is a solution of (1.35) for sufficiently large x.

Now suppose there is a finite $\alpha > 0$ such that (1.35) has a solution for $x \geq \alpha$. It follows from the form of the equation that $v^2 \in L^1(\alpha, \infty)$ and $v(x)$ is positive, monotone decreasing, absolutely continuous function. Differentiating with respect to x shows that $v(x)$ satisfies (1.34) for almost all x. Hence if we put $u(x) = e^{\int_\alpha^x v(t)dt}$, then u satisfies (1.29) for almost all $x \geq \alpha$ and $u(x) \geq 1$. Equation (1.29) is therefore non-oscillatory. □

Introduce now the notation

$$y(x) = xv(x), \quad d(x) = x\int_x^\infty c(t)dt,$$

in terms of which (1.35) becomes

$$y(x) = x \int_x^\infty y^2(t) \frac{dt}{t^2} + d(x). \tag{1.36}$$

The following comparison lemma will be needed for the proof of Theorem 1.2.6.

LEMMA 1.2.9. *Consider the ordinary differential equations*

$$U'' + C(x)U = 0, \tag{1.37}$$

and

$$u'' + c(x)u = 0. \tag{1.38}$$

Set $D(x) = x \int_x^\infty C(t)dt$ and $d(x) = x \int_x^\infty c(t)dt$. If equation (1.37) is non-oscillatory and $D(x) \geq d(x)$ for $x \geq \alpha$, then equation (1.38) is also non-oscillatory.

Proof: By Lemma 1.2.8, the integral equation

$$Y(x) = x \int_x^\infty Y^2(t) \frac{dt}{t^2} + D(x),$$

has a solution $Y(x)$ for $x \geq \beta$ for some $\beta > 0$. We now consider equation (1.36) for $x \geq \gamma = \max\{\alpha, \beta\}$, and define successive approximations by writing

$$y_0(x) = Y(x), \quad y_n(x) = x \int_x^\infty y_{n-1}^2(t) \frac{dt}{t^2} + d(x).$$

We have

$$y_1(x) = x \int_x^\infty Y^2(t) \frac{dt}{t^2} + d(x) \leq x \int_x^\infty Y^2(t) \frac{dt}{t^2} + D(x) = Y(x) = y_0(x).$$

Since

$$y_n(x) - y_{n-1}(x) = x \int_x^\infty [y_{n-1}^2(t) - y_{n-2}^2(t)] \frac{dt}{t^2},$$

we see that $y_{n-1}(x) \geq y_n(x) \geq d(x)$ for all x and all n. Hence $\lim y_n(x) = y(x)$ exists and satisfies (1.36). It then follows from Lemma 1.2.8 that equation (1.38) is non-oscillatory as claimed. $\square$

Proof of Theorem 1.2.6: To prove (1) we apply Lemma 1.2.9. Indeed $D(x) = \frac{1}{4}$ corresponds to $C(x) = \frac{1}{4}x^{-2}$ and $U(x) = x^{1/2} \log(x)$ so that the corresponding equation (1.37) in non-oscillatory. Since, $\limsup_{t \to \infty} t \int_t^\infty c(s)ds < \frac{1}{4}$, it follows from Theorem 1.2.6 that equation (1.29) is non-oscillatory.

(2) We shall show that if the equation (1.29) is non-oscillatory, then we necessarily have that $\liminf_{t \to \infty} t \int_t^\infty c(t) \leq \frac{1}{4}$. Since (1.29) is assumed to be non-oscillatory, equation (1.36) has a solution $y(x)$ for x sufficiently large. Define

$$y_* := \liminf_{t \to \infty} y(t), \quad d_* := \liminf_{t \to \infty} t \int_t^\infty c(t).$$

Since for every $\epsilon > 0$ $(y_*)^2 - \epsilon \leq t \int_t^\infty y^2(s) \frac{ds}{s^2}$ for x sufficiently large, it follows from (1.36) that $y_* \geq (y_*)^2 + d_*$, which is possible only if $d_* \leq \frac{1}{4}$. The proof is now complete. $\square$

The following summarizes the connection between the oscillatory behavior of equation (1.29) and the existence of positive solutions of $(\mathcal{B}_P)$ on a finite interval.

COROLLARY 1.2.1. *Let P be a positive locally integrable function on $\mathbb{R}$.*
(1) *If $\liminf_{r \to 0} \log r \int_0^r sP(s)ds > -\infty$, then for every $R > 0$, there exists $\alpha := \alpha(R) > 0$ such that the scaled function $P_\alpha(r) := \alpha^2 P(\alpha r)$ is a HI-potential on $(0, R)$.*
(2) *If $\lim_{r \to 0} \log r \int_0^r sP(s)ds = -\infty$, then there are no $\alpha, c > 0$, for which $P_{\alpha,c}(r) := cP(\alpha r)$ is a HI-potential on $(0, R)$.*

Proof: The change of variable $s = -\log r$, $z(s) = \varphi(e^{-s})$ maps a solution φ of the equation $\mathcal{B}_P$ (i.e., $\varphi'' + \frac{1}{r}\varphi' + P(r)\varphi = 0$) to a solution of the

$(\mathcal{B}'_P)$ $\qquad\qquad z'' + e^{-2s}P(e^{-s})z(s) = 0.$

It is clear that the equation $z''(s) + a(s)z(s) = 0$ where $a(s) = e^{-2s}P(e^{-s})$ is non-oscillatory (i.e., has a positive solution on some interval (b, ∞)) if and only if $(\mathcal{B}_P)$ has a positive solution on some interval $(0, R)$. The criteria of the preceeding corollary, coupled with the scaling property in Proposition 1.1.2, yield the result.

1.3. The class of Bessel pairs

We shall say that a couple of C^1-functions (V, W) is a *n-dimensional Bessel pair on $(0, R)$*, provided there exists a scalar $c > 0$ such that the ordinary differential equation

$(\mathcal{B}_{V,cW})$ $\qquad y''(r) + (\frac{n-1}{r} + \frac{V_r(r)}{V(r)})y'(r) + \frac{cW(r)}{V(r)}y(r) = 0,$

has a positive solution on the interval $(0, R)$. The *weight* of such a pair is then defined as

(1.39) $\qquad \beta(V, W; R) = \sup\{c;\ (\mathcal{B}_{V,cW}) \text{ has a positive solution on } (0, R)\}.$

Note that we can rewrite $(\mathcal{B}_{V,cW})$ as

(1.40) $\qquad\qquad (r^{n-1}V(r)y')' + cr^{n-1}W(r)y = 0,$

which means that (V, W) is a *n-dimensional Bessel pair on $(0, R)$* if and only if the pair $(\tilde{V}, \tilde{W}) := (r^{n-1}V, r^{n-1}W)$ is a 1-*dimensional Bessel pair –or simply a Bessel pair on $(0, R)$–* meaning that the ODE

(1.41) $\qquad\qquad (\tilde{V}(r)y')' + c\tilde{W}(r)y = 0$

has a positive solution on the interval $(0, R)$.

A simple change of variables in the corresponding ODEs, gives the following relationship between the HI-potentials defined in the last section and Bessel pairs.

PROPOSITION 1.3.1. *Assume $n \geq 3$. The following assertions are then equivalent:*
(1) *P is a HI-potential on $(0, R)$ with $\beta(P, R) = 1$,*
(2) *For any $0 \leq \lambda \leq n-2$, the pair $\left(r^{-\lambda}, (\frac{n-\lambda-2}{2})^2 r^{-\lambda-2} + r^{-\lambda}P(r)\right)$ is a n-dimensional Bessel pair on $(0, R)$.*
(3) *For any $1 \leq \alpha \leq n-1$, the couple $\left(r^\alpha, \frac{(\alpha-1)^2}{4}r^{\alpha-2} + r^\alpha P(r)\right)$ is a Bessel pair on $(0, R)$.*

Proof: It follows from a straightforward calculation that $y(r)$ is a solution of $(\mathcal{B}_P)$ if and only if $\varphi := r^{-\frac{n-\lambda-2}{2}} y(r)$ is a solution of

$$\varphi'' + \frac{n-\lambda-1}{r}\varphi' + \left(P(r) + \frac{(n-\lambda-2)^2}{4r^2}\right)\varphi = 0.$$

Note that the above equation is the corresponding ordinary differential equation $(\mathcal{B}_{V,W})$ for the n-dimensional Bessel pair $(V,W) = \left(r^{-\lambda}, (\frac{n-\lambda-2}{2})^2 r^{-\lambda-2} + r^{-\lambda}P(r)\right)$. Therefore 1) and 2) are equivalent. That they are equivalent to 3) follows from (1.40). □

COROLLARY 1.3.1. *(Explicit Bessel pairs) Assume $n \geq 3$, and $0 \leq \lambda \leq n-2$. We then have the following:*

(1) *For any $R > 0$, $\left(r^{-\lambda}, r^{-\lambda-2}\right)$ is a n-dimensional Bessel pair on $(0, R)$, and*

(1.42) $$\beta\left(r^{-\lambda}, r^{-\lambda-2}\right) = \left(\frac{n-\lambda-2}{2}\right)^2.$$

(2) *For any $R > 0$, the couple $\left(r^{-\lambda}, (\frac{n-\lambda-2}{2})^2 r^{-\lambda-2} + \frac{z_0^2}{R^2} r^{-\lambda}\right)$ is a n-dimensional Bessel pair on $(0, R)$, and*

(1.43) $$\beta\left(r^{-\lambda}, \left(\frac{n-\lambda-2}{2}\right)^2 r^{-\lambda-2} + \frac{z_0^2}{R^2} r^{-\lambda}\right) = 1.$$

(3) *For any $R > 0$, the couple*

(1.44) $$(V, W_\rho) := \left(r^{-\lambda}, \left(\frac{n-\lambda-2}{2}\right)^2 r^{-\lambda-2} + \frac{1}{4} r^{-2-\lambda} \left(\log\frac{\rho}{r}\right)^{-2}\right),$$

where $\rho > Re$, is a n-dimensional Bessel pair on $(0, R)$, and

(1.44) $$\beta(V, W_\rho) = 1.$$

(4) *The couple $\left(r^{-\lambda}, (\frac{n-\lambda-2}{2})^2 r^{-\lambda-2} + r^{-\lambda-\alpha}\right)$ is a n-dimensional Bessel pair on some $(0, R_\alpha)$, whenever $0 \leq \alpha < 2$.*

Proof: Statements 1), 2), and 3) follow directly from Theorem 1.1.3 and Proposition 1.3.1. To prove 4) notice that by Corollary 1.2.1, r^α is a Bessel potential for any $\alpha < 2$. The proof of 4) follows form Proposition 1.3.1. □

We now make an important connection between Bessel pairs and the oscillatory behavior of the following related equations. For that, we rewrite again $(\mathcal{B}_{V,W})$ as

$$(r^{n-1} V(r) y')' + r^{n-1} W(r) y = 0,$$

and then by setting $s = \frac{1}{r}$ and $x(s) = y(r)$, we see that y is a solution of $(\mathcal{B}_{V,W})$ on an interval $(0, \delta)$ if and only if x is a positive solution for the equation

(1.45) $$\left(s^{-(n-3)} V(\tfrac{1}{s}) x'(s)\right)' + s^{-(n+1)} W(\tfrac{1}{s}) x(s) = 0 \quad \text{on} \quad (\tfrac{1}{\delta}, \infty).$$

As in the previous section, the fact that (V, W) is a Bessel pair or not is closely related to the oscillatory behavior of the equation (1.45).

THEOREM 1.3.1. *Let V and W be positive C^1-functions on $(0, R)$. Assume*

(1.46) $$\int_0^R \frac{1}{\tau^{n-1} V(\tau)} d\tau = +\infty \quad \text{and} \quad \int_0^R \tau^{n-1} V(\tau) d\tau < \infty.$$

(1) If

(1.47) $$\limsup_{r\to 0} r^{2(n-1)} V(r) W(r) \Big(\int_r^R \frac{1}{\tau^{n-1} V(\tau)} d\tau\Big)^2 < \frac{1}{4},$$

then (V, W) is a n-dimensional Bessel pair on $(0, \rho)$ for some $\rho > 0$.

(2) On the other hand, if

(1.48) $$\liminf_{r\to 0} r^{2(n-1)} V(r) W(r) \Big(\int_r^R \frac{1}{\tau^{n-1} V(\tau)} d\tau\Big)^2 > \frac{1}{4},$$

then there is no interval $(0, \rho)$ on which (V, W) is a n-dimensional Bessel pair.

Proof: The proof follows from Theorem 1.2.5 applied to the ordinary differential equation (1.45). □

The above integral criterium allows to show the following extension of Proposition 1.3.1.

THEOREM 1.3.2. *Let V be a strictly positive C^1-function on $(0, R)$ such that*

(1.49) $$\frac{rV_r(r)}{V(r)} + \lambda \geq 0 \text{ on } (0, R) \text{ and } \lim_{r\to 0} \frac{rV_r(r)}{V(r)} + \lambda = 0,$$

where $\lambda \leq n - 2$. Then for any HI-potential P on $(0, R)$ and any $c \leq \beta(P; R)$, the couple $(V, W_{\lambda,c})$ is a n-dimensional Bessel pair, where

(1.50) $$W_{\lambda,c}(r) = V(r)\Big(\big(\frac{n-\lambda-2}{2}\big)^2 r^{-2} + cP(r)\Big).$$

Moreover, $\beta(V, W_{\lambda,c}; R) = 1$ for all $c \leq \beta(P; R)$.

Proof: Write $\frac{V_r(r)}{V(r)} = -\frac{\lambda}{r} + f(r)$ where $f(r) \geq 0$ on $(0, R)$ and $\lim_{r\to 0} rf(r) = 0$. In order to prove that $\big(V(r), V(r)\big(\big(\frac{n-\lambda-2}{2}\big)^2 r^{-2} + cP(r)\big)\big)$ is a n-dimensional Bessel pair, we need to show that the equation

(1.51) $$y'' + \Big(\frac{n-\lambda-1}{r} + f(r)\Big) y' + \Big(\big(\frac{n-\lambda-2}{2}\big)^2 r^{-2} + cP(r)\Big) y(r) = 0,$$

has a positive solution on $(0, R)$. But first we note that the equation

$$x'' + \big(\frac{n-\lambda-1}{r}\big) x' + \Big(\big(\frac{n-\lambda-2}{2}\big)^2 r^{-2} + cP(r)\Big) x(r) = 0,$$

has a positive solution on $(0, R)$ whenever $c \leq \beta(P; R)$. Since $f(r) \geq 0$ and since, by Lemma 1.1.2, $x'(r) \leq 0$, we get that x is a positive subsolution for the equation (1.51) on $(0, R)$, and thus it has a positive solution of $(0, R)$. This means that $\beta(V, W_{\lambda,c}; R) \geq 1$.

For the reverse inequality, we shall use the criterium in Theorem 1.3.1. Indeed apply (1.47) to $V(r)$ and $W_1(r) = C\frac{V(r)}{r^2}$ to get

$$\lim_{r\to 0} r^{2(n-1)}V(r)W_1(r)\Big(\int_r^R \frac{1}{\tau^{n-1}V(\tau)}d\tau\Big)^2 = C\lim_{r\to 0} r^{2(n-2)}V^2(r)\Big(\int_r^R \frac{1}{\tau^{n-1}V(\tau)}d\tau\Big)^2$$

$$= C\Big(\lim_{r\to 0} r^{(n-2)}V(r)\int_r^R \frac{1}{\tau^{n-1}V(\tau)}d\tau\Big)^2$$

$$= C\Big(\lim_{r\to 0} \frac{\frac{1}{r^{n-1}V(r)}}{\frac{(n-2)r^{n-3}V(r)+r^{n-2}V_r(r)}{r^{2(n-2)}V^2(r)}}\Big)^2$$

$$= C\Big(\lim_{r\to 0} \frac{1}{(n-2)+r\frac{V_r(r)}{V(r)}}\Big)^2$$

$$= \frac{C}{(n-\lambda-2)^2}.$$

For $(V, CV(r^{-2}+cP))$ to be a n-dimensional Bessel pair, it is necessary that $\frac{C}{(n-\lambda-2)^2} \leq \frac{1}{4}$, and the proof for the best constant is complete. □

COROLLARY 1.3.2. *Let V and W be positive C^1-functions on $(0, +\infty)$. Assume that*

(1.52) $$\lim_{r\to 0} r\frac{V_r(r)}{V(r)} = -\lambda \text{ and } \lambda \leq n-2.$$

(1) *If* $\limsup_{r\to 0} r^2 \frac{W(r)}{V(r)} < (\frac{n-\lambda-2}{2})^2$, *then (V,W) is a n-dimensional Bessel pair on some interval $(0, \rho)$.*
(2) *On the other hand, if* $\liminf_{r\to 0} r^2 \frac{W(r)}{V(r)} > (\frac{n-\lambda-2}{2})^2$, *then there is no $R > 0$ such that (V,W) is a n-dimensional Bessel pair on the interval $(0, R)$.*

Proof: To prove 1) assume

$$\limsup_{r\to 0} r^2 \frac{W(r)}{V(r)} < (\frac{n-\lambda-2}{2})^2.$$

With an argument similar to that of Theorem 1.3.2 we have

$$\limsup_{r\to 0} r^{2(n-1)}V(r)W_1(r)\Big(\int_r^R \frac{1}{\tau^{n-1}V(\tau)}d\tau\Big)^2 < \frac{1}{4},$$

and 1) then follows from Theorem 1.3.1. Proof of 2) follows from a similar argument. □

By applying the above to $V \equiv 1$ and $W(r) := \frac{(n-2)^2}{4}r^{-2} + \alpha r^{-a}$, we get the following result.

COROLLARY 1.3.3. *If $n \geq 2$, $a \geq 2$ and $\alpha > 0$, then there is no $R > 0$ such that the couple $\Big(1, \frac{(n-2)^2}{4}r^{-2} + \alpha r^{-a}\Big)$ is a n-dimensional Bessel pair on $(0, R)$.*

1.4. Further comments

The book by Agarwal-Bohner-Li [**12**] is a good reference on the oscillatory theory of first and second order differential equations. It addresses delay and ordinary differential equations as well as non-linear differential systems. Another good source on the oscillatory behaviour of ordinary differential equations and Sturm theory is Hartman's book [**178**]. Theorems 1.2.4 and 1.2.6 were proved by E. Hille in [**180**]. The criterium at infinity for studying the oscillatory behavior of equation (1.21) (Theorem 1.2.5) is more recent and is due to Sugie-Kita-Yamaoka [**254**]. See [**13**] and [**186**] for more recent results about the oscillatory behaviour of solutions of second order differential equations. The notions of HI-potential (originally named Bessel potential) and Bessel pairs were introduced by Ghoussoub-Moradifam [**162, 163**] in their work that connected improvements of Hardy inequalities – studied in the next chapters– to the oscillatory behaviour of associated differential equations.

It is important to relate the above notion of Bessel pairs to other notions introduced for the same purpose of extending Hardy's inequality. Say that (V, W) is a *Muckenhought pair* [**225**] on the interval $(0, R)$ if

$$(1.53) \qquad \gamma(V, W, R) := \sup_{0 < r < R} \left(\int_0^r W(t)\, dt \right) \left(\int_r^R \frac{1}{V(t)}\, dt \right) < \infty.$$

Problem: Describe the relationship between Bessel pairs and Muckenhought pairs, as well as the constants $\beta(V, W, R)$ and $\gamma(V, W, R)$.

CHAPTER 2

The Classical Hardy Inequality and Its Improvements

A C^1-function P is a HI-potential on an interval $(0, R)$ if and only if the following improved Hardy inequality holds on the ball B_R of radius R in $\mathbb{R}^n$:

$\int_{B_R} |\nabla u|^2 dx - (\frac{n-2}{2})^2 \int_{B_R} \frac{u^2}{|x|^2} dx \geq c \int_{B_R} P(|x|) u^2 dx$ for all $u \in H_0^1(B_R)$.

A characterization of the best possible constant $c := c(P, R)$ satisfying the above inequality is given by the HIP-constant $\beta(P, R)$ introduced in the last chapter. We show that if $\Omega \subset B_R$ is a smooth domain containing 0 in its interior, then for every $0 \leq \lambda \leq \beta(P, R)$, the best constant

$$\mu_\lambda(P, \Omega) := \inf_{\substack{u \in H_0^1(\Omega) \\ u \neq 0}} \frac{\int_\Omega |\nabla u|^2 \, dx - \lambda \int_\Omega P(|x|) u^2 \, dx}{\int_\Omega \frac{|u|^2}{|x|^2} \, dx}$$

is always equal to $\frac{(n-2)^2}{4}$ and is never attained in $H_0^1(\Omega)$.

2.1. One dimensional Poincaré inequalities

A natural proof of the Hardy inequality consists of associating to any smooth radial positive functions $u \in C_c^2(B_R)$, where B_R is the ball of radius R in $\mathbb{R}^n$, the function $v(r) = u(r) r^{(n-2)/2}$ where $r = |x|$. Denoting ω_n the volume of the unit sphere, one can estimate the quantity

$$H(u) := \int_\Omega |\nabla u|^2 dx - (\frac{n-2}{2})^2 \int_\Omega \frac{u^2}{|x|^2} dx,$$

as follows:

$$\begin{aligned} H(u) &= \omega_n \int_0^R |\frac{n-2}{2} r^{-n/2} v(r) - r^{1-n/2} v'(r)|^2 r^{n-1} dr - (\frac{n-2}{2})^2 \omega_n \int_0^R \frac{v^2(r)}{r} dr \\ &= \omega_n (\frac{n-2}{2})^2 \int_0^R v^2 [(1 - \frac{2 v'(r) r}{(n-2) v(r)})^2 - 1] \frac{dr}{r} \\ &= \omega_n \int_0^R v'(r)^2 r \, dr - \omega_n (\frac{n-2}{2}) \int_0^R v(r) v'(r) dr \\ &= \omega_n \int_0^R v'(r)^2 r \, dr. \end{aligned}$$

This is clearly non-negative and we are essentially done for the classical Hardy inequality. For an improvement, one needs to establish sharper lower estimates for $\int_0^R v'(r)^2 r \, dr$, which leads to one dimensional Poincaré and Poincaré-Wirtinger inequalities of the following type.

PROPOSITION 2.1.1. For $a < b$, let k be a differentiable function on (a, b), and let φ be a strictly positive differentiable function on (a, b). Then, every $h \in C^1([a, b])$ with

$$(2.1) \quad -\infty < \lim_{r \to a} k(r)|h(r)|^2 \frac{\varphi'(r)}{\varphi(r)} = \lim_{r \to b} k(r)|h(r)|^2 \frac{\varphi'(r)}{\varphi(r)} < \infty,$$

satisfies the following inequality:

$$(2.2) \quad \int_a^b |h'(r)|^2 k(r) dr \geq \int_a^b -|h(r)|^2 \left(\frac{k'(r)\varphi'(r) + k(r)\varphi''(r)}{\varphi(r)} \right) dr.$$

Moreover, assuming (2.1), equality holds if and only if $h(r) = \varphi(r)$ for all $r \in (a, b)$.

Proof: Define $\psi(r) = h(r)/\varphi(r)$, $r \in [a, b]$. Then

$$\int_a^b |h'(r)|^2 k(r) dr = \int_a^b |\psi(r)|^2 |\varphi'(r)|^2 k(r) dr + \int_a^b 2\varphi(r)\varphi'(r)\psi(r)\psi'(r)k(r) dr$$

$$+ \int_a^b |\varphi(r)|^2 |\psi'(r)|^2 k(r) dr$$

$$= \int_a^b |\psi(r)|^2 |\varphi'(r)|^2 k(r) dr - \int_a^b |\psi(r)|^2 (k\varphi\varphi')'(r) dr$$

$$+ \int_a^b |\varphi(r)|^2 |\psi'(r)|^2 k(r) dr$$

$$= \int_a^b |\psi(r)|^2 (|\varphi'(r)|^2 k(r) - (k\varphi\varphi')'(r)) dr + \int_a^b |\varphi(r)|^2 |\psi'(r)|^2 k(r) dr.$$

Hence, we have

$$\int_a^b |h'(r)|^2 k(r) dr = \int_a^b -|h(r)|^2 \left(\frac{k'(r)\varphi'(r) + k(r)\varphi''(r)}{\varphi} \right) dr + \int_a^b |\varphi(r)|^2 |\psi'(r)|^2 k(r) dr$$

$$\geq \int_a^b -|h(r)|^2 \left(\frac{k'(r)\varphi'(r) + k(r)\varphi''(r)}{\varphi(r)} \right) dr,$$

and therefore (2.2) holds. Note that the last inequality is obviously an idendity if and only if $h(r) = \varphi(r)$ for all $r \in (a, b)$. The proof is complete. □

By applying Proposition 2.1.1 to the weight $k(r) = r$, we obtain the following generalization of the 2-dimensional Poincaré inequality.

COROLLARY 2.1.1. (Generalized 2-dimensional Poincaré inequality) *For* $0 \leq a < b$, *let* φ *be a strictly positive differentiable function on* (a, b). *Then, every* $h \in C^1([a, b])$ *with*

$$(2.3) \quad -\infty < \lim_{r \to a} r|h(r)|^2 \frac{\varphi'(r)}{\varphi(r)} = \lim_{r \to b} r|h(r)|^2 \frac{\varphi'(r)}{\varphi(r)} < \infty,$$

satisfies the following inequality:

$$(2.4) \quad \int_a^b |h'(r)|^2 r \, dr \geq \int_a^b -|h(r)|^2 \left(\frac{\varphi'(r) + r\varphi''(r)}{\varphi(r)} \right) dr.$$

Moreover, under assumption (2.3), equality holds if and only if $h(r) = \varphi(r)$ *for all* $r \in (a, b)$.

By applying Proposition 2.1.1 to the weight $k(r) = 1$, we obtain the following generalization of the 2-dimensional Poincaré-Wirtinger inequality.

COROLLARY 2.1.2. (Generalized Poincaré-Wirtinger inequality) *For $a < b$, let φ be a strictly positive real valued differentiable function on (a,b). Then, every $h \in C^1([a,b])$ with*

(2.5) $$-\infty < \lim_{r \to a} |h(r)|^2 \frac{\varphi'(r)}{\varphi(r)} = \lim_{r \to b} |h(r)|^2 \frac{\varphi'(r)}{\varphi(r)} < \infty,$$

satisfies the following inequality:

(2.6) $$\int_a^b |h'(r)|^2 dr \geq \int_a^b -|h(r)|^2 \frac{\varphi''(r)}{\varphi(r)} dr.$$

Moreover, under assumption (2.5), equality holds if and only if $h(r) = \varphi(r)$ for all $r \in (a,b)$.

REMARK 2.1.1. Note that all inequalities presented above hold when we replace (2.1), (2.3), and (2.5) with the following weaker conditions:

$$\liminf_{r \to b} k(r)|h(r)|^2 \frac{\varphi'(r)}{\varphi(r)} \geq \limsup_{r \to a} k(r)|h(r)|^2 \frac{\varphi'(r)}{\varphi(r)},$$

$$\liminf_{r \to b} r|h(r)|^2 \frac{\varphi'(r)}{\varphi(r)} \geq \limsup_{r \to a} r|h(r)|^2 \frac{\varphi'(r)}{\varphi(r)},$$

$$\liminf_{r \to b} |h(r)|^2 \frac{\varphi'(r)}{\varphi(r)} \geq \limsup_{r \to a} |h(r)|^2 \frac{\varphi'(r)}{\varphi(r)},$$

respectively, provided both sides in the above inequalities are not equal to ∞.

In the sequel, we shall repeatedly use the notion of a *symmetric-decreasing rearrangement* of a measurable function, which is defined as follows: Starting with a general set $A \subset \mathbb{R}^n$, denote by χ_A^* the characteristic function of a ball of volume $|A|$ centered at the origin. The symmetric-decreasing rearrangement of a measurable function $u : \Omega \to \mathbb{R}$ is then defined as:

$$u^*(x) = \int_0^{+\infty} \chi^*_{\{|u|>t\}}(x)\,dt.$$

Note that u^* is clearly symmetric-decreasing, and that $\|u^*\|_p = \|u\|_p$ for any p, since the rearrangement does not change the values of u, while only changing the places where these values occur. What is less obvious is that

(2.7) $$\int_\Omega |\nabla u^*|^2\,dx \leq \int_\Omega |\nabla u|^2\,dx,$$

a proof of which can be found in [**203**]

2.2. HI-potentials and improved Hardy inequalities on balls

The following theorem describes the crucial role of HI-potentials in improving the classical Hardy inequality.

THEOREM 2.2.1. *Let P be a decreasing non-negative C^1-function on $(0, R)$. The following properties are then equivalent:*

(1) *P is a HI-potential on $(0, R)$.*
(2) *There exists $c > 0$ such that for all $n \geq 2$, the following inequality holds for any $u \in H_0^1(B_R)$, where B_R is a ball of radius R in $\mathbb{R}^n$:*

(H_{cP}) $$\int_{B_R} |\nabla u|^2 dx - \left(\frac{n-2}{2}\right)^2 \int_{B_R} \frac{u^2}{|x|^2} dx \geq c \int_{B_R} P(|x|) u^2 dx.$$

Moreover, the largest constant c for which (H_{cP}) holds is equal to $\beta(P, R)$.

We prove that 1) implies 2) in Theorem 2.2.1 by establishing the following.

PROPOSITION 2.2.1. *Let Ω be a bounded smooth domain in $\mathbb{R}^n$ ($n \geq 2$) with $0 \in \Omega$, and set $R = (|\Omega|/\omega_n)^{1/n}$. Suppose P is a continuous function on $(0, R)$ and that φ is a positive C^2-function on $(0, R)$ such that*

(2.8) $$0 \leq P(r) \leq -\frac{\varphi'(r) + r\varphi''(r)}{r\varphi(r)} \quad \text{for all } 0 < r < R,$$

(2.9) $$\liminf_{r \to 0} r\frac{\varphi'(r)}{\varphi(r)} \geq 0 \quad \text{and} \quad \limsup_{r \to R} \frac{\varphi'(r)}{\varphi(r)} < \infty,$$

(2.10) $$\left(\frac{n-2}{2}\right)^2 \frac{1}{r^2} + P(r) \text{ is a decreasing function of } r.$$

Then, we have for all $u \in H_0^1(\Omega)$,

(2.11) $$\int_\Omega |\nabla u|^2 dx - \left(\frac{n-2}{2}\right)^2 \int_\Omega \frac{u^2}{|x|^2} dx \geq \int_\Omega P(|x|) u^2 dx.$$

Proof: We first prove the inequality for smooth radial positive functions on the ball $\Omega = B_R$. For such $u \in C_0^2(B_R)$, define $v(r) = u(r) r^{(n-2)/2}$ where $r = |x|$. In view of Corollary 2.1.1, we can estimate $H(u) := \int_\Omega |\nabla u|^2 dx - \left(\frac{n-2}{2}\right)^2 \int_\Omega \frac{u^2}{|x|^2} dx$ as follows:

$$\begin{aligned}
H(u) &= \omega_n \int_0^R \left|\frac{n-2}{2} r^{-n/2} v(r) - r^{1-n/2} v'(r)\right|^2 r^{n-1} dr - \left(\frac{n-2}{2}\right)^2 \omega_n \int_0^R \frac{v^2(r)}{r} dr \\
&= \omega_n \left(\frac{n-2}{2}\right)^2 \int_0^R v^2 \left[\left(1 - \frac{2v'(r) r}{(n-2) v(r)}\right)^2 - 1\right] \frac{dr}{r} \\
&= \omega_n \int_0^R v'(r)^2 r - \omega_n \left(\frac{n-2}{2}\right) \int_0^R v(r) v'(r) dr \\
&= \omega_n \int_0^R v'(r)^2 r \\
&\geq \omega_n \int_0^R -v^2(r) \left(\frac{\varphi'(r) + r\varphi''(r)}{\varphi(r)}\right) dr \\
&= \omega_n \int_0^R -u^2(r) \left(\frac{\varphi'(r) + r\varphi''(r)}{\varphi(r)}\right) r^{n-2} dr \\
&= -\int_\Omega u^2(x) \left(\frac{\varphi'(|x|) + |x|\varphi''(|x|)}{|x|\varphi(|x|)}\right) dx \\
&\geq \int_\Omega P(|x|) u^2 dx.
\end{aligned}$$

Hence, inequality (2.11) holds for radial smooth positive functions. If now u is a non-radial function on a general domain Ω, we use the symmetrization argument as follows. Let B_R be a ball having the same volume as Ω with $R = (|\Omega|/\omega_n)^{1/n}$ and let u^* be the symmetric decreasing rearrangement of the function $|u|$. As mentioned above, if $u \in H_0^1(\Omega)$ then $u^* \in H_0^1(B_R)$ and has the same L^p-norm as u. Moreover, the symmetrization $u \to u^*$ decreases the Dirichlet energy, while increasing the integrals $\int_\Omega ((\frac{n-2}{2})^2 \frac{1}{|x|^2} + P(|x|)) u^2 dx$, since the weight $(\frac{n-2}{2})^2 \frac{1}{|x|^2} + P(|x|)$ is a decreasing function of $|x|$. Hence, (2.11) holds for every $u \in H_0^1(\Omega)$. $\square$

In order to prove that 2) implies 1) in Theorem 2.2.1, we shall need the following result.

LEMMA 2.2.2. *Let P be a continuous positive function on $(0, R)$, and let B_R be a ball of radius R in $\mathbb{R}^n$ $(n \geq 2)$. Assume that*

$$\int_{B_R} \left(|\nabla u|^2 - \left(\tfrac{n-2}{2}\right)^2 \tfrac{u^2}{|x|^2} - P(|x|)u^2\right) dx \geq 0 \text{ for all } u \in H_0^1(B_R).$$

Then, there exists a C^2-supersolution u to the equation

(2.12)
$$\begin{cases} -\Delta u - \left(\tfrac{n-2}{2}\right)^2 \tfrac{u}{|x|^2} - P(|x|)u = 0 & \text{in } B_R, \\ u > 0 & \text{in } B_R \setminus \{0\}, \\ u = 0 & \text{in } \partial B_R. \end{cases}$$

Proof: Define

$$\lambda_1(P) := \inf\left\{\frac{\int_{B_R} |\nabla \psi|^2 - \left(\tfrac{n-2}{2}\right)^2 \tfrac{|\psi|^2}{|x|^2} - P|\psi|^2}{\int_{B_R} |\psi|^2}; \ \psi \in C_0^\infty(B_R \setminus \{0\})\right\}.$$

By our assumption $\lambda_1(P) \geq 0$. Let (φ_k, λ_1^k) be the first eigenpair for the problem

$$(L - \lambda_1(P) - \lambda_1^k)\varphi_n = 0 \text{ on } B_R \setminus B_{\frac{R}{k}}$$
$$\varphi_k(r) = 0 \text{ on } \partial(B_R \setminus B_{\frac{R}{k}}),$$

where $L = -\Delta - \left(\tfrac{n-2}{2}\right)^2 \tfrac{1}{|x|^2} - P$, and $B_{\frac{R}{k}}$ is a ball of radius $\tfrac{R}{k}$, $n \geq 2$. The eigenfunctions can be chosen in such a way that $\varphi_k > 0$ on $B_R \setminus B_{\frac{R}{k}}$ and $\varphi_R(b) = 1$, for some $b \in B_R$ with $\tfrac{R}{2} < |b| < R$.

Note that $\lambda_1^k \downarrow 0$ as $k \to \infty$. Harnak's inequality yields that for any compact subset K, $\tfrac{\max_K \varphi_k}{\min_K \varphi_k} \leq C(K)$ with the latter constant being independent of φ_k. Also standard elliptic estimates yield that the family (φ_k) have uniformly bounded derivatives on the compact sets $B_R \setminus B_{\frac{R}{k}}$. Therefore, there exists a subsequence $(\varphi_{k_{l_2}})_{l_2}$ of $(\varphi_k)_k$ such that $(\varphi_{k_{l_2}})_{l_2}$ converges to some $\varphi_2 \in C^2(B_R \setminus B_{\frac{R}{k}})$. Now consider $(\varphi_{k_{l_2}})_{l_2}$ on $B_R \setminus B_{\frac{R}{k}}$. Again there exists a subsequence $(\varphi_{k_{l_3}})_{l_3}$ of $(\varphi_{k_{l_2}})_{l_2}$ which converges to $\varphi_3 \in C^2(B_R \setminus B_{\frac{R}{3}})$, and $\varphi_3(x) = \varphi_2(x)$ for all $x \in B_R \setminus B_{\frac{R}{2}}$. By repeating this argument we get a supersolution $\varphi \in C^2(B_R \setminus \{0\})$ i.e. $L\varphi \geq 0$, such that $\varphi > 0$ on $B_R \setminus \{0\}$. □

Proof of Theorem 2.2.1: That 1) implies 2) follows immediately from Proposition 2.2.1 and Lemma 1.1.2. Indeed, if P is a HI-potential, then there exists a positive solution φ for (B_P), which by Lemma 1.1.2, satisfies the boundary conditions that are needed for Proposition 2.2.1. Note that this implication is valid for any smooth bounded domain provided P is assumed to be non-decreasing on $(0, R)$. Otherwise this condition is not needed if the domain is simply a ball of radius R.

To show that 2) implies 1), we assume that inequality (H_P) holds on a ball B_R of radius R, then apply Lemma 2.2.2 to obtain a C^2-supersolution for equation (2.12). Now take the surface average of u, that is

(2.13) $$w(r) = \frac{1}{n\omega_w r^{n-1}} \int_{\partial B_r} u(x) dS = \frac{1}{n\omega_n} \int_{|z|=1} u(rz) dz > 0,$$

where ω_n denotes the volume of the unit ball in $\mathbb{R}^n$. We may assume that the unit ball is contained in B_R (otherwise we just use a smaller ball). By a standard calculation we get

$$(2.14) \qquad w''(r) + \frac{n-1}{r}w'(r) \leq \frac{1}{n\omega_n r^{n-1}} \int_{\partial B_r} \Delta u(x) dS.$$

Since $u(x)$ is a supersolution of (2.12), w satisfies the inequality:

$$(2.15) \quad w''(r) + \frac{n-1}{r}w'(r) + (\frac{n-2}{2})^2 \frac{w(r)}{r^2} \leq -P(r)w(r), \quad for \quad 0 < r < R.$$

Now define

$$(2.16) \qquad \varphi(r) = r^{\frac{n-2}{2}} w(r), \quad for \ 0 < r < R.$$

Using (2.15), a straightforward calculation shows that φ satisfies

$$(2.17) \qquad \varphi''(r) + \frac{\varphi'(r)}{r} \leq -\varphi(r)P(r) \quad for \ 0 < r < R.$$

By Proposition 1.1.1 we conclude that the equation $y''(r) + \frac{1}{r}y' + P(r)y = 0$ has a positive solution φ on $(0, R)$, and that P is therefore a HI-potential.

To establish that $\beta(P, R)$ is the best constant, note that by the sufficient condition we have that the best constant c is larger than $\beta(P, R)$. On the other hand, the necessary condition yields that $y'' + \frac{1}{r}y' + cP(r)y = 0$ has a positive solution on $(0, R)$, which means that $c \leq \beta(P, R)$. The proof of the theorem is now complete. $\square$

2.3. Improved Hardy inequalities on domains with 0 in their interior

The following direct application of Theorem 2.2.1 will be frequently used in the sequel.

THEOREM 2.3.1. *Let Ω be a smooth bounded domain in $\mathbb{R}^n$ ($n \geq 2$) containing 0 in its interior, and let $R_0 = (|\Omega|/\omega_n)^{1/n}$ and $R_1 = \sup_{x \in \Omega} |x|$. If P is a HI-potential on $(0, R_1)$, then there exists $c > 0$ such for all $u \in H_0^1(\Omega)$, we have*

$$\int_\Omega |\nabla u|^2 dx - (\frac{n-2}{2})^2 \int_\Omega \frac{u^2}{|x|^2} dx \geq c \int_\Omega P(|x|) u^2 dx,$$

and the best constant $c := c(P, \Omega) \geq \beta(P, R_1)$.

Moreover, if P is a decreasing HI-potential, then $c(P, \Omega) \geq \beta(P, R_0)$.

Proof: This follows immediately from Theorem 2.2.1 applied to the ball B_{R_1} which contains Ω. On the other hand, if P is decreasing, then it suffices to apply Proposition 2.2.1. Note that $R_0 \leq R_1$ and consequently $\beta(P, R_0) \geq \beta(P, R_1)$. $\square$

We now give some of the applications of Theorem 2.3.1 by exploiting the richness of the class of HI-potentials as described in Chapter 1.

COROLLARY 2.3.1. *Let Ω be a smooth bounded domain in $\mathbb{R}^n$ ($n \geq 2$) containing 0 in its interior.*

(1) *For any $0 \leq \alpha < 2$, there is $c > 0$ such that the following inequality holds:*

$$(2.18) \qquad \int_\Omega |\nabla u|^2 dx - (\frac{n-2}{2})^2 \int_\Omega \frac{u^2}{|x|^2} dx \geq c \int_\Omega \frac{u^2}{|x|^\alpha} dx \quad for \ u \in H_0^1(\Omega).$$

2.3. IMPROVED HARDY INEQUALITIES ON DOMAINS WITH 0 IN THEIR INTERIOR

Moreover, the best constant $c(r^{-\alpha}, \Omega)$ is at least as large as $\dfrac{z_\alpha^2 \omega_n^{\frac{2-\alpha}{n}}}{|\Omega|^{\frac{2-\alpha}{n}}}$, where z_α is the first root of the largest positive solution of $y'' + \frac{1}{r}y' + \frac{1}{r^\alpha}y = 0$.

(2) *In particular, the following inequality holds:*

(2.19) $\quad \int_\Omega |\nabla u|^2 dx - (\frac{n-2}{2})^2 \int_\Omega \frac{u^2}{|x|^2} dx \geq c \int_\Omega u^2 dx \quad$ *for $u \in H_0^1(\Omega)$,*

and the best constant is at least as large as $\dfrac{z_0^2 \omega_n^{2/n}}{|\Omega|^{2/n}}$, where z_0 is the first root of the Bessel function J_0.

(3) *If $\alpha \geq 2$, then for any ball B_R, there is no $c > 0$ for which inequality $(H_{cr^{-\alpha}})$ holds on B_R, and in particular, $(\frac{n-2}{2})^2$ is the best constant for the classical Hardy inequality.*

Proof: The first two assertions follow immediately from Theorem 2.3.1 applied to the HI-potentials $P(r) = r^{-\alpha}$, and $P(r) = 1$ respectively.

The last assertion follows from Theorem 2.2.1 and the fact that if $\alpha \geq 2$, then there is no $c > 0$ such that $(\frac{n-2}{2})^2 r^{-2} + cr^{-\alpha}$ is a HI-potential which follows from Corollary 1.3.3.

COROLLARY 2.3.2. *Let Ω be a smooth domain in $\mathbb{R}^n$ with $0 \in \Omega$ and $n \geq 2$, then we have the following inequalities:*

- *For every $k \geq 1$, and $\rho = (\sup_{x \in \Omega} |x|)(e^{e^{e^{\cdot^{\cdot^{e}}}}{}^{(k-times)}})$, the following holds for all $u \in H_0^1(\Omega)$,*

(2.20) $\quad \int_\Omega |\nabla u|^2 dx \geq (\frac{n-2}{2})^2 \int_\Omega \frac{|u|^2}{|x|^2} dx + \frac{1}{4} \sum_{j=1}^k \int_\Omega \frac{|u|^2}{|x|^2} \left(\prod_{i=1}^j \log^{(i)} \frac{\rho}{|x|}\right)^{-2} dx,$

and $\frac{1}{4}$ is the best constant for which the above holds.

- *For any $D \geq \sup_{x \in \Omega} |x|$, the following inequality holds for all $u \in H_0^1(\Omega)$,*

(2.21)
$$\int_\Omega |\nabla u|^2 dx \geq (\frac{n-2}{2})^2 \int_\Omega \frac{u^2}{|x|^2} dx + \frac{1}{4} \sum_{i=1}^\infty \int_\Omega \frac{1}{|x|^2} X_1^2(\frac{|x|}{D}) X_2^2(\frac{|x|}{D}) \ldots X_i^2(\frac{|x|}{D}) |u|^2 dx,$$

and $\frac{1}{4}$ is again the best constant for which the inequality holds.

Proof: Both inequalities follow again from Theorem 2.3.1 applied to the HI-potentials $P_1(r) = \frac{1}{r^2} \sum_{j=1}^k (\prod_{i=1}^j \log^{(i)} \frac{\rho}{r})^{-2}$ and $P_2(r) = \frac{1}{r^2} \sum_{j=1}^{+\infty} X_1^2(\frac{r}{D}) X_2^2(\frac{r}{D}) \ldots X_{j-1}^2(\frac{r}{D}) X_j^2(\frac{r}{D})$, respectively. Note that in both cases, Theorem 2.2.1 characterizes the best constant when Ω is a ball, as $\beta(P, R)$ which is equal to $\frac{1}{4}$ by Theorem 1.1.3.

REMARK 2.3.2. The above corollary yields Hardy inequalities as well as their corresponding best constants, in the critical dimension $n = 2$, where the leading term $(\frac{n-2}{2})^2 \int_\Omega \frac{|u|^2}{|x|^2} dx$ disappears.

We can also use Theorem 2.2.1 in conjunction with the integral criteria for HI-potentials in Corollary 1.2.1 to deduce the following general result.

COROLLARY 2.3.3. *Let P be a non-negative C^1-function on $(0, +\infty)$.*

- *If $\liminf_{r \to 0} \log r \int_0^r sP(s) ds > -\infty$, then for any bounded domain in $\mathbb{R}^n$, $n \geq 2$, there exists $\alpha := \alpha(\Omega) > 0$ such that an improved Hardy inequality (H_{P_α}) holds on Ω for the scaled potential $P_\alpha(x) := \alpha^2 P(\alpha x)$.*

- If $\lim_{r\to 0} \log r \int_0^r sP(s)ds = -\infty$, then for any bounded domain in $\mathbb{R}^n$, there is no $\beta, c > 0$ for which $(H_{P_{\beta,c}})$ holds on Ω with $P_{\beta,c} = cP(\beta x)$.

Here is another immediate application of the characterization in Theorem 2.2.1 in conjunction with Lemma 1.1.2.

COROLLARY 2.3.4. *Assume $n \geq 2$, then there is no positive $P \in C^1(0, +\infty)$ such that the inequality*

$$\int_{\mathbb{R}^n} |\nabla u|^2 dx - \left(\frac{n-2}{2}\right)^2 \int_{\mathbb{R}^n} \frac{u^2}{|x|^2} dx \geq \int_{\mathbb{R}^n} P(|x|) u^2 dx,$$

holds for all $u \in \mathcal{D}^{1,2}(\mathbb{R}^n)$. . □

2.4. Attainability of the best Hardy constant on domains with 0 in their interior

We now tackle the question of evaluating the best constant, which is given by the following minimization problem

$$(2.22) \qquad \mu_\lambda(P, \Omega) := \inf_{\substack{u \in H_0^1(\Omega) \\ u \neq 0}} \frac{\int_\Omega |\nabla u|^2 \, dx - \lambda \int_\Omega P(|x|) u^2 \, dx}{\int_\Omega \frac{|u|^2}{|x|^2} \, dx},$$

where $\lambda \in \mathbb{R}$ is a parameter, P is a HI-potential and Ω is a smooth bounded domain in $\mathbb{R}^n$ ($n \geq 2$). It is clear from the previous results, that for $0 \leq \lambda \leq \beta(P, R)$ where $R = \sup_{x \in \Omega} |x|$, we have

$$(2.23) \qquad \frac{(n-2)^2}{4} \leq \mu_\lambda(P, \Omega) \leq \mu_0(\Omega),$$

where $\mu_0(\Omega)$ is the best constant in the classical Hardy inequality for functions in $H_0^1(\Omega)$. We shall see that the determination of $\mu_\lambda(P, \Omega)$ and its attainability depend heavily on the location of the singularity point 0 with respect to the domain Ω. We shall deal here with the case where 0 is in the interior of the domain Ω.

We start by showing that the Sobolev inequality can actually be derived from Hardy's except for the value of the best constant, which will be again studied in Chapter 13. In this sense, Hardy's inequality is stronger than Sobolev's.

COROLLARY 2.4.1. *For any $n \geq 3$, there exists a constant $C(n) > 0$ such that for all $u \in \mathcal{D}^{1,2}(\mathbb{R}^n)$, we have*

$$(2.24) \qquad \int_{\mathbb{R}^n} |u|^{\frac{2n}{n-2}} dx \leq C(n) \int_{\mathbb{R}^n} |\nabla u|^2 dx.$$

Proof: We shall derive the inequality for radial decreasing functions, which is sufficient for the proof of Proposition 2.4.1 below, but also for the general case in view of the properties of the symmetric rearrangement defined above. The argument goes as follows: If u is radial and decreasing, then for any $y \in \mathbb{R}^n$ we have

$$\|u\|_p^p = \int_{\mathbb{R}^n} |u|^p \, dx \geq u(y)^p |y|^n |B_1^n|,$$

where $|B_1^n|$ is the volume of the unit ball in $\mathbb{R}^n$. Now take this to the power $1 - \frac{2}{p}$, multiply by $|u(y)|^2 |y|^{\frac{n(2-p)}{p}}$ and integrate over y to obtain

$$\int_{\mathbb{R}^n} \frac{|u(y)|^2}{|y|^{\frac{n(p-2)}{p}}} \, dy \geq |B_1^n|^{1-\frac{2}{p}} \|u\|_p^2.$$

It now suffices to take $p := \frac{2n}{n-2}$ and use Hardy's inequality to conclude. $\square$

THEOREM 2.4.1. *Let Ω be a smooth bounded domain of $\mathbb{R}^n$ such that $0 \in \Omega$ and $n \geq 2$. Then, for any HI-potential P on $(0, R)$, where $R = \sup_{x \in \Omega} |x|$, we have for any $0 \leq \lambda \leq \beta(P, R)$,*

(2.25) $$\mu_\lambda(P, \Omega) = \frac{(n-2)^2}{4},$$

and the latter is never attained in $H_0^1(\Omega)$.

Proof: It is clear that $\mu_0(\Omega) \geq \mu_0(\mathbb{R}^n) = \frac{(n-2)^2}{4}$. The reverse inequality follows from the fact that for any $c > 0$, the function $P(r) = \frac{c}{r^2}$ cannot be a HI-potential on any interval $(0, R)$ by Corollary 1.3.3. Since 0 is in the interior of Ω, we have $\mu_0(\Omega) \leq \mu_0(B_\rho) = \frac{(n-2)^2}{4}$ whenever $B_\rho \subset \Omega$.

We now establish that this constant is never attained in $H_0^1(\Omega)$. Indeed, if $\mu_\lambda(P, \Omega)$ is achieved by some $u \in H_0^1(\Omega)$, then u would satisfy the corresponding Euler-Lagrange equation, that is, it would be at least a weak solution of the following problem:

(2.26) $$\begin{cases} \Delta u + \left(\frac{n-2}{2}\right)^2 \frac{u}{|x|^2} + P(|x|)u = 0 & \text{in } \Omega, \\ u > 0 & \text{in } \Omega \setminus \{0\}, \\ u = 0 & \text{in } \partial\Omega. \end{cases}$$

The proof will then follow from the following proposition.

PROPOSITION 2.4.1. *Let Ω be a smooth bounded domain in $\mathbb{R}^n$ with $n \geq 2$ such that $0 \in \Omega$. Let P be radial and non-negative function in $C_{loc}^{0,\alpha}(\Omega \setminus \{0\})$ for some $\alpha \in (0, 1)$. Then Equation (2.26) has no $H_0^1(\Omega)$ solution.*

Proof: Suppose that u is a $H_0^1(\Omega)$ positive solution of (2.26). By standard elliptic regularity we know that $u \in C_{loc}^{2,\alpha}(\Omega \setminus \{0\})$. Taking again the surface average of u,

$$v(r) = \frac{1}{n\omega_n r^{n-1}} \int_{\partial B_r} u(x) dS = \frac{1}{n\omega_n} \int_{|\omega|=} u(r\omega) d\omega > 0,$$

where ω_n denotes the volume of the unit ball in $\mathbb{R}^n$, and where we assume without loss that the unit ball B_1 is contained in Ω. We have as in the proof of Theorem 2.2.1,

(2.27) $$v''(r) + \frac{n-1}{r} v'(r) + \frac{\left(\frac{n-2}{2}\right)^2}{r^2} v(r) = -P(r)v(r), \quad 0 < r \leq 1.$$

By setting $w(r) = r^{(n-2)/2} v(r) > 0$ for $r > 0$, we get that

(2.28) $$(rw')' = -rP(r)w(r) \leq 0 \quad 0 < r \leq 1.$$

We now show that if $u \in H_0^1(\Omega)$ then $\liminf_{r \downarrow 0} w(r) = 0$, which will clearly contradict that w is necessarily decreasing on $(0,1)$, a fact established in Lemma 1.1.2. Indeed, otherwise, there exist constants C_0, r_0 such that $w(r) > C_0 > 0$ for $0 < r \leq r_0$. But if $u \in H_0^1(\Omega)$, then $u \in L^{2n/(n-2)}(B_{r_0})$ by Corollary 2.4.1. It follows from the definitions of w and v and from Hölder's inequality that for $t \in (0, r_0]$,

$$C \leq t^{-n/2} \int_{\partial B_t} u \, dS \leq (n\omega_n)^{\frac{n+2}{2n}} t^{\frac{n-2}{2n}} \left(\int_{\partial B_t} u^{\frac{2n}{n-2}} \right)^{\frac{n-2}{2n}}.$$

Integrating this from 0 to $r \leq r_0$ and using once more Hölder's inequality we easily end up with $C \leq \|u\|_{L^{\frac{2n}{n-2}}}(B_r)$ for some positive constant C independent of r. This is clearly a contradiction, and the proposition is proved. □

2.5. Further comments

Considering the vast amount of literature about the classical Hardy inequality and its improvements, it is hard to believe that the whole thing started with an attempt to give an easy proof for the following inequality established by Hilbert,

$$(2.29) \qquad \sum_{m,n=1}^{\infty} \frac{a_m b_n}{m+n} \leq \pi \left(\sum_{m=1}^{\infty} a_m^2 \right)^{\frac{1}{2}} \left(\sum_{m=1}^{\infty} b_m^2 \right)^{\frac{1}{2}}.$$

See Masmoudi [215] and A. Kufner, L. Maligranda, and L-E Persson [196] for historical accounts. The books of B. Opic and A. Kufner [231] and the one by A. Kufner, L-E Persson [195] are standard references on the subject. This chapter is mostly focused on certain types of improvement involving radially symmetric potentials, on the corresponding best constants, and on whether they are attainable or not. A first application of improved Hardy inequalities to PDEs appears in Chapter 5, but the importance of the approach we adopt will be more transparent once we tackle the fourth order counterpart of the Hardy inequality in Part II of this book.

A good account on Schwarz symmetrization can be found in Lieb and Loss [203]. The classical Hardy inequality on domains of $\mathbb{R}^n$ ($n \geq 2$) having 0 in their interior has been well understood for some time, including the facts that $\frac{(n-2)^2}{4}$ is the best constant and that it is never attained in $H_0^1(\Omega)$. Actually, I. Herbst has established in [179] the sharp constant in Hardy-Rellich inequalities of arbitrary order.

One novelty here lies with improvements via the addition of extra terms to the classical potential $\frac{(n-2)^2}{4} \frac{1}{|x|^2}$ and the new phenomena –described in the next chapter– which appear in the case where the singularity (at 0) is on the boundary of the domain Ω. The first such improved Hardy inequality on bounded domains was given by Brezis-Vázquez [65] in order to study the stability of certain singular solutions of nonlinear elliptic equations. They established that

$$\int_\Omega |\nabla u|^2 \geq \left(\frac{n-2}{2} \right)^2 \int_\Omega \frac{u^2}{|x|^2} + \lambda_\Omega \int_\Omega u^2 \quad \text{for every } u \in H_0^1(\Omega),$$

calculated the best constant λ_Ω, and showed that it is optimal when Ω is a ball, yet still not achieved in $H_0^1(\Omega)$. This led them to ask ([65] Problem 2) whether the two terms on the right-hand side of the above inequality are simply the first two terms of an infinite series of correcting terms. This question was addressed by several authors. In particular, Gazzola-Grunau-Mitidieri [151] and Adimurthi-Chaudhuri-Ramaswamy [5] improved the Hardy inequality by adding arbitrary large number of positive terms (Equation 2.20). Filippas-Tertikas [142] showed that the Hardy inequality can be improved by adding an infinite number of terms to the right hand side of the inequality (Equation (2.21)).

Ghoussoub-Moradifam [162] eventually extended and unified the above results by introducing the Hardy improving potentials and establishing the connection between such improvements and the oscillatory theory of second order linear differential equations. They also proved that the best constant in the improvement of Adimurthi et al. is actually equal to $\frac{1}{4}$ which was conjectured by Chaudhuri in [88] where he had shown the estimate $\frac{1}{4} \leq c \leq \frac{1}{2}$.

Missing from this chapter is the Hardy-Leray inequality for vector fields in $\mathbb{R}^n$ [201], which states that

$$(2.30) \qquad \int_{\mathbb{R}^n} \frac{|\mathbf{u}|^2}{|x|^2} dx \leq \frac{4}{(n-2)^2} \int_{\mathbb{R}^n} |\nabla \mathbf{u}|^2 dx$$

for all vector fields $\mathbf{u}$ in $C_0^\infty(\mathbb{R}^n)$ as well as its improvements in the case of divergence-free fields by O. Costin and V. Maz'ya [94], who showed that the best constant becomes

$$(2.31) \qquad \frac{4}{(n-2)^2}\left(1 - \frac{8}{(n+2)^2}\right).$$

Open Problem (1): Since the best constant in the Hardy-Leray inequality as well as the one for divergence-free fields are not attained, establish improved Hardy-Leray inequalities of the form

$$(2.32) \qquad \int_\Omega |\nabla \mathbf{u}|^2 dx - \frac{(n-2)^2}{4} \int_\Omega \frac{|\mathbf{u}|^2}{|x|^2} dx \geq \int_\Omega P(x)|\mathbf{u}|^2 dx$$

on bounded domains Ω of $\mathbb{R}^n$. The same question applies to the case of divergence-free vector fields.

CHAPTER 3

Improved Hardy Inequality with Boundary Singularity

If P is a HI-potential on an interval $(0, R)$ and $0 \leq \lambda \leq \beta(P, R)$, then for domains Ω having 0 on their boundary and such that $\sup\limits_{x \in \Omega} |x| \leq R$, the best constant in the improved Hardy inequality

$$\mu_\lambda(P, \Omega) := \inf_{\substack{u \in H_0^1(\Omega) \\ u \neq 0}} \frac{\int_\Omega |\nabla u|^2 \, dx - \lambda \int_\Omega P(|x|) u^2 \, dx}{\int_\Omega \frac{|u|^2}{|x|^2} \, dx}$$

can be anywhere between $\frac{(n-2)^2}{4}$ and $\frac{n^2}{4}$, and it is attained in $H_0^1(\Omega)$ whenever $\mu_\lambda(P, \Omega) < \frac{n^2}{4}$. Moreover, $\mu_\lambda(P, \Omega)$ is not attained if it is equal to $\frac{n^2}{4}$ and if –for example– Ω contains a half-ball centered at zero. Furthermore, $\mu_\lambda(P, \Omega)$ can be equal to $\frac{n^2}{4}$ for any $0 \leq \lambda \leq \beta(P, R)$ for certain domains having 0 on their boundaries, such as those that lie on one side of a half-space going through 0.

3.1. Improved Hardy inequalities on conical domains with vertex at 0

To any sub-domain Σ of the n-dimensional sphere $\mathbb{S}^{n-1}$, we associate a conical domain $\mathcal{C}_\Sigma \subset \mathbb{R}^n$ having 0 as a vertex, as well as a (half) cylinder $\mathcal{Z}_\Sigma \subset \mathbb{R}^{n+1}$ by setting

$$\mathcal{C}_\Sigma := \{ t\sigma \mid t > 0, \, \sigma \in \Sigma \} \quad \mathcal{Z}_\Sigma := \mathbb{R}_+ \times \Sigma.$$

If Σ is a smooth, then $\mathcal{C}_\Sigma$ is a dilation-invariant domain in $\mathbb{R}^n$. In particular, if Σ is the sphere $\mathbb{S}^{n-1}$, then $\mathcal{C}_\Sigma = \mathbb{R}^n \setminus \{0\}$ and if Σ is the half-sphere $\mathbb{S}_+^{n-1}$, then $\mathcal{C}_\Sigma$ is the half-space $\mathbb{R}_+^n$. The map τ from $\mathbb{R}^n \setminus \{0\}$ to $\mathbb{R}^{n+1}$, defined by

$$\tau(x) := \left(-\log |x|, \frac{x}{|x|} \right)$$

is a homeomorphism mapping $\mathcal{C}_\Sigma$ onto $\mathcal{Z}_\Sigma$. It induces the Emden-Fowler transform T from $C_c^\infty(\mathcal{C}_\Sigma)$ to $C_c^\infty(\mathcal{Z}_\Sigma)$ defined via the formula

$$u(x) = |x|^{\frac{2-n}{2}} (Tu)(\tau(x)).$$

The divergence theorem readily implies the representations

$$(3.1) \qquad \int_{\mathcal{C}_\Sigma} |\nabla u|^2 \, dx = \frac{(n-2)^2}{4} \int_0^\infty \int_\Sigma |Tu|^2 \, ds d\sigma + \int_0^\infty \int_\Sigma |\nabla_{s,\sigma} Tu|^2 \, ds d\sigma$$

and

$$(3.2) \qquad \int_{\mathcal{C}_\Sigma} \frac{|u|^2}{|x|^2} \, dx = \int_0^\infty \int_\Sigma |Tu|^2 \, ds d\sigma,$$

where $\nabla_{s,\sigma} = (\partial_s, \nabla_\sigma)$ denotes the gradient on $\mathbb{R}_+ \times \mathbb{S}^{n-1}$, with ∇_σ being the gradient on the manifold $\mathbb{S}^{n-1}$. Now associate to Σ the following two constants:

$$(3.3) \quad \mu_0(\mathcal{C}_\Sigma) := \inf_{\substack{u \in C_c^\infty(\mathcal{C}_\Sigma) \\ u \neq 0}} \frac{\int_{\mathcal{C}_\Sigma} |\nabla u|^2 \, dx}{\int_{\mathcal{C}_\Sigma} \frac{|u|^2}{|x|^2} \, dx},$$

and

$$(3.4) \quad \lambda_1(\mathcal{Z}_\Sigma) := \inf_{\substack{v \in C_c^\infty(\mathcal{Z}_\Sigma) \\ v \neq 0}} \frac{\int_0^\infty \int_\Sigma |\nabla_{s,\sigma} v|^2 \, ds d\sigma}{\int_0^\infty \int_\Sigma |v|^2 \, ds d\sigma}.$$

If $n \geq 3$, we shall consider the space $\mathcal{D}^{1,2}(\mathcal{C}_\Sigma)$, which is the closure of $C_c^\infty(\mathcal{C}_\Sigma)$ with respect to the norm $(\int_\Omega |\nabla u|^2 \, dx)^{1/2}$, as a closed subspace of $\mathcal{D}^{1,2}(\mathbb{R}^n)$. Note that if $\Sigma = \mathbb{S}^{n-1}$, then by a standard density result, we have,

$$\mathcal{D}^{1,2}(\mathcal{C}_{\mathbb{S}^{n-1}}) = \mathcal{D}^{1,2}(\mathbb{R}^n \setminus \{0\}) = \mathcal{D}^{1,2}(\mathbb{R}^n).$$

In the next proposition we relate the Hardy inequality on $\mathcal{C}_\Sigma$ to the Poincaré inequality for maps supported be the cylinder $\mathcal{Z}_\Sigma$.

PROPOSITION 3.1.1. Let $\mathcal{C}_\Sigma$ be the cone generated by a domain $\Sigma \subset \mathbb{S}^{n-1}$. Then, the best constant for the ratio (3.3) is equal to

$$(3.5) \quad \mu_0(\mathcal{C}_\Sigma) = \frac{(n-2)^2}{4} + \lambda_1(\Sigma),$$

and is never attained in $\mathcal{D}^{1,2}(\mathcal{C}_\Sigma)$. In particular,

$$(3.6) \quad \mu_0(\mathbb{R}_+^n) = \frac{n^2}{4}.$$

Proof: By (3.1) and (3.2) we have that

$$\mu_0(\mathcal{C}_\Sigma) - \frac{(n-2)^2}{4} = \inf_{\substack{v \in C_c^\infty(\mathcal{Z}_\Sigma) \\ v \neq 0}} \frac{\int_0^\infty \int_\Sigma |\nabla_{s,\sigma} v|^2 \, ds d\sigma}{\int_0^\infty \int_\Sigma |v|^2 \, ds d\sigma} = \lambda_1(\mathcal{Z}_\Sigma).$$

The result follows by noticing that $\lambda_1(\mathcal{Z}_\Sigma) = \lambda_1(\Sigma)$, where

$$\lambda_1(\Sigma) := \inf_{\substack{v \in C_c^\infty(\Sigma) \\ v \neq 0}} \frac{\int_\Sigma |\nabla_\sigma v|^2 \, d\sigma}{\int_\Sigma |v|^2 \, d\sigma}.$$

On the other hand, that $\mu_0(\mathcal{C}_\Sigma)$ is not achieved in $\mathcal{D}^{1,2}(\mathcal{C}_\Sigma)$ is an immediate consequence of the fact that the Dirichlet eigenvalue problem of $-\Delta$ in the half-cylinder $\mathcal{Z}_\Sigma$ is never achieved. □

The eigenvalues $\lambda_1(\Sigma)$ and therefore the Hardy constants $\mu_0(\mathcal{C}_\Sigma)$ are explicitly known in the following special cases: If $\Sigma = \mathbb{S}_+^{n-1}$ is a half-sphere then $\lambda_1(\mathbb{S}_+^{n-1}) = n - 1$, which means that the Hardy constant of a half space is given by

$$(3.7) \quad \mu_0(\mathbb{R}_+^n) = \frac{n^2}{4}.$$

If $n = 2$ and if $\mathcal{C}_{\Sigma_\theta} \subset \mathbb{R}^2$ is a cone of amplitude $\theta \in (0, 2\pi]$, then $\lambda_1(\Sigma_\theta)$ coincides with the Dirichlet eigenvalue on the interval $(0, \theta)$, hence

$$(3.8) \quad \mu_0(\mathcal{C}_{\Sigma_\theta}) = \frac{\pi^2}{\theta^2} \geq \frac{1}{4}.$$

We now show the following Hardy inequality.

THEOREM 3.1.1. *Let $\mathcal{C}_{R,\Sigma} = \{t\sigma \mid t \in (0, R), \sigma \in \Sigma\}$ be the conical domain where $\Sigma \subset \mathbb{S}^{n-1}$, $n \geq 2$ and $R > 0$, and let P be a HI-potential on $(0, R)$. Then the following holds for all $u \in C_c^\infty(\mathcal{C}_{1,\Sigma})$,*

$$(3.9) \quad \int_{\mathcal{C}_{R,\Sigma}} |\nabla u|^2 \, dx - \mu_0(\mathcal{C}_\Sigma) \int_{\mathcal{C}_{R,\Sigma}} \frac{|u|^2}{|x|^2} \, dx \geq \beta(P, R) \int_{\mathcal{C}_{R,\Sigma}} P(|x|) |u|^2 \, dx.$$

In particular,

$$(3.10) \quad \int_{\mathcal{C}_{R,\Sigma}} |\nabla u|^2 \, dx - \mu_0(\mathcal{C}_\Sigma) \int_{\mathcal{C}_{R,\Sigma}} \frac{|u|^2}{|x|^2} \, dx \geq \frac{z_0^2}{|\operatorname{diam}(\mathcal{C}_{R,\Sigma})|^2} \int_{\mathcal{C}_{R,\Sigma}} |u|^2 \, dx,$$

where z_0 is the first zero of the Bessel function J_0.

Proof: By homogeneity, it suffices to prove the result for $R = 1$. Fix $u \in C_c^\infty(\mathcal{C}_{1,\Sigma})$ and compute in polar coordinates $t = |x|$, $\sigma = x/|x|$:

$$\int_{\mathcal{C}_{1,\Sigma}} |\nabla u|^2 \, dx = \int_0^1 \int_\Sigma \left|\frac{\partial u}{\partial t}\right|^2 t^{n-1} dt d\sigma + \int_0^1 \int_\Sigma |\nabla_\sigma u|^2 t^{n-3} dt d\sigma,$$

$$\int_{\mathcal{C}_{1,\Sigma}} \frac{|u|^2}{|x|^2} \, dx = \int_0^1 \int_\Sigma |u|^2 t^{n-3} dt d\sigma.$$

Since for every fixed $t \in (0, 1)$, we have

$$\int_\Sigma |\nabla_\sigma u|^2 t^{n-3} d\sigma \geq \lambda_1(\Sigma) \int_\Sigma |u|^2 t^{n-3} d\sigma,$$

then in view of Proposition 3.1.1, we only need to show that

$$(3.11) \quad \int_0^1 \left|\frac{\partial u}{\partial t}\right|^2 t^{n-1} dt - \frac{(n-2)^2}{4} \int_0^1 |u|^2 t^{n-3} dt \geq \beta(P, 1) \int_0^1 P(t)|u|^2 t^{n-1} dt$$

for any fixed $\sigma \in \Sigma$. For that, we put $w(t, \sigma) = t^{\frac{n-2}{2}} u(t\sigma)$ and use Corollary 2.1.1 to obtain

$$\int_0^1 \left|\frac{\partial u}{\partial t}\right|^2 t^{n-1} dt - \mu_0(\mathbb{R}^n) \int_0^1 |u|^2 t^{n-3} dt = \int_0^1 \left|\frac{\partial w}{\partial t}\right|^2 t \, dt + (2-n) \int_0^1 \frac{\partial w}{\partial t} w \, dt$$

$$= \int_0^1 \left|\frac{\partial w}{\partial t}\right|^2 t \, dt + \frac{(2-n)}{2} \int_0^1 \frac{\partial w^2}{\partial t} \, dt$$

$$= \int_0^1 \left|\frac{\partial w}{\partial t}\right|^2 t \, dt \geq \beta(P, 1) \int_0^1 P(t) w^2 t \, dt$$

$$= \beta(P, 1) \int_0^1 P(t)|u|^2 t^{n-1} dt.$$

This gives (3.11) and the inequality is proved for conical domains. $\square$

COROLLARY 3.1.1. *Let Ω be a smooth bounded domain of $\mathbb{R}^n$ ($n \geq 2$) such that $0 \in \partial\Omega$, and let P be a HI-potential on $(0, R)$, where $R \geq \sup_{x \in \Omega} |x|$. If Ω is contained in a half-space, then the following holds:*

$$(3.12) \quad \int_\Omega |\nabla u|^2 \, dx - \frac{n^2}{4} \int_\Omega \frac{|u|^2}{|x|^2} \, dx \geq \beta(P, R) \int_\Omega P(|x|)|u|^2 \, dx \qquad \forall u \in H_0^1(\Omega).$$

In particular,

$$(3.13) \quad \int_\Omega |\nabla u|^2 \, dx - \frac{n^2}{4} \int_\Omega \frac{|u|^2}{|x|^2} \, dx \geq \frac{z_0^2}{|\text{diam}(\Omega)|^2} \int_\Omega |u|^2 \, dx \qquad \forall u \in H_0^1(\Omega),$$

where z_0 is the first zero of the Bessel function J_0.

Proof: Assume that Ω is a bounded domain of $\mathbb{R}^n$ that is contained in a half-space while $0 \in \partial\Omega$. If $R > 0$ is the diameter of Ω, we have $\Omega \subset B_R^+(0)$, where $B_R^+(0)$ is a half ball of radius R centered at the origin. Take in the above theorem $\Sigma := \mathbb{S}_+^{n-1}$ to be a half sphere in $\mathbb{S}^{n-1}$ so that $\mathcal{C}_\Sigma$ is a half-space. We deduce from Theorem 3.1.1 that

$$\int_{B_R^+} |\nabla u|^2 \, dx - \frac{n^2}{4} \int_{B_R^+} \frac{|u|^2}{|x|^2} \, dx \geq \beta(P, R) \int_{B_R^+} P(|x|)|u|^2 \, dx,$$

for all $u \in C_c^\infty(\Omega)$ and the proof of Theorem 3.1.1 is complete. $\square$

COROLLARY 3.1.2. *Let Ω be a Lipschitz domain of class C^2 in $\mathbb{R}^n$ such that $0 \in \partial\Omega$ and let P be a HI-potential on $(0, R)$ for some $R > 0$. Then, there exists $r_0 := r_0(\Omega, P) > 0$ and $c_0 := c_0(\Omega, P) > 0$ such that for all $0 < r < r_0$ the following inequality holds for all $u \in H_0^1(\Omega \cap B_r(0))$,*

$$(3.14) \quad \int_{\Omega \cap B_r(0)} |\nabla u|^2 \, dx - \frac{n^2}{4} \int_{\Omega \cap B_r(0)} \frac{|u|^2}{|x|^2} \, dx \geq c_0 \int_{\Omega \cap B_r(0)} P(|x|)|u|^2 \, dx.$$

Proof: It suffices to notice that for r_0 small enough, the subdomain $\Omega \cap B_r(0)$ is contained in a half-space and therefore Corollary 3.1.1 applies.

3.2. Attainability of the Hardy constants on domains having 0 on the boundary

Let now Ω be a smooth bounded domain in $\mathbb{R}^n$ with $n \geq 2$, such that 0 belongs to the boundary of Ω. We shall see that the situation is quite different from the case where 0 is in the interior of the domain, as the geometry of the domain will play a role in both the value of the best constant and its attainability in $H_0^1(\Omega)$. For each $\lambda \in \mathbb{R}$ and $P \geq 0$ we consider the best constant

$$(3.15) \quad \mu_\lambda(P, \Omega) := \inf_{\substack{u \in H_0^1(\Omega) \\ u \neq 0}} \frac{\int_\Omega |\nabla u|^2 \, dx - \lambda \int_\Omega P(|x|) u^2 \, dx}{\int_\Omega \frac{|u|^2}{|x|^2} \, dx}.$$

It is clear that the map $\lambda \to \mu_\lambda(P, \Omega)$ is non-increasing on $\mathbb{R}$. We start by noting the following observations.

LEMMA 3.2.1. *Suppose P is a HI-potential on $(0, R)$, and let Ω be a smooth bounded domain in $\mathbb{R}^n$ with $0 \in \partial\Omega$ and $\Omega \subset B_R$. Then,*

(1) If B is a ball of diameter $R > 0$ with $0 \in \partial B$, then $\mu_\lambda(P, B) = \frac{n^2}{4}$ for all $0 \leq \lambda \leq \beta(P, R)$.
(2) $\frac{(n-2)^2}{4} < \mu_\lambda(P, \Omega) \leq \frac{n^2}{4}$ whenever $0 \leq \lambda \leq \beta(P, R)$.

Proof: Note that 1) follows from Corollary 3.1.1, which yields that $\mu_\lambda(P, B) \geq \frac{n^2}{4}$, and the fact that $\mu_\lambda(P, B) \leq \mu_0(B) = \mu_0(\mathbb{R}^n_+) = \frac{n^2}{4}$.

2) The inequality $\mu_\lambda(P, \Omega) \leq \frac{n^2}{4}$ follows from the fact that if Ω is a smooth domain with $0 \in \partial\Omega$, then one can always find a ball $B \subset \Omega$ with $0 \in \partial B$. This then yields that $\mu_\lambda(P, \Omega) \leq \mu_\lambda(P, B) \leq \frac{n^2}{4}$.

The inequality $\frac{(n-2)^2}{4} \leq \mu_\lambda(P, \Omega)$ follows from the main result of Chapter 2. To prove that it is a strict inequality, assume not; that is, $\mu_0(P, \Omega) = \frac{(n-2)^2}{4} < \frac{n^2}{4}$. It then follows from Lemma 3.2.5 below that there exists a $u \in H^1_0(\Omega)$ that achieves $\mu_0(P, \Omega)$, which is not possible since then a null extension of u outside Ω would achieve the Hardy constant on $\mathbb{R}^n$, hence contradicting a result of the last section. $\square$

In the sequel we investigate for which value of λ-positive or negative- $\mu_\lambda(P, \Omega) = \frac{n^2}{4}$, and whether it can be attained or not. Here is the main result of this section.

THEOREM 3.2.2. *Let Ω be a smooth bounded domain of $\mathbb{R}^n$ such that $0 \in \partial\Omega$ and let P be a continuous positive function on $(0, R]$ where $R \geq \sup_{x \in \Omega} |x|$ such that $rP(r) = O(1)$ whenever $r \to 0$. There exists then $\lambda^*(\Omega, P) \in \mathbb{R}$ such that*

(1) *For all $\lambda > \lambda^*(\Omega, P)$, the best constant $\mu_\lambda(\Omega, P) < \frac{n^2}{4}$ and is attained in $H^1_0(\Omega)$.*
(2) *For all $\lambda \leq \lambda^*(\Omega, P)$, the best constant $\mu_\lambda(\Omega, P) = \frac{n^2}{4}$ and the latter is not attained in $H^1_0(\Omega)$ for any $\lambda < \lambda^*(\Omega, P)$.*

The proof will rely on the following lemmas.

LEMMA 3.2.3. *Let Ω be a Lipschitz domain of class C^2 in $\mathbb{R}^n$ such that $0 \in \partial\Omega$ and $\Omega \subset (0, R)$, and let P be a positive continuous function on $(0, R]$. Then, there exists λ_0 such that $\mu_{\lambda_0}(P, \Omega) \geq \frac{n^2}{4}$.*

Proof: For $\delta > 0$ small, consider a radially symmetric cut-off function χ in $C^\infty(B_\delta(0))$ satisfying
$$0 \leq \chi \leq 1, \quad \chi \equiv 0 \text{ in } \mathbb{R}^n \setminus B_{\frac{\delta}{2}}(0), \quad \chi \equiv 1 \text{ in } B_{\frac{\delta}{4}}(0).$$

Write any $u \in H^1_0(\Omega)$ as $u = \chi u + (1 - \chi)u$, and use that $\inf_{\frac{\delta}{4} \leq r \leq R} P(r) \geq c_0 > 0$ to get

(3.16) $$\int_\Omega |x|^{-2} |u|^2 \, dx \leq \int_\Omega |x|^{-2} |\chi u|^2 \, dx + c_1 \int_\Omega P(|x|) |u|^2 \, dx,$$

where the constant c_1 depends only on δ and P. Since $\chi u \in H^1_0(\Omega \cap B_\delta(0))$, then Corollary 3.1.2 implies that for δ sufficiently small,

(3.17) $$\frac{n^2}{4} \int_\Omega |x|^{-2} |\chi u|^2 \, dx \leq \int_\Omega |\nabla(\chi u)|^2 \, dx.$$

Note now that
$$\int_\Omega |\nabla(\chi u)|^2\, dx \le \int_\Omega |\nabla u|^2\, dx + \frac{1}{2}\int_\Omega \nabla(\chi^2)\cdot \nabla(u^2)\, dx + c_2 \int_\Omega P(|x|)|u|^2\, dx,$$
and by integration by parts,
$$\int_\Omega |\nabla(\chi u)|^2\, dx \le \int_\Omega |\nabla u|^2\, dx - \frac{1}{2}\int_\Omega \Delta(\chi^2)|u|^2\, dx + c_3 \int_\Omega P(|x|)|u|^2\, dx.$$
Combining this with (3.16) and (3.17) we infer that there exists a positive constant c such that
$$\frac{n^2}{4}\int_\Omega |x|^{-2}|u|^2\, dx \le \int_\Omega |\nabla u|^2\, dx + c\int_\Omega P(|x|)|u|^2\, dx \quad \forall u \in H_0^1(\Omega).$$
This implies that $\mu_{-c}(\Omega, P) \ge \frac{n^2}{4}$. □

LEMMA 3.2.4. *Let Ω be a Lipschitz domain of class C^2 in $\mathbb{R}^n$ such that $0 \in \partial\Omega$ and let P be a positive function on $(0,R)$ for some $R > 0$ such that $rP(r) = O(1)$ whenever $r \to 0$. Then*

$$(3.18) \qquad \sup_{\lambda \in \mathbb{R}} \mu_\lambda(\Omega, P) \le \frac{n^2}{4}.$$

Proof: Note that $\mu_\lambda(\Omega, P) \le \mu_\lambda(B_r, P)$ for any ball B_r contained in Ω in such a way that $0 \in \partial B_r$. Since $\mu_0(B_r) = \frac{n^2}{4}$, we can find for each $\delta > 0$, a $u_\delta \in C_c^\infty(B_r)$ such that
$$\int_{B_r} |\nabla u_\delta|^2\, dx \le \left(\frac{n^2}{4} + \delta\right) \int_{B_r} |x|^{-2} u_\delta^2\, dx.$$
It follows that
$$\mu_\lambda(\Omega, P) \le \frac{\int_{B_r} |\nabla u_\delta|^2\, dx - \lambda \int_{B_r} P(|x|)u_\delta^2\, dx}{\int_{B_r} |x|^{-2} u_\delta^2\, dx}$$
$$\le \frac{\int_{B_r} |\nabla u_\delta|^2\, dx}{\int_{B_r} |x|^{-2} u_\delta^2\, dx} + \frac{|\lambda|\int_{B_r} P(|x|)u_\delta^2\, dx}{\int_{B_r} |x|^{-2} u_\delta^2\, dx}$$
$$\le \left(\frac{n^2}{4} + \delta\right) + cr|\lambda|.$$
Taking the limit in δ and then in r, we get our claim. □

LEMMA 3.2.5. *Let Ω be a smooth bounded domain of $\mathbb{R}^n$ such that $0 \in \partial\Omega$ and $n \ge 2$, and let P be a positive continuous function on $(0,R)$, where $R \ge \sup_{x \in \Omega} |x|$. If $\mu_\lambda(P, \Omega) < \frac{n^2}{4}$, then it is attained in $H_0^1(\Omega)$.*

Proof: We first consider the best constant $\mu_\lambda(\Omega, P)$ in the case of the HI-potential $P \equiv 1$, that is the minimization problem

$$(3.19) \qquad \mu_\lambda(\Omega) := \inf_{\substack{u \in H_0^1(\Omega) \\ u \ne 0}} \frac{\int_\Omega |\nabla u|^2\, dx - \lambda \int_\Omega |u|^2\, dx}{\int_\Omega \frac{|u|^2}{|x|^2}\, dx},$$

3.2. ATTAINABILITY OF THE HARDY CONSTANTS

and note that Lemma 3.2.3 combined with Lemma 3.2.4 yield that if Ω is any smooth bounded domain of $\mathbb{R}^n$ such that $0 \in \partial\Omega$, then

$$\sup_{\lambda \in \mathbb{R}} \mu_\lambda(\Omega) = \frac{n^2}{4}. \tag{3.20}$$

Assume now that $\mu_\lambda(P,\Omega) < \frac{n^2}{4}$ and let $(u_k)_k \in H_0^1(\Omega)$ be a minimizing sequence for $\mu_\lambda(P,\Omega)$. We can normalize it in such a way that

$$\int_\Omega |\nabla u_k|^2\, dx = 1, \tag{3.21}$$

$$1 - \lambda \int_\Omega P(|x|) u_k^2\, dx = \mu_\lambda(P,\Omega) \int_\Omega |x|^{-2} u_k^2\, dx + o(1). \tag{3.22}$$

We can assume that $u_k \rightharpoonup u$ weakly in $H_0^1(\Omega)$, $|x|^{-1} u_k \rightharpoonup |x|^{-1} u$, and $P(|x|) u_k \rightharpoonup P(|x|) u$ weakly in $L^2(\Omega)$, and by the Rellich Theorem, that $u_k \to u$ strongly in $L^2(\Omega)$. Putting $\theta_k := u_k - u$, we get from (3.21) and (3.22)

$$\int_\Omega |\nabla \theta_k|^2\, dx + \int_\Omega |\nabla u|^2\, dx = 1 + o(1), \tag{3.23}$$

and

$$1 - \lambda \int_\Omega P(|x|) u^2\, dx = \mu_\lambda(P,\Omega) \left(\int_\Omega |x|^{-2} |\theta_k|^2\, dx + \int_\Omega \frac{|u|^2}{|x|^2}\, dx \right) + o(1). \tag{3.24}$$

By (3.20), we can find for any fixed positive $\delta < \frac{n^2}{4} - \mu_\lambda(P,\Omega)$, a $\lambda_\delta \in \mathbb{R}$ such that $\mu_{\lambda_\delta}(\Omega) \geq \frac{n^2}{4} - \delta$ in such a way that

$$\int_\Omega |\nabla \theta_k|^2\, dx \geq \left(\frac{n^2}{4} - \delta\right) \int_\Omega |x|^{-2} |\theta_k|^2\, dx + \lambda_\delta \int_\Omega |\theta_k|^2\, dx.$$

Since $\theta_k \to 0$ in $L^2(\Omega)$, it follows that

$$\int_\Omega |\nabla \theta_k|^2\, dx + o(1) \geq \left(\frac{n^2}{4} - \delta\right) \int_\Omega |x|^{-2} |\theta_k|^2\, dx.$$

This, combined with the definition of $\mu_\lambda(P,\Omega)$, (3.23) and (3.24), yield

$$\mu_\lambda(P,\Omega) \int_\Omega \frac{|u|^2}{|x|^2}\, dx \leq \int_\Omega |\nabla u|^2\, dx - \lambda \int_\Omega P(|x|) u^2\, dx$$

$$\leq 1 - \int_\Omega |\nabla \theta_k|^2\, dx - \lambda \int_\Omega P(|x|) |u|^2\, dx + o(1)$$

$$\leq 1 - \left(\frac{n^2}{4} - \delta\right) \int_\Omega |x|^{-2} |\theta_k|^2\, dx - \lambda \int_\Omega P(|x|) u^2\, dx + o(1)$$

$$\leq \left(\mu_\lambda(P,\Omega) - \frac{n^2}{4} + \delta\right) \int_\Omega |x|^{-2} |\theta_k|^2\, dx + \mu_\lambda(P,\Omega) \int_\Omega \frac{|u|^2}{|x|^2}\, dx + o(1).$$

Therefore $\int_\Omega |x|^{-2} |\theta_k|^2\, dx \to 0$ since $\mu_\lambda(P,\Omega) - \frac{n^2}{4} + \delta < 0$. In particular,

$$\mu_\lambda(P,\Omega) \int_\Omega \frac{|u|^2}{|x|^2}\, dx = \int_\Omega |\nabla u|^2\, dx - \lambda \int_\Omega P(|x|) |u|^2\, dx,$$

and $u \neq 0$ by (3.24). Thus u achieves $\mu_\lambda(P,\Omega)$. $\square$

Proof of Theorem 3.2.2: Note again that Lemmas 3.2.3 and 3.2.4 yield that

$$\sup_{\lambda \in \mathbb{R}} \mu_\lambda(P, \Omega) = \frac{n^2}{4}.$$

Since $\lambda \to \mu_\lambda(P, \Omega)$ is non-increasing, we can set

(3.25) $$\lambda^*(\Omega, P) = \sup\left\{\lambda \in \mathbb{R};\ \mu_\lambda(\Omega, P) = \frac{n^2}{4}\right\}.$$

Note that Lemma 3.2.3 also yields that $\lambda^*(\Omega, P) \in \mathbb{R}$. Assertion 1) follows from the definition of $\lambda^*(\Omega, P)$ and from Lemma 3.2.5. Since the mapping $\lambda \to \mu_\lambda(\Omega, P)$ is constant on $(-\infty, \lambda^*(\Omega, P))$, it is easy to show that $\mu_\lambda(\Omega, P)$ cannot be achieved for λ in such an interval.

3.3. Best Hardy constant for domains contained in a half-space

The results of the preceding section deal mostly with the case when $\lambda^*(\Omega, P) \leq 0$. We now study a case when $\lambda^*(\Omega, P) > 0$. This is obviously related to whether P is an HI-potential.

THEOREM 3.3.1. *Let Ω be a smooth bounded domain of $\mathbb{R}^n$ such that $0 \in \partial\Omega$ and $n \geq 2$, and let P be a HI-potential on $(0, R)$, where $R = \sup_{x \in \Omega} |x|$. If Ω is contained in a half-space going through 0, then*

(3.26) $$\lambda^*(\Omega, P) \geq \beta(P, R),$$

and for any $\lambda \leq \beta(P, R)$, the best constant $\mu_\lambda(P, \Omega)$ is equal to $\frac{n^2}{4}$ and is not attained in $H_0^1(\Omega)$.

Proof: That $\lambda^*(\Omega, P) \geq \beta(P, R)$ follows from Theorem 3.1.1. For the rest of the proof, we need the following lemma.

LEMMA 3.3.2. *Consider a half-ball $B_R^+ = B_R \cap \mathcal{C}_{\mathbb{S}_+^{n-1}}$ centered at 0 and let P be a positive and continuous function on $(0, R]$. Then any weak solution to*

(3.27) $$-\Delta u = \frac{n^2}{4}|x|^{-2}u + \lambda P(|x|)u \quad \text{on } B_R^+$$

is necessarily trivial.

Proof: For any domain $\Omega \subset \mathbb{R}^n$, recall the notation $\widehat{H}^1(\Omega) := H^1(\Omega) \cap L^2(\Omega; |x|^{-2}dx)$ and $\mathcal{D}'(\Omega)$ for the space of distributions on Ω. Test (3.27) with the negative and the positive part of u to conclude that u has constant sign, and by the maximum principle we have that $u > 0$ in B_R^+. We shall now prove that this is not possible.

For that, we consider $\Phi > 0$ to be the first eigenfunction of the Laplacian $-\Delta_\sigma$ on $\mathbb{S}_+^{n-1}$. In other words, since $\lambda_1(\mathbb{S}_+^{n-1}) = n - 1$, the function Φ solves

(3.28) $$\begin{cases} -\Delta_\sigma \Phi = (n-1)\Phi & \text{in } \mathbb{S}_+^{n-1} \\ \Phi = 0,\ \dfrac{\partial \Phi}{\partial \eta} \leq 0 & \text{on } \partial\mathbb{S}_+^{n-1}, \end{cases}$$

where $\eta \in T_\sigma(\mathbb{S}^{n-1})$ is the exterior normal to $\mathbb{S}_+^{n-1}$ at $\sigma \in \partial\mathbb{S}_+^{n-1}$.

3.3. BEST HARDY CONSTANT FOR DOMAINS CONTAINED IN A HALF-SPACE

We associate to the solution u –by density and via the trace theorem– the following radially symmetric function ψ on $\mathbb{D}_R \setminus \{0\}$, where $\mathbb{D}_R \subset \mathbb{R}^2$ is the open 2-dimensional disc of radius R centered at 0.

$$(3.29) \qquad \psi(z) = |z|^{\frac{n-2}{2}} \int_{\mathbb{S}^{n-1}_+} u(|z|\sigma) \Phi(\sigma) \, d\sigma = |z|^{\frac{n-2}{2}} \int_{|z|\mathbb{S}^{n-1}_+} u(\tau) \Phi_{|z|}(\tau) \, d\tau,$$

where $\Phi_r(\tau) = \Phi(\frac{\tau}{r})$ for all $\tau \in r\mathbb{S}^{n-1}_+$. In polar coordinates $(r, \sigma) \in (0, \infty) \times \mathbb{S}^{n-1}$, we have that

$$u_{rr} = -(n-1)r^{-1}u_r - r^{-2}\Delta_\sigma u,$$

and direct computations based on (3.27), (3.28) and (3.29) lead to ψ satisfying

$$-\Delta \psi = \lambda P(|x|)\psi \quad \text{in } \mathcal{D}'(\mathbb{D}_R \setminus \{0\}).$$

We now claim that $\psi \in \widehat{H}^1(\mathbb{D}_R)$. Indeed, for $r = |z|$,

$$|\psi'| \leq cr^{\frac{n-2}{2}-1} \int_{\mathbb{S}^{n-1}_+} |u(r\sigma)| d\sigma + cr^{\frac{n-2}{2}} \int_{\mathbb{S}^{n-1}_+} |\nabla u(r\sigma)| d\sigma,$$

and by Hölder's inequality,

$$\int_{\mathbb{D}_R} \left(r^{\frac{n-2}{2}-1} \int_{\mathbb{S}^{n-1}_+} |u(r\sigma)| d\sigma \right)^2 dz = c\int_0^R \int_{\mathbb{S}^{n-1}_+} r^{n-3} u^2 d\sigma \leq c\int_\Omega |x|^{-2} u^2 dx < \infty$$

$$\int_{\mathbb{D}_R} \left(r^{\frac{n-2}{2}} \int_{\mathbb{S}^{n-1}_+} |\nabla u(r\sigma)| d\sigma \right)^2 dz \leq c\int_0^R r^{n-1} \int_{\mathbb{S}^{n-1}_+} |\nabla u|^2 d\sigma \leq c\int_\Omega |\nabla u|^2 dx < \infty.$$

Finally, $\psi \in L^2(\mathbb{D}_R; |z|^{-2}dz)$ since

$$\int_{\mathbb{D}_R} |z|^{-2} |\psi|^2 dz = 2\pi \int_0^R r^{-1} |\psi|^2 dz \leq c\int_0^R r^{n-3} \int_{\mathbb{S}^{n-1}_+} |u|^2 d\sigma = c\int_\Omega \frac{|u|^2}{|x|^2} dx < \infty.$$

We now show that there exists $R_\lambda > 0$ such that for any $R \in (0, R_\lambda)$, we have $\psi = 0$ on $\partial \mathbb{D}_R$. Indeed, let $R_\lambda < 1/3$ be small enough, in such a way that

$$(3.30) \qquad \lambda < \inf \left\{ \frac{\int_{\mathbb{D}_{R_\lambda}} |\nabla v|^2}{\int_{\mathbb{D}_{R_\lambda}} P(|x|)|v|^2}; v \in H^1_0(\mathbb{D}_{R_\lambda}) \right\}.$$

Pick any $R < R_\lambda$ and assume that $\psi(x) \neq 0$ for some $x \in \partial \mathbb{D}_R$. Since ψ is radially symmetric, we can assume that $\psi \geq \varepsilon$ on $\partial \mathbb{D}_R$ for some $\varepsilon > 0$.

For any $\delta \in (1/2, 1)$ introduce the following radially symmetric function on $\mathbb{D}_R$,

$$\varphi_\delta(z) = |\log |z||^{-\delta}.$$

By direct computation one can easily check that $\varphi_\delta \in \widehat{H}^1(\mathbb{D}_R)$, and in particular that

$$(3.31) \qquad (2\delta - 1) \int_{\mathbb{D}_R} |z|^{-2} |\varphi_\delta|^2 = 2\pi + o(1) \quad \text{as } \delta \to \frac{1}{2}.$$

Since $\delta > 1/2$ then φ_δ is a smooth solution to

$$(3.32) \qquad \Delta \varphi_\delta \geq \tfrac{3}{4} |z|^{-2} |\log |z||^{-2+\delta} = \tfrac{3}{4} |z|^{-2} |\log |z||^{-2} \varphi_\delta \text{ on } \mathbb{D}_R \setminus \{0\}.$$

It is standard (see Lemma (3.4.1) below) to infer that φ_δ solves (3.32) weakly in $\widehat{H}^1(\mathbb{D}_R)$. Now set

$$v := \varepsilon \varphi_\delta - \psi \in \widehat{H}^1(\mathbb{D}_R),$$

and notice that $v \leq 0$ on $\partial \mathbb{D}_R$ since $R < 1/3$. Note also that

$$\begin{aligned}\Delta v &\geq \frac{3}{4}|z|^{-2}|\log|z||^{-2}(\varepsilon\varphi_\delta) + \lambda P(|x|)\psi \\ &= \left[\frac{3}{4}|z|^{-2}|\log|z||^{-2} + \lambda\right]\varepsilon\varphi_\delta P(|x|) - \lambda P(|x|)v\end{aligned}$$

weakly (i.e., on the dual of $\hat{H}^1(\mathbb{D}_R)$). Use $v^+ := \max\{v,0\} \in H_0^1(\mathbb{D}_R) \cap \hat{H}^1(\mathbb{D}_R)$ as a test function to get

$$-\int_{\mathbb{D}_R}|\nabla v^+|^2 \geq \int_{\mathbb{D}_R}\left[\frac{3}{4}|z|^{-2}|\log|z||^{-2} + \lambda\right]\varepsilon\varphi_\delta P(|x|)v^+ - \lambda\int_{\mathbb{D}_R}P(|x|)|v^+|^2.$$

Since $\lambda \geq 0$, we infer that

$$\int_{\mathbb{D}_R}|\nabla v^+|^2 dx \leq \lambda \int_{\mathbb{D}_R}P(|x|)|v^+|^2 dx,$$

and hence $v^+ \equiv 0$ on $\mathbb{D}_R$ in view of (3.30). Thus $\psi \geq \varepsilon\varphi_\delta$ on $\mathbb{D}_R$ and therefore

$$+\infty > \int_{\mathbb{D}_R}|z|^{-2}|\psi|^2 dz \geq \varepsilon\int_{\mathbb{D}_R}|z|^{-2}|\varphi_\delta|^2 dz,$$

which contradicts (3.31).

It follows that $\psi \equiv 0$ in a neighborhood of 0, and hence $u \equiv 0$ in $B_r \cap C_{\mathbb{S}_+^{n-1}}$, for $r > 0$ small enough, which contradicts that $u > 0$ everywhere on B_R^+. $\square$

In order to complete the proof of Theorem 3.3.1, we assume that Ω is a smooth bounded domain containing a half-ball centered at $0 \in \partial\Omega$. If $\mu_\lambda(P,\Omega) = \frac{n^2}{4}$ and if the latter is achieved at some $u \in \hat{H}^1(\Omega)$, then u is a weak solution of

$$(3.33) \qquad -\Delta u = \frac{n^2}{4}|x|^{-2}u + \lambda P(|x|)u \quad \text{on } \Omega.$$

Since Ω is contained in a half-ball $B_R^+(0)$ for some $R > 0$, then u is necessarily equal to zero by the preceding lemma.

REMARK 3.3.3. Note that the same argument shows that if Ω contains a half-ball $B_\rho^+(0)$ for some $\rho > 0$, and if $\mu_\lambda(P,\Omega) = \frac{n^2}{4}$, then the latter is not attained in $H_0^1(\Omega)$.

In the case where $P \equiv 1$, Theorem 3.3.1 yields in particular the following result.

COROLLARY 3.3.1. If Ω be a smooth bounded domain of $\mathbb{R}^n$ such that $0 \in \partial\Omega$, then
(1) $\frac{(n-2)^2}{4} < \mu_0(\Omega) \leq \frac{n^2}{4}$.
(2) $\mu_0(\Omega)$ is attained in $H_0^1(\Omega)$ if and only if $\mu_0(\Omega) < \frac{n^2}{4}$.
(3) if Ω is contained in a half-space going through 0, and $R = \sup_{x \in \Omega}|x|$, then $\lambda^*(\Omega) \geq \frac{z_0^2}{R^2}$. In other words, for any $0 \leq \lambda \leq \frac{z_0^2}{R^2}$ we have $\mu_\lambda(\Omega) = \frac{n^2}{4}$, and the latter is not attained in $H_0^1(\Omega)$.

Now we show that there exist indeed domains Ω with $0 \in \partial\Omega$ such that $\mu_0(\Omega) < \frac{n^2}{4}$ and therefore the best constant is attained in $H_0^1(\Omega)$.

PROPOSITION 3.3.1. For any $\delta > 0$ there exists $\rho_\delta > 0$ such that if Ω is a smooth domain with $0 \in \partial\Omega$ and
$$\Omega \supseteq \{x \in \mathbb{R}^n \mid x \cdot \nu > -\delta|x|\,,\ \alpha < |x| < \beta\,\}$$
for some $\nu \in \mathbb{S}^{n-1}$, $\beta > \alpha > 0$ with $\beta/\alpha > \rho_\delta$, then $\mu_0(\Omega) < \frac{n^2}{4}$. In particular the Hardy constant $\mu_0(\Omega)$ is achieved.

Proof: Since the cone $\mathcal{C}_\delta = \{x \in \mathbb{R}^n \mid x \cdot \nu > -\delta|x|\,\}$ contains a half-space, then its Hardy constant is smaller than $\frac{n^2}{4}$. Hence, there exists $u \in C_c^\infty(\mathcal{C}_\delta)$ such that
$$\frac{\int_{\mathcal{C}_\delta} |\nabla u|^2\, dx}{\int_{\mathcal{C}_\delta} \frac{|u|^2}{|x|^2}\, dx} < \frac{n^2}{4}\,.$$
Assuming that the support of u is contained in an annulus of radii $\beta > \alpha > 0$, the claim in Corollary 3.3.1 then holds for $\rho := \beta/\alpha$. $\square$

3.4. The Poisson equation on the punctured disc

Before ending this chapter we include the following lemma, which was used in the proof of Theorem 3.3.1

LEMMA 3.4.1. Let $\psi \in \widehat{H}^1(\mathbb{D}_R)$ and $f \in L^1_{\mathrm{loc}}(\mathbb{D}_R)$ for some $R > 0$. If ψ solves
(3.34) $$-\Delta\psi \geq f \quad \text{in } \mathcal{D}'(\mathbb{D}_R \setminus \{0\})$$
then $-\Delta\psi \geq f$ in $\mathcal{D}'(\mathbb{D}_R)$.

Proof: Note first that since
$$+\infty > \int_{\mathbb{D}_R} |z|^{-2}|\psi|^2\, dz = \int_0^R \frac{1}{r}\left(r^{-1}\int_{\partial B_r} |\psi|^2\, d\sigma\right) dr,$$
there exists then a sequence $r_k \to 0$, $r_k \in (0, R)$ such that
(3.35) $$r_k^{-1} \int_{\partial B_{r_k}} |\psi|^2\, d\sigma \to 0, \quad \text{and} \quad r_k^{-2} \int_{\partial B_{r_k^2}} |\psi|^2\, d\sigma \to 0$$
as $k \to \infty$. Consider the following cut-off functions:
$$\eta_k(z) = \begin{cases} 0 & \text{if } |z| \leq r_k^2 \\ \dfrac{\log |z|/r_k^2}{|\log r_k|} & \text{if } r_k^2 < |z| < r_k \\ 1 & \text{if } r_k \leq |z| \leq R, \end{cases}$$
and let φ be any nonnegative function in $C_c^\infty(\mathbb{D}_R)$. Test (3.34) with $\eta_k\varphi$ to get
$$\int_{\mathbb{D}_R} \nabla\psi \cdot \nabla(\eta_k\varphi)\, dz \geq \int_{\mathbb{D}_R} f\, \eta_k\varphi\, dz.$$
Since $\psi \in H^1(\mathbb{D}_R)$ and since $\eta_k \rightharpoonup 1$ weak* in L^∞, it follows that
$$\int f\, \eta_k\varphi\, dz = \int_{\mathbb{D}_R} f\varphi\, dz + o(1)\,, \quad \int \eta_k \nabla\psi \cdot \nabla\varphi\, dz = \int_{\mathbb{D}_R} \nabla\psi \cdot \nabla\varphi\, dz + o(1)$$
as $k \to \infty$. Therefore
(3.36) $$\int_{\mathbb{D}_R} \nabla\psi \cdot \nabla\varphi\, dz + \int \varphi \nabla\psi \cdot \nabla\eta_k\, dz \geq \int_{\mathbb{D}_R} f\varphi\, dz + o(1)\,.$$

Note that $\nabla\eta_k$ vanishes outside the annulus $A_k := \{r_k^2 < |z| < r_k\}$, and that η_k is harmonic on A_k. Thus

$$\int_{\mathbb{D}_R} \varphi\nabla\psi \cdot \nabla\eta_k \, dz = \int_{A_k} \nabla(\psi\varphi) \cdot \nabla\eta_k \, dz - \int_{A_k} \psi\nabla\varphi \cdot \nabla\psi \, dz$$
$$= \mathcal{R}_k - \int_{A_k} \psi\nabla\varphi \cdot \nabla\eta_k \, dz,$$

where

$$\mathcal{R}_k := -r_k^{-2} \int_{\partial B_{r_k^2}} (\nabla\eta_k \cdot z)\psi\varphi \, d\sigma + r_k^{-1} \int_{\partial B_{r_k}} (\nabla\eta_k \cdot z)\psi\varphi \, d\sigma.$$

Now

$$|\mathcal{R}_k| \leq c \, (r_k |\log r_k|)^{-1} \int_{\partial B_{r_k}} |\psi| \, d\sigma + c \, (r_k^2 |\log r_k|)^{-1} \int_{\partial B_{r_k^2}} |\psi| \, d\sigma,$$

where $c > 0$ is a constant that does not depend on k, and by Hölder inequality and (3.35)

$$(r_k |\log r_k|)^{-1} \int_{\partial B_{r_k}} |\psi| \, d\sigma \leq c \, |\log r_k|^{-1} \left(r_k^{-1} \int_{\partial B_{r_k}} |\psi|^2 \, d\sigma \right)^{1/2} = o(1).$$

For the same reasons,

$$(r_k^2 |\log r_k|)^{-1} \int_{\partial B_{r_k^2}} |\psi| \, d\sigma \leq c \, |\log r_k|^{-1} \left(r_k^{-2} \int_{\partial B_{r_k^2}} |\psi|^2 \, d\sigma \right)^{1/2} = o(1)$$

and hence $\mathcal{R}_k = o(1)$. Since $\psi \in L^2(\mathbb{D}_R; |z|^{-2} \, dz)$ it follows that

$$\left| \int_{A_k} \psi\nabla\varphi \cdot \nabla\eta_k \, dz \right| \leq |\log r_k|^{-1} \int |z|^{-1} \psi |\nabla\varphi| \, dz = o(1).$$

In conclusion, $\int_{\mathbb{D}_R} \varphi\nabla\psi \cdot \nabla\eta_k \, dz = o(1)$ and therefore (3.36) gives that $\int_{\mathbb{D}_R} \nabla\psi \cdot \nabla\varphi \, dz \geq \int_{\mathbb{D}_R} f\varphi \, dz$. Since φ was an arbitrary nonnegative function in $C_c^\infty(\mathbb{D}_R)$, this shows that $-\Delta\psi \geq f$ in the distributional sense on $\mathbb{D}_R$, as desired.

3.5. Further comments

Hardy inequalities on domains having 0 on their boundary were studied by Fall-Musina [133], Fall [134] and Cazacu [107]. They showed that the best constant $\mu_\lambda(\Omega)$ can then be anywhere between $\frac{(n-2)^2}{4}$ and $\frac{n^2}{4}$. Motivated by the work of Ghoussoub-Kang [157] on the Hardy-Sobolev inequalities, they showed that $\mu_\lambda(\Omega)$ is attained in $H_0^1(\Omega)$ whenever it is strictly below $\frac{n^2}{4}$. The proofs here are adaptations of theirs, including Lemma 3.4.1, which is taken from [133]. That all these results also hold in the presence of an improving potential (i.e., Theorems 3.2.5 and 3.3.1) are part of an unpublished work by the authors [164] and appear here for the very first time.

It has been shown by Fall [134] that unlike the Hardy-Sobolev case (see Chapter 16), the strict local concavity of Ω at 0 does not necessarily imply that $\mu_0(\Omega) < \frac{n^2}{4}$. The following question is therefore in order.

Open problem (2): Assuming 0 is on the boundary of a bounded smooth domain Ω of $\mathbb{R}^n$. Find a necessary and sufficient geometric condition on Ω, which insures that the best constant $\mu_\lambda(\Omega) < \frac{n^2}{4}$ and therefore is attained.

On the other hand, show that $\mu_{\lambda^*}(\Omega, P)$ – where $\lambda^* := \lambda^*(\Omega, P)$ – is never attained in $H_0^1(\Omega)$. Note that this is indeed the case if $\lambda^* = \lambda^*(\Omega, P) \leq 0$ **[134]** and – as seen in this chapter – if $\lambda^* = \lambda^*(\Omega, P) > 0$ provided that Ω is contained in a half-space going through 0.

CHAPTER 4

Weighted Hardy Inequalities

Two positive functions V and W constitute an n-dimensional Bessel pair on an interval $(0, R)$ if and only if for some $c > 0$, the following inequalities hold for all $u \in C_c^\infty(B_R)$, where B_R is a ball of radius R in $\mathbb{R}^n$ and $n \geq 1$:

$$\int_B V(|x|)|\nabla u|^2 dx \geq c \int_B W(|x|)u^2 dx.$$

This characterization allows one to improve, extend, and unify many results –old and new– about Hardy-type inequalities and their best constants, including those corresponding to power-weights à la Caffarelli-Kohn-Nirenberg.

This chapter also explores Hardy-type inequalities for $H^1(\Omega)$-functions, i.e., those which do not necessarily have compact support in Ω. In this case, a penalizing term appears in order to account for the boundary contribution. In particular, there exists $c > 0$ and $\theta > 0$ such the following inequalities hold:

$$\int_\Omega V(|x|)|\nabla u|^2 dx \geq c \int_\Omega W(|x|)u^2 dx - \theta \int_{\partial\Omega} u^2 ds \quad \text{for all } u \in H^1(\Omega).$$

4.1. Bessel pairs and weighted Hardy inequalities

The following theorem is the main result of this chapter.

THEOREM 4.1.1. *Let V and W be positive C^1-functions on some interval $(0, R)$ such that for some $n \geq 1$, we have*

(4.1) $$\int_0^R \frac{1}{r^{n-1}V(r)} dr = +\infty \quad \text{and} \quad \int_0^R r^{n-1} V(r) dr < +\infty.$$

The following two statements are then equivalent:

(1) *(V, W) is a n-dimensional Bessel pair on $(0, R)$.*
(2) *There exists $c > 0$ such that for all $u \in C_c^\infty(B_R)$, where B_R is a ball of radius R in $\mathbb{R}^n$, we have*

($H_{V,cW}$) $$\int_{B_R} V(|x|)|\nabla u|^2 dx \geq c \int_{B_R} W(|x|)u^2 dx.$$

Moreover, the largest c for which the inequality holds is equal to $\beta(V, W; R)$.

For the proof of Theorem 4.1.1, we shall need the following lemmas.

LEMMA 4.1.2. *Let Ω be a smooth bounded domain in $\mathbb{R}^n$ ($n \geq 1$) containing 0. Set $R =: \sup_{x \in \partial\Omega} |x|$ and assume $\varphi \in C^1(0, R)$ is a positive solution of the ordinary differential equation*

(4.2) $$y'' + \left(\frac{n-1}{r} + \frac{V_r(r)}{V(r)}\right)y' + \frac{W(r)}{V(r)}y = 0 \quad \text{on } (0, R),$$

where $V, W \geq 0$ on $(0, R)$ while $\int_0^R \frac{1}{r^{n-1}V(r)} dr = +\infty$ and $\int_0^R r^{n-1}V(r) dr < +\infty$. Setting $\psi(x) = \frac{u(x)}{\varphi(|x|)}$ for any $u \in C_c^\infty(\Omega)$, we then have the following properties:

(1) $\int_0^R r^{n-1} V(r) \left(\frac{\varphi'(r)}{\varphi(r)}\right)^2 dr < \infty$ *and* $\lim_{r \to 0} r^{n-1} V(r) \frac{\varphi'(r)}{\varphi(r)} = 0.$

45

(2) $\int_\Omega V(|x|)(\varphi'(|x|))^2\psi^2(x)dx < \infty$.
(3) $\int_\Omega V(|x|)\varphi^2(|x|)|\nabla\psi|^2(x)dx < \infty$.
(4) $|\int_\Omega V(|x|)\varphi'(|x|)\varphi(|x|)\psi(x)\frac{x}{|x|}\cdot\nabla\psi(x)dx| < \infty$.
(5) $\lim_{r\to 0}|\int_{\partial B_r} V(|x|)\varphi'(|x|)\varphi(|x|)\psi^2(x)ds| = 0$, where $B_r \subset \Omega$ is a ball of radius r centered at 0.

Proof: 1) Setting $x(r) = r^{n-1}V(r)\frac{\varphi'(r)}{\varphi(r)}$, we have for $0 < r < R$,

$$r^{n-1}V(r)x'(r) + x^2(r) = \frac{r^{2(n-1)}V^2(r)}{\varphi}(\varphi''(r) + (\frac{n-1}{r} + \frac{V_r(r)}{V(r)})\varphi'(r))$$
$$= -\frac{r^{2(n-1)}V(r)W(r)}{\varphi(r)} \leq 0.$$

Dividing by $r^{n-1}V(r)$ and integrating once, we obtain

$$(4.3) \qquad x(r) \geq \int_r^R \frac{|x(s)|^2}{s^{n-1}V(s)}ds + x(R).$$

To prove that $\lim_{r\to 0} G(r) < \infty$, where $G(r) := \int_r^R \frac{x^2(s)}{s^{n-1}V(s)}ds$, we assume the contrary and use (4.3) to write that

$$(-r^{n-1}V(r))G'(r))^{\frac{1}{2}} \geq G(r) + x(R).$$

Thus, for r sufficiently small we have $-r^{n-1}V(r)G'(r) \geq \frac{1}{2}G^2(r)$ and hence, $(\frac{1}{G(r)})' \geq \frac{1}{2r^{n-1}V(r)}$, which contradicts the fact that $G(r)$ goes to infinity as r tends to zero.

Also in view of (4.3), we have that $x_0 := \lim_{r\to 0} x(r)$ exists, and since $\lim_{r\to 0} G(r) < \infty$, we necessarily have $x_0 = 0$ and 1) is proved.

For assertion 2), we use 1) to see that

$$\int_\Omega V(|x|)\varphi'(|x|)^2\psi^2(x)dx \leq \|u\|_\infty^2 \int_\Omega V(|x|)\frac{\varphi'(|x|)^2}{\varphi^2(|x|)}dx < \infty.$$

3) Note that

$$|\nabla\psi(x)| \leq \frac{|\nabla u(x)|}{\varphi(|x|)} + |u(x)|\frac{|\varphi'(|x|)|}{\varphi^2(|x|)} \leq \frac{C_1}{\varphi(|x|)} + C_2\frac{|\varphi'(|x|)|}{\varphi^2(|x|)} \quad \text{for all } x \in \Omega,$$

where $C_1 = \max_{x\in\Omega}|\nabla u|$ and $C_2 = \max_{x\in\Omega}|u|$. Hence,

$$\int_\Omega V(|x|)\varphi^2(|x|)|\nabla\psi|^2(x)dx \leq \int_\omega V(|x|)\frac{(C_1\varphi(|x|) + C_2\varphi'(|x|))^2}{\varphi^2(|x|)}dx$$
$$= \int_\Omega C_1^2 V(|x|)dx + \int_\Omega 2C_1C_2 \frac{|\varphi'(|x|)|}{\varphi(|x|)}V(|x|)dx$$
$$+ \int_\Omega C_2^2 \frac{\varphi'(|x|)^2}{\varphi(|x|)^2}V(|x|)dx$$
$$\leq L_1 + 2C_1C_2\big(\int_\Omega V(|x|)\frac{\varphi'(|x|)^2}{\varphi(|x|)^2}dx\big)^{\frac{1}{2}}\big(\int_\Omega V(|x|)dx\big)^{\frac{1}{2}} + L_2,$$
$$< \infty,$$

which proves 3).

4) now follows from 2) and 3) since

$$V(|x|)|\nabla u|^2 = V(|x|)\varphi'(|x|)^2\psi^2(x)$$
$$+ 2V(|x|)\varphi'(|x|)\varphi(|x|)\psi(x)\frac{x}{|x|}\cdot\nabla\psi(x) + V(|x|)\varphi^2(|x|)|\nabla\psi|^2.$$

Finally, 5) follows from 1) since

$$\left|\int_{\partial B_r} V(|x|)\varphi'(|x|)\varphi(|x|)\psi^2(x)ds\right| \le ||u||_\infty^2 \left|\int_{\partial B_r} V(|x|)\frac{\varphi'(|x|)}{\varphi(|x|)}ds\right|$$

$$= ||u||_\infty^2 V(r)\frac{|\varphi'(r)|}{\varphi(r)} \int_{\partial B_r} 1 ds$$

$$= n\omega_n ||u||_\infty^2 r^{n-1} V(r)\frac{|\varphi'(r)|}{\varphi(r)}.$$

LEMMA 4.1.3. *Let V and W be positive C^1-functions on $(0, R)$, and let $B := B_R$ be a ball with radius R in $\mathbb{R}^n$ ($n \ge 1$) centered at zero. Assuming*

$\int_B \left(V(|x|)|\nabla u|^2 - W(|x|)u^2\right) dx \ge 0$ *for all* $u \in C_c^\infty(B)$,

then there exists a C^2-supersolution to the following linear elliptic equation

(4.4) $\quad -\text{div}(V(|x|)\nabla u) - W(|x|)u = 0, \quad \text{in } B,$

(4.5) $\quad\quad\quad\quad\quad\quad\quad\quad\quad\quad\quad u > 0 \quad \text{in } B \setminus \{0\},$

(4.6) $\quad\quad\quad\quad\quad\quad\quad\quad\quad\quad\quad u = 0 \quad \text{in } \partial B.$

Proof: Define

$$\lambda_1(V) := \inf\left\{\frac{\int_B V(|x|)|\nabla \psi|^2 - W(|x|)|\psi|^2}{\int_B |\psi|^2}; \ \psi \in C_c^\infty(B \setminus \{0\})\right\}.$$

By our assumption $\lambda_1(V) \ge 0$. Let (φ_k, λ_1^k) be the first eigenpair for the problem

$$(L - \lambda_1(V) - \lambda_1^k)\varphi_k = 0 \text{ on } B \setminus B_{\frac{R}{k}}$$

$$\varphi_k = 0 \text{ on } \partial(B \setminus B_{\frac{R}{k}}),$$

where $Lu = -\text{div}(V(|x|)\nabla u) - W(|x|)u$, and $B_{\frac{R}{k}}$ is a ball of radius $\frac{R}{k}$, $k \ge 2$. The eigenfunctions can be chosen in such a way that $\varphi_k > 0$ on $B \setminus B_{\frac{R}{k}}$ and $\varphi_k(b) = 1$, for some $b \in B$ with $\frac{R}{2} < |b| < R$.

Note that $\lambda_1^k \downarrow 0$ as $k \to \infty$. Harnak's inequality yields that for any compact subset K, $\frac{\max_K \varphi_k}{\min_K \varphi_k} \le C(K)$ with the later constant being independent of φ_k. Also standard elliptic estimates also yields that the family (φ_n) have also uniformly bounded derivatives on the compact sets $B \setminus B_{\frac{R}{k}}$.

Therefore, there exists a subsequence $(\varphi_{k_{l_2}})_{l_2}$ of $(\varphi_k)_k$ such that $(\varphi_{k_{l_2}})_{l_2}$ converges to some $\varphi_2 \in C^2(B \setminus B_{\frac{R}{2}})$. Now consider $(\varphi_{k_{l_2}})_{l_2}$ on $B \setminus B_{\frac{R}{3}}$. Again there exists a subsequence $(\varphi_{k_{l_3}})_{l_3}$ of $(\varphi_{k_{l_2}})_{l_2}$ which converges to $\varphi_3 \in C^2(B \setminus B_{\frac{R}{3}})$, and $\varphi_3(x) = \varphi_2(x)$ for all $x \in B \setminus B_{\frac{R}{2}}$. By repeating this argument, we get a supersolution $\varphi \in C^2(B \setminus \{0\})$, i.e., $L\varphi \ge 0$, such that $\varphi > 0$ on $B \setminus \{0\}$. $\square$

Proof of Theorem 4.1.1: To prove that 1) implies 2), let $\varphi \in C^1(0, R]$ be a solution of $(B_{V,W})$ such that $\varphi(x) > 0$ for all $x \in (0, R)$. Define $\frac{u(x)}{\varphi(|x|)} = \psi(x)$. Then

$$|\nabla u|^2 = \varphi'(|x|)^2 \psi^2(x) + 2\varphi'(|x|)\varphi(|x|)\psi(x)\frac{x}{|x|}.\nabla\psi + \varphi^2(|x|)|\nabla\psi|^2.$$

Hence,

$$V(|x|)|\nabla u|^2 \ge V(|x|)\varphi'(|x|)^2\psi^2(x) + 2V(|x|)\varphi'(|x|)\varphi(|x|)\psi(x)\frac{x}{|x|}.\nabla\psi(x).$$

Thus, we have
$$\int_B V(|x|)|\nabla u|^2 dx \geq \int_B V(|x|)\varphi'(|x|)^2 \psi^2(x) dx$$
$$+ \int_B 2V(|x|)\varphi'(|x|)\varphi(|x|)\psi(x)\frac{x}{|x|}\cdot\nabla\psi dx.$$

Let B_ϵ be a ball of radius ϵ centered at the origin. Integrate by parts to get
$$\int_B V(|x|)|\nabla u|^2 dx \geq \int_B V(|x|)\varphi'(|x|)^2 \psi^2(x) dx + \int_{B_\epsilon} 2V(|x|)\varphi'(|x|)\varphi(|x|)\psi(x)\frac{x}{|x|}\cdot\nabla\psi dx$$
$$+ \int_{B\setminus B_\epsilon} 2V(|x|)\varphi'(|x|)\varphi(|x|)\psi(x)\frac{x}{|x|}\cdot\nabla\psi dx$$
$$= \int_{B_\epsilon} V(|x|)\varphi'(|x|)^2 \psi^2(x) dx + \int_{B_\epsilon} 2V(|x|)\varphi'(|x|)\varphi(|x|)\psi(x)\frac{x}{|x|}\cdot\nabla\psi dx$$
$$- \int_{B\setminus B_\epsilon} (V(|x|)\varphi''(|x|)\varphi(|x|)\, dx$$
$$- \int_{B\setminus B_\epsilon} \left\{\frac{(n-1)V(|x|)}{r} + V_r(|x|))\varphi'(|x|)\varphi(|x|)\right\}\psi^2(x)\Big\} dx$$
$$+ \int_{\partial(B\setminus B_\epsilon)} V(|x|)\varphi'(|x|)\varphi(|x|)\psi^2(x) ds.$$

Let $\epsilon \to 0$ and use Lemma 4.1.2 and the fact that φ is a solution of (4.2) to get
$$\int_B V(|x|)|\nabla u|^2 dx \geq -\int_B [V(|x|)\varphi''(|x|) + (\frac{(n-1)V(|x|)}{r} + V_r(|x|))\varphi'(|x|)]\frac{u^2(x)}{\varphi(|x|)} dx$$
$$= \int_B W(|x|)u^2(x) dx.$$

To show that 2) implies 1), we assume that inequality $(H_{V,W})$ holds on a ball B of radius R, and then apply Lemma 4.1.3 to obtain a C^2-supersolution for the equation (4.4). Take the surface average of u, that is

$$(4.7) \qquad y(r) = \frac{1}{n\omega_w r^{n-1}}\int_{\partial B_r} u(x) dS = \frac{1}{n\omega_n}\int_{|\omega|=1} u(r\omega) d\omega > 0,$$

where ω_n denotes the volume of the unit ball in $\mathbb{R}^n$. We may assume that the unit ball is contained in B (otherwise we just use a smaller ball). We then have

$$(4.8) \qquad y''(r) + \frac{n-1}{r}y'(r) = \frac{1}{n\omega_n r^{n-1}}\int_{\partial B_r} \Delta u(x) dS.$$

Since u is a supersolution of (4.4),
$$\int_{\partial B_r} \text{div}(V(|x|)\nabla u) ds - \int_{\partial B_r} W(|x|) u\, dx \geq 0,$$
and therefore,
$$V(r)\int_{\partial B_r} \Delta u\, dS - V_r(r)\int_{\partial B_r} \nabla u.x\, ds - W(r)\int_{\partial B_r} u(x) ds \geq 0.$$

It follows that
$$(4.9) \qquad V(r)\int_{\partial B_r} \Delta u\, dS - V_r(r) y'(r) - W(r) y(r) \geq 0,$$

and in view of (4.7), we see that y satisfies the inequality

$$(4.10) \quad V(r)y''(r) + (\frac{(n-1)V(r)}{r} + V_r(r))y'(r) \leq -W(r)y(r) \quad \text{for} \quad 0 < r < R,$$

whch means that y is a positive supersolution for $(B_{V,W})$. It follows that $(B_{V,W})$ has actually a positive solution on $(0,R)$, and the proof of Theorem 4.1.1 is now complete.

4.2. Improved weighted Hardy-type inequalities on bounded domains

We start by applying Theorem 4.1.1 to improve the following class of weighted Hardy inequalities established by Caffarelli-Kohn-Nirenberg:

(4.11) $\quad \left(\int_{\mathbb{R}^n} |x|^{-bp} u^p dx \right)^{\frac{2}{p}} \leq C_{a,b} \int_{\mathbb{R}^n} |x|^{-2a} |\nabla u|^2 dx$ for all $u \in C_c^\infty(\mathbb{R}^n)$,

where for $n \geq 3$,

(4.12) $\quad -\infty < a < \frac{n-2}{2}, \ a \leq b \leq a+1, \text{ and } p = \frac{2n}{n-2+2(b-a)}$.

For $n = 2$

(4.13) $\quad -\infty < a < 0, \ a < b \leq a+1, \text{ and } p = \frac{2}{b-a}$,

and for $n = 1$

(4.14) $\quad -\infty < a < -\frac{1}{2}, \ a+\frac{1}{2} < b \leq a+1, \text{ and } p = \frac{2}{-1+2(b-a)}$.

Let $\mathcal{D}_a^{1,2}$ be the completion of $C_c^\infty(\mathbb{R}^n)$ for the inner product

$$(u,v) = \int_{\mathbb{R}^n} |x|^{-2a} \nabla u . \nabla v \, dx$$

and let

(4.15) $\quad S(a,b) = \inf_{u \in \mathcal{D}_a^{1,2} \setminus \{0\}} \frac{\int_{\mathbb{R}^n} |x|^{-2a} |\nabla u|^2 dx}{\left(\int_{\mathbb{R}^n} |x|^{-bp} |u|^p dx \right)^{2/p}}$

denote the corresponding best constant.

We shall deal with these inequalities in their full generality in Part V of this book. For now, we shall give a necessary and sufficient condition for the improvement of (4.11) with $b = a+1$ and $n \geq 1$. Our results cover also the critical case when $a = \frac{n-2}{2}$ which is not covered by the methods of Caffarelli-Kohn-Nirenberg.

THEOREM 4.2.1. *Let P be positive C^1-functions on $(0,R)$, and let B_R be a ball with radius R in $\mathbb{R}^n$ ($n \geq 1$). Assuming $a \leq \frac{n-2}{2}$, then the following two statements are then equivalent:*

(1) *P is a HI-potential on $(0,R)$.*
(2) *There exists $c > 0$ such that the following inequality holds for all $u \in C_c^\infty(B)$,*

(4.16) $\quad \int_B |x|^{-2a} |\nabla u|^2 dx \geq \left(\frac{n-2a-2}{2}\right)^2 \int_B |x|^{-2a-2} u^2 dx + c \int_B |x|^{-2a} P(|x|) u^2 dx$.

Moreover, $\left(\frac{n-2a-2}{2}\right)^2$ is the best constant for the first term and $\beta(P;R)$ is the largest constant c for which (4.16) holds.

On the other hand, there is no strictly positive $P \in C^1(0,\infty)$, such that the following inequality holds for all $u \in C_c^\infty(\mathbb{R}^n)$,

(4.17) $\quad \int_{\mathbb{R}^n} |x|^{-2a} |\nabla u|^2 dx \geq \left(\frac{n-2a-2}{2}\right)^2 \int_{\mathbb{R}^n} |x|^{-2a-2} u^2 dx + c \int_{\mathbb{R}^n} P(|x|) u^2 dx$.

Proof: It suffices to use Theorems 4.1.1 and 4.2.3 with $V(r) = r^{-2a}$ to get that P is a HI-potential if and only if the pair $(r^{-2a}, W_{a,c}(r))$ is a Bessel pair on $(0, R)$ for some $c > 0$, where

$$W_{a,c}(r) = (\frac{n-2a-2}{2})^2 r^{-2-2a} + cr^{-2a} P(r).$$

For the last part, assume that (4.17) holds for some P. Then it follows from Theorem 4.2.1 that for $V(r) := cr^{2a} P(r)$ the equation $y''(r) + \frac{1}{r} y' + V(r) y = 0$ has a positive solution on $(0, \infty)$, which is not possible in view of Lemma 1.1.2. □

By applying the above to various examples of HI-potentials, we can deduce several old and new inequalities.

COROLLARY 4.2.1. *Let Ω be a bounded smooth domain in $\mathbb{R}^n$ with $n \geq 1$ and $a \leq \frac{n-2}{2}$. Then, for any $b < 2$ there exists $c > 0$ such that for all $u \in C_c^\infty(\Omega)$*

(4.18) $\quad \int_\Omega |x|^{-2a} |\nabla u|^2 dx \geq (\frac{n-2a-2}{2})^2 \int_\Omega |x|^{-2a-2} u^2 dx + c \int_\Omega |x|^{-2a-b} u^2 dx.$

Moreover, when Ω is a ball B of radius R the best constant c for which (4.18) holds is equal to the weight $\beta(r^{-b}; R) = \frac{z_b^{2-b}}{R^{2-b}}$ of the HI-potential $W(r) = r^{-b}$ on $(0, R]$. In particular,

(4.19) $\quad \int_\Omega |x|^{-2a} |\nabla u|^2 dx \geq (\frac{n-2a-2}{2})^2 \int_\Omega |x|^{-2a-2} u^2 dx + z_0^2 \frac{\omega_n^{2/n}}{|\Omega|^{2/n}} \int_\Omega |x|^{-2a} u^2 dx,$

where $z_0 = 2.4048...$ is the first zero of the Bessel function $J_0(z)$.

Proof: It suffices to apply Theorem 4.2.1 with the function $W(r) = r^{-b}$ which is a HI-potential whenever $b < 2$.

COROLLARY 4.2.2. *Let Ω be a bounded smooth domain in $\mathbb{R}^n$ with $n \geq 1$ and $a \leq \frac{n-2}{2}$. Then*

(1) *For every integer k, and $\rho = (\sup_{x \in \Omega} |x|)(e e^{e^{\cdot^{\cdot^{e^{((k-1)-times)}}}}})$, we have for $u \in H_0^1(\Omega)$,*

(4.20) $\quad \int_\Omega \frac{|\nabla u|^2}{|x|^{2a}} dx \geq (\frac{n-2a-2}{2})^2 \int_\Omega \frac{u^2}{|x|^{2a+2}} dx$

$\quad + \frac{1}{4} \sum_{j=1}^k \int_\Omega \frac{|u|^2}{|x|^{2a+2}} (\prod_{i=1}^j \log^{(i)} \frac{\rho}{|x|})^{-2} dx.$

(2) *For any $D \geq \sup_{x \in \Omega} |x|$, we have for $u \in H_0^1(\Omega)$,*

(4.21) $\quad \int_\Omega \frac{|\nabla u|^2}{|x|^{2a}} dx \geq (\frac{n-2a-2}{2})^2 \int_\Omega \frac{u^2}{|x|^{2a+2}} dx$

$\quad + \frac{1}{4} \sum_{i=1}^\infty \int_\Omega \frac{1}{|x|^{2a+2}} X_1^2(\frac{|x|}{D}) X_2^2(\frac{|x|}{D}) ... X_i^2(\frac{|x|}{D}) u^2 dx,$

In both cases, $\frac{1}{4}$ is the best constant which is not attained in $H_0^1(\Omega)$.

Proof: As seen in Chapter 1, $W_{k,\rho}(r) = \sum_{j=1}^k \frac{1}{r^2} (\prod_{i=1}^j \log^{(i)} \frac{\rho}{|x|})^{-2} dx$ is a HI-potential on $(0, R)$ where $R = \sup_{x \in \Omega} |x|$, and $\beta(W_{k,\rho}; R) = \frac{1}{4}$. The same applies to $\tilde{W}_{k,\rho}(r)$. □

We now apply the integral criteria introduced in Chapter 1 to identify new weights V and W for which inequality $(H_{V,W})$ holds. This will lead to new Hardy-type inequalities.

THEOREM 4.2.2. *Let V and W be positive C^1-functions on $(0,R)$, and let B_R be a ball with radius R in $\mathbb{R}^n$ ($n \geq 1$). Assume*

(4.22) $\quad \int_0^R \frac{1}{r^{n-1}V(r)} dr = +\infty \quad$ and $\quad \int_0^R r^{n-1} v(r) dr < \infty.$

(1) *Suppose*

(4.23) $\quad \limsup_{r \to 0} r^{2(n-1)} V(r) W(r) \Big(\int_r^R \frac{1}{\tau^{n-1} V(\tau)} d\tau \Big)^2 < \frac{1}{4},$

then (V,W) is a n-dimensional Bessel pair on $(0,\rho)$ for some $\rho > 0$ and consequently, inequality $(H_{V,W})$ holds for all $u \in C_c^\infty(B_\rho)$, where B_ρ is a ball of radius ρ.

(2) *On the other hand, if*

(4.24) $\quad \liminf_{r \to 0} r^{2(n-1)} V(r) W(r) \Big(\int_r^R \frac{1}{\tau^{n-1} V(\tau)} d\tau \Big)^2 > \frac{1}{4},$

then there is no interval $(0,\rho)$ on which (V,W) is a n-dimensional Bessel pair and consequently, there is no smooth bounded domain Ω containing 0 on which inequality $(H_{V,W})$ holds.

An immediate application of Theorem 4.2.2 and Theorem 4.1.1 is the following general Hardy inequality.

THEOREM 4.2.3. *Let V be a positive C^1-function on $(0,R)$, and let Ω be a smooth domain in $\mathbb{R}^n$ ($n \geq 1$) containing 0 and such that $R = \sup_{x \in \Omega} |x|$. Assume that for some $\lambda \in \mathbb{R}$*

(4.25) $\quad \frac{rV_r(r)}{V(r)} + \lambda \geq 0$ on $(0,R)$ and $\lim_{r \to 0} \frac{rV_r(r)}{V(r)} + \lambda = 0.$

(1) *If $\lambda \leq n-2$, and P is a HI-potential on $(0,R)$, then for all $u \in C_c^\infty(\Omega)$,*
(4.26)
$\int_\Omega V(|x|) |\nabla u|^2 dx \geq (\frac{n-\lambda-2}{2})^2 \int_\Omega \frac{V(|x|)}{|x|^2} u^2 dx + \beta(P;R) \int_\Omega V(|x|) P(|x|) u^2 dx,$

and both $(\frac{n-\lambda-2}{2})^2$ and $\beta(P;R)$ are the best constants.

(2) *In particular, $\beta(V, r^{-2}V; R) = (\frac{n-\lambda-2}{2})^2$ is the best constant in the following inequality*

(4.27) $\quad \int_\Omega V(|x|) |\nabla u|^2 dx \geq (\frac{n-\lambda-2}{2})^2 \int_\Omega \frac{V(|x|)}{|x|^2} u^2 dx \quad$ for all $u \in C_c^\infty(\Omega).$

Applied to $V_1(r) = r^{-m} W_{k,\rho}(r)$ and $V_2(r) = r^{-m} \tilde{W}_{k,\rho}(r)$ where

$$W_{k,\rho}(r) = \sum_{j=1}^k \frac{1}{r^2} \Big(\prod_{i=1}^j \log^{(i)} \frac{\rho}{r} \Big)^{-2}$$

and

$$\tilde{W}_{k;\rho}(r) = \sum_{j=1}^k \frac{1}{r^2} X_1^2(\frac{r}{\rho}) X_2^2(\frac{r}{\rho}) \ldots X_{j-1}^2(\frac{r}{\rho}) X_j^2(\frac{r}{\rho}),$$

and noting that in both cases the corresponding λ is equal to $2m+2$, we get the following weighted Hardy inequalities.

COROLLARY 4.2.3. *Let Ω be a smooth bounded domain in $\mathbb{R}^n$ ($n \geq 1$) containing 0, and $m \leq \frac{n-4}{2}$. Then, the following inequalities hold:*

$$(4.28) \quad \int_\Omega \frac{W_{k,\rho}(x)}{|x|^{2m}} |\nabla u|^2 dx \geq \left(\frac{n-2m-4}{2}\right)^2 \int_\Omega \frac{W_{k,\rho}(x)}{|x|^{2m+2}} u^2 dx,$$

$$(4.29) \quad \int_\Omega \frac{\tilde{W}_{k,\rho}(x)}{|x|^{2m}} |\nabla u|^2 dx \geq \left(\frac{n-2m-4}{2}\right)^2 \int_\Omega \frac{\tilde{W}_{k,\rho}(x)}{|x|^{2m+2}} u^2 dx.$$

Moreover, $\left(\frac{n-2m-4}{2}\right)^2$ is the best constant in both inequalities.

4.3. Weighted Hardy-type inequalities on $\mathbb{R}^n$

In this section we deal with Hardy-type inequalities on the whole of $\mathbb{R}^n$. Theorem 4.1.1 already yields that inequality $(H_{V,W})$ holds for all $u \in C_c^\infty(\mathbb{R}^n)$ if and only if the equation $(B_{V,W})$ has a positive solution on $(0, \infty)$. The latter equation is therefore non-oscillatory, which by Theorem 1.2.5 implies some conditions on V and W. This will again be a very useful fact for computing best constants.

Recall that Theorem 1.2.5 states that if $a(r)$ and $b(r)$ are positive real valued functions, such that $\int_d^\infty \frac{1}{a(\tau)} d\tau = \infty$ for some $d > 0$, and the limit

$$L := \lim_{r \to \infty} a(r) b(r) \left(\int_r^\infty \frac{1}{a(r)} dr\right)^2$$

exists, then for the equation $(a(r) y')' + b(r) y(r) = 0$, to be non-oscillatory, it is necessary that $L \leq \frac{1}{4}$.

We now show the following result.

THEOREM 4.3.1. *Let $a, b > 0$, and α, β, m be real numbers.*

- *If $\alpha\beta > 0$, and $m \leq \frac{n-2}{2}$, then for all $u \in C_c^\infty(\mathbb{R}^n)$,*

$$(4.30) \quad \int_{\mathbb{R}^n} \frac{(a+b|x|^\alpha)^\beta}{|x|^{2m}} |\nabla u|^2 dx \geq \left(\frac{n-2m-2}{2}\right)^2 \int_{\mathbb{R}^n} \frac{(a+b|x|^\alpha)^\beta}{|x|^{2m+2}} u^2 dx,$$

and $\left(\frac{n-2m-2}{2}\right)^2$ is the best constant in the inequality.
- *If $\alpha\beta < 0$, and $2m - \alpha\beta \leq n-2$, then for all $u \in C_c^\infty(\mathbb{R}^n)$,*

$$(4.31) \quad \int_{\mathbb{R}^n} \frac{(a+b|x|^\alpha)^\beta}{|x|^{2m}} |\nabla u|^2 dx \geq \left(\frac{n-2m+\alpha\beta-2}{2}\right)^2 \int_{\mathbb{R}^n} \frac{(a+b|x|^\alpha)^\beta}{|x|^{2m+2}} u^2 dx,$$

and $\left(\frac{n-2m+\alpha\beta-2}{2}\right)^2$ is the best constant in the inequality.

Proof: Letting $V(r) = \frac{(a+br^\alpha)^\beta}{r^{2m}}$, then

$$r \frac{V'(r)}{V(r)} = -2m + \frac{b\alpha\beta r^\alpha}{a+br^\alpha} = -2m + \alpha\beta - \frac{a\alpha\beta}{a+br^\alpha}.$$

Hence, in the case $\alpha, \beta > 0$ and $2m \leq n-2$, (4.30) follows directly from Theorem 4.2.3. The same holds for (4.31) since it also follows directly from Theorem 4.2.3 in the case where $\alpha < 0$, $\beta > 0$ and $2m - \alpha\beta \leq n-2$.

For the remaining two other cases, we will use Theorem 4.1.1. Indeed, in this case the equation $(B_{V,W})$ becomes

(4.32) $$y'' + \left(\frac{n-2m-1}{r} + \frac{b\alpha\beta r^{\alpha-1}}{a+br^\alpha}\right)y' + \frac{c}{r^2}y = 0,$$

and the best constant in inequalities (4.30) and (4.31) is the largest c such that the above equation has a positive solution on $(0, +\infty)$. Note that by Lemma 1.1.2, we have that $y' < 0$ on $(0, +\infty)$. Hence, if $\alpha < 0$ and $\beta < 0$, then $y(r) = r^{-(\frac{n-2m-2}{2})}$ is a positive solution of the equation

$$y'' + \frac{n-2m-1}{r}y' + \frac{(\frac{n-2m-2}{2})^2}{r^2}y = 0$$

which is a positive super-solution for (4.32) and therefore the latter ODE has a positive solution on $(0, +\infty)$, from which we conclude that (4.30) holds. To prove now that $(\frac{n-2m-2}{2})^2$ is the best constant in (4.30), we use the fact that if the equation (4.32) has a positive solution on $(0, +\infty)$, then the equation is necessarily non-oscillatory. By rewriting (4.32) as

(4.33) $$\left(r^{n-2m-1}(a+br^\alpha)^\beta y'\right)' + cr^{n-2m-3}(a+br^\alpha)^\beta y = 0,$$

and by noting that

$$\int_d^\infty \frac{1}{r^{n-2m-1}(a+br^\alpha)^\beta} < \infty,$$

and

$$\lim_{r\to\infty} cr^{2(n-2m-2)}(a+br^\alpha)^{2\beta}\left(\int_r^\infty \frac{1}{r^{n-2m-1}(a+br^\alpha)^\beta}dr\right)^2 = \frac{c}{(n-2m-2)^2},$$

we can use Theorem 1.2.5 recalled above to conclude that for equation (4.33) to be non-oscillatory it is necessary that $\frac{c}{(n-2m-2)^2} \leq \frac{1}{4}$. Thus, $\frac{(n-2m-2)^2}{4}$ is the best constant in the inequality (4.30).

A very similar argument applies in the case where $\alpha > 0$, $\beta < 0$, and $2m < n-2$, to obtain that inequality (4.31) holds for all $u \in C_c^\infty(\mathbb{R}^n)$ and that $(\frac{n-2m+\alpha\beta-2}{2})^2$ is indeed the best constant. □

Note that the above two inequalities can be improved on smooth bounded domains by using Theorem 4.2.3. The following result is a bit more involved.

THEOREM 4.3.2. *Let $a, b > 0$, and α, β be real numbers.*

(1) *If $\alpha\beta < 0$ and $-\alpha\beta \leq n-2$, then for all $u \in C_c^\infty(\mathbb{R}^n)$*

(4.34) $$\int_{\mathbb{R}^n}(a+b|x|^\alpha)^\beta|\nabla u|^2 dx \geq b^{\frac{2}{\alpha}}\left(\frac{n-\alpha\beta-2}{2}\right)^2 \int_{\mathbb{R}^n}(a+b|x|^\alpha)^{\beta-\frac{2}{\alpha}}u^2 dx,$$

and $b^{\frac{2}{\alpha}}(\frac{n-\alpha\beta-2}{2})^2$ is the best constant in the inequality.

(2) *If $\alpha\beta > 0$ and $n \geq 2$, then there exists a constant $C > 0$ such that for all $u \in C_c^\infty(\mathbb{R}^n)$*

(4.35) $$\int_{\mathbb{R}^n}(a+b|x|^\alpha)^\beta|\nabla u|^2 dx \geq C \int_{\mathbb{R}^n}(a+b|x|^\alpha)^{\beta-\frac{2}{\alpha}}u^2 dx.$$

Moreover, $b^{\frac{2}{\alpha}}(\frac{n-2}{2})^2 \leq C \leq b^{\frac{2}{\alpha}}(\frac{n+\alpha\beta-2}{2})^2$.

Proof: Letting $V(r) = (a + br^\alpha)^\beta$, then we have
$$r\frac{V'(r)}{V(r)} = \frac{b\alpha\beta r^\alpha}{a + br^\alpha} = \alpha\beta - \frac{a\alpha\beta}{a + br^\alpha}.$$

Inequality (4.49) and its best constant in the case when $\alpha < 0$ and $\beta > 0$, then follow immediately from Theorem 4.2.3 with $\lambda = -\alpha\beta$. The proof of the remaining cases will use Theorem 4.1.1 as well as the integral criteria for the oscillatory behavior of solutions for ODEs of the form $(B_{V,W})$.

Assuming still that $\alpha\beta < 0$, then with an argument similar to that of Theorem 4.3.1 above, one can show that $y(r) = r^{-(\frac{n+\alpha\beta-2}{2})}$, a positive solution of the equation $y'' + (\frac{n+\alpha\beta-1}{r})y' + \frac{(n+\alpha\beta-2)^2}{4r^2}y = 0$ on $(0, +\infty)$, is a positive supersolution for the equation
$$y'' + \left(\frac{n-1}{r} + \frac{V'(r)}{V(r)}\right)y' + \frac{b^{\frac{2}{\alpha}}(n + \alpha\beta - 2)^2}{4(a + br^\alpha)^{\frac{2}{\alpha}}}y = 0.$$

Theorem 4.1.1 then yields that the inequality (4.34) holds for all $u \in C_c^\infty(\mathbb{R}^n)$. To prove now that $b^{\frac{2}{\alpha}}(\frac{n+\alpha\beta-2}{2})^2$ is the best constant in (4.34) it is enough to show that if the following equation

(4.36) $$\left(r^{n-1}(a + br^\alpha)^\beta y'\right)' + Cr^{n-1}(a + br^\alpha)^{\beta - \frac{2}{\alpha}}y = 0$$

has a positive solution on $(0, +\infty)$, then $C \leq b^{\frac{2}{\alpha}}(\frac{n+\alpha\beta-2}{2})^2$. If now $\alpha > 0$ and $\beta < 0$, then we have
$$\lim_{r\to\infty} cr^{2(n-1)}(a + br^\alpha)^{2\beta - \frac{2}{\alpha}} \left(\int_r^\infty \frac{1}{r^{n-1}(a + br^\alpha)^\beta}dr\right)^2 = \frac{C}{b^{\frac{2}{\alpha}}(n + \alpha\beta - 2)^2}.$$

Hence, by Theorem 1.2.5 again, the non-oscillatory aspect of the equation holds for $C \leq \frac{b^{\frac{2}{\alpha}}(n+\alpha\beta-2)^2}{4}$ which completes the proof of the first part.

Now assume $\alpha\beta > 0$. With a very similar argument one can conclude that $C \leq b^{\frac{2}{\alpha}}(\frac{n+\alpha\beta-2}{2})^2$. On the other hand since $\alpha\beta > 0$, we have that
$$\alpha\beta - \frac{a\alpha\beta}{a + br^\alpha} \geq 0$$
and therefore $y := r^{-\frac{n-2}{2}}$ is a positive super solution of
$$y'' + \left(\frac{n-1}{r} + \frac{V'(r)}{V(r)}\right)y' + \frac{b^{\frac{2}{\alpha}}(n - 2)^2}{4(a + br^\alpha)^{\frac{2}{\alpha}}}y = 0.$$

It follows that $C \geq b^{\frac{2}{\alpha}}(\frac{n-2}{2})^2$. $\square$

4.4. Hardy inequalities for functions in $H^1(\Omega)$

In this section, we explore Hardy-type inequalities for $H^1(\Omega)$-functions, i.e., for functions which do not necessarily have compact support in Ω. In this case, one expects a penalizing term to account for the boundary. In this direction, we have the following theorem which is the main result of this section.

THEOREM 4.4.1. *Let V and W be positive C^1-functions on $(0, R)$, and let B_R be a ball with radius R in $\mathbb{R}^n$ ($n \geq 1$). Assume that for some $0 < a < R$*

(4.37) $$\int_0^a \frac{1}{r^{n-1}V(r)}dr = +\infty \quad \text{and} \quad \int_0^a r^{n-1}V(r)dr < +\infty.$$

The following statements are then equivalent:

(1) (V, W) is a n-dimensional Bessel pair on $(0, R)$.
(2) There exists $c > 0$ and $\theta > 0$ such that for all $u \in C^\infty(\bar{B})$,

(4.38) $\quad \int_B V(|x|)|\nabla u|^2 dx \geq c \int_B W(|x|)u^2 dx - \theta \int_{\partial B} u^2$ for all $u \in H^1(B)$.

Moreover, $\theta := -V(R)\frac{\varphi'(R)}{\varphi(R)}$, where φ is the corresponding positive solution of $(B_{(V,cW)})$ on $(0, R]$, and $\beta(V, W, R)$ is the largest c for which the inequality holds.

The above theorem allows to extend all Hardy-type inequalities on $H_0^1(\Omega)$ to corresponding inequalities on $H^1(\Omega)$. For instance one can get the following general form of the Caffarelli-Kohn-Nirenberg inequalities (in the case where $b = a + 1$)

COROLLARY 4.4.1. *Assume B is the ball of radius R in $\mathbb{R}^n$. If $a \leq n - 2$, then for all $u \in H^1(B)$,*

(4.39) $\quad \int_B |x|^{-a}|\nabla u|^2 dx \geq \left(\frac{n-a-2}{2}\right)^2 \int_B |x|^{-a-2} u^2 dx$

$$- \frac{(n-a-2)R^{-a-1}}{2} \int_{\partial B} u^2 dx.$$

Proof: It is easy to see that $\varphi(r) = r^{\frac{n-a-2}{2}}$ is a positive supersolution of $B_{(r^{-a}, (\frac{n-a-2}{2})^2 r^{-2-a})}$. This means that θ can be taken to be equal to $\frac{(n-a-2)R^{-a-1}}{2}$ in the above theorem. $\square$

More generally, one has the following Hardy inequality on $H^1(\Omega)$.

THEOREM 4.4.2. *Let Ω be a smooth domain Ω containing 0 such that $R = \sup_{x \in \Omega} |x|$, and let V be a strictly positive C^1-function on $(0, R]$. Assume that for some $\lambda \in \mathbb{R}$*

(4.40) $\quad \frac{rV_r(r)}{V(r)} + \lambda \geq 0$ on $(0, R)$ and $\lim_{r \to 0} \frac{rV_r(r)}{V(r)} + \lambda = 0$.

If $\lambda \leq n - 2$, and if P is a HI-potential on $(0, R)$, then we have for all $u \in H^1(\Omega)$,

$$\int_\Omega V(|x|)|\nabla u|^2 dx \geq \left(\frac{n-\lambda-2}{2}\right)^2 \int_\Omega \frac{V(|x|)}{|x|^2} u^2 dx + \beta(P; R) \int_\Omega V(|x|)P(|x|)u^2 dx$$

$$+ V(R)\left(\frac{\varphi'(R)}{\varphi(R)} + \frac{n-\lambda-2}{2R}\right)\int_{\partial\Omega} u^2 dx,$$

where φ is the positive solution corresponding to the equation (B_P).

Proof: If φ is a positive solution of (B_P) on $(0, R)$, then $y(r) = r^{\frac{n-\lambda-2}{2}}\varphi(r)$ is a positive super-solution of $B_{(V, V(\frac{n-\lambda-2}{2})^2 r^{-2} + P)}$, which means that the couple $(V, V(\frac{n-\lambda-2}{2})^2 r^{-2} + P)$ is a Bessel pair. Now apply Theorem 4.4.1 and Proposition 1.2.1 to complete the proof. $\square$

COROLLARY 4.4.2. *Let z_0 be the first root of the Bessel function J_0. Then, for any $0 < \mu < z_0$, and any $R > 0$, the following inequality holds for all $u \in H^1(B_R)$,*

(4.41) $\quad \int_{B_R} |\nabla u|^2 dx \geq \left(\frac{n-2}{2}\right)^2 \int_{B_R} \frac{u^2}{|x|^2} dx + \frac{\mu^2}{R^2}\int_{B_R} u^2 dx + \frac{\mu J_0'(\mu)}{R J_0(\mu)}\int_{\partial B_R} u^2 dx,$

where B_R is the ball of radius R in $\mathbb{R}^n$ $(n \geq 2)$.

Proof: Note that $\varphi(r) = J_0(\frac{\mu r}{R})$ is a positive solution for the equation $(B_{V,W})$ on $(0, R)$, where $V = 1$ and $W = \left(\frac{n-2}{2}\right)^2 r^{-2} + \frac{\mu^2}{R^2}$. $\square$

COROLLARY 4.4.3. *Let B_R be the ball of radius R in $\mathbb{R}^n$ $(n \geq 2)$.*

4. WEIGHTED HARDY INEQUALITIES

(1) For $k \geq 1$, let ρ be such that $\rho \geq R(e^{e^{\cdot^{\cdot^{e}}}}$ $^{e(k-times)})$ and $R\frac{\varphi_k'(R)}{\varphi_k(R)} \geq -\frac{n}{2}$, where $\varphi_k = \left(\prod_{i=1}^{j} \log^{(k)} \frac{\rho}{|x|}\right)^{\frac{1}{2}}$. Then the following inequality holds for all $u \in H^1(B_R)$,

$$\int_{B_R} |\nabla u|^2 dx \geq \left(\frac{n-2}{2}\right)^2 \int_{B_R} \frac{|u|^2}{|x|^2} dx + \frac{1}{4}\sum_{j=1}^{k} \int_{B_R} \frac{|u|^2}{|x|^2} \left(\prod_{i=1}^{j} \log^{(i)} \frac{\rho}{|x|}\right)^{-2} dx$$

$$- \frac{n}{2R} \int_{\partial B_R} \times u^2 dx.$$

(2) For $k \geq 1$, we have for $u \in H^1(B_R)$,

$$\int_{B_R} |\nabla u|^2 dx \geq \int_{B_R} \frac{u^2}{|x|^2} dx + \frac{1}{4}\sum_{i=1}^{k} \int_{B_R} X_1^2\left(\frac{|x|}{R}\right) X_2^2\left(\frac{|x|}{R}\right)\ldots X_i^2\left(\frac{|x|}{R}\right) \frac{u^2}{|x|^2} dx - \frac{k}{2R}\int_{\partial B_R} u^2 dx.$$

Proof: The first inequality is an immediate application of Theorem 4.4.1 applied to the HI-potential $W_{k;\rho}$. The second follows from an application to the HI-potential $\tilde{W}_{k;R}(r) = \Sigma_{j=1}^{k} \frac{1}{r^2} X_1^2\left(\frac{r}{R}\right) X_2^2\left(\frac{r}{R}\right)\ldots X_{j-1}^2\left(\frac{r}{R}\right) X_j^2\left(\frac{r}{R}\right)$ where the functions X_i are defined in Chapter 1. Note that $\varphi_k = \left(X_1\left(\frac{r}{R}\right)X_2\left(\frac{r}{R}\right)\ldots X_{j-1}\left(\frac{r}{R}\right)X_k\left(\frac{r}{R}\right)\right)^{\frac{1}{2}}$ is a positive solution for the corresponding equation, and that $R\frac{\varphi_k'(R)}{\varphi_k(R)} = -\frac{k}{2}$. $\square$

Proof of Theorem 4.4.1: It is a slight perturbation of the proof of Theorem 4.1.1, and hence we provide a sketch. First we prove that 1) implies 2). Let $\varphi \in C^1(0, R]$ be a solution of $(B_{V,W})$ such that $\varphi(x) > 0$ for all $x \in (0, R)$. Define $\frac{u(x)}{\varphi(|x|)} = \psi(x)$. Then,

$$|\nabla u|^2 = \varphi'(|x|)^2 \psi^2(x) + 2\varphi'(|x|)\varphi(|x|)\psi(x)\frac{x}{|x|}\cdot\nabla\psi + \varphi^2(|x|)|\nabla\psi|^2,$$

and

$$\int_B V(|x|)|\nabla u|^2 dx \geq \int_B V(|x|)\varphi'(|x|)^2\psi^2(x)dx + \int_B 2V(|x|)\varphi'(|x|)\varphi(|x|)\psi(x)\frac{x}{|x|}\cdot\nabla\psi dx.$$

Set B_ϵ to be a ball of radius ϵ, then integrate by parts to get as in Theorem 4.1.1,

$$\int_B V(|x|)|\nabla u|^2 dx \geq \int_{B_\epsilon} V(|x|)\varphi'(|x|)^2\psi^2(x)dx + \int_{B_\epsilon} 2V(|x|)\varphi'(|x|)\varphi(|x|)\psi(x)\frac{x}{|x|}\cdot\nabla\psi dx$$

$$- \int_{B\setminus B_\epsilon} V(|x|)\varphi''(|x|)\varphi(|x|)\psi^2(x)\,dx$$

$$- \int_{B\setminus B_\epsilon} \left(\frac{(n-1)V(|x|)}{r} + V_r(|x|)\varphi'(|x|)\varphi(|x|)\right)\psi^2(x)\,dx$$

$$+ \int_{\partial(B\setminus B_\epsilon)} V(|x|)\varphi'(|x|)\varphi(|x|)\psi^2(x)ds.$$

Let $\epsilon \to 0$ and use Lemma 4.1.2 and the fact that φ is a solution of $(B_{V,W})$ to get

$$\int_B V(|x|)|\nabla u|^2 dx \geq -\int_B \left\{V(|x|)\varphi''(|x|) + \left(\frac{(n-1)V(|x|)}{r} + V_r(|x|)\right)\varphi'(|x|)\right\}\frac{u^2(x)}{\varphi(|x|)}dx$$

$$+ \int_{\partial B} V(|x|)\varphi'(|x|)\varphi(|x|)\psi^2(x)ds$$

$$= \int_B W(|x|)u^2(x)dx + \theta \int_{\partial B} u^2 ds.$$

To show that 2) implies 1), we assume that for some $\theta < 0$, we have

$$\int_B \left(V(|x|)|\nabla u|^2 - W(|x|)u^2\right) dx - \theta \int_{\partial B} u^2 ds \geq 0 \text{ for all } u \in H^1(B),$$

Similarly to Lemma 4.1.3, one can show that there exists a C^2-supersolution to the following linear elliptic equation

$$\begin{align}
(4.42) \quad -\text{div}(V(|x|)\nabla u) - W(|x|)u &= 0, \quad \text{in } B, \\
(4.43) \quad u &> 0 \quad \text{in } B \setminus \{0\}, \\
(4.44) \quad V\nabla u.\nu &= \theta u \quad \text{in } \partial B.
\end{align}$$

Take again the surface average of u,

$$(4.45) \quad y(r) = \frac{1}{n\omega_w r^{n-1}} \int_{\partial B_r} u(x) dS = \frac{1}{n\omega_n} \int_{|\omega|=1} u(r\omega) d\omega > 0,$$

where ω_n denotes the volume of the unit ball in $\mathbb{R}^n$. We may assume that the unit ball is contained in B (otherwise we just use a smaller ball). It is easy to see that $V(R)\frac{y'(R)}{y(R)} = \theta$. We have

$$(4.46) \quad y''(r) + \frac{n-1}{r} y'(r) = \frac{1}{n\omega_n r^{n-1}} \int_{\partial B_r} \Delta u(x) dS.$$

Since $u(x)$ is a supersolution of (4.42),

$$\int_{\partial B_r} \text{div}(V(|x|)\nabla u) ds - \int_{\partial B} W(|x|) u dx \geq 0,$$

and therefore,

$$V(r) \int_{\partial B_r} \Delta u dS - V_r(r) \int_{\partial B_r} \nabla u.x ds - W(r) \int_{\partial B_r} u(x) ds \geq 0.$$

It follows that

$$(4.47) \quad V(r) \int_{\partial B_r} \Delta u dS - V_r(r) y'(r) - W(r) y(r) \geq 0,$$

and in view of (4.45), we see that y satisfies the inequality

$$(4.48) \quad V(r) y''(r) + \left(\frac{(n-1)V(r)}{r} + V_r(r)\right) y'(r) \leq -W(r) y(r), \quad \text{for } 0 < r < R,$$

and hence it is a positive supersolution y for $(B_{V,W})$ with $V(R)\frac{y'(R)}{y(R)} = \theta$. It follows from Proposition 1.2.1 that $(B_{V,W})$ has actually a positive solution φ on $(0, R)$ such that $V(R)\frac{\varphi'(R)}{\varphi(R)} = \theta$, and the proof of theorem 4.4.1 is now complete. $\square$

4.5. Further comments

The notion of a Bessel pair was introduced by Ghoussoub-Moradifam [163] in order to establish the general Hardy inequality on $H_0^1(\Omega)$ with radial weights as stated in Theorem 4.1.1. Though unbeknown to the authors when they published [162] and [163], this characterization is closely related to the sufficient condition involving PDEs described in the book of B. Opic and A. Kufner [231]. The novelty here is that we have a necessary and sufficient condition, which allows for a useful formulation for the best constants involved. Moreover, by concentrating on the radial case, one can use Sturm's theory to exhibit many new interesting examples of Bessel pairs for which the weighted Hardy inequality holds. The appearance of radial Bessel pairs in the study of Hardy-Rellich inequalities in Part III is another validation to this approach. It should be interesting to compare this notion to other *Muckenhought*-type characterizations of Hardy admissible pairs [**225**].

The full Caffarelli-Kohn-Nirenberg inequalities [**76**] will be considered in Part V of this book. They can be obtained by interpolating between the Hardy and the Sobolev inequalities. The case when $b = a+1$ can be seen as weighted Hardy inequalities. In this case, Catrina and Wang [**85**] showed that for $n \geq 3$, $S(a, a+1) = (\frac{n-2a-2}{2})^2$ and that $S(a, a+1)$ is not achieved while $S(a, b)$ is always achieved for $a < b < a+1$. For the case $n = 2$, they also showed that $S(a, a+1) = a^2$, and that $S(a, a+1)$ is not achieved, while for $a < b < a+1$, $S(a, b)$ is again achieved. For $n = 1$, $S(a, a+1) = (\frac{1+2a}{2})^2$ is also not achieved.

Corollary 4.2.1 is an extension by Ghoussoub-Moradifam [**163**] of a result established by Brezis and Vázquez [**65**] in the case where $a = 0$. Corollary 4.2.2 is also an extension by Ghoussoub-Moradifam of a result established by Adimurthi et al. [**5**] and Filippas-Tertikas [**142**] in the case where $a = 0$, and of another result by Wang and Willem ([**270**] Theorem 2) in the case $k = 1$.

All the results in Section 4.4 are due to Moradifam [**223**] who extended the methods developed in [**163**] to obtain weighted Hardy inequalities on $H^1(\Omega)$.

Open Problem (3): Take any $\alpha \in \mathbb{R} \setminus \{\alpha^*(n)\}$ where $\alpha^*(n) = -\frac{n-2}{2}$ and $n \geq 3$. Find the best constant $C(\alpha, n) > 0$ such that the following inequality holds

$$(4.49) \qquad \int_{\mathbb{R}^n} (1+|x|^2)^\alpha |\nabla u|^2 dx \geq c(\alpha, n) \int_{\mathbb{R}^n} (1+|x|^2)^{\alpha-1} u^2 dx,$$

for $u \in C_c^\infty(\mathbb{R}^n)$. Note that for $\alpha < -\frac{n-2}{2}$ one assumes that $\int_{\mathbb{R}^n} (1+|x|^2)^{\alpha-1} u(x)\, dx = 0$.

Here are the cases known to Blanchet-Bonforte-Dolbeault-Grillo-Vasquez [**50**]:
- If $0 < \alpha \leq 1$, then $C(\alpha, n) = \frac{(n+2\alpha-2)^2}{4}$.
- If $\alpha = n$, then $C(\alpha, n) = 2n(n-1)$.

Note that Theorem 4.3.2 which is due to Ghoussoub-Moradifam [**163**] deals with a more general setting and yields the following cases:
- If $-\frac{n-2}{2} \leq \alpha < 0$, then $C(\alpha, n) = \frac{(n-2\alpha-2)^2}{4}$.
- If $\alpha > 0$, then $\frac{(n-2)^2}{4} \leq C(\alpha, n) \leq \frac{(n+2\alpha-2)^2}{4}$.

The remaining cases are still open. We refer to [**50**] for more details.

CHAPTER 5

The Hardy Inequality and Second Order Nonlinear Eigenvalue Problems

We present here various applications of Hardy-type inequalities to second order nonlinear elliptic eigenvalue problems. The associated bifurcation diagram generally depends on the regularity of the extremal solution, i.e., the one corresponding to the largest parameter for which the boundary value problem is solvable. Whether the solution is regular or singular depends on the dimension and on the given nonlinearity.

The general approach to showing regularity of the extremal is to use the spectral properties of the linearized equation at neighboring solutions in order to obtain estimates that translate into uniform bounds, which then allow passing to the limit. On the other hand, in order to show the optimality of the regularity result, one generally finds an explicit singular and semi-stable weak solution on a ball. Both directions are closely related to Hardy-type inequalities, which will be also crucial in section 5.4 for showing the uniqueness of solutions for small λ.

5.1. Second order nonlinear eigenvalue problems

Consider the following second order nonlinear eigenvalue problem with Dirichlet boundary conditions,

$$(P_\lambda) \qquad \begin{cases} -\Delta u = \lambda f(u) & \text{in } \Omega \\ u = 0 & \text{on } \partial\Omega, \end{cases}$$

where $\lambda \geq 0$ is a parameter, Ω is a bounded domain in $\mathbb{R}^n$, $n \geq 2$, and where f satisfies one of the following two conditions:

(R): f is smooth, increasing, convex on $\mathbb{R}$ with $f(0) = 1$ and f is superlinear at infinity, that is $\lim_{t \to \infty} \frac{f(t)}{t} = +\infty$.

(S): f is smooth, increasing, convex on $[0, 1)$ with $f(0) = 1$ and $\lim_{t \nearrow 1} f(t) = +\infty$.

We recall without proofs various properties one comes to expect when studying (P_λ). For more details, we refer to [83], [54], [112], [130][152]. Under either condition (R) or (S) on the nonlinearity f, the following holds true.

- There exists a finite positive critical parameter λ^* such that for all $0 < \lambda < \lambda^*$, a *minimal solution* u_λ of (P_λ) exists. By minimal solution, we mean here that if v is another solution of (P_λ) then $v \geq u_\lambda$ in Ω.
- For each $0 < \lambda < \lambda^*$ the minimal solution u_λ is *semi-stable* in the sense that

(5.1) $$\int_\Omega \lambda f'(u_\lambda) \psi^2 dx \leq \int_\Omega |\nabla \psi|^2 dx, \qquad \forall \psi \in H_0^1(\Omega),$$

and is unique among all the weak semi-stable solutions.
- The map $\lambda \mapsto u_\lambda(x)$ is increasing on $(0, \lambda^*)$ for each $x \in \Omega$. This allows one to define $u^*(x) := \lim_{\lambda \nearrow \lambda^*} u_\lambda(x)$, the so-called *extremal solution*, which can be shown to be a weak solution of (P_{λ^*}). In addition one can show that u^* is the unique weak solution of (P_{λ^*}).
- There are no solutions of (P_λ) (even in a very weak sense) for $\lambda > \lambda^*$.

A question which has attracted a lot of attention is when the extremal function u^* is a classical solution of (P_{λ^*}). The regularity of u^* is of interest since in that case, one can then apply the bifurcation theorem of Crandall-Rabinowitz [106] to start a second branch of solutions emanating from (λ^*, u^*).

This analysis depends closely on the spectral properties of the solutions around –and at– λ^*. Note that a solution u of (P_λ) or (S_λ) is semi-stable (resp., stable) if the linearized operator

$$(5.2) \qquad L_{\lambda,u} := -\Delta - \lambda f'(u)$$

is non-negative (resp., positive definite), meaning that

$$(5.3) \qquad \mu_1(L_{\lambda,u}) := \inf \left\{ \frac{\int_\Omega |\nabla \varphi|^2 \, dx - \lambda \int_\Omega f'(u)\varphi^2 \, dx}{\int_\Omega \varphi^2 \, dx}; \varphi \in H_0^1(\Omega) \right\}$$

is non-negative (resp., positive). As mentioned above, one can show that $\mu_1(L_{\lambda, u_\lambda}) > 0$, that is u_λ is stable for each $\lambda < \lambda^*$. Moreover, the convexity of f yields that $\mu_1(L_{\lambda, u_\lambda})$ is decreasing in λ, and therefore $\mu_1^* = \lim_{\lambda \to \lambda^*} \mu_1(L_{\lambda, u_\lambda}) \geq 0$. Moreover, $\mu_1^* = \mu_1(L_{\lambda^*, u^*})$.

Now whether $\mu_1^* > 0$ or $\mu_1^* = 0$ turn out to depend on whether u^* is a classical or a singular solution, which itself depends on the dimension and so, given a nonlinearity f, we say that N is the associated *critical dimension* provided the extremal solution u^* associated with (P_{λ^*}) is a classical solution for any bounded smooth domain $\Omega \subset \mathbb{R}^n$ and any $n \leq N - 1$, while there exists a domain $\Omega \subset \mathbb{R}^N$ such that the associated extremal solution u^* is not a classical solution.

The general approach to showing N is the critical dimension for a particular nonlinearity f is to use the semi-stability of the minimal solutions u_λ to obtain various estimates which translate to uniform L^∞ bounds and then passing to the limit. These estimates generally depend on the ambient space dimension.

On the other hand, in order to show the optimality of the regularity result one generally finds an explicit singular extremal solution u^* on a radial domain. Here the crucial tool is the fact that if there exists a semi-stable singular solution in $H_0^1(\Omega)$, then it has to be the extremal solution. A proof of the following proposition can be found in [65], [130] or [161].

PROPOSITION 5.1.1. *Let Ω be a bounded domain Ω in $\mathbb{R}^n$, and assume f is a non-linearity satisfying (R) or (S). Let u be a $H_0^1(\Omega)$-weak solution of (P_λ) for some $\lambda > 0$, such that $\|f(u)\|_\infty = +\infty$. Then, the following assertions are equivalent:*
  (1) *u is semi-stable.*
  (2) *$\lambda = \lambda^*$ and $u = u^*$.*

In practice –as shown below– one considers an explicit singular solution on the unit ball and applies the Hardy inequality (resp., an improved Hardy inequality) to show its semi-stability (resp., stability) in the right dimension. We also remark that one cannot remove the $H_0^1(\Omega)$ condition as counterexamples can be found.

5.2. The role of dimensions in the regularity of extremal solutions

We now illustrate the role of Hardy's inequality with the following example.

THEOREM 5.2.1. *Consider the nonlinear eigenvalue problem (P_λ) on a bounded domain $\Omega \subset \mathbb{R}^n$, $n \geq 2$ with the nonlinearity $f(t) = (1-t)^{-p}$, $p > 0$. Then,*

(1) *If $n < \frac{2+6p+4\sqrt{p^2+p}}{1+p}$, then u^* is a semi-stable regular solution such that $\mu_1^* = 0$.*

(2) *If Ω the unit ball and $n \geq \frac{2+6p+4\sqrt{p^2+p}}{1+p}$, then the extremal solution is singular and stable, that is $\mu_1^* > 0$. It is explicitly given by $u^*(x) = 1 - |x|^{\frac{2}{1+p}}$ at $\lambda^* = \frac{2(n-2)(1+p)+4}{(1+p)^2}$.*

In other words, the critical dimension for $f(t) = (1-t)^{-p}$ is the smallest integer above or equal to $N(p) = \frac{2+6p+4\sqrt{p^2+p}}{1+p}$.

Proof: 1) We first prove that in order to establish regularity of u^*, it suffices to show that $(1-u^*)^{-p}$ is in $L^{\frac{(1+p)n}{2p}}(\Omega)$. Indeed, standard regularity theory would then implies that $u \in C^{0,\frac{2}{1+p}}(\bar{\Omega}) \cap H^1(\Omega)$ (see for example Gilbarg-Trudinger [**169**]). If now $u^*(x_0) = 1$ at some $x_0 \in \Omega$, then the Hölder continuity of u implies that

$$(5.4) \qquad |1 - u^*(x)| = |u^*(x_0) - u^*(x)| \leq C|x - x_0|^{\frac{2}{1+p}}.$$

Choosing $\delta > 0$ small so that $B_\delta(x_0) \subset \Omega$, it follows that

$$\int_{B_\delta(x_0)} \frac{1}{|1-u^*|^{\frac{(1+p)n}{2}}} \geq \frac{1}{C} \int_{B_\delta(x_0)} \frac{1}{|x-x_0|^n} = +\infty,$$

which contradicts the integrability assumption on $(1-u^*)$. Therefore, $u^* < 1$ on Ω.

Now we show that this integrability assumption is indeed satisfied for $n < \frac{2+6p+4\sqrt{p^2+p}}{1+p}$. For that, we shall prove that $(1-u)^{-p} \in L^q(\Omega)$ for $q \leq \frac{n(1+p)}{2p}$.

Take $\psi := (1-u)^{-t} - 1$, and $\varphi := (1-u)^{-2t-1} - 1$. By putting ψ into the stability condition and testing (P_λ) on φ. We obtain

$$\frac{\lambda p}{t^2} \int_\Omega \left(\frac{1}{(1-u)^{p+2t+1}} - \frac{2}{(1-u)^{p+t+1}} + \frac{1}{(1-u)^{p+1}} \right) dx \leq \int_\Omega \frac{|\nabla u|^2}{(1-u)^{2t+2}} dx,$$

and

$$\int_\Omega \frac{|\nabla u|^2}{(1-u)^{2t+2}} dx \leq \frac{\lambda}{2t+1} \int_\Omega \left[\frac{1}{(1-u)^{p+2t+1}} - \frac{1}{(1-u)^p} \right] dx,$$

respectively. After dropping a couple of positive terms we have

$$\left(\frac{p}{t^2} - \frac{1}{2t+1} \right) \int_\Omega \frac{dx}{(1-u)^{2t+p+1}} \leq \frac{2p}{t^2} \int_\Omega \frac{dx}{(1-u)^{t+p+1}}.$$

Hölder's inequality then yields

$$(5.5) \qquad \left(\frac{2}{t^2} - \frac{1}{2t+1} \right) \left\| \frac{1}{1-u} \right\|_{L^{2t+p+1}} \leq C,$$

where C depends on p, t, and Ω but not on u. Since

$$\left(\frac{2}{t^2} - \frac{1}{2t+1} \right) > 0, \quad \text{for } t < p + \sqrt{p^2 + p},$$

we have

$$(1-u)^{-p} \in L^q(\Omega) \text{ for all } q < \frac{3p+1+2\sqrt{p^2+p}}{p} = \frac{(1+p)N(p)}{2p}.$$

Note that since u^* is a classical solution, μ_1^* cannot be strictly positive, since then one can use the implicit function theorem to find stable solutions beyond λ^*.

To prove 2) note that $u(x) := 1 - |x|^{\frac{2}{1+p}}$ is clearly a $H_0^1(B_1)$-weak solution of (P_λ). It corresponds to $\bar{\lambda}(n) = \frac{2(n-2)(1+p)+4}{(1+p)^2}$. Since now $\|u\|_\infty = 1$, it suffices to show that it is semi-stable so that Proposition 5.1.1 applies to yield that it is indeed the extremal solution. In other words, we need to prove that for all $\varphi \in H_0^1(B_1)$,

$$(5.6) \qquad \int_{B_1} |\nabla \varphi|^2 dx \geq \int_{B_1} \frac{p\bar{\lambda}(n)\varphi^2}{(1-u)^{p+1}} dx = p\bar{\lambda}(n) \int_{B_1} \frac{\varphi^2}{|x|^2} dx.$$

But Hardy's inequality gives for $n \geq 2$ that for any $\varphi \in H_0^1(B_1)$,

$$\int_{B_1} |\nabla \varphi|^2 dx \geq \frac{(n-2)^2}{4} \int_{B_1} \frac{\varphi^2}{|x|^2} dx,$$

which means that (5.6) holds whenever $p\bar{\lambda}(n) \leq \frac{(n-2)^2}{4}$, equivalently if $n \geq \frac{2+6p+4\sqrt{p^2+p}}{1+p}$.

Now, if $p\bar{\lambda}(n) < \frac{(n-2)^2}{4}$, then the Hardy inequality yields immediately that $\mu_1^* > 0$. But if $p\bar{\lambda}(n) = \frac{(n-2)^2}{4}$, then to prove that $\mu_1^* > 0$, one requires the improved Hardy inequality (2.19), which states that

$$(5.7) \qquad \int_{B_1} |\nabla u|^2 dx - \left(\frac{n-2}{2}\right)^2 \int_{B_1} \frac{u^2}{|x|^2} dx \geq z_0^2 \int_{B_1} u^2 dx \quad \text{for } u \in H_0^1(\Omega),$$

where z_0 is the first root of the Bessel function J_0. Actually, one can show that $\mu_1^* = z_0^2$.

Note that if $p\bar{\lambda}(n) > \frac{(n-2)^2}{4}$, then the fact that $\frac{(n-2)^2}{4}$ is the best constant in the Hardy inequality means that $u(x) := 1 - |x|^{\frac{2}{1+p}}$ —which is still a solution for $\lambda = \frac{2(n-2)(1+p)+4}{(1+p)^2}$— is an unstable solution. $\square$

REMARK 5.2.2. If $f(t) = e^t$ one can similarly use Proposition 5.1.1 and the classical Hardy inequality to show that the extremal solution is singular and is explicitly given by $u^*(x) = -2\log(|x|)$ as long as the dimension $n \geq 10$. Again, if $n = 10$, one then requires the improved Hardy inequality (5.7) to show that in this case the extremal solution is indeed stable.

5.3. Asymptotic behavior of stable solutions near the extremals

We now establish pointwise upper and lower estimates on the minimal solutions u_λ in terms of λ, λ^* and the extremal solution u^*. For simplicity we restrict our attention to the nonlinearities $f(u) = e^u$ and $f(u) = (1-u)^{-2}$.

THEOREM 5.3.1. Let u^* denote the extremal solution of (P_λ) on a smooth bounded domain Ω in $\mathbb{R}^n$.

(1) If $f(u) = (1-u)^{-2}$, then for $0 < \lambda < \lambda^*$,

$$(5.8) \qquad u_\lambda(x) \leq \left(\frac{\lambda}{\lambda^*}\right)^{\frac{1}{3}} u^*(x) \quad \text{for a.e. } x \in \Omega.$$

Moreover, if Ω is the unit ball in $\mathbb{R}^n$ and $n \geq \frac{14+4\sqrt{6}}{3}$, then for $0 < \lambda < \lambda^* = \frac{6n-8}{9}$ we have for a.e. $x \in \Omega$,

$$1 - |x|^{\frac{2}{3}} - \frac{3(\lambda^* - \lambda)}{(6n-8)} \left(|x|^{\frac{-n}{2}+1+\frac{\sqrt{9n^2-84n+100}}{6}} - 1 \right) \leq u_\lambda(x) \leq \left(\frac{\lambda}{\lambda^*}\right)^{\frac{1}{3}} (1 - |x|^{\frac{2}{3}}). \quad (5.9)$$

(2) If $f(u) = e^u$, then for $0 < \lambda < \lambda^*$,

$$u_\lambda(x) \leq \log\left(\frac{\lambda^*}{\lambda^* - \lambda + \lambda e^{-u^*}}\right) \quad \text{for a.e. } x \in \Omega. \quad (5.10)$$

Moreover, if Ω is the unit ball in $\mathbb{R}^n$ with $n \geq 10$, then for $0 < \lambda < \lambda^* = 2n - 4$ we have for a.e. $x \in \Omega$,

$$\log(\frac{1}{|x|^2}) - \frac{(\lambda^* - \lambda)}{(2n-4)} \left(|x|^{\frac{-n}{2}+1+\frac{\sqrt{n^2-12n+20}}{2}} - 1 \right) \leq u_\lambda(x) \leq \log\left(\frac{\lambda^*}{\lambda^* - \lambda + \lambda|x|^2}\right). \quad (5.11)$$

Proof: The upper estimates follow easily from the minimality of u_λ and the fact that $x \mapsto \left(\frac{\lambda}{\lambda^*}\right)^{\frac{1}{3}} u^*(x)$ (resp., $x \mapsto \log\left(\frac{\lambda^*}{\lambda^* - \lambda + \lambda e^{-u^*}}\right)$) is a supersolution of (P_λ) in the case that $f(u) = (1-u)^{-2}$ (resp., $f(u) = e^u$).

For the lower bound, recall first that $\lambda \mapsto u_\lambda$ is differentiable (thanks to the Implicit Function Theorem) and increasing on $(0, \lambda^*)$, and so if one defines $v_\lambda(x) := \frac{d}{d\lambda} u_\lambda(x)$, then v_λ is positive and solves the linear equation

$$(L_\lambda) \quad \begin{cases} -\Delta v = f(u_\lambda) + \lambda f'(u_\lambda) v & \text{in } \Omega, \\ v = 0 & \text{on } \partial\Omega, \end{cases}$$

where f is given by either e^u or $(1-u)^{-2}$. Define now the following notion.

DEFINITION 5.3.2. An extremal solution u^* associated with (P_λ) is said to be *super-stable* provided there exists $\varepsilon > 0$ such that

$$(\lambda^* + \varepsilon) \int_\Omega f'(u^*) \psi^2 \, dx \leq \int_\Omega |\nabla \psi|^2 \, dx \quad \text{for } \psi \in H_0^1(\Omega).$$

Note that if u^* is a super-stable extremal solution then the first eigenvalue $\mu_1(\lambda^*, u^*)$ of the linearized equation at u^* is strictly positive but the converse is however not true.

LEMMA 5.3.3. *Assume Ω is a smooth bounded domain in $\mathbb{R}^n$. Then,*

(1) *For $0 < \lambda < \lambda^*$, v_λ is the unique H_0^1-weak solution of (L_λ).*
(2) *$\lambda \mapsto v_\lambda$ is increasing on $(0, \lambda^*)$ and therefore $v^*(x) := \lim_{\lambda \to \lambda^*} v_\lambda(x)$ is defined for a.e. $x \in \Omega$.*
(3) *$\lambda \mapsto u_\lambda$ is convex on $(0, \lambda^*)$ and therefore for $0 < \lambda < \lambda^*$,*

$$u_\lambda(x) \geq u^*(x) + (\lambda - \lambda^*) v^*(x) \quad \text{for all } x \in \Omega. \quad (5.12)$$

(4) *If u^* is super-stable, then v^* is the unique H_0^1-weak solution of (L_{λ^*}).*

Proof: (1) One can use the fact that $\mu_1(\lambda, u_\lambda) \geq 0$, and a standard minimization argument to show the existence of an H_0^1-solution to (L_λ). Using the fact that $\mu_1(\lambda, u_\lambda) > 0$ one can see that the solution is unique.

(2) Let $0 < \lambda < \lambda^*$ and $\varepsilon > 0$ small. Note that

$$\begin{aligned} -\Delta(v_{\lambda+\varepsilon} - v_\lambda) &= f(u_{\lambda+\varepsilon}) - f(u_\lambda) + \varepsilon f'(u_{\lambda+\varepsilon}) v_{\lambda+\varepsilon} \\ &\quad + \lambda f'(u_{\lambda+\varepsilon}) v_{\lambda+\varepsilon} - \lambda f'(u_\lambda) v_\lambda \\ &= g(x) + \lambda f'(u_\lambda)(v_{\lambda+\varepsilon} - v_\lambda), \end{aligned}$$

where
$$g(x) := f(u_{\lambda+\varepsilon}) - f(u_\lambda) + \varepsilon f'(u_{\lambda+\varepsilon})v_{\lambda+\varepsilon} + \lambda\big(f'(u_{\lambda+\varepsilon})v_{\lambda+\varepsilon} - f'(u_\lambda)v_{\lambda+\varepsilon}\big)$$
is in $H^1(\Omega)$ and is positive. Now set $w := v_{\lambda+\varepsilon} - v_\lambda$ in such a way that w solves
$$\begin{aligned} -\Delta w &= g(x) + \lambda f'(u_\lambda)w && \text{on } \Omega, \\ w &= 0 && \text{on } \partial\Omega. \end{aligned}$$
Testing this equation on w^- gives
$$-\int_\Omega gw^- \geq \mu_1(\lambda, u_\lambda)\int_\Omega (w^-)^2,$$
and hence $w^- = 0$ a.e. in Ω. By the maximum principle one then get that $w > 0$ in Ω and hence that $\lambda \to v_\lambda$ is increasing. We can therefore define the limit $v^*(x) := \lim_{\lambda \to \lambda^*} v_\lambda(x)$, which exists a.e. x in Ω, though it might be infinite on a large set.

(3) The convexity of $\lambda \mapsto u_\lambda$ follows from the fact that $\lambda \mapsto v_\lambda$ is increasing. We can therefore write $u_\lambda \geq u_t + (\lambda - t)v_t$ for $0 < \lambda, t < \lambda^*$ and a.e. $x \in \Omega$. The claim now follows by letting t go to λ^*.

(4) Since u^* is super-stable one has
$$(\lambda + \varepsilon)\int_\Omega f'(u_\lambda)\psi^2\, dx \leq \int_\Omega |\nabla\psi|^2\, dx \qquad \forall \psi \in H_0^1.$$
Using this and testing (L_λ) on v_λ gives
$$\varepsilon \int_\Omega f'(u_\lambda)v_\lambda^2\, dx \leq \int_\Omega f(u_\lambda)v_\lambda\, dx.$$
Since f is either $f(u) = e^u$ or $f(u) = (1-u)^{-2}$, the left hand side is necessarily bounded. From this and again by testing (P_λ) on v_λ one sees that v_λ is bounded in H_0^1. Passing to limits, one sees that v^* is a H_0^1–weak solution of (L_{λ^*}). The uniqueness follows from the fact that $\mu_1(\lambda^*, u^*) > 0$. □

We now complete the proof of Theorem 5.3.1. For that we assume that Ω is the unit ball in $\mathbb{R}^n$. It is then easy to show using Hardy's inequality that the explicit extremal solutions for (P_λ) –given above– are super-stable provided $n > 10$ (resp., $n > \frac{14+4\sqrt{6}}{3} = 7.93...$) when $f(u) = e^u$ (resp., $f(u) = (1-u)^{-2}$). An easy calculation also shows that
$$v^*(x) = \frac{1}{2n-4}\left(|x|^{\frac{-n}{2}+1+\frac{\sqrt{n^2-12n+20}}{2}} - 1\right),$$
(when $f(u) = e^u$) resp.,
$$v^*(x) = \frac{3}{6n-8}\left(|x|^{\frac{-n}{2}+1+\frac{\sqrt{9n^2-84n+100}}{6}} - 1\right),$$
(when $f(u) = (1-u)^{-2}$) are H_0^1–weak solutions of (P_{λ^*}) in the respective cases, assuming the dimension restrictions above. Using this and the earlier convexity result gives the desired lower bounds for $n > 10$ (resp., $n > \frac{14+4\sqrt{6}}{3}$). To obtain the result for the critical dimensions one passes to the limit in λ. We omit the details. □

5.4. The bifurcation diagram for small parameters

In this section we show how the Hardy inequality can be used to address the issue of uniqueness for the solution of

$$
(5.13) \quad \begin{cases} -\Delta u = \frac{\lambda}{(1-u)^2} & \text{in } \Omega, \\ 0 < u < 1 & \text{in } \Omega, \\ u = 0 & \text{on } \partial\Omega, \end{cases}
$$

for $\lambda > 0$ small, which is a statement on the bifurcation diagram around $\lambda = 0$. Our goal is to show that if (for example) Ω is a star-shaped domain with respect to 0 and if $n \geq 3$, then for λ small, the minimal solution u_λ is the unique solution of (5.13). By setting $v = u - u_\lambda$, equation (5.13) yields

$$
(5.14) \quad \begin{cases} -\Delta v = \lambda g_\lambda(x, v) & \text{in } \Omega \\ 0 \leq v < 1 - u_\lambda & \text{in } \Omega \\ v = 0 & \text{on } \partial\Omega, \end{cases}
$$

where

$$
(5.15) \quad g_\lambda(x, s) = \frac{1}{(1 - u_\lambda(x) - s)^2} - \frac{1}{(1 - u_\lambda(x))^2}.
$$

It then suffices to prove that the solutions of (5.14) must be trivial for λ small enough.

We shall actually prove the result in the following general setting. For any bounded domain Ω in $\mathbb{R}^n$, we consider the – possibly empty – set

$$
H(\Omega) = \Big\{ h \in C^1(\bar{\Omega}, \mathbb{R}^n) : \operatorname{div}(h) \equiv 1 \text{ and } \langle h, \nu \rangle \geq 0 \text{ on } \partial\Omega \Big\},
$$

and the corresponding parameter

$$
M(\Omega) := \inf \Big\{ \sup_{x \in \Omega} \bar{\mu}(h, x) : h \in H(\Omega) \Big\},
$$

where

$$
\bar{\mu}(h, x) = \frac{1}{2} \sup_{|\xi|=1} \langle (Dh(x) + Dh(x)^T)\xi, \xi \rangle.
$$

THEOREM 5.4.1. *Let Ω be a bounded domain in $\mathbb{R}^n$ containing 0 such that $M(\Omega) < \frac{1}{2}$. If the dimension $n \geq 3$, then for λ small the minimal solution u_λ is the unique solution of problem (5.13).*

Note that the above condition on Ω is clearly satisfied whenever it is a star-shaped domain, as it suffices to take $h(\mathbf{x}) = \frac{\mathbf{x}}{n}$. The proof will make crucial use of the following extension of the Pohozaev identity due to Pucci and Serrin [**241**].

PROPOSITION 5.4.1. *Let v be a solution of the boundary value problem*

$$
\begin{cases} -\Delta v = f(x, v) & \text{in } \Omega, \\ v = 0 & \text{on } \partial\Omega. \end{cases}
$$

Then, for any $a \in \mathbb{R}$ and any $h \in C^2(\Omega; \mathbb{R}^n) \cap C^1(\bar{\Omega}; \mathbb{R}^n)$, the following identity holds

$$
(5.16) \quad \begin{aligned} & \int_\Omega \Big[\operatorname{div}(h) F(x, v) - a v f(x, v) + \langle \nabla_x F(x, v), h \rangle \Big] dx \\ & = \int_\Omega \Big[(\tfrac{1}{2}\operatorname{div}(h) - a)|\nabla v|^2 - \langle Dh \nabla v, \nabla v \rangle \Big] dx + \frac{1}{2} \int_{\partial\Omega} |\nabla v|^2 \langle h, \nu \rangle d\sigma, \end{aligned}
$$

where $F(x, s) = \int_0^s f(x, t)\, dt$.

Proof of Theorem 5.4.1: As mentioned above, it suffices to show that equation (5.14) has only trivial solutions for λ small. For that note first that if we set

$$G_\lambda(x,s) := \int_0^s g_\lambda(x,t)\,dt = \frac{1}{1-u_\lambda(x)-s} - \frac{1}{1-u_\lambda(x)} - \frac{s}{(1-u_\lambda(x))^2},$$

then easy calculations show that

$$\frac{G_\lambda(x,s)}{g_\lambda(x,s)} = \frac{1 - u_\lambda(x) - s - \frac{(1-u_\lambda(x)-s)^2(1-u_\lambda(x)+s)}{(1-u_\lambda(x))^2}}{1 - \frac{(1-u_\lambda(x)-s)^2}{(1-u_\lambda(x))^2}}$$

and

$$\frac{\nabla_x G_\lambda(x,s)}{g_\lambda(x,s)} = \frac{1 - \frac{(1-u_\lambda(x)-s)^2(1-u_\lambda(x)+2s)}{(1-u_\lambda(x))^3}}{1 - \frac{(1-u_\lambda(x)-s)^2}{(1-u_\lambda(x))^2}}\nabla u_\lambda(x),$$

which then yield that for some $C_0 > 0$ and provided λ is away from λ^*,
(5.17)
$$\left|\frac{G_\lambda(x,s)}{g_\lambda(x,s)}\right| \leq C_0|1 - u_\lambda(x) - s| \text{ and } \left|\frac{\nabla_x G_\lambda(x,s)}{g_\lambda(x,s)} - \nabla u_\lambda\right| \leq C_0|1 - u_\lambda(x) - s|^2|\nabla u_\lambda|.$$

It follows that for any (x,s) satisfying $|1 - u_\lambda(x) - s| \leq \delta|x|$, there holds

(5.18) $\begin{aligned}L_\lambda(x,s) :&= G_\lambda(x,s) - av(x)g_\lambda(x,s) + \langle \nabla_x G_\lambda(x,s), h(x)\rangle \\ &\leq g_\lambda(x,s)\big[C_0\delta|x| + \alpha C_0\delta|h| - a(1 - u_\lambda(x) - \delta|x|) \\ &\quad + \langle \nabla u_\lambda(x), h(x)\rangle + C_0\delta^2|x|^2|\nabla u_\lambda(x)||h(x)|\big] \\ &\leq 0,\end{aligned}$

provided λ and δ are sufficiently small.

Now for a solution v of (5.14), the Pohozaev identity (5.16) with $h \in H(\Omega)$ yields

(5.19)
$$\begin{aligned}\lambda \int_\Omega L_\lambda(x,v(x))\,dx &= \int_\Omega \left[(\frac{1}{2} - a)|\nabla v|^2 - \frac{1}{2}\langle(Dh + Dh^T)\nabla v, \nabla v\rangle\right]dx \\ &\quad + \frac{1}{2}\int_{\partial\Omega}|\nabla v|^2\langle h, \nu\rangle\,d\sigma \\ &\geq \int_\Omega (\frac{1}{2} - a - \bar\mu(h,x))|\nabla v|^2\,dx.\end{aligned}$$

Fix $0 < a < \frac{1}{2} - M(\Omega)$ and choose $h \in H(\Omega)$ such that

$$\frac{1}{2} - a - \sup_{x\in\Omega}\bar\mu(h,x) > 0.$$

It then follows from (5.19) and (5.18) that

(5.20) $\displaystyle\lambda \int_{\{0\leq v\leq 1-u_\lambda-\delta|x|\}} L_\lambda(x,v(x))\,dx \geq (\frac{1}{2} - a - \sup_{x\in\Omega}\bar\mu(h,x))\int_\Omega |\nabla v|^2\,dx.$

On the other hand, there exists a constant $C_\delta > 0$ such that

$$\begin{aligned}
L_\lambda(x, v(x)) &= \frac{v^2(x)}{(1 - u_\lambda(x) - v(x))(1 - u_\lambda(x))^2} + \frac{av^2(x)[v(x) - 2 + 2u_\lambda(x)]}{(1 - u_\lambda(x) - v(x))^2(1 - u_\lambda(x))^2} \\
&\quad + \frac{v^2(x)(3 - 3u_\lambda(x) - 2v(x))}{(1 - u_\lambda(x) - v(x))^2(1 - u_\lambda(x))^3} <\nabla u_\lambda(x), h(x)> \\
&\leq C_\delta \frac{v^2(x)}{|x|^2}
\end{aligned}$$

for $x \in \{0 \leq v \leq 1 - u_\lambda - \delta|x|\}$ uniformly for λ away from λ^*.

If now $N \geq 3$, then Hardy's inequality combined with (5.20) implies

$$\frac{(N-2)^2}{4}\left(\frac{1}{2} - a - \sup_{x \in \Omega} \bar{\mu}(h, x)\right) \int_{\{0 \leq v \leq 1 - u_\lambda - \delta|x|\}} \frac{v^2 dx}{|x|^2} \leq \lambda C_\delta \int_{\{0 \leq v \leq 1 - u_\lambda - \delta|x|\}} \frac{v^2 dx}{|x|^2}.$$

We can therefore conclude that for λ sufficiently small, $v \equiv 0$ for $x \in \{0 \leq v \leq 1 - u_\lambda - \delta|x|\}$ with $\delta > 0$ small. Since we can assume δ and λ sufficiently small to have

$$1 - u_\lambda - \delta|x| \geq \frac{1}{2} \quad \text{in} \quad \{x \in \Omega : |x| \geq \frac{1}{2}\text{dist}(0, \partial\Omega)\},$$

we then have

$$v \equiv 0 \quad \text{in} \quad \{x \in \Omega : v(x) \leq \frac{1}{2}\} \cap \{x \in \Omega : |x| \geq \frac{1}{2}\text{dist}(0, \partial\Omega)\}.$$

Since $v = 0$ on $\partial\Omega$ and the domain $\{x \in \Omega : |x| \geq \frac{1}{2}\text{dist}(0, \partial\Omega)\}$ is connected, the continuity of v gives that

$$v \equiv 0 \quad \text{in} \quad \{x \in \Omega : |x| \geq \frac{1}{2}\text{dist}(0, \partial\Omega)\}.$$

Therefore, the maximum principle for elliptic equations implies $v \equiv 0$ in Ω, which completes the proof of Theorem 5.4.1.

REMARK 5.4.1. Examples of dumbell shaped domains $\Omega \subset \mathbb{R}^n$ which satisfy condition $M(\Omega) < \frac{1}{2}$ are given for $n \geq 3$ in Schaaf [247]. When $n \geq 4$, there even exist topologically nontrivial domains with this property. Let us stress that in both cases Ω is not starlike, which means that the assumption $M(\Omega) < \frac{1}{2}$ on a domain Ω is more general than being star-shaped.

5.5. Further comments

The Hardy inequality and its various improvements have been used in many contexts, such as in the study of the stability of solutions of semi-linear elliptic and parabolic equations [65, 72], the analysis of the asymptotic behavior of the heat equation with singular potentials [267], the study of the stability of eigenvalues in elliptic problems such as Schrödinger operators [144], and more recently in the analysis of compressible Euler equations [191]. We have elected here to show their impact on the recent progress in nonlinear eigenvalue problems.

Such problems corresponding to nonlinearities f of type (R) have been studied for a long time. See, for example Brezis-Vasquez [65] for the case where $f(t) = e^t$. For general nonlinearities on general domains, Nedev [227] showed that u^* is regular for $n \leq 3$ [227]. This was improved by Cabré [68] to $n \leq 4$ provided the domain is convex. In the case where Ω is a ball, Cabré and Capella [71] proved that u^* is a regular solution in dimensions $n \leq 9$, again regardless of the nonlinearity as long as it satisfies (R).

The case when $f(t) = (1-t)^{-2}$ had already been considered by Mignot-Puel [**219**]. The book of Esposito-Ghoussoub-Guo [**130**] contains a much more penetrating analysis of this case and more. The results in section 5.3 are due to Cowan-Ghoussoub [**101**]. See also [**84**] for many interesting properties concerning the mapping $\lambda \mapsto u_\lambda(x)$. The uniqueness of the solution for λ small is due to Esposito-Ghoussoub [**129**] and is based on ideas of R. Schaaf [**247**].

Open problem (4): Suppose that f is a non-linearity of type (R). What is then the critical dimension for the non-linear eigenvalue problem (P_λ), that is what is the maximal dimension n^* such that the extremal solution for the corresponding (P_λ) on any bounded domain is regular? Note that the above mentioned results show that $n^* \geq 4$ (at least when the domain is convex).

Open problem (5): Suppose f is a non-linearity of type (S). Show that the critical dimension n^* is then equal to 2. Note that by considering non-linearities of the form $f(u) = \frac{1}{(1-u)^p}$, Theorem 5.2.1 shows that $n^* \leq 2$.

Part 2

Hardy-Rellich Type Inequalities

CHAPTER 6

Improved Hardy-Rellich Inequalities on $H_0^2(\Omega)$

If (V,W) is an n-dimensional Bessel pair on $(0,R)$, then for any radial function $u \in C_c^\infty(B_R)$ where B_R is a ball of radius R in $\mathbb{R}^n$, $n \geq 1$, we have

$$\int_B V(|x|)|\Delta u|^2 dx \geq \int_B W(|x|)|\nabla u|^2 dx + (n-1)\int_B \left(\frac{V(|x|)}{|x|^2} - \frac{V_r(|x|)}{|x|}\right)|\nabla u|^2 dx.$$

Moreover, if

$$W(r) - \frac{2V(r)}{r^2} + \frac{2V_r(r)}{r} - V_{rr}(r) \geq 0 \quad for \ \ 0 \leq r \leq R,$$

then the above inequality holds true for all u –radial or not– in $C_c^\infty(B_R)$.

By combining this with the main inequality of Chapter 3, one obtain various improvements of the Hardy-Rellich inequality for $H_0^2(\Omega)$. In particular, If Ω is a domain such that $0 \in \Omega \subset B_R \subset \mathbb{R}^n$ with $n \geq 4$, $\lambda < n-2$, and P is a HI-potential on $(0, R)$ such that $\frac{P_r(r)}{P(r)} = \frac{\lambda}{r} + f(r)$, where $f(r) \geq 0$ and $\lim_{r \to 0} rf(r) = 0$, then the following inequality holds for all $u \in H_0^2(\Omega)$,

$$(6.1) \quad \int_\Omega |\Delta u|^2 dx \geq \frac{n^2(n-4)^2}{16}\int_\Omega \frac{u^2}{|x|^4}dx + \left(\frac{n^2}{4} + \frac{(n-\lambda-2)^2}{4}\right)\beta(P;R)\int_\Omega \frac{P(|x|)}{|x|^2}u^2 dx.$$

6.1. General Hardy-Rellich inequalities for radial functions

Let $0 \in \Omega \subset \mathbb{R}^n$ be a smooth domain, and denote

$$C_{c,r}^k(\Omega) = \{v \in C_r^k(\Omega) : v \text{ is radial and supp } v \subset \Omega\},$$

$$H_{0,r}^m(\Omega) = \{u \in H_0^m(\Omega) : u \text{ is radial}\}.$$

We start by considering a general inequality for radial functions.

THEOREM 6.1.1. *Let (V,W) be a pair of non-negative C^1-functions on $(0, R)$, and let B_R be the ball of radius R in $\mathbb{R}^n$, $n \geq 1$. Assume*

$$(6.2) \quad \int_0^R \frac{1}{r^{n-1}V(r)}dr = \infty \quad \text{and} \quad \lim_{r \to 0} r^\alpha V(r) = 0 \text{ for some } \alpha < n-2.$$

The following statements are then equivalent:

(1) *(V,W) is a n-dimensional Bessel pair on $(0, R)$.*
(2) *There exists $c > 0$ such that for all radial functions $u \in C_{c,r}^\infty(B_R)$,*

$$(HR_{V,cW}) \quad \int_{B_R} V(x)|\Delta u|^2 dx \geq c\int_{B_R} W(x)|\nabla u|^2 dx + (n-1)\int_{B_R}\left(\frac{V(x)}{|x|^2} - \frac{V_r(|x|)}{|x|}\right)|\nabla u|^2 dx.$$

Moreover, the best constant is given by

$$(6.3) \quad \beta(V,W;R) = \sup\{c;\ (HR_{V,cW}) \text{ holds for radial functions}\}.$$

Proof: Assume $u \in C_{c,r}^\infty(B)$ and observe that

$$\frac{1}{n\omega_n} \int_B V(x)|\Delta u|^2 dx = \int_0^R V(r) u_{rr}^2 r^{n-1} dr + (n-1)^2 \int_0^R V(r) \frac{u_r^2}{r^2} r^{n-1} dr$$
$$+ 2(n-1) \int_0^R V(r) u u_r r^{n-2} dr.$$

Setting $\nu = u_r$, we have

$$\int_B V(x)|\Delta u|^2 dx = \int_B V(x)|\nabla \nu|^2 dx + (n-1) \int_B \left(\frac{V(|x|)}{|x|^2} - \frac{V_r(|x|)}{|x|} \right) |\nu|^2 dx.$$

Thus, inequality (HR$_{V,W}$) for radial functions is equivalent to

$$\int_B V(x)|\nabla \nu|^2 dx \geq \int_B W(x)\nu^2 dx.$$

Letting $y(r) = \nu(x)$ where $|x| = r$, we then have

(6.4) $$\int_0^R V(r) y'(r)^2 r^{n-1} dr \geq \int_0^R W(r) y^2(r) r^{n-1} dr.$$

It therefore follows from Theorem 4.4.1 that 1) and 2) are equivalent. □

By applying the above theorem to the n-dimensional Bessel pair

$$V(x) = |x|^{-2m} \quad \text{and} \quad W_m(x) = V(x)\left[\left(\frac{n-2m-2}{2} \right)^2 |x|^{-2} + P(r) \right]$$

where P is a HI-potential, and by using Theorem 4.2.3, we get the following result in the case of radial functions.

COROLLARY 6.1.1. *Suppose $n \geq 1$, $m < \frac{n-2}{2}$, $B_R \subset \mathbb{R}^n$, and let P be a HI-potential on $(0, R)$. Then we have for all $u \in C_{c,r}^\infty(B_R)$,*

(6.5) $$\int_{B_R} \frac{|\Delta u|^2}{|x|^{2m}} dx \geq \left(\frac{n+2m}{2} \right)^2 \int_{B_R} \frac{|\nabla u|^2}{|x|^{2m+2}} dx + \beta(P;R) \int_{B_R} P(r) \frac{|\nabla u|^2}{|x|^{2m}} dx.$$

Moreover, $\left(\frac{n+2m}{2} \right)^2$ and $\beta(P;R)$ are the best constants.

We now give a few immediate applications of the above in the case where $m = 0$ and $n \geq 3$. Actually, we shall see in the next section that many of the following results are still true in the non-radial case – though sometimes under more stringent conditions on the weights and with different constants – at least in lower dimensions.

COROLLARY 6.1.2. *Assume P is a HI-potential on $(0, R)$, and let $B_R \subset \mathbb{R}^n$ with $n \geq 3$, then for all $u \in C_{c,r}^\infty(B_R)$ we have*

(6.6) $$\int_{B_R} |\Delta u|^2 dx \geq \frac{n^2}{4} \int_{B_R} \frac{|\nabla u|^2}{|x|^2} dx + \beta(P;R) \int_{B_R} P(x) |\nabla u|^2 dx,$$

Moreover, $\frac{n^2}{4}$ and $\beta(P;R)$ are the best constants.

In particular, the following hold for smooth bounded domains Ω in $\mathbb{R}^n$ with $0 \in \Omega \subset B_R$,

- For $\alpha < 2$,

(6.7) $$\int_\Omega |\Delta u|^2 dx \geq \frac{n^2}{4} \int_\Omega \frac{|\nabla u|^2}{|x|^2} dx + \frac{z_\alpha^{2-\alpha}}{R^{2-\alpha}} \int_\Omega \frac{|\nabla u|^2}{|x|^\alpha} dx \quad \text{for } u \in C_{c,r}^\infty(B_R),$$

6.1. GENERAL HARDY-RELLICH INEQUALITIES FOR RADIAL FUNCTIONS

and for $\alpha = 0$,

(6.8) $\quad \int_\Omega |\Delta u|^2 dx \geq \frac{n^2}{4} \int_\Omega \frac{|\nabla u|^2}{|x|^2} dx + \frac{z_0^2}{R^2} \int_\Omega |\nabla u|^2 dx \quad$ for $u \in C_{c,r}^\infty(B_R)$,

the constants being optimal when Ω is the ball B_R.

- For $k \geq 1$ and $\rho = R(e^{e^{e^{\cdot^{\cdot^{e(k-times)}}}}})$, we have

(6.9) $\quad \int_\Omega |\Delta u|^2 dx \geq \frac{n^2}{4} \int_\Omega \frac{|\nabla u|^2}{|x|^2} dx + \frac{1}{4} \sum_{j=1}^k \int_\Omega \frac{|\nabla u|^2}{|x|^2} \Big(\prod_{i=1}^j log^{(i)} \frac{\rho}{|x|}\Big)^{-2} dx,$

- For $D \geq R$, we have

(6.10)
$$\int_\Omega |\Delta u|^2 dx \geq \frac{n^2}{4} \int_\Omega \frac{|\nabla u|^2}{|x|^2} dx + \frac{1}{4} \sum_{i=1}^\infty \int_\Omega \frac{|\nabla u|^2}{|x|^2} X_1^2(\frac{|x|}{D}) X_2^2(\frac{|x|}{D}) ... X_i^2(\frac{|x|}{D}) dx,$$

Moreover, all constants appearing in the above two inequalities are optimal.

We now combine Theorem 6.1.1 with Theorem 4.1.1 to get the following inequalities for radial functions.

THEOREM 6.1.2. *Let P be a HI-potential on $(0, R)$ such that $\frac{P_r(r)}{P(r)} = \frac{\lambda}{r} + f(r)$, where $f(r) \geq 0$ and $\lim_{r \to 0} rf(r) = 0$. If $\lambda < n - 2$, and B is a ball of radius R in $\mathbb{R}^n$ with $n \geq 4$, then the following holds for any radial function $u \in C_{c,r}^\infty(B_R)$,*

(6.11) $\quad \int_B |\Delta u|^2 dx \geq \frac{n^2(n-4)^2}{16} \int_B \frac{u^2}{|x|^4} dx$
$\qquad + \Big(\frac{n^2}{4} + \frac{(n-\lambda-2)^2}{4}\Big) \beta(P; R) \int_B \frac{P(x)}{|x|^2} u^2 dx.$

Proof: Use first inequality (6.6) with the HI-potential P, then Theorem 4.2.1 with the Bessel pair $(|x|^{-2}, |x|^{-2}\frac{(n-4)^2}{4}|x|^{-2} + P)$ to the term $\int_B \frac{|\nabla u|^2}{|x|^2} dx$, and finally apply Theorem 4.2.3 with the Bessel pair $(P, \frac{(n-\lambda-2)^2}{4})|x|^{-2}P)$ to the term $\int_B P(x)|\nabla u|^2 dx$, to obtain

$$\int_B |\Delta u|^2 dx \geq \frac{n^2}{4} \int_B \frac{|\nabla u|^2}{|x|^2} dx + \beta(P, R) \int_B P(x)|\nabla u|^2 dx$$
$$\geq \frac{n^2}{4} \frac{(n-4)^2}{4} \int_B \frac{u^2}{|x|^4} dx + \frac{n^2}{4} \beta(P, R) \int_B \frac{P(x)}{|x|^2} u^2 + \beta(P, R) \int P(x)|\nabla u|^2 dx$$
$$\geq \frac{n^2(n-4)^2}{16} \int_B \frac{u^2}{|x|^4} dx + \Big(\frac{n^2}{4} + \frac{(n-\lambda-2)^2}{4}\Big) \beta(P, R) \int_B \frac{P(x)}{|x|^2} u^2 dx.$$

The following is immediate from Theorem 6.1.2 and from the fact that $\lambda = 2$ for the HI-potential under consideration.

COROLLARY 6.1.3. *Let Ω be a smooth bounded domain containing 0 in $\mathbb{R}^n$ ($n \geq 4$) and $R = \sup_{x \in \Omega} |x|$. Then, the following holds for all $u \in C_{c,r}^\infty(\Omega)$:*

(1) If $k \geq 1$ and $\rho = R(e^{e^{e^{\cdot^{\cdot^{\cdot^{e}}}}}})$ ($e(k-times)$), then

(6.12) $$\int_\Omega |\Delta u|^2 dx \geq \frac{n^2(n-4)^2}{16} \int_\Omega \frac{u^2}{|x|^4} dx$$
$$+ (1 + \frac{n(n-4)}{8}) \sum_{j=1}^k \int_\Omega \frac{u^2}{|x|^4} (\prod_{i=1}^j \log^{(i)} \frac{\rho}{|x|})^{-2} dx.$$

(2) If $D \geq R$, then

(6.13) $$\int_\Omega |\Delta u|^2 dx \geq \frac{n^2(n-4)^2}{16} \int_\Omega \frac{u^2}{|x|^4} dx$$
$$+ (1 + \frac{n(n-4)}{8}) \sum_{i=1}^\infty \int_\Omega \frac{u^2}{|x|^4} X_1^2(\frac{|x|}{D}) X_2^2(\frac{|x|}{D}) \ldots X_i^2(\frac{|x|}{D}) dx.$$

THEOREM 6.1.3. *Let P_1 and P_2 be two HI-potentials on $(0, R)$, and let $B_R \subset \mathbb{R}^n$ with $n \geq 4$. If $a < 2$, then for all $u \in C_{c,r}^\infty(B_R)$*

(6.14)
$$\int_B |\Delta u|^2 dx \geq \frac{n^2(n-4)^2}{16} \int_B \frac{u^2}{|x|^4} dx + \frac{n^2}{4} \beta(P_1; R) \int_B P_1(x) \frac{u^2}{|x|^2} dx$$
$$+ \frac{z_a^{2-a}}{R^{2-a}} (\frac{n-a-2}{2})^2 \int_B \frac{u^2}{|x|^{a+2}} dx + \frac{z_a^{2-a}}{R^{2-a}} \beta(P_2; R) \int_B P_2(x) \frac{u^2}{|x|^a} dx.$$

Proof: We first use inequality (6.7) with the HI-potential $|x|^{-a}$ where $a < 2$, then Theorem 4.2.1 with the Bessel pair $(|x|^{-2}, |x|^{-2}(\frac{(n-4)^2}{4}|x|^{-2} + P))$, then again Theorem 4.2.1 with the Bessel pair $(|x|^{-a}, |x|^{-a}((\frac{n-a-2}{2})^2|x|^{-2} + P))$ to obtain

$$\int_B |\Delta u|^2 dx \geq \frac{n^2}{4} \int_B \frac{|\nabla u|^2}{|x|^2} dx + \frac{z_a^{2-a}}{R^{2-a}} \int_B \frac{|\nabla u|^2}{|x|^a} dx$$
$$\geq \frac{n^2(n-4)^2}{16} \int_B \frac{u^2}{|x|^4} dx + \frac{n^2}{4} \beta(P_1; R) \int_B P_1(x) \frac{u^2}{|x|^2} dx$$
$$+ \frac{z_a^{2-a}}{R^{2-a}} \int_B \frac{|\nabla u|^2}{|x|^a} dx$$
$$\geq \frac{n^2(n-4)^2}{16} \int_B \frac{u^2}{|x|^4} dx + \frac{n^2}{4} \beta(P_1; R) \int_B P_1(x) \frac{u^2}{|x|^2} dx$$
$$+ \frac{z_a^{2-a}}{R^{2-a}} (\frac{n-a-2}{2})^2 \int_B \frac{u^2}{|x|^{a+2}} dx$$
$$+ \frac{z_a^{2-a}}{R^{2-a}} \beta(P_2; R) \int_B P_2(x) \frac{u^2}{|x|^a} dx.$$

6.2. General Hardy-Rellich inequalities for non-radial functions

We now deal with the case of non-necessarily radial functions in H_0^2. We first prove the following result.

THEOREM 6.2.1. *Let (V, W) be a n-dimensional Bessel pair such that $\beta(V, W; R) \geq 1$, and let B_R be a ball with radius R in $\mathbb{R}^n$ ($n \geq 1$). Assume*

(6.15) $\int_0^R \frac{1}{r^{n-1} V(r)} dr = \infty,$ $\quad \lim_{r \to 0} r^\alpha V(r) = 0$ *for some* $\alpha < n - 2,$

and

(6.16) $$W(r) - \frac{2V(r)}{r^2} + \frac{2V_r(r)}{r} - V_{rr}(r) \geq 0 \quad for \ 0 \leq r \leq R.$$

There exists then $c \geq 1$ such that the following holds for all $u \in C_c^\infty(B)$,

(HR$_{V,cW}$)
$$\int_B V(x)|\Delta u|^2 dx \geq c \int_B W(x)|\nabla u|^2 dx + (n-1)\int_B \left(\frac{V(x)}{|x|^2} - \frac{V_r(|x|)}{|x|}\right)|\nabla u|^2 dx.$$

Moreover,

(6.17) $$\beta(V, W; R) = \sup\{c;\ (\text{HR}_{V,cW})\ holds\}.$$

Before proceeding with the proof, we shall give a few implications while relating it to the radial case of the last section. For that we need to check the condition (6.16) on the various Bessel pairs that we consider.

Starting with the Bessel pair
$$V(x) = |x|^{-2m} \quad \text{and} \quad W_{m,c}(x) = V(x)\left[\left(\tfrac{n-2m-2}{2}\right)^2|x|^{-2} + cP(x)\right]$$
where P is a HI-potential, we see that in order to satisfy (6.16), we need to have

(6.18) $$\frac{-(n+4) - 2\sqrt{n^2 - n + 1}}{6} \leq m \leq \frac{-(n+4) + 2\sqrt{n^2 - n + 1}}{6}.$$

We can therefore deduce that under (6.18), the following remain valid for non-radial functions.

COROLLARY 6.2.1. *Suppose $m < \frac{n-2}{2}$, n satisfies (6.18) and $B_R \subset \mathbb{R}^n$. If P is a HI-potential on $(0, R)$, then for all $u \in H_0^2(B_R)$ we have*

(6.19) $$\int_{B_R} \frac{|\Delta u|^2}{|x|^{2m}} \geq \left(\frac{n+2m}{2}\right)^2 \int_{B_R} \frac{|\nabla u|^2}{|x|^{2m+2}} dx + \beta(P; R) \int_{B_R} P(x)\frac{|\nabla u|^2}{|x|^{2m}} dx.$$

Moreover, $\left(\frac{n+2m}{2}\right)^2$ and $\beta(P; R)$ are the best constants.

We shall see however that this inequality remains true without condition (6.18), but with a constant that is sometimes different from $\left(\frac{n+2m}{2}\right)^2$ in the cases where (6.18) is not valid.

In the simplest case $V \equiv 1$ and $W_{m,c}(x) = \left(\frac{n-2}{2}\right)^2|x|^{-2} + W(x)$, condition (6.16) reduces to $\left(\frac{n-2}{2}\right)^2|x|^{-2} + W(x) \geq 2|x|^{-2}$, which is then guaranteed if $n \geq 5$. Therefore, we can now state that the above inequalities hold for all functions in $H_0^2(B_R)$.

COROLLARY 6.2.2. *Assume P is a HI-potential on $(0, R)$, and let $B_R \subset \mathbb{R}^n$. If $n \geq 5$, then inequalities (6.6), (6.7), ... (6.12) hold for all $u \in H_0^2(B_R)$.*

As mentioned before, we shall show in the next section, that the above inequalities still hold without condition (6.18), that is for dimensions $n = 3, 4$. However, in this case the best constant will be different from $\frac{n^2}{4}$. We actually show that the best constant is equal to 3 in dimension 4 and to $\frac{25}{36}$ in dimension 3, which are obviously different from the corresponding best constants in the radial case. □

We now proceed with the proof of Theorem 6.2.1, which will rely on the decomposition of a function into its spherical harmonics as follows: A function $u \in C_c^\infty(\Omega)$ could be extended by zero outside Ω, and could therefore be considered as a function in $C_c^\infty(\mathbb{R}^n)$. We can then decompose u into spherical harmonics in the following way:

$$u(x) = \sum_{k=0}^\infty u_k(x) := \sum_{k=0}^\infty f_k(|x|)\varphi_k(x)$$

where $(\varphi_k)_k$ are the orthonormal eigenfunctions of the Laplace-Beltrami operator with eigenvalues $c_k = k(n+k-2)$, $k \geq 0$, and f_k are the corresponding components, which belong to $C_c^\infty(\Omega)$ and satisfy $f_k(r) = O(r^k)$ and $f'(r) = O(r^{k-1})$ as $r \to 0$. In particular,

(6.20) $\quad \varphi_0 = 1$ and $f_0 = \frac{1}{n\omega_n r^{n-1}} \int_{\partial B_r} u\, ds = \frac{1}{n\omega_n} \int_{|x|=1} u(rx) ds.$

We also have for any $k \geq 0$, and any continuous real valued functions V and W on $(0, \infty)$,

(6.21) $\quad \int_{\mathbb{R}^n} V(|x|) |\Delta u_k|^2 dx = \int_{\mathbb{R}^n} V(|x|) \left(\Delta f_k(|x|) - c_k \frac{f_k(|x|)}{|x|^2}\right)^2 dx,$

and

(6.22) $\quad \int_{\mathbb{R}^n} W(|x|) |\nabla u_k|^2 dx = \int_{\mathbb{R}^n} W(|x|) |\nabla f_k|^2 dx + c_k \int_{\mathbb{R}^n} W(|x|) |x|^{-2} f_k^2 dx.$

Proof of Theorem 6.2.1: Assume that the equation $(B_{V,W})$ has a positive solution on $(0, R]$. We prove that the inequality $(HR_{V,W})$ holds for all $u \in C_c^\infty(B)$ by frequently using the following inequality:

(6.23) $\quad \int_0^R V(r) |x'(r)|^2 r^{n-1} dr \geq \int_0^R W(r) x^2(r) r^{n-1} dr$ for all $x \in C^1(0, R]$.

Indeed, for all $n \geq 1$ and $k \geq 0$ we have

$$\frac{1}{n\omega_n} \int_{\mathbb{R}^n} V(x) |\Delta u_k|^2 dx$$
$$= \frac{1}{n\omega_n} \int_{\mathbb{R}^n} V(x) \left(\Delta f_k(|x|) - c_k \frac{f_k(|x|)}{|x|^2}\right)^2 dx$$
$$= \int_0^R V(r) \left(f_k''(r) + \frac{n-1}{r} f_k'(r) - c_k \frac{f_k(r)}{r^2}\right)^2 r^{n-1} dr$$
$$= \int_0^R V(r) (f_k''(r))^2 r^{n-1} dr + (n-1)^2 \int_0^R V(r) (f_k'(r))^2 r^{n-3} dr$$
$$+ c_k^2 \int_0^R V(r) f_k^2(r) r^{n-5} dr + 2(n-1) \int_0^R V(r) f_k''(r) f_k'(r) r^{n-2} dr$$
$$- 2c_k \int_0^R V(r) f_k''(r) f_k(r) r^{n-3} dr$$
$$- 2c_k(n-1) \int_0^R V(r) f_k'(r) f_k(r) r^{n-4} dr.$$

6.2. GENERAL HARDY-RELLICH INEQUALITIES FOR NON-RADIAL FUNCTIONS

Integrate by parts and use (6.20) for $k = 0$ to get

(6.24)
$$\frac{1}{n\omega_n} \int_{\mathbb{R}^n} V(x)|\Delta u_k|^2 dx = \int_0^R V(r)f_k''(r)^2 r^{n-1} dr$$
$$+ (n - 1 + 2c_k) \int_0^R V(r)f_k'(r)^2 r^{n-3} dr$$
$$+ (2c_k(n - 4) + c_k^2) \int_0^R V(r) r^{n-5} f_k^2(r) dr$$
$$- (n - 1) \int_0^R V_r(r) r^{n-2} f_k'(r)^2 dr$$
$$- c_k(n - 5) \int_0^R V_r(r) f_k^2(r) r^{n-4} dr - c_k \int_0^R V_{rr}(r) f_k^2(r) r^{n-3} dr.$$

Now define $g_k(r) = \frac{f_k(r)}{r}$ and note that $g_k(r) = O(r^{k-1})$ for all $k \geq 1$. We have

$$\int_0^R V(r) f_k'(r)^2 r^{n-3} = \int_0^R V(r) g_k'(r)^2 r^{n-1} dr + \int_0^R 2V(r) g_k(r) g_k'(r) r^{n-2} dr$$
$$+ \int_0^R V(r) g_k^2(r) r^{n-3} dr$$
$$= \int_0^R V(r) g_k'(r)^2 r^{n-1} dr - (n - 3) \int_0^R V(r) g_k^2(r) r^{n-3} dr$$
$$- \int_0^R V_r(r) g_k^2(r) r^{n-2} dr.$$

Thus,

(6.25)
$$\int_0^R V(r) f_k'(r)^2 r^{n-3} \geq \int_0^R W(r) f_k^2(r) r^{n-3} dr$$
$$- (n - 3) \int_0^R V(r) f_k^2(r) r^{n-5} dr - \int_0^R V_r(r) f_k^2(r) r^{n-4} dr.$$

Substituting $2c_k \int_0^R V(r) f_k'(r)^2 r^{n-3}$ in (6.24) by its lower estimate in the last inequality (6.25), we get

$$\frac{1}{n\omega_n} \int_{\mathbb{R}^n} V(x)|\Delta u_k|^2 dx \geq \int_0^R W(r) f_k'(r)^2 r^{n-1} dr + \int_0^R W(r) f_k(r)^2 r^{n-3} dr$$
$$+ (n - 1) \int_0^R V(r) f_k'(r)^2 r^{n-3} dr + c_k(n - 1) \int_0^R V(r) f_k(r)^2 r^{n-5} dr$$
$$- (n - 1) \int_0^R V_r(r) r^{n-2} f_k'(r)^2 dr - c_k(n - 1) \int_0^R V_r(r) r^{n-4} f_k(r)^2 dr$$
$$+ c_k(c_k - (n - 1)) \int_0^R V(r) r^{n-5} f_k^2(r) dr$$
$$+ c_k \int_0^R \left[(W(r) - \frac{2V(r)}{r^2} + \frac{2V_r(r)}{r} - V_{rr}(r)) \right] f_k^2(r) r^{n-3} dr.$$

The proof is now complete since the last term is non-negative by condition (6.16). Note also that because of this condition, the formula for the best constant requires that $\beta(V, W; R) \geq 1$, since if W satisfies (6.16) then cW satisfies it for any $c \geq 1$.

6.3. Optimal Hardy-Rellich inequalities with power weights $|x|^m$

The general Theorem 6.2.1 allowed us to deduce inequality (6.26) below for a restricted interval of powers m. We shall now prove that the same holds for all $m < \frac{n-2}{2}$, though with an optimal constant $a_{n,m}$ that will be evaluated in section 6.5, and which can sometimes be different from $(\frac{n+2m}{2})^2$.

THEOREM 6.3.1. *Suppose $n \geq 1$, $m < \frac{n-2}{2}$, $B_R \subset \mathbb{R}^n$ a ball of radius R, and let P be a HI-potential on $(0, R)$. Then, we have for all $u \in C_c^\infty(B_R)$,*

$$(6.26) \quad \int_{B_R} \frac{|\Delta u|^2}{|x|^{2m}} dx \geq a_{n,m} \int_{B_R} \frac{|\nabla u|^2}{|x|^{2m+2}} dx + \beta(P; R) \int_{B_R} P(|x|) \frac{|\nabla u|^2}{|x|^{2m}} dx,$$

where

$$a_{n,m} = \inf \left\{ \frac{\int_{B_R} \frac{|\Delta u|^2}{|x|^{2m}} dx}{\int_{B_R} \frac{|\nabla u|^2}{|x|^{2m+2}} dx}; u \in C_c^\infty(B_R) \setminus \{0\} \right\}.$$

Moreover, $\beta(P; R)$ and $a_{m,n}$ are the best constants.

Proof: Assuming the inequality

$$\int_{B_R} \frac{|\Delta u|^2}{|x|^{2m}} dx \geq a_{n,m} \int_{B_R} \frac{|\nabla u|^2}{|x|^{2m+2}} dx,$$

holds for all $u \in C_c^\infty(B_R)$, we shall prove that it can be improved by any HI-potential P. We will use the following inequality repeatedly in the proof which follows directly from Theorem 4.2.1 with n=1. For $\alpha \geq 1$, we have for all $f \in C^\infty(0, R)$,

$$(6.27) \quad \int_0^R r^\alpha (f'(r))^2 dr \geq (\frac{\alpha-1}{2})^2 \int_0^R r^{\alpha-2} f^2(r) dr$$
$$+ \beta(P; R) \int_0^R r^\alpha P(r) f^2(r) dr,$$

and both $(\frac{\alpha-1}{2})^2$ and $\beta(P; R)$ are best constants.

6.3. OPTIMAL HARDY-RELLICH INEQUALITIES WITH POWER WEIGHTS $|x|^m$

Decompose $u \in C_c^\infty(B_R)$ into its spherical harmonics $\Sigma_{k=0}^\infty u_k$, where $u_k(x) = f_k(|x|)\varphi_k(x)$. We evaluate $I_k = \frac{1}{nw_n} \int_{\mathbb{R}^n} \frac{|\Delta u_k|^2}{|x|^{2m}} dx$ in the following way:

$$I_k = \int_0^R r^{n-2m-1}(f_k''(r))^2 dr + [(n-1)(2m+1) + 2c_k] \int_0^R r^{n-2m-3}(f_k')^2 dr$$

$$+ c_k[c_k + (n-2m-4)(2m+2)] \int_0^R r^{n-2m-5}(f_k(r))^2 dr$$

$$\geq \beta(P,R) \int_0^R r^{n-2m-1} P(r)(f_k')^2 dr + [(\frac{n+2m}{2})^2 + 2c_k] \int_0^R r^{n-2m-3}(f_k')^2 dr$$

$$+ c_k[c_k + (n-2m-4)(2m+2)] \int_0^R r^{n-2m-5}(f_k(r))^2 dr$$

$$\geq \beta(P,R) \int_0^R r^{n-2m-1} P(r)(f_k')^2 dr + a_{n,m} \int_0^R r^{n-2m-3}(f_k')^2 dr$$

$$+ \beta(P,R)[(\frac{n+2m}{2})^2 + 2c_k - a_{n,m}] \int_0^R r^{n-2m-3} P(r)(f_k)^2 dr$$

$$+ ((\frac{n-2m-4}{2})^2[(\frac{n+2m}{2})^2 + 2c_k - a_{n,m}]$$

$$+ c_k[c_k + (n-2m-4)(2m+2)]) \int_0^R r^{n-2m-5}(f_k(r))^2 dr.$$

Now by (6.44) in section 5.5, we have

$$(\frac{n-2m-4}{2})^2[(\frac{n+2m}{2})^2 + 2c_k - a_{n,m}] + c_k[c_k + (n-2m-4)(2m+2)] \geq c_k a_{n,m},$$

for all $k \geq 0$. Hence, we have

$$I_k \geq a_{n,m} \int_0^R r^{n-2m-3}(f_k')^2 dr + a_{n,m} c_k \int_0^R r^{n-2m-5}(f_k(r))^2 dr$$

$$+ \beta(P,R) \int_0^R r^{n-2m-1} P(r)(f_k')^2 dr$$

$$+ \beta(P,R)[(\frac{n+2m}{2})^2 + 2c_k - a_{n,m}] \int_0^R r^{n-2m-3} P(r)(f_k)^2 dr$$

$$\geq a_{n,m} \int_0^R r^{n-2m-3}(f_k')^2 dr + a_{n,m} c_k \int_0^R r^{n-2m-5}(f_k(r))^2 dr$$

$$+ \beta(P,R) \int_0^R r^{n-2m-1} P(r)(f_k')^2 dr + \beta(P,R) c_k \int_0^R r^{n-2m-3} P(r)(f_k)^2 dr$$

$$= a_{n,m} \int_{B_R} \frac{|\nabla u|^2}{|x|^{2m+2}} dx + \beta(P,R) \int_{B_R} P(|x|) \frac{|\nabla u|^2}{|x|^{2m}} dx.$$

As was done in Theorem 6.1.2, we can now combine Theorem 6.3.1 with Theorem 4.1.1 to get the following inequality.

THEOREM 6.3.2. *Let P be a HI-potential on $(0, R)$, such that $\frac{P_r(r)}{P(r)} = \frac{\lambda}{r} + f(r)$, where $f(r) \geq 0$ and $\lim_{r \to 0} rf(r) = 0$. If $\lambda < n - 2$, and B is a ball of radius R in*

$\mathbb{R}^n$ with $n \geq 3$, then the following Hardy-Rellich inequality holds $u \in C_c^\infty(B_R)$,
(6.28)
$$\int_B |\Delta u|^2 dx \geq \frac{C(n)(n-4)^2}{4} \int_B \frac{u^2}{|x|^4} dx + \left[C(n) + \frac{(n-\lambda-2)^2}{4}\right] \beta(P;R) \int_B \frac{P(x)}{|x|^2} u^2 dx,$$
where $C(3) = \frac{25}{36}$, $C(4) = 3$ and $C(n) = \frac{n^2}{4}$ for all $n \geq 5$.

Proof: Set $C(n) = a_{n,0}$, and use first Theorem 6.3.1 with the HI-potential P, then Theorem 4.2.1 with the Bessel pair $(|x|^{-2}, |x|^{-2}(\frac{(n-4)^2}{4}|x|^{-2} + P)$, then Theorem 4.2.3 with the Bessel pair $(P, \frac{(n-\lambda-2)^2}{4})|x|^{-2}P)$ to obtain

$$\int_B |\Delta u|^2 dx \geq C(n) \int_B \frac{|\nabla u|^2}{|x|^2} dx + \beta(P,R) \int_B P(x) |\nabla u|^2 dx$$

$$\geq C(n) \frac{(n-4)^2}{4} \int_B \frac{u^2}{|x|^4} dx + C(n)\beta(P,R) \int_B \frac{P(x)}{|x|^2} u^2 + \beta(P,R) \int P(x) |\nabla u|^2 dx$$

$$\geq C(n) \frac{(n-4)^2}{4} \int_B \frac{u^2}{|x|^4} dx + (C(n) + \frac{(n-\lambda-2)^2}{4}) \beta(P,R) \int_B \frac{P(x)}{|x|^2} u^2 dx.$$

Note that $C(n) = a_{n,0}$ are evaluated in section 6.5. $\square$

THEOREM 6.3.3. *Let P_1 and P_2 be two HI-potential s on $(0,R)$, and let B_R be a ball of radius R in $\mathbb{R}^n$ with $n \geq 3$. If $a < 2$, then for all $u \in C_0^1(B_R)$*

(6.29)
$$\int_B |\Delta u|^2 dx \geq \frac{C(n)(n-4)^2}{4} \int_B \frac{u^2}{|x|^4} dx + C(n)\beta(P_1;R) \int_B P_1(x) \frac{u^2}{|x|^2} dx$$
$$+ \frac{z_a^{2-a}}{R^{2-a}} (\frac{n-a-2}{2})^2 \int_B \frac{u^2}{|x|^{a+2}} dx + \frac{z_a^{2-a}}{R^{2-a}} \beta(P_2;R) \int_B P_2(x) \frac{u^2}{|x|^a} dx.$$

Proof: We first use Theorem 6.3.1 with the HI-potential $|x|^{-a}$ where $a < 2$, then Theorem 4.2.1 with the Bessel pair $(|x|^{-2}, |x|^{-2}[\frac{(n-4)^2}{4}|x|^{-2} + P_1])$, then again Theorem 4.2.1 with the Bessel pair $(|x|^{-2a}, |x|^{-2a}[(\frac{n-2a-2}{2})^2|x|^{-2} + P_2])$ to obtain

$$\int_B |\Delta u|^2 dx \geq C(n) \int_B \frac{|\nabla u|^2}{|x|^2} dx + \frac{z_a^{2-a}}{R^{2-a}} \int_B \frac{|\nabla u|^2}{|x|^{-a}} dx$$

$$\geq C(n) \frac{(n-4)^2}{4} \int_B \frac{u^2}{|x|^4} dx + C(n)\beta(P_1;R) \int_B P_1(x) \frac{u^2}{|x|^2} dx$$

$$+ \frac{z_a^{2-a}}{R^{2-a}} \int_B \frac{|\nabla u|^2}{|x|^{-a}} dx$$

$$\geq C(n) \frac{(n-4)^2}{4} \int_B \frac{u^2}{|x|^4} dx + C(n)\beta(P_1;R) \int_B P_1(x) \frac{u^2}{|x|^2} dx$$

$$+ \frac{z_a^{2-a}}{R^{2-a}} (\frac{n-a-2}{2})^2 \int_B \frac{u^2}{|x|^{a+2}} dx$$

$$+ \frac{z_a^{2-a}}{R^{2-a}} \beta(P_2;R) \int_B P_2(x) \frac{u^2}{|x|^a} dx.$$

In the following we relate $\int_\Omega V(|x|)|\Delta u|^2 dx$ directly to $\int_\Omega W(|x|)|u|^2 dx$ without going through an inequality involving $\int_\Omega Z(|x|)|\nabla u|^2 dx$. This will allow us to get better constants.

6.3. OPTIMAL HARDY-RELLICH INEQUALITIES WITH POWER WEIGHTS $|x|^m$

THEOREM 6.3.4. *Let Ω be a smooth domain in $\mathbb{R}^n$ with $n \geq 1$ and $0 \in \Omega$. Set $R =: \sup_{x \in \Omega} |x|$ and let $V \in C^2(0, R)$ be a non-negative function that satisfies the following conditions:*

(6.30) $\quad V_r(r) \leq 0 \quad$ and $\quad \int_0^R \frac{1}{r^{n-3}V(r)} dr = -\int_0^R \frac{1}{r^{n-4}V_r(r)} dr = +\infty.$

There exist then $\lambda_1, \lambda_2 \in \mathbb{R}$ such that

(6.31) $\quad \frac{rV_r(r)}{V(r)} + \lambda_1 \geq 0 \quad$ on $(0, R)$ and $\lim_{r \to 0} \frac{rV_r(r)}{V(r)} + \lambda_1 = 0,$

(6.32) $\quad \frac{rV_{rr}(r)}{V_r(r)} + \lambda_2 \geq 0 \quad$ on $(0, R)$ and $\lim_{r \to 0} \frac{rV_{rr}(r)}{V_r(r)} + \lambda_2 = 0,$

and for $r \in (0, R)$,

(6.33) $\quad \left[\frac{1}{2}(n - \lambda_1 - 2)^2 + 3(n-3)\right] V(r) - (n-5)rV_r(r) - r^2 V_{rr}(r) \geq 0.$

Then, the following inequality holds:

$$\int_\Omega V(|x|)|\Delta u|^2 dx \geq \left(\frac{(n - \lambda_1 - 2)^2}{4} + (n-1)\right) \frac{(n - \lambda_1 - 4)^2}{4} \int_\Omega \frac{V(|x|)}{|x|^4} u^2 dx$$

(6.34)
$$- \frac{(n-1)(n - \lambda_2 - 2)^2}{4} \int_\Omega \frac{V_r(|x|)}{|x|^3} u^2 dx.$$

Proof: Using Theorem 4.2.3 and condition (6.33), we can evaluate the integral $H := \frac{1}{n\omega_n} \int_{\mathbb{R}^n} V(|x|)|\Delta u_k|^2 dx$ as follows:

$$H = \int_0^R V(r)(f_k''(r))^2 r^{n-1} dr + (n - 1 + 2c_k) \int_0^R V(r)(f_k'(r))^2 r^{n-3} dr$$

$$+ (2c_k(n-4) + c_k^2) \int_0^R V(r) r^{n-5} f_k^2(r) dr - (n-1) \int_0^R V_r(r) r^{n-2} (f_k')^2(r) dr$$

$$- c_k(n-5) \int_0^R V_r(r) f_k^2(r) r^{n-4} dr - c_k \int_0^R V_{rr}(r) f_k^2(r) r^{n-3} dr$$

$$\geq \int_0^R V(r)(f_k''(r))^2 r^{n-1} dr + (n-1) \int_0^R V(r)(f_k'(r))^2 r^{n-3} dr$$

$$- (n-1) \int_0^R V_r(r) r^{n-2} (f_k')^2(r) dr$$

$$+ c_k \int_0^R \left(\left(\frac{1}{2}(n - \lambda_1 - 2)^2 + 3(n-3)\right) V(r) - (n-5)rV_r(r) - r^2 V_{rr}(r)\right) f_k^2(r) r^{n-5} dr.$$

The rest of the proof follows from the above inequality combined with Theorem 4.2.3. $\square$

REMARK 6.3.5. Let $V(r) = r^{-2m}$ with $m \leq \frac{n-4}{2}$. Then in order to satisfy condition (6.33) we must have $-1 - \frac{\sqrt{1+(n-1)^2}}{2} \leq m \leq \frac{n-4}{2}$. Under this assumption the inequality (6.34) gives the following weighted second order Rellich inequality:

$$\int_B \frac{|\Delta u|^2}{|x|^{2m}} dx \geq \frac{(n+2m)^2(n-4-2m)^2}{16} \int_B \frac{u^2}{|x|^{2m+4}} dx.$$

In the following theorem we will show that the constant appearing in the above inequality is optimal. Moreover, we will see that if $m < -1 - \frac{\sqrt{1+(n-1)^2}}{2}$, then the best constant is strictly less than $(\frac{(n+2m)(n-4-2m)}{4})^2$. This shows that inequality (6.34) is actually sharp.

6. IMPROVED HARDY-RELLICH INEQUALITIES ON $H_0^2(\Omega)$

THEOREM 6.3.6. *Let $m \leq \frac{n-4}{2}$, $B_R \subset \mathbb{R}^n$ a ball of radius R, and let P be a HI-potential on $(0, R)$ such that $\frac{P(r)}{P_r(r)} = -\frac{\lambda}{r} + f(r)$, where $f(r) \geq 0$ and $\lim_{r\to 0} r f(r) = 0$. Then the following inequality holds for all $u \in C_c^\infty(B_R)$*

$$\int_{B_R} \frac{|\Delta u|^2}{|x|^{2m}} dx \geq \beta_{n,m} \int_{B_R} \frac{u^2}{|x|^{2m+4}} dx$$
(6.35)
$$+ \beta(P; R)\left[\frac{(n+2m)^2}{4} + \frac{(n-2m-\lambda-2)^2}{4}\right] \int_{B_R} \frac{P(|x|)}{|x|^{2m+2}} u^2 dx,$$

where

(6.36)
$$\beta_{n,m} = \inf_{u \in C_c^\infty(B)\setminus\{0\}} \frac{\int_B \frac{|\Delta u|^2}{|x|^{2m}} dx}{\int_B \frac{u^2}{|x|^{2m+4}} dx}.$$

Proof: Again we will repeatedly use inequality (6.27) in the proof. Decomposing $u \in C_c^\infty(B_R)$ into spherical harmonics $\Sigma_{k=0}^\infty u_k$, where $u_k(x) = f_k(|x|)\varphi_k(x)$, we can estimate the integral $H := \frac{1}{n\omega_n} \int_{\mathbb{R}^n} \frac{|\Delta u_k|^2}{|x|^{2m}} dx$ as follows:

$$H = \int_0^R r^{n-2m-1}(f_k''(r))^2 dr + [(n-1)(2m+1) + 2c_k]\int_0^R r^{n-2m-3} f_k'(r)^2 dr$$

$$+ c_k[c_k + (n-2m-4)(2m+2)]\int_0^R r^{n-2m-5}(f_k(r))^2 dr$$

$$\geq \left(\frac{n+2m}{2}\right)^2 \int_0^R r^{n-2m-3} f_k'(r)^2 dr + \beta(P; R)\int_0^R r^{n-2m-1} P(r) f_k'(r)^2 dr$$

$$+ c_k[c_k + 2\left(\frac{n-\lambda-4}{2}\right)^2 + (n-2m-4)(2m+2)]\int_0^R r^{n-2m-5}(f_k(r))^2 dr,$$

where we have used the fact that $c_k \geq 0$ to get the above inequality. We now have

$$H \geq \beta_{n,m} \int_0^R r^{n-2m-5}(f_k)^2 dr$$

$$+ \beta(P; R)\frac{(n+2m)^2}{4} \int_0^R r^{n-2m-3} P(r)(f_k)^2 dr$$

$$+ \beta(P; R) \int_0^R r^{n-2m-1} P(r)(f_k')^2 dr$$

$$\geq \beta_{n,m} \int_0^R r^{n-2m-5}(f_k)^2 dr$$

$$+ \beta(P; R)\left(\frac{(n+2m)^2}{4} + \frac{(n-2m-\lambda-2)^2}{4}\right) \int_0^R r^{n-2m-3} P(r)(f_k)^2 dr$$

$$\geq \frac{\beta_{n,m}}{n\omega_n} \int_B \frac{u_k^2}{|x|^{2m+4}} dx$$

$$+ \frac{\beta(P; R)}{n\omega_n}\left[\frac{(n+2m)^2}{4} + \frac{(n-2m-\lambda-2)^2}{4}\right] \int_B \frac{P(|x|)}{|x|^{2m+2}} u_k^2 dx,$$

by Theorem 4.2.3. Hence, (6.35) holds and the proof is complete. $\square$

6.4. HIGHER ORDER RELLICH INEQUALITIES

THEOREM 6.3.7. *Assume* $-1 < m \leq \frac{n-4}{2}$, $B_R \subset \mathbb{R}^n$ *is a ball of radius* R, *and let* P *be a HI-potential on* $(0, R)$. *Then, for all* $u \in C_c^\infty(B_R)$:

$$(6.37) \quad \int_B \frac{|\Delta u|^2}{|x|^{2m}} dx \geq \frac{(n+2m)^2(n-2m-4)^2}{16} \int_B \frac{u^2}{|x|^{2m+4}} dx$$
$$+ \beta(P;R) \frac{(n+2m)^2}{4} \int_B \frac{P(x)}{|x|^{2m+2}} u^2 dx + \frac{z_{2m}^{2-2m}}{R^{2-2m}} \|u\|_{H_0^1}.$$

Proof: Decomposing again $u \in C_c^\infty(B_R)$ into its spherical harmonics $\Sigma_{k=0}^\infty u_k$ where $u_k(x) = f_k(|x|) \varphi_k(x)$, we calculate again $H := \frac{1}{n\omega_n} \int_{\mathbb{R}^n} \frac{|\Delta u_k|^2}{|x|^{2m}} dx$ as follows:

$$H = \int_0^R r^{n-2m-1} (f_k''(r))^2 dr + [(n-1)(2m+1) + 2c_k] \int_0^R r^{n-2m-3} (f_k')^2 dr$$
$$+ c_k [c_k + (n-2m-4)(2m+2)] \int_0^R r^{n-2m-5} (f_k(r))^2 dr$$
$$\geq (\frac{n+2m}{2})^2 \int_0^R r^{n-2m-3} (f_k')^2 dr + \frac{z_{2m}^{2-2m}}{R^{2-2m}} \int_0^R r^{n-1} (f_k')^2 dr$$
$$+ c_k \int_0^R r^{n-2m-3} (f_k')^2 dr$$
$$\geq \frac{(n+2m)^2(n-2m-4)^2}{16} \int_0^R r^{n-2m-5} (f_k)^2 dr$$
$$+ \beta(P;R) \frac{(n+2m)^2}{4} \int_0^R P(r) r^{n-2m-3} (f_k)^2 dr$$
$$+ \frac{z_{2m}^{2-2m}}{R^{2-2m}} \int_0^R r^{n-1} (f_k')^2 dr + c_k \frac{z_{2m}^{2-2m}}{R^{2-2m}} \int_0^R r^{n-3} (f_k)^2 dr$$
$$= \frac{(n+2m)^2(n-2m-4)^2}{16 n\omega_n} \int_{\mathbb{R}^n} \frac{u_k^2}{|x|^{2m+4}} dx$$
$$+ \frac{\beta(P;R)}{n\omega_n} \frac{(n+2m)^2}{4} \int_{\mathbb{R}^n} \frac{P(x)}{|x|^{2m+2}} u_k^2 dx + \frac{z_{2m}^{2-2m}}{R^{2-2m}} \|u_k\|_{W_0^{1,2}}.$$

Hence (6.37) holds.

6.4. Higher order Rellich inequalities

We shall now use the results obtained in the previous section to derive higher order Rellich inequalities with corresponding improvements. Let P be a HI-potential, $\beta_{n,m}$ as defined in Theorem 6.3.6 and

$$\sigma_{n,m} = \beta(P;R) \Big[\frac{(n+2m)^2}{4} + \frac{(n-2m-\lambda-2)^2}{4} \Big].$$

For the sake of convenience we make the following convention: $\prod_{i=1}^{0} a_i = 1$.

THEOREM 6.4.1. *Let P be a HI-potential on $(0, R)$ such that* $\frac{P(r)}{P_r(r)} = -\frac{\lambda}{r} + f(r)$, *where* $f(r) \geq 0$ *and* $\lim_{r \to 0} r f(r) = 0$. *Assume* $m \in \mathbb{N}$, $1 \leq l \leq m$, *and* $2k + 4m \leq$

n. Then the following inequality holds for all $u \in C_c^\infty(B_R)$

$$(6.38)\int_{B_R} \frac{|\Delta^m u|^2}{|x|^{2k}} dx \geq \prod_{i=0}^{l-1} \beta_{n,k+2i} \int_{B_R} \frac{|\Delta^{m-l} u|^2}{|x|^{2k+4l}} dx$$

$$+ \sum_{i=0}^{l-1} \sigma_{n,k+2i} \prod_{j=1}^{l-1} \beta_{n,k+2j-2} \int_{B_R} \frac{P(x)|\Delta^{m-i-1}u|^2}{|x|^{2k+4i+2}} dx.$$

Proof: Follows directly from Theorem 6.3.6. □

THEOREM 6.4.2. *Let P be a HI-potential on $(0,R)$ such that $\frac{P(r)}{P_r(r)} = -\frac{\lambda}{r} + f(r)$, where $f(r) \geq 0$ and $\lim_{r \to 0} rf(r) = 0$. Assume $m \in \mathbb{N}$, $1 \leq l \leq m$, and $2k + 4m + 2 \leq n$. Then the following inequality holds for all $u \in C_c^\infty(B_R)$:*

$$(6.39)$$

$$\int_{B_R} \frac{|\nabla \Delta^m u|^2}{|x|^{2k}} dx \geq (\frac{n-2k-2}{2})^2 \prod_{i=0}^{l-1} \beta_{n,k+2i+1} \int_{B_R} \frac{|\Delta^{m-l} u|^2}{|x|^{2k+4l+2}} dx$$

$$+ (\frac{n-2k-2}{2})^2 \sum_{i=0}^{l-1} \sigma_{n,k+2i+1} \prod_{j=1}^{l-1} \beta_{n,k+2j-1} \int_{B_R} \frac{P(x)|\Delta^{m-i-1}u|^2}{|x|^{2k+4i+4}} dx$$

$$+ \beta(P;R) \int_{B_R} P(x) \frac{|\Delta^m u|^2}{|x|^{2k}} dx.$$

Proof: Follows directly from Theorem 4.2.1 and the previous theorem. □

THEOREM 6.4.3. *Let P be a HI-potential on $(0,R)$ such that $\frac{P(r)}{P_r(r)} = -\frac{\lambda}{r} + f(r)$, where $f(r) \geq 0$ and $\lim_{r \to 0} rf(r) = 0$. Assume $m \in \mathbb{N}$, $1 \leq l \leq m-1$, and $2k + 4m \leq n$. Then the following inequality holds for all $u \in C_c^\infty(B_R)$:*

$$(6.40)$$

$$\int_{B_R} \frac{|\Delta^m u|^2}{|x|^{2k}} dx \geq a_{n,k}(\frac{n-2k-4}{2})^2 \prod_{i=0}^{l-1} \beta_{n,k+2i+2} \int_{B_R} \frac{|\Delta^{m-l-1} u|^2}{|x|^{2k+4l+4}} dx$$

$$+ a_{n,k}(\frac{n-2k-4}{2})^2 \sum_{i=0}^{l-1} \sigma_{n,k+2i+2} \prod_{j=1}^{l-1} \beta_{n,k+2j} \int_{B_R} \frac{P(x)|\Delta^{m-i-2}u|^2}{|x|^{2k+4i+6}} dx$$

$$+ \beta(P;R) a_{n,k} \int_{B_R} P(x) \frac{|\Delta^{m-1} u|^2}{|x|^{2k+2}} dx + \beta(P;R) \int_{B_R} P(x) \frac{|\nabla \Delta^{m-1} u|^2}{|x|^{2k}} dx,$$

where $a_{n,m}$ is defined in Theorem 6.3.1.

Proof: Follows directly from Theorem 6.3.1 and the previous theorem. □

THEOREM 6.4.4. *Let P be a HI-potential on $(0,R)$ such that $\frac{P(r)}{P_r(r)} = -\frac{\lambda}{r} + f(r)$, where $f(r) \geq 0$ and $\lim_{r \to 0} rf(r) = 0$. Assume $m \in \mathbb{N}$, $1 \leq l \leq m$, and $2k + 4m \leq n$. Then the following inequality holds for all $u \in C_c^\infty(B_R)$:*

$$\int_{B_R} \frac{|\Delta^m u|^2}{|x|^{2k}} dx \geq \prod_{i=1}^{l} \frac{a_{n,k+2i-2}(n-2k-4i)^2}{4} \int_{B_R} \frac{|\Delta^{m-l} u|^2}{|x|^{2k+4l}} dx$$

$$+ \beta(P;R) \sum_{i=1}^{l} \prod_{j=1}^{i-1} \frac{a_{n,k+2j-2}(n-2k-4j)^2}{4} \int_{B_R} P(x) \frac{|\nabla \Delta^{m-i} u|^2}{|x|^{2k+4i-4}} dx$$

$$+ \beta(P;R) \sum_{i=1}^{l} a_{n,k+2i-2} \prod_{j=1}^{i-1} \frac{a_{n,k+2j-2}(n-2k-4j)^2}{4} \int_{B_R} P(x) \frac{|\Delta^{m-i} u|^2}{|x|^{2k+4i-2}} dx,$$

where $a_{n,m}$ are the best constants in inequality (6.26).

Proof: Follows directly from Theorem 6.3.1. □

6.5. Calculations of best constants

We start by the evaluation of the constants $a_{n,m}$.

THEOREM 6.5.1. *Suppose $n \geq 1$ and $m \leq \frac{n-2}{2}$. Then for any $R > 0$,*

$$a_{n,m} = \inf\left\{\frac{\int_{B_R} \frac{|\Delta u|^2}{|x|^{2m}} dx}{\int_{B_R} \frac{|\nabla u|^2}{|x|^{2m+2}} dx}; u \in C_c^\infty(B_R) \setminus \{0\}\right\}$$

$$= \min_{k \geq 0}\left\{\frac{\left(\frac{(N-4-2m)(N+2m)}{4} + c_k\right)^2}{\left(\frac{N-4-2m}{2}\right)^2 + c_k}\right\},$$

where $c_k = k(n+k-2)$. In particular

(1) For $n = 1$
  - if $m \in (-\infty, -\frac{3}{2}) \cup [-\frac{7}{6}, -\frac{1}{2}]$, then
  $$a_{1,m} = \left(\frac{1+2m}{2}\right)^2$$
  - if $-\frac{3}{2} < m < -\frac{7}{6}$, then
  $$a_{1,m} = \min\left\{\left(\frac{n+2m}{2}\right)^2, \frac{\left(\frac{(n-4-2m)(n+2m)}{4} + 2\right)^2}{\left(\frac{n-4-2m}{2}\right)^2 + 2}\right\}.$$

(2) If $m = \frac{n-4}{2}$, then
$$a_{m,n} = \min\{(n-2)^2, n-1\}.$$

(3) If $n \geq 2$ and $m \leq \frac{-(n+4)+2\sqrt{n^2-n+1}}{6}$, then $a_{n,m} = \left(\frac{n+2m}{2}\right)^2$.

(4) If $2 \leq n \leq 3$ and $\frac{-(n+4)+2\sqrt{n^2-n+1}}{6} < m \leq \frac{n-2}{2}$, or $n \geq 4$ and $\frac{n-4}{2} < m \leq \frac{n-2}{2}$, then
$$a_{n,m} = \frac{\left(\frac{(n-4-2m)(n+2m)}{4} + n - 1\right)^2}{\left(\frac{n-4-2m}{2}\right)^2 + n - 1}.$$

(5) If $n \geq 4$, $\frac{-(n+4)+2\sqrt{n^2-n+1}}{6} < m < \frac{n-4}{2}$, define $k^* = [(\frac{\sqrt{3}}{3} - \frac{1}{2})(n-2)]$.
  - If $k^* \leq 1$, then
  $$a_{n,m} = \frac{\left(\frac{(n-4-2m)(n+2m)}{4} + n - 1\right)^2}{\left(\frac{n-4-2m}{2}\right)^2 + n - 1}.$$
  - For $k^* > 1$ the interval $(m_0^1 := \frac{-(n+4)+2\sqrt{n^2-n+1}}{6}, m_0^2 := \frac{n-4}{2})$ can be divided in $2k^* - 1$ subintervals. For $1 \leq k \leq k^*$ define
  $$m_k^1 := \frac{2(n-5) - \sqrt{(n-2)^2 - 12k(k+n-2)}}{6},$$
  $$m_k^2 := \frac{2(n-5) + \sqrt{(n-2)^2 - 12k(k+n-2)}}{6}.$$

If $m \in (m_0^1, m_1^1] \cup [m_1^2, m_0^2)$, then
$$a_{n,m} = \frac{(\frac{(n-4-2m)(n+2m)}{4} + n - 1)^2}{(\frac{n-4-2m}{2})^2 + n - 1}.$$

- For $k \geq 1$ and $m \in (m_k^1, m_{k+1}^1] \cup [m_{k+1}^2, m_k^2)$, then

$$a_{n,m} = \min\{\frac{(\frac{(n-4-2m)(n+2m)}{4} + k(n+k-2))^2}{(\frac{n-4-2m}{2})^2 + k(n+k-2)}, \frac{(\frac{(n-4-2m)(n+2m)}{4} + (k+1)(n+k-1))^2}{(\frac{n-4-2m}{2})^2 + (k+1)(n+k-1)}\}.$$

For $m \in (m_{k^*}^1, m_{k^*}^2)$,

$$a_{n,m} = \min\{\frac{(\frac{(n-4-2m)(n+2m)}{4} + k^*(n+k^*-2))^2}{(\frac{n-4-2m}{2})^2 + k^*(n+k^*-2)}, \frac{(\frac{(n-4-2m)(n+2m)}{4} + (k^*+1)(n+k^*-1))^2}{(\frac{n-4-2m}{2})^2 + (k^*+1)(n+k^*-1)}\}.$$

Proof: Letting $V(r) = r^{-2m}$ then,
$$W(r) - \frac{2V(r)}{r^2} + \frac{2V_r(r)}{r} - V_{rr}(r) = ((\frac{n-2m-2}{2})^2 - 2 - 4m - 2m(2m+1))r^{-2m-2}.$$

In order to satisfy condition (6.16) we should have

(6.41) $$\frac{-(n+4) + 2\sqrt{n^2 - n + 1}}{6} \leq m \leq \frac{-(n+4) + 2\sqrt{n^2 - n + 1}}{6}.$$

So, by Theorem 6.2.1 under the above condition we have $a_{n,m} = (\frac{n+2m}{2})^2$ as in the radial case.

Decomposing again $u \in C_c^\infty(B_R)$ into spherical harmonics; $u = \Sigma_{k=0}^\infty u_k$, where $u_k(x) = f_k(|x|)\varphi_k(x)$, one has

(6.42)
$$\int_{\mathbb{R}^n} \frac{|\Delta u_k|^2}{|x|^{2m}} dx = \int_{\mathbb{R}^n} |x|^{-2m}(f_k''(|x|))^2 dx$$
$$+ ((n-1)(2m+1) + 2c_k) \int_{\mathbb{R}^n} |x|^{-2m-2}(f_k')^2 dx$$
$$+ c_k(c_k + (n-4-2m)(2m+2)) \int_{\mathbb{R}^n} |x|^{-2m-4}(f_k)^2 dx$$
$$\geq \left((\frac{n+2m}{2})^2 + 2c_k\right) \int_{\mathbb{R}^n} |x|^{-2m-2}(f_k')^2 dx$$
$$+ c_k(c_k + (n-4-2m)(2m+2)) \int_{\mathbb{R}^n} |x|^{-2m-4}(f_k)^2 dx$$

(6.43) $$\int_{\mathbb{R}^n} \frac{|\nabla u_k|^2}{|x|^{2m+2}} dx = \int_{\mathbb{R}^n} |x|^{-2m-2}(f_k')^2 dx + c_k \int_{\mathbb{R}^n} |x|^{-2m-4}(f_k)^2 dx.$$

Hence
$$a_{m,N} \leq \frac{C_1 \frac{\int_0^\infty r^{N-3-2m}(f_k')^2 dr}{\int_0^\infty r^{N-5-2m}(f_k)^2 dr} + C_2}{\frac{\int_0^\infty r^{N-3-2m}(f_k')^2 dr}{\int_0^\infty r^{N-5-2m}(f_k)^2 dr} + c_k}$$

where $C_1 = \left[\left(\frac{N+2m}{2}\right)^2 + 2c_k\right]$ and

$C_2 = c_k[c_k - (N-3-2m)(N-4-2m) + (N-1)(N-4-2m)].$

6.5. CALCULATIONS OF BEST CONSTANTS

However, since $C_2 - c_k C_1 \leq 0$, the real function
$$\omega(y) := \frac{C_1 y + C_2}{y + c_k} = \frac{C_1(y + c_k) + C_2 - c_k C_1}{y + c_k}$$
is increasing for positive y. Hence, from the Hardy inequality
$$\int_0^\infty r^{N-3-2m}(f_k')^2 \, dr \geq \left(\frac{N-4-2m}{2}\right)^2 \int_0^\infty r^{N-5-2m}(f_k')^2 \, dr,$$
we conclude that
$$a_{m,N} \leq A(k, N, m) := \frac{\left(\frac{(N-4-2m)(N+2m)}{4} + c_k\right)^2}{\left(\frac{N-4-2m}{2}\right)^2 + c_k}.$$

One can also show that indeed
$$(6.44) \qquad a_{n,m} = \min\{A(k, m, n); \, k \in \mathbb{N}\}.$$

Note that when $m = \frac{n-4}{2}$ and $n + k > 2$, then $c_k \neq 0$. Actually, this also holds for $n + k \leq 2$, in which case one deduces that if $m = \frac{n-4}{2}$, then
$$a_{n,m} = \min\{(n-2)^2 = \left(\frac{n+2m}{2}\right)^2, (n-1) = c_1\}$$
which is statement 2).

Now we consider the function
$$f(x) = \frac{\left(\frac{(n-4-2m)(n+2m)}{4} + x\right)^2}{\left(\frac{n-4-2m}{2}\right)^2 + x}.$$

It is easy to check that $f'(x) = 0$ at x_1 and x_2, where

$$(6.45) \qquad x_1 = -\frac{(n-4-2m)(n+2m)}{4}$$

$$(6.46) \qquad x_2 = \frac{(n-4-2m)(-n+6m+8)}{4}.$$

Observe that for for $n \geq 2$, $\frac{n-8}{6} \leq \frac{n-4}{2}$. So, for $m \leq \frac{n-8}{6}$ both x_1 and x_2 are negative and hence $a_{n,m} = \left(\frac{n+2m}{2}\right)^2$. Also note that
$$\frac{-(n+4) - 2\sqrt{n^2 - n + 1}}{6} \leq \frac{n-8}{6} \quad \text{for all } n \geq 1.$$
Hence, under the condition in 3) we have $a_{n,m} = \left(\frac{n+2m}{2}\right)^2$.

Also for $n = 1$ if $m \leq -\frac{3}{2}$ both critical points are negative and we have $a_{1,m} \leq \left(\frac{1+2m}{2}\right)^2$. Comparing $A(0, m, n)$ and $A(1, m, n)$ we see that $A(1, m, n) \geq A(0, m, n)$ if and only if (6.41) holds.

For $n = 1$ and $-\frac{3}{2} < m < -\frac{7}{6}$ both x_1 and x_2 are positive. Consider the equations
$$x(x-1) = x_1 = \frac{(2m+3)(2m+1)}{4},$$
and
$$x(x-1) = x_2 = -\frac{(2m+3)(6m+7)}{4}.$$

By simple calculations we can see that all four solutions of the above two equations are less that two. Since, $A(1,m,1) < A(0,m,1)$ for $m < -\frac{7}{6}$, we have $a_{1,m} \leq \min\{A(1,m,1), A(2,m,1)\}$ and 1) follows.

For $n \geq 2$ and $\frac{n-4}{2} < m < \frac{n-2}{2}$ we have $x_1 > 0$ and $x_2 < 0$. Consider the equation
$$x(x+n-2) = x_1 = -\frac{(n-4-2m)(n+2m)}{4}.$$
Then $\frac{2m+4-n}{2}$ and $-\frac{(2m+n)}{2}$ are solutions of the above equation and both are less than one. Since, for $n \geq 4$
$$\frac{n-2}{2} > \frac{-(n+4)+2\sqrt{n^2-n+1}}{6},$$
and $A(1,m,n) \leq A(0,m,n)$ for $m \geq \frac{-(n+4)+2\sqrt{n^2-n+1}}{6}$, the best constant is equal to what 4) claims.

5) also follows from a similar elementary argument. □

We now compute the constants

(6.47) $$\beta_{n,m} = \inf\left\{\frac{\int_B \frac{|\Delta u|^2}{|x|^{2m}}dx}{\int_B \frac{u^2}{|x|^{2m+4}}dx}; u \in C_c^\infty(B)\setminus\{0\}\right\}.$$

THEOREM 6.5.2. *If* $m \leq \frac{n-4}{2}$, *then*
$$\beta_{n,m} = \frac{(n+2m)^2(n-4-2m)^2}{16} + \gamma_{n,m},$$
where
$$\gamma_{n,m} := \min_{k=0,1,2,\ldots}\{k(n+k-2)[k(n+k-2) + \frac{(n+2m)(n-2m-4)}{2}]\}.$$
Consequently, the values of $\beta_{n,m}$ are as follows:

(1) If $-1 - \frac{\sqrt{1+(n-1)^2}}{2} \leq m \leq \frac{n-4}{2}$, then
$$\beta_{n,m} = (\frac{(n+2m)(n-4-2m)}{4})^2.$$

(2) If $\frac{n}{2} - 3 \leq m \leq -1 - \frac{\sqrt{1+(n-1)^2}}{2}$, then
$$\beta_{n,m} = (\frac{(n+2m)(n-4-2m)}{4})^2 + (n-1)[(n-1) + \frac{(n+2m)(n-2m-4)}{2}].$$

(3) If $k := \frac{n-2m-4}{2} \in N$, then
$$\beta_{n,m} = (\frac{(n+2m)(n-4-2m)}{4})^2 + k(n+k-2)[k(n+k-2) + \frac{(n+2m)(n-2m-4)}{2}].$$

(4) If $k < \frac{n-2m-4}{2} < k+1$ for some $k \in N$, then
$$\beta_{n,m} = \frac{(n+2m)^2(n-2m-4)^2}{16} + a(m,n,k),$$
where
$$a(m,n,k) = \min\{k(n+k-2)[k(n+k-2) + \frac{(n+2m)(n-2m-4)}{2}],$$
$$(k+1)(n+k-1)[(k+1)(n+k-1) + \frac{(n+2m)(n-2m-4)}{2}]\}.$$

Proof: Decompose $u \in C_c^\infty(B_R)$ into spherical harmonics $\Sigma_{k=0}^\infty u_k$, where $u_k(x) = f_k(|x|)\varphi_k(x)$. we have

$$\frac{1}{n\omega_n}\int_{\mathbb{R}^n} \frac{|\Delta u_k|^2}{|x|^{2m}}dx = \int_0^R r^{n-2m-1}(f_k''(r))^2 dr + [(n-1)(2m+1) + 2c_k]\int_0^R r^{n-2m-3}(f_k')^2 dr$$

$$+ c_k[c_k + (n-2m-4)(2m+2)]\int_0^R \times r^{n-2m-5}(f_k(r))^2 dr$$

$$\geq ((\frac{(n+2m)(n-4-2m)}{4})^2$$

$$+ c_k[c_k + \frac{(n+2m)(n-2m-4)}{2}])\int_0^R r^{n-2m-5}(f_k(r))^2 dr,$$

by the Hardy inequality. Hence,

$$\beta_{n,m} \geq B(n,m,k) := \frac{(n+2m)^2(n-4-2m)^2}{16} + \gamma_{n,m}.$$

To prove that $\beta_{n,m}$ is the best constant, let k be such that

$$\beta_{n,m} = \frac{(n+2m)(n-4-2m)}{4})^2$$

$$+ k(n+k-2)[k(n+k-2) + \frac{(n+2m)(n-2m-4)}{2}].$$

Set

$$u(x) = |x|^{-\frac{n-4}{2}+m+\epsilon}\varphi_k(x)\varphi(|x|),$$

where $\varphi_k(x)$ is an eigenfunction corresponding to the eigenvalue c_k and $\varphi(r)$ is a smooth cutoff function, such that $0 \leq \varphi \leq 1$, with $\varphi \equiv 1$ in $[0, \frac{1}{2}]$. We have

$$\frac{\int_{B_R} \frac{|\Delta u|^2}{|x|^{2m}} dx}{\int_{B_R} \frac{u^2}{|x|^{2m+4}} dx} = \left[-\frac{(n+2m)(n-4-2m)}{4} - c_k + \epsilon(2+2m+\epsilon)\right]^2 + O(1).$$

Let now $\epsilon \to 0$ to obtain the result. Thus the inequality

$$\int_{B_R} \frac{|\Delta u|^2}{|x|^{2m}} \geq \beta_{n,m} \int_{B_R} \frac{u^2}{|x|^{2m+4}} dx,$$

holds for all $u \in C_c^\infty(B_R)$.

To calculate explicit values of $\beta_{n,m}$ we need to find the minimum point of the function

$$f(x) = x(x + \frac{(n+2m)(n-2m-4)}{2}), \quad x \geq 0.$$

Observe that

$$f'(-\frac{(n+2m)(n-2m-4)}{4}) = 0.$$

To find a minimizer $k \in N$, we should solve the equation

$$k^2 + (n-2)k + \frac{(n+2m)(n-2m-4)}{4} = 0.$$

The roots of the above equation are $x_1 = \frac{n+2m}{2}$ and $x_2 = \frac{n-2m-4}{2}$. 1) follows from Theorem 6.3.4. It is easy to see that if $m \leq -1 - \frac{\sqrt{1+(n-1)^2}}{2}$, then $x_1 < 0$. Hence, for $m \leq -1 - \frac{\sqrt{1+(n-1)^2}}{2}$ the minimum of the function f is attained in x_2. Note that if $m \leq -1 - \frac{\sqrt{1+(n-1)^2}}{2}$, then $B(n,m,1) \leq B(n,m,0)$. Therefore claims 2), 3), and 4) follow. $\square$

6.6. Further comments

The classical Hardy-Rellich inequality was first proved by Rellich [**242**]. It states that for dimensions $n \geq 5$, one has

(6.48) $\quad \int_\Omega |\Delta u|^2 dx \geq \frac{n^2(n-4)^2}{16} \int_\Omega \frac{u^2}{|x|^4} dx \quad$ for $u \in H_0^2(\Omega)$.

The sharp constant in Hardy-Rellich inequalities of arbitrary order were established by I. Herbst in [**179**]. Subsequently many authors such as [**18**], [**46**], [**110**] studied the version with weights

(6.49) $\quad \int_{\mathbb{R}^n} |x|^\alpha |\Delta u|^2 \, dx \geq \mu_{n,\alpha} \int_{\mathbb{R}^n} |x|^{\alpha-4} |u|^2 \, dx \quad$ for any $u \in C_c^2(\mathbb{R}^n \setminus \{0\})$,

under some restrictions on α.

In view of the natural question of what happens at the critical dimension $n = 4$, and no doubt motivated by the fact that Hardy's inequality can be improved once restricted to a smooth bounded domain Ω in $\mathbb{R}^n$, there was a flurry of activity about possible improvements of the following type:

(6.50) $\quad \int_\Omega |\Delta u|^2 dx - \frac{n^2(n-4)^2}{16} \int_\Omega \frac{u^2}{|x|^4} dx \geq \int_\Omega W(x) u^2 dx.$

Hardy-Rellich inequalities in dimension $n = 4$ were first obtained by Adimurthi-Grossi-Santra in [**8**], as well as various other improved Hardy-Rellich inequalities for specific logarithmic potentials W. Tertikas and Zographopoulos [**262**] followed suit and added to the list of improving potentials. They also considered the intermediate step of showing inequalities of the form

(6.51) $\quad \int_{B_R} |\Delta u|^2 dx \geq \frac{n^2}{4} \int_{B_R} \frac{|\nabla u|^2}{|x|^2} dx,$

for all radial functions $u \in C_{c,r}^\infty(B_R)$.

The results presented in this chapter are mostly due to Ghoussoub-Moradifam [**163**]. They extend, improve, and unify the above results. They first showed that a general HI-potential could be added to (6.51) before adding another HI-potential in the final step that relates $\int_{B_R} \frac{|\nabla u|^2}{|x|^2} dx$ to $\int_{B_R} \frac{u^2}{|x|^4} dx$, hence creating a myriad of new and improved Hardy-Rellich inequalities. This leads to optimal improvements for (6.51) such as the already encompassing inequality for all $u \in H_0^2(B)$:

(6.52) $\quad \int_B |\Delta u|^2 dx - C(n) \int_B \frac{|\nabla u|^2}{|x|^2} dx \geq c(W, R) \int_B W(x) |\nabla u|^2 dx.$

The case when $W \equiv \tilde{W}_{k,\rho}$ and $n \geq 5$ was established previously by Tertikas-Zographopoulos [**262**], who used the decomposition of a function into spherical harmonics to initiate the computation of the best constants $a_{n,m}$ and $\beta_{n,m}$ before it was completed in [**163**]. See also Adimurthi, M. Grossi, and S. Santra [**8**], R. Brown [**67**], X. Cabré [**107**] and E. B. Davies, A.M. Hinz [**110**] for more about Hardy-Rellich type inequalities.

The role of condition (6.16) in symmetry breaking was singled out in Ghoussoub-Moradifam [**163**]. Note that it is a sufficient but not necessary condition, which guarantees that the best constant is the same for the radial and for the non-radial case. For example, it was shown in [**163**] that the best constants for the radial case of (6.52) are $C(n) = \frac{n^2}{4}$ for all $n \geq 2$, and that they are equal to those corresponding to the same inequality for general H_0^2-functions as long as

$n \geq 5$. On the other hand, there is symmetry breaking in dimensions 3 and 4 since the best constants for radial functions are different from the general case, where $C(3) = \frac{25}{36}$ and $C(4) = 3$. In a related context and using different methods, Beckner [41] also computed the values of the constants $C(n)$, when $W \equiv 0$. An older paper by Yafaev [277] also contains various results about symmetry breaking for Hardy-Rellich inequalities of arbitrary order in the supercritical range. However, the following question remains of interest.

Open problem (6): Devise a necessary and sufficient condition on a Bessel pair (V, W) that prevents symmetry breaking in Hardy-Rellich type inequalities involving V, W as weights.

Finally, we note that just as in the case of Hardy's inequalities– the best constants in the Hardy-Rellich inequalities are not attained whenever $0 \in \Omega$ while this can happen in the case when $0 \in \partial\Omega$, and in particular on conical domains.

Open problem (7): Assuming that $0 \in \partial\Omega$, where Ω is a smooth bounded domain in $\mathbb{R}^n$, develop a counterpart of Chapter 3 for Hardy-Rellich inequalities. In particular, evaluate best constants in the various inequalities and study how their attainability depend on the domain.

CHAPTER 7

Weighted Hardy-Rellich Inequalities on $H^2(\Omega) \cap H_0^1(\Omega)$

We show here that if (V, W) is a n-dimensiona Bessel pair on an interval $(0, R)$, and if B_R is a ball of radius R in $\mathbb{R}^n$, $n \geq 1$, then for any $0 < c \leq \beta(V, W, R)$, there exists $\theta > 0$ such the following inequalities hold for all radial functions $u \in H^2(B)$,

$$\int_B V(|x|)|\Delta u|^2 dx \geq c \int_B W(|x|)|\nabla u|^2 dx + (n-1) \int_B \left(\frac{V(|x|)}{|x|^2} - \frac{V_r(|x|)}{|x|}\right)|\nabla u|^2 dx$$
$$+ [(n-1) - \theta] V(R) \int_{\partial B} |\nabla u|^2 \, dx.$$

The latter inequality holds for all functions –radial or not– in $H^2(B) \cap H_0^1(\Omega)$, provided

$$cW(r) - \frac{2V(r)}{r^2} + \frac{2V_r(r)}{r} - V_{rr}(r) \geq 0 \text{ on } (0, R).$$

This leads to various classes of weighted Hardy-Rellich inequalities on $H^2(\Omega) \cap H_0^1(\Omega)$.

7.1. Inequalities between Hessian and Dirichlet energies on $H^2(\Omega) \cap H_0^1(\Omega)$

We start by considering a general inequality for radial functions. Let $0 \in \Omega \subset \mathbb{R}^n$ be a smooth domain, and denote $C_r^k(\bar{\Omega}) = \{v \in C^k(\bar{\Omega}) : \text{v is radial}\}$.

THEOREM 7.1.1. *Let (V, W) be a Bessel pair on $(0, R)$, and let B_R be a ball with radius R in $\mathbb{R}^n$ ($n \geq 1$) and centered at zero. Assume*

(7.1) $\quad \int_0^R \frac{1}{r^{n-1}V(r)} dr = \infty \quad \text{and} \quad \lim_{r \to 0} r^\alpha V(r) = 0 \text{ for some } \alpha < n - 2.$

Then, there exists $c > 0$ and $\theta > 0$ so that the following hold:

(1) *For all radial functions $u \in C^\infty(\bar{B})$,*

(7.2)
$$\int_B V(|x|)|\Delta u|^2 dx \geq c \int_B W(|x|)|\nabla u|^2 dx + (n-1) \int_B \left(\frac{V(|x|)}{|x|^2} - \frac{V_r(|x|)}{|x|}\right)|\nabla u|^2 dx$$
$$+ [(n-1) - \theta] V(R) \int_{\partial B} |\nabla u|^2 \, dx.$$

Moreover, the inequality holds for any $c \leq \beta(V, W, R)$ and any $\theta \geq -R\frac{\varphi'(R)}{\varphi(R)}$, where φ is a positive solution of $(\mathcal{B}_{(V, cW)})$ on $(0, R]$.

(2) *If the following condition is also satisfied:*

(7.3) $\quad cW(r) - \frac{2V(r)}{r^2} + \frac{2V_r(r)}{r} - V_{rr}(r) \geq 0 \quad \text{for } 0 \leq r \leq R,$

then inequality (7.2) holds for all $u \in H^2(B) \cap H_0^1(B)$.

Proof: 1) Assume $u \in C_r^\infty(\bar{B})$ and observe that

$$\int_B V(x)|\Delta u|^2 dx = n\omega_n \{ \int_0^R V(r) u_{rr}^2 r^{n-1} dr + (n-1)^2 \int_0^R V(r) \frac{u_r^2}{r^2} r^{n-1} dr$$
$$+ 2(n-1) \int_0^R V(r) u u_r r^{n-2} dr \}.$$

Setting $\nu = u_r$, we then have

$$\int_B V(x)|\Delta u|^2 dx = \int_B V(x)|\nabla \nu|^2 dx + (n-1) \int_B \left(\frac{V(|x|)}{|x|^2} - \frac{V_r(|x|)}{|x|} \right) |\nu|^2 dx$$
$$+ (n-1) V(R) \int_{\partial B} |\nu|^2 ds.$$

Thus, inequality (7.2) for radial functions follows from inequality (4.38), i.e.,

$$\int_B V(|x|)|\nabla w|^2 dx \geq c \int_B W(|x|) w^2 dx - \theta \int_{\partial B} w^2 d\sigma,$$

once applied to $w = \nu \in H^1(B)$.

2) For the non-radial case, we shall again use the decomposition of a function into its spherical harmonics $u(x) = \Sigma_{k=0}^\infty u_k$, where $u_k(x) = f_k(|x|) \varphi_k(x)$ and $(\varphi_k(x))_k$ are the orthogonal eigenfunctions of the Laplace-Beltrami operator with corresponding eigenvalues $c_k = k(N+k-2)$, $k \geq 0$. The functions f_k belong to $C^\infty([0,R])$, $f_k(R) = 0$, and satisfy $f_k(r) = O(r^k)$ and $f'(r) = O(r^{k-1})$ as $r \to 0$. In particular,

(7.4) $\qquad \varphi_0 = 1$ and $f_0 = \frac{1}{n\omega_n r^{n-1}} \int_{\partial B_r} u ds = \frac{1}{n\omega_n} \int_{|x|=1} u(rx) ds.$

We also have for any $k \geq 0$, and any continuous real valued functions V and W on $(0, \infty)$,

(7.5) $\qquad \int_{\mathbb{R}^n} V(|x|)|\Delta u_k|^2 dx = \int_{\mathbb{R}^n} V(|x|) \left(\Delta f_k(|x|) - c_k \frac{f_k(|x|)}{|x|^2} \right)^2 dx,$

and

(7.6) $\qquad \int_{\mathbb{R}^n} W(|x|)|\nabla u_k|^2 dx = \int_{\mathbb{R}^n} W(|x|)|\nabla f_k|^2 dx + c_k \int_{\mathbb{R}^n} W(|x|)|x|^{-2} f_k^2 dx.$

We shall use repeatedly the following inequality: for all $x \in C^1(0, R]$,

(7.7) $\qquad \int_0^R V(r)|x'(r)|^2 r^{n-1} dr \geq \int_0^R W(r) x^2(r) r^{n-1} dr + V(R) \frac{\varphi'(R)}{\varphi(R)} R^{n-1} x(R)^2.$

7.1. INEQUALITIES BETWEEN HESSIAN AND DIRICHLET TYPE ENERGIES

Indeed, for all $n \geq 1$ and $k \geq 0$ we have

$$\frac{1}{nw_n}\int_{\mathbb{R}^n} V(x)|\Delta u_k|^2 dx = \frac{1}{nw_n}\int_{\mathbb{R}^n} V(x)\Big(\Delta f_k(|x|) - c_k \frac{f_k(|x|)}{|x|^2}\Big)^2 dx$$

$$= \int_0^R V(r)\Big(f_k''(r) + \frac{n-1}{r}f_k'(r) - c_k \frac{f_k(r)}{r^2}\Big)^2 r^{n-1} dr$$

$$= \int_0^R V(r) f_k''(r)^2 r^{n-1} dr + (n-1)^2 \int_0^R V(r) f_k'(r)^2 r^{n-3} dr$$

$$+ c_k^2 \int_0^R V(r) f_k^2(r) r^{n-5} + 2(n-1)\int_0^R V(r) f_k''(r) f_k'(r) r^{n-2}$$

$$- 2c_k \int_0^R V(r) f_k''(r) f_k(r) r^{n-3} dr$$

$$- 2c_k(n-1)\int_0^R V(r) f_k'(r) f_k(r) r^{n-4} dr.$$

Integrate by parts and use (7.4) for $k = 0$ to get

(7.8)
$$\frac{1}{n\omega_n}\int_{\mathbb{R}^n} V(x)|\Delta u_k|^2 dx = \int_0^R V(r)(f_k''(r))^2 r^{n-1} dr$$

$$+ (n-1+2c_k)\int_0^R V(r)(f_k'(r))^2 r^{n-3} dr$$

$$+ (2c_k(n-4)+c_k^2)\int_0^R V(r) r^{n-5} f_k^2(r) dr$$

$$- (n-1)\int_0^R V_r(r) r^{n-2} f_k'(r)^2 dr$$

$$- c_k(n-5)\int_0^R V_r(r) f_k^2(r) r^{n-4} dr - c_k \int_0^R V_{rr}(r) f_k^2(r) r^{n-3} dr$$

$$+ (n-1)V(R)(f_k'(R))^2 R^{n-2}.$$

Now define $g_k(r) = \frac{f_k(r)}{r}$ and note that $g_k(r) = O(r^{k-1})$ for all $k \geq 1$. We have

$$\int_0^R V(r) f_k'(r)^2 r^{n-3} = \int_0^R V(r) g_k'(r)^2 r^{n-1} dr + \int_0^R 2V(r) g_k(r) g_k'(r) r^{n-2} dr$$

$$+ \int_0^R V(r) g_k^2(r) r^{n-3} dr$$

$$= \int_0^R V(r) g_k'(r)^2 r^{n-1} dr - (n-3)\int_0^R V(r) g_k^2(r) r^{n-3} dr$$

$$- \int_0^R V_r(r) g_k^2(r) r^{n-2} dr.$$

Thus,

(7.9)
$$\int_0^R V(r) f_k'(r)^2 r^{n-3} \geq \int_0^R W(r) f_k^2(r) r^{n-3} dr - (n-3)\int_0^R V(r) f_k^2(r) r^{n-5} dr$$

$$- \int_0^R V_r(r) f_k^2(r) r^{n-4} dr.$$

Substituting $2c_k \int_0^R V(r)(f_k'(r))^2 r^{n-3}$ in (7.8) by its lower estimate in the last inequality (7.9), we get

(7.10)
$$\frac{1}{n\omega_n}\int_{\mathbb{R}^n} V(x)|\Delta u_k|^2 dx \geq \int_0^R W(r) f_k'(r)^2 r^{n-1} dr + \int_0^R W(r) f_k(r)^2 r^{n-3} dr$$
$$+ (n-1)\int_0^R V(r) f_k'(r)^2 r^{n-3} dr + c_k(n-1)\int_0^R V(r) f_k(r)^2 r^{n-5} dr$$
$$- (n-1)\int_0^R V_r(r) r^{n-2} (f_k')^2(r) dr - c_k(n-1)\int_0^R V_r(r) r^{n-4} f_k^2(r) dr$$
$$+ c_k(c_k - (n-1))\int_0^R V(r) r^{n-5} f_k^2(r) dr$$
$$+ c_k \int_0^R \left(W(r) - \frac{2V(r)}{r^2} + \frac{2V_r(r)}{r} - V_{rr}(r)\right) f_k^2(r) r^{n-3} dr$$
$$+ (n-1) V(R) f_k'(R)^2 R^{n-2} + V(R) \frac{\varphi'(R)}{\varphi(R)} R^{n-1} f_k'(R)^2.$$

The proof is now complete since the last two terms are non-negative by our assumptions. $\square$

As in the last chapter, we see that in the simplest case $V \equiv 1$, condition (7.3) which insures that the above inequality holds for all –and not necessarily radial– functions, requires that the dimension $n \geq 5$. More generally, when $V(x) = |x|^{-2m}$, then (7.3) is satisfied if

(7.11)
$$\frac{-(n+4) - 2\sqrt{n^2 - n + 1}}{6} \leq m \leq \frac{-(n+4) + 2\sqrt{n^2 - n + 1}}{6}.$$

Also to satisfy the condition (7.1) we need to have $m > -\frac{n}{2}$. Thus for m satisfying (7.11) the inequality

(7.12)
$$\int_{B_R} \frac{|\Delta u|^2}{|x|^{2m}} \geq \left(\frac{n+2m}{2}\right)^2 \int_{B_R} \frac{|\nabla u|^2}{|x|^{2m+2}} dx$$

holds for all $u \in H^2(B_R)$. Moreover, $(\frac{n+2m}{2})^2$ is the best constant. We shall now see that - just like in the last chapter - this inequality remains true without condition (7.11), but with a constant that is sometimes different from $(\frac{n+2m}{2})^2$. For example, if $m = 0$, then the best constant is 3 in dimension 4 and $\frac{25}{36}$ in dimension 3.

Recall from the last chapter the definition of the best constants

(7.13)
$$a_{n,m} = \inf \left\{ \frac{\int_{B_R} \frac{|\Delta u|^2}{|x|^{2m}} dx}{\int_{B_R} \frac{|\nabla u|^2}{|x|^{2m+2}} dx} ; u \in H_0^2(\Omega) \setminus \{0\} \right\},$$

which have been computed explicitly in Section 6.5. We start by proving the following result.

LEMMA 7.1.2. *Assume* $-\frac{n}{2} \leq m < \frac{n-2}{2}$ *and* Ω *be a smooth domain in* $\mathbb{R}^n$, $n \geq 1$. *Then,*

(7.14)
$$a_{n,m} = \inf \left\{ \frac{\int_{B_R} \frac{|\Delta u|^2}{|x|^{2m}} dx}{\int_{B_R} \frac{|\nabla u|^2}{|x|^{2m+2}} dx} ; u \in H^2(\Omega) \cap H_0^1(\Omega) \setminus \{0\} \right\}.$$

7.1. INEQUALITIES BETWEEN HESSIAN AND DIRICHLET TYPE ENERGIES

Proof: Decomposing again $u \in C^\infty(\bar{B}_R)$ into spherical harmonics; $u = \Sigma_{k=0}^\infty u_k$, where $u_k(x) = f_k(|x|)\varphi_k(x)$, we have by (7.6)

$$(7.15) \quad \int_{\mathbb{R}^n} \frac{|\nabla u_k|^2}{|x|^{2m+2}} dx = \int_{\mathbb{R}^n} |x|^{-2m-2}(f_k')^2 dx + c_k \int_{\mathbb{R}^n} |x|^{-2m-4}(f_k)^2 dx.$$

On the other hand, for $V(x) = |x|^{-2m}$ the boundary term in equality (7.10)

$$(n-1)V(R)f_k'(R)^2 R^{n-2} + V(R)\frac{\varphi'(R)}{\varphi(R)}R^{n-1}f_k'(R)^2,$$

simplifies to $(\frac{n}{2}+m)R^{n-2m-2}f_k'(R)^2$, since the solution for the corresponding ODE is $\varphi(r) = r^{-\frac{n-2m-2}{2}}$. By the assumption $(\frac{n}{2}+m) \geq 0$, the exact same proof as in Theorem 6.5.1 yields that

$$\inf\left\{\frac{\int_{B_R}\frac{|\Delta u|^2}{|x|^{2m}}dx}{\int_{B_R}\frac{|\nabla u|^2}{|x|^{2m+2}}dx}; H^2(\Omega)\setminus\{0\}\right\} \geq \inf\left\{\frac{\int_{B_R}\frac{|\Delta u|^2}{|x|^{2m}}dx}{\int_{B_R}\frac{|\nabla u|^2}{|x|^{2m+2}}dx}; u \in H_0^2(\Omega)\setminus\{0\}\right\}.$$

The proof is complete since the reverse inequality holds trivially. □

THEOREM 7.1.3. *Suppose $-\frac{n}{2} \leq m < \frac{n-2}{2}$ and B_R is a ball of radius R in $\mathbb{R}^n$ with $n \geq 3$. Let P be a HI-potential on $(0,R)$ in such a way that the corresponding positive solution φ for the equation $(\mathcal{B}_P)$ on $(0,R]$ satisfies:*

$$(7.16) \quad R\frac{\varphi'(R)}{\varphi(R)} \geq -\frac{n}{2} + m.$$

Then, for all $u \in H^2(B_R) \cap H_0^1(B_R)$ we have

$$(7.17) \quad \int_{B_R}\frac{|\Delta u|^2}{|x|^{2m}} \geq a_{n,m}\int_{B_R}\frac{|\nabla u|^2}{|x|^{2m+2}}dx + \beta(P;R)\int_{B_R}P(x)\frac{|\nabla u|^2}{|x|^{2m}}dx,$$

where $a_{n,m}$ are defined in (7.14). Moreover $\beta(P;R)$ and $a_{m,n}$ are the best constants.

Proof: Assuming the inequality $\int_{B_R}\frac{|\Delta u|^2}{|x|^{2m}} \geq a_{n,m}\int_{B_R}\frac{|\nabla u|^2}{|x|^{2m+2}}dx$, holds for all $u \in C^\infty(\bar{B}_R)$, we shall prove that it can be improved by any HI-potential P. We shall repeatedly use the following inequality which follows directly from the inequality (4.39) with $n = 1$, and $\alpha = -a \geq 1$,

$$(7.18) \quad \int_0^R r^\alpha f'(r)^2 dr \geq (\frac{\alpha-1}{2})^2 \int_0^R r^{\alpha-2}f^2(r)dr$$

$$+ \beta(P;R)\int_0^R r^\alpha P(r)f^2(r)dr + (\frac{\varphi'(R)}{\varphi(R)}+\frac{\alpha-1}{2R})R^\alpha f(R)^2,$$

for all $f \in C^\infty(0,R]$, where both $(\frac{\alpha-1}{2})^2$ and $\beta(P;R)$ are best constants. Decompose $u \in C^\infty(\bar{B}_R)$ into its spherical harmonics $\Sigma_{k=0}^\infty u_k$, where $u_k(x) = f_k(|x|)\varphi_k(x)$.

We evaluate $I_k = \frac{1}{nw_n} \int_{\mathbb{R}^n} \frac{|\Delta u_k|^2}{|x|^{2m}} dx$ in the following way

$$\begin{aligned}
I_k &\geq \int_0^R r^{n-2m-1} f_k''(r)^2 dr + [(n-1)(2m+1) + 2c_k] \int_0^R r^{n-2m-3} (f_k')^2 dr \\
&\quad + c_k[c_k + (n-2m-4)(2m+2)] \int_0^R r^{n-2m-5} f_k(r)^2 dr \\
&\quad + (n-1) R^{n-2m-2} f_k'(R)^2 \\
&\geq \beta(P,R) \int_0^R r^{n-2m-1} P(x)(f_k')^2 dr + [(\frac{n+2m}{2})^2 + 2c_k] \int_0^R r^{n-2m-3}(f_k')^2 dr \\
&\quad + c_k[c_k + (n-2m-4)(2m+2)] \int_0^R r^{n-2m-5} f_k(r)^2 dr \\
&\geq \beta(P,R) \int_0^R r^{n-2m-1} P(x)(f_k')^2 dr + a_{n,m} \int_0^R r^{n-2m-3}(f_k')^2 dr \\
&\quad + \beta(P,R)[(\frac{n+2m}{2})^2 + 2c_k - a_{n,m}] \int_0^R r^{n-2m-3} P(x) f_k^2 dr \\
&\quad + ((\frac{n-2m-4}{2})^2[(\frac{n+2m}{2})^2 + 2c_k - a_{n,m}] + c_k[c_k \\
&\quad + (n-2m-4)(2m+2)]) \int_0^R r^{n-2m-5} f_k(r)^2 dr.
\end{aligned}$$

Now by (6.44) in the previous chapter, we have

$$(\frac{n-2m-4}{2})^2[(\frac{n+2m}{2})^2 + 2c_k - a_{n,m}] + c_k[c_k + (n-2m-4)(2m+2)] \geq c_k a_{n,m},$$

for all $k \geq 0$. Hence, we have

$$\begin{aligned}
I_k &\geq a_{n,m} \int_0^R r^{n-2m-3}(f_k')^2 dr + a_{n,m} c_k \int_0^R r^{n-2m-5} f_k(r)^2 dr \\
&\quad + \beta(P,R) \int_0^R r^{n-2m-1} P(x)(f_k')^2 dr \\
&\quad + \beta(P,R)[(\frac{n+2m}{2})^2 + 2c_k - a_{n,m}] \int_0^R r^{n-2m-3} P(x)(f_k)^2 dr \\
&\geq a_{n,m} \int_0^R r^{n-2m-3}(f_k')^2 dr + a_{n,m} c_k \int_0^R r^{n-2m-5} f_k(r)^2 dr \\
&\quad + \beta(P,R) \int_0^R r^{n-2m-1} P(x)(f_k')^2 dr + \beta(P,R) c_k \int_0^R r^{n-2m-3} P(x) f_k^2 dr \\
&= a_{n,m} \int_{B_R} \frac{|\nabla u|^2}{|x|^{2m+2}} dx + \beta(P,R) \int_{B_R} P(x) \frac{|\nabla u|^2}{|x|^{2m}} dx.
\end{aligned}$$

We shall now give a few immediate applications of the above in the case where $m = 0$ and $n \geq 3$.

THEOREM 7.1.4. *Let P be a HI-potential on $(0, R)$ in such a way that the corresponding positive solution φ for the equation $(\mathcal{B}_{1,P})$ on $(0, R]$ satisfies $R\frac{\varphi'(R)}{\varphi(R)} \geq$*

7.1. INEQUALITIES BETWEEN HESSIAN AND DIRICHLET TYPE ENERGIES

$-\frac{n}{2}$ for $n \geq 3$. Then, for all $u \in H^2(B_R) \cap H_0^1(B_R)$ we have

$$(7.19) \quad \int_{B_R} |\Delta u|^2 dx \geq C(n) \int_{B_R} \frac{|\nabla u|^2}{|x|^2} dx + \beta(W; R) \int_{B_R} W(x) |\nabla u|^2 dx,$$

where $C(3) = \frac{25}{36}$, $C(4) = 3$ and $C(n) = \frac{n^2}{4}$ for all $n \geq 5$. Moreover, $C(n)$ and $\beta(W; R)$ are best constants.

COROLLARY 7.1.1. *The following hold for any smooth bounded domain Ω in $\mathbb{R}^n$ with $R = \sup_{x \in \Omega} |x|$:*

(1) *Let z_0 be the first zero of the Bessel function $J_0(z)$ and choose $0 < \mu_n < z_0$ so that $\mu_n \frac{J_0'(\mu_n)}{J_0(\mu_n)} = -\frac{n}{2}$. Then, for any $u \in H^2(\Omega) \cap H_0^1(\Omega)$,*

$$(7.20) \quad \int_\Omega |\Delta u|^2 dx \geq C(n) \int_\Omega \frac{|\nabla u|^2}{|x|^2} dx + \frac{\mu_n^2}{R^2} \int_\Omega |\nabla u|^2 dx.$$

(2) *For $k \geq 1$, choose $\rho \geq R(e^{e^{e^{\cdot^{\cdot^{e(k-times)}}}}})$ large so that $R\frac{\varphi'(R)}{\varphi(R)} \geq -\frac{n}{2}$, where $\varphi = \big(\prod_{i=1}^j \log^{(i)} \frac{\rho}{|x|}\big)^{\frac{1}{2}}$. Then, for any $u \in H^2(\Omega) \cap H_0^1(\Omega)$,*

$$(7.21) \quad \int_\Omega |\Delta u|^2 dx \geq C(n) \int_\Omega \frac{|\nabla u|^2}{|x|^2} dx + \frac{1}{4} \sum_{j=1}^k \int_\Omega \frac{|\nabla u|^2}{|x|^2} \big(\prod_{i=1}^j \log^{(i)} \frac{\rho}{|x|}\big)^{-2} dx,$$

(3) *For $n \geq 1$, we have for any $u \in H^2(\Omega) \cap H_0^1(\Omega)$,*
$$(7.22)$$
$$\int_\Omega |\Delta u|^2 dx \geq C(n) \int_\Omega \frac{|\nabla u|^2}{|x|^2} dx + \frac{1}{4} \sum_{i=1}^n \int_\Omega \frac{|\nabla u|}{|x|^2} X_1^2(\frac{|x|}{R}) X_2^2(\frac{|x|}{R})...X_i^2(\frac{|x|}{R}) dx.$$

The following Theorem will be needed in the next chapter.

THEOREM 7.1.5. *Let B be the unit ball in $\mathbb{R}^n$ ($n \geq 5$). Then the inequality*

$$(7.23) \quad \int_B |\Delta u|^2 dx \geq \int_B \frac{|\nabla u|^2}{|x|^2 - \frac{n}{2(n-1)} |x|^{\frac{n}{2}+1}} dx + (n-1) \int_B \frac{|\nabla u|^2}{|x|^2} dx,$$

holds for all $u \in C_0^\infty(\bar{B})$.

We need the following lemma.

LEMMA 7.1.6. *For every $u \in C^1([0,1])$, the following inequality holds:*

$$(7.24) \quad \int_0^1 |u'(r)|^2 r^{n-1} dr \geq \int_0^1 \frac{u^2}{r^2 - \frac{n}{2(n-1)} r^{\frac{n}{2}+1}} r^{n-1} dr - (n-1) u(1)^2.$$

Proof: Let $\varphi := r^{-\frac{n}{2}+1} - \frac{n}{2(n-1)}$ and $k(r) := r^{n-1}$. Define $\psi(r) = u(r)/\varphi(r)$, $r \in [0,1]$. Then,

$$\begin{aligned}
\int_0^1 |u'(r)|^2 k(r)dr &= \int_0^1 |\psi(r)|^2 |\varphi'(r)|^2 k(r)dr \\
&\quad + \int_0^1 2\varphi(r)\varphi'(r)\psi(r)\psi'(r)k(r)dr + \int_0^1 \varphi(r)^2 \psi'(r)^2 k(r)dr \\
&= \int_0^1 |\psi(r)|^2 [\varphi'(r)^2 k(r) - (k\varphi\varphi')'(r)]dr + \int_0^1 |\varphi(r)|^2 |\psi'(r)|^2 \\
&\quad \times k(r)dr \\
&\quad + \psi^2(1)\varphi'(1)\varphi(1) \\
&\geq \int_0^1 |\psi(r)|^2 [\varphi'(r)^2 k(r) - (k\varphi\varphi')'(r)]dr + \psi^2(1)\varphi'(1)\varphi(1).
\end{aligned}$$

Note that $\psi^2(1)\varphi'(1)\varphi(1) = u^2(1)\frac{\varphi'(1)}{\varphi(1)} = -(n-1)u^2(1)$. Hence, we have

$$(7.25) \quad \int_0^1 |u'(r)|^2 k(r)dr \geq \int_0^1 -u^2(r)\frac{k'(r)\varphi'(r) + k(r)\varphi''(r)}{\varphi}dr - (n-1)u^2(1).$$

We now get (7.24) by simplifying the above inequality. $\square$

Proof of Theorem 7.1.5: We again decompose u into spherical harmonics

$$u = \Sigma_{k=0}^\infty u_k \text{ where } u_k(x) = f_k(|x|)\varphi_k(x)$$

with $f_k \in C^\infty([0,1])$, $f_k(1) = 0$, $f_k(r) = O(r^k)$ and $f'(r) = O(r^{k-1})$ as $r \to 0$.

As before, we have for any $k \geq 0$, and any continuous real valued W on $(0,1)$,

$$(7.26) \quad \int_B |\Delta u_k|^2 dx = \int_B \left(\Delta f_k(|x|) - c_k \frac{f_k(|x|)}{|x|^2}\right)^2 dx,$$

and

$$(7.27) \quad \int_B W(|x|)|\nabla u_k|^2 dx = \int_B W(|x|)|\nabla f_k|^2 dx + c_k \int_B W(|x|)|x|^{-2} f_k^2 dx.$$

We shall repeatedly use that for all $x \in C^1([0,1])$ with $x(1) = 0$,

$$(7.28) \quad \int_0^1 |x'(r)|^2 r^{n-1} dr \geq \frac{(n-2)^2}{4} \int_0^1 \frac{x^2(r)}{r^2 - \frac{n}{2(n-1)}r^{\frac{n}{2}+1}} r^{n-1} dr,$$

For all $n \geq 5$ and $k \geq 0$ we have

$$\begin{aligned}
\frac{1}{n\omega_n}\int_B |\Delta u_k|^2 dx &= \frac{1}{n\omega_n}\int_B \left(\Delta f_k(|x|) - c_k \frac{f_k(|x|)}{|x|^2}\right)^2 dx \\
&= \int_0^1 \left(f_k''(r) + \frac{n-1}{r}f_k'(r) - c_k \frac{f_k(r)}{r^2}\right)^2 r^{n-1} dr \\
&= \int_0^1 f_k''(r)^2 r^{n-1} dr + (n-1)^2 \int_0^1 f_k'(r)^2 r^{n-3} dr \\
&\quad + c_k^2 \int_0^1 f_k(r)^2 r^{n-5} + 2(n-1)\int_0^1 f_k''(r)f_k'(r)r^{n-2} \\
&\quad - 2c_k \int_0^1 f_k''(r)f_k(r)r^{n-3} dr - 2c_k(n-1)\int_0^1 f_k'(r)f_k(r)r^{n-4} dr.
\end{aligned}$$

Integrate by parts and use (6.20) for $k = 0$ to get

(7.29) $$\frac{1}{n\omega_n}\int_B |\Delta u_k|^2 dx \geq \int_0^1 f_k''(r)^2 r^{n-1} dr$$

(7.30) $$+(n-1+2c_k)\int_0^1 f_k'(r)^2 r^{n-3} dr$$

$$+(2c_k(n-4)+c_k^2)\int_0^1 r^{n-5} f_k(r)^2 dr + (n-1)f_k'(1)^2.$$

Now define $g_k(r) = \frac{f_k(r)}{r}$ and note that $g_k(r) = O(r^{k-1})$ for all $k \geq 1$. We have

$$\int_0^1 f_k'(r)^2 r^{n-3} = \int_0^1 g_k'(r)^2 r^{n-1} dr + \int_0^1 2g_k(r)g_k'(r) r^{n-2} dr + \int_0^1 g_k(r)^2 r^{n-3} dr$$

$$= \int_0^1 g_k'(r)^2 r^{n-1} dr - (n-3)\int_0^1 g_k(r)^2 r^{n-3} dr.$$

It follows that

(7.31) $$\int_0^1 f_k'(r)^2 r^{n-3} \geq \frac{(n-2)^2}{4}\int_0^1 \frac{f_k(r)^2}{r^2 - \frac{n}{2(n-1)}r^{\frac{n}{2}+1}} r^{n-3} dr$$

$$-(n-3)\int_0^1 f_k^2(r) r^{n-5} dr.$$

Substituting $2c_k \int_0^1 f_k'(r)^2 r^{n-3}$ in (7.30) by its lower estimate in the last inequality (7.31), and using Lemma 7.1.6 we get

$$\frac{1}{n\omega_n}\int_B |\Delta u_k|^2 dx \geq \frac{(n-2)^2}{4}\int_0^1 \frac{f_k(r)^2}{r^2 - \frac{n}{2(n-1)}r^{\frac{n}{2}+1}} r^{n-1} dr$$

$$+ 2c_k \frac{(n-2)^2}{4}\int_0^1 \frac{f_k(r)^2}{r^2 - \frac{n}{2(n-1)}r^{\frac{n}{2}+1}} r^{n-3} dr$$

$$+ (n-1)\int_0^1 f_k'(r)^2 r^{n-3} dr + c_k(n-1)\int_0^1 f_k(r)^2 r^{n-5} dr$$

$$+ c_k(c_k - (n-1))\int_0^1 r^{n-5} f_k(r)^2 dr + c_k \int_0^1 \frac{(n-2)^2}{4(r^2 - \frac{n}{2(n-1)}r^{\frac{n}{2}+1})} - \frac{2}{r^2} dr$$

$$\geq \frac{(n-2)^2}{4}\int_0^1 \frac{f_k'(r)^2}{r^2 - \frac{n}{2(n-1)}r^{\frac{n}{2}+1}} r^{n-1} dr$$

$$+ c_k \frac{(n-2)^2}{4}\int_0^1 \frac{f_k(r)^2}{r^2 - \frac{n}{2(n-1)}r^{\frac{n}{2}+1}} r^{n-3} dr$$

$$+ (n-1)\int_0^1 f_k'(r)^2 r^{n-3} dr + c_k(n-1)\int_0^1 f_k(r)^2 r^{n-5} dr.$$

The proof is complete in view of (7.27). $\square$

7.2. Hardy-Rellich inequalities on $H^2(\Omega) \cap H_0^1(\Omega)$

We can clearly combine Theorem 7.1.4 for H^2-functions, with Theorems 4.2.1 and 4.4.2 for H_0^1-functions to obtain inequalities of the following form.

THEOREM 7.2.1. *Let P be a HI-potential on $(0, R)$, such that $\frac{P_r(r)}{P(r)} = \frac{\lambda}{r} + f(r)$, where $f(r) \geq 0$ and $\lim_{r \to 0} rf(r) = 0$. If $\lambda < n-2$, and B_R is a ball of radius*

R in $\mathbb{R}^n$ with $n \geq 3$. Assume the positive solution φ of equation (B_P) satisfies $R\frac{\varphi'(R)}{\varphi(R)} \geq -\frac{n}{2}$. Then the following holds for all $u \in H^2(B_R) \cap H^1_0(B_R)$,

$$(7.32) \quad \int_{B_R} |\Delta u|^2 dx \geq \frac{C(n)(n-4)^2}{4} \int_{B_R} \frac{u^2}{|x|^4} dx$$
$$+ (C(n) + \frac{(n-\lambda-2)^2}{4})\beta(P;R) \int_{B_R} \frac{P(x)}{|x|^2} u^2 dx,$$

where $C(3) = \frac{25}{36}$, $C(4) = 3$ and $C(n) = \frac{n^2}{4}$ for all $n \geq 5$.

Similarly, by combining Corollary 7.1.1 for H^2-functions, with Theorems 4.2.1 and 4.4.2 for H^1_0-functions, we obtain

COROLLARY 7.2.1. *For $u \in H^2(B_R) \cap H^1_0(B_R)$, we have*

$$(7.33) \quad \int_{B_R} |\Delta u|^2 dx \geq \frac{C(n)(n-4)^2}{4} \int_{B_R} \frac{u^2}{|x|^4} dx$$
$$+ (C(n) + \frac{(n-2)^2}{4})\frac{\mu_n^2}{R^2} \int_{B_R} \frac{u^2}{|x|^2} dx,$$

where μ_n is defined in Corollary 7.1.1.

Proof: Use first Corollary 7.1.1 for H^2-functions, then Theorem 4.2.1 with the Bessel pair $(|x|^{-2}, |x|^{-2}(\frac{(n-4)^2}{4}|x|^{-2} + \frac{z_0^2}{R^2}))$, then Theorem 4.4.2 with the Bessel pair $(1, \frac{(n-\lambda-2)^2}{4})|x|^{-2})$ to obtain

$$\int_B |\Delta u|^2 dx \geq C(n) \int_B \frac{|\nabla u|^2}{|x|^2} dx + \frac{\mu_n^2}{R^2} \int_{B_R} |\nabla u|^2 dx$$
$$\geq C(n)\frac{(n-4)^2}{4} \int_B \frac{u^2}{|x|^4} dx + C(n)\frac{z_0^2}{R^2} \int_B \frac{u^2}{|x|^2} dx + \frac{\mu_n^2}{R^2} \int_{B_R} |\nabla u|^2 dx$$
$$\geq C(n)\frac{(n-4)^2}{4} \int_B \frac{u^2}{|x|^4} dx + (C(n)z_0^2 + \frac{(n-2)^2}{4}\mu_n^2)\frac{1}{R^2} \int_B \frac{u^2}{|x|^2} dx.$$

COROLLARY 7.2.2. *Let $P_1(x)$ and $P_2(x)$ be two radial HI-potentials on a ball B of radius R in $\mathbb{R}^n$ with $n \geq 4$. Then, for all $u \in H^2(B) \cap H^1_0(B)$*

$$\int_B |\Delta u|^2 dx \geq \frac{C(n)(n-4)^2}{4} \int_B \frac{u^2}{|x|^4} dx + C(n)\beta(P_1;R) \int_B P_1(x)\frac{u^2}{|x|^2} dx$$
$$+ \frac{\mu_n^2}{R^2}(\frac{n-2}{2})^2 \int_B \frac{u^2}{|x|^2} dx + \frac{\mu_n^2}{R^2}\beta(P_2;R) \int_B P_2(x) u^2 dx.$$

Proof: We again first use Corollary 7.1.1, then Theorem 4.4.2 with the Bessel pair $(|x|^{-2}, |x|^{-2}(\frac{(n-4)^2}{4}|x|^{-2} + P))$, then again Theorem 4.4.2 with the Bessel pair

$(1, (\frac{n-2}{2})^2|x|^{-2} + P)$ to obtain

$$\begin{aligned}
\int_B |\Delta u|^2 dx &\geq C(n) \int_B \frac{|\nabla u|^2}{|x|^2} dx + \frac{\mu_n^2}{R^2} \int_B |\nabla u|^2 dx \\
&\geq \frac{C(n)(n-4)^2}{4} \int_B \frac{u^2}{|x|^4} dx + C(n)\beta(P_1; R) \int_B P_1(x) \frac{u^2}{|x|^2} dx \\
&\quad + \frac{\mu_n^2}{R^2} \int_B |\nabla u|^2 dx \\
&\geq \frac{C(n)(n-4)^2}{4} \int_B \frac{u^2}{|x|^4} dx + C(n)\beta(P_1; R) \int_B P_1(x) \frac{u^2}{|x|^2} dx \\
&\quad + \frac{\mu_n^2}{R^2}(\frac{n-2}{2})^2 \int_B \frac{u^2}{|x|^2} dx + \frac{\mu_n^2}{R^2} \beta(P_2; R) \int_B P_2(x) u^2 dx.
\end{aligned}$$

The following theorem is a counterpart of Theorem 6.3.4 for $H^2(\Omega) \cap H_0^1(\Omega)$.

THEOREM 7.2.2. *Let Ω be a smooth domain in R^n with $n \geq 1$ and let $V \in C^2(0, R)$ ($R =: \sup_{x \in \Omega} |x|$) be a non-negative function that satisfies the following conditions:*

(7.34) $\quad V_r(r) \leq 0 \quad and \quad \int_0^R \frac{1}{r^{n-3}V(r)} dr = -\int_0^R \frac{1}{r^{n-4}V_r(r)} dr = +\infty.$

There exist $\lambda_1, \lambda_2 \in R$ such that

(7.35) $\quad \frac{rV_r(r)}{V(r)} + \lambda_1 \geq 0 \ on \ (0, R) \ and \ \lim_{r \to 0} \frac{rV_r(r)}{V(r)} + \lambda_1 = 0,$

(7.36) $\quad \frac{rV_{rr}(r)}{V_r(r)} + \lambda_2 \geq 0 \ on \ (0, R) \ and \ \lim_{r \to 0} \frac{rV_{rr}(r)}{V_r(r)} + \lambda_2 = 0,$

and

(7.37) $\quad [\frac{1}{2}(n-\lambda_1-2)^2 + 3(n-3)] V(r) - (n-5)rV_r(r) - r^2 V_{rr}(r) \geq 0$ on $(0, R)$.

If $\lambda_1 \leq n$, then the following inequality holds:

(7.38)
$$\int_\Omega V(|x|)|\Delta u|^2 dx \geq \frac{1}{16}[(n-\lambda_1-2)^2 + 4(n-1)](n-\lambda_1-4)^2 \int_\Omega \frac{V(|x|)}{|x|^4} u^2 dx \\
- \frac{(n-1)(n-\lambda_2-2)^2}{4} \int_\Omega \frac{V_r(|x|)}{|x|^3} u^2 dx.$$

Proof: By Theorem 4.4.2 and condition (7.37), we can estimate $H := \frac{1}{n\omega_n} \int_{\mathbb{R}^n} V(x)|\Delta u_k|^2 dx$ as follows:

$$\begin{aligned}
H &= \int_0^R V(r) f_k''(r)^2 r^{n-1} dr + (n-1+2c_k) \int_0^R V(r) f_k'(r)^2 r^{n-3} dr \\
&+ (2c_k(n-4) + c_k^2) \int_0^R V(r) r^{n-5} f_k^2(r) dr - (n-1) \int_0^R V_r(r) r^{n-2} (f_k')^2(r) dr \\
&- c_k(n-5) \int_0^R V_r(r) f_k^2(r) r^{n-4} dr - c_k \int_0^R V_{rr}(r) f_k^2(r) r^{n-3} dr \\
&+ (n-1) V(R) f_k'(R)^2 R^{n-2} \\
&\geq \int_0^R V(r) f_k''(r)^2 r^{n-1} dr + (n-1) \int_0^R V(r) f_k'(r)^2 r^{n-3} dr \\
&- (n-1) \int_0^R V_r(r) r^{n-2} (f_k')^2(r) dr \\
&+ c_k \int_0^R \left(\left(\frac{1}{2}(n-\lambda_1-2)^2 + 3(n-3) \right) V(r) - (n-5) r V_r(r) - r^2 V_{rr}(r) \right) \\
&\times f_k^2(r) r^{n-5} dr \\
&+ (n-1) V(R) f_k'(R)^2 R^{n-2}.
\end{aligned}$$

The rest of the proof follows from the above inequality combined with Theorem 4.4.2. □

REMARK 7.2.3. Let $V(r) = r^{-2m}$ with $-\frac{n}{2} \leq m \leq \frac{n-4}{2}$. Then in order to satisfy condition (7.37) we must have $-1 - \frac{\sqrt{1+(n-1)^2}}{2} \leq m \leq \frac{n-4}{2}$. Since $-1 - \frac{\sqrt{1+(n-1)^2}}{2} \leq -\frac{n}{2}$, then if $-\frac{n}{2} \leq m \leq \frac{n-4}{2}$, inequality (7.38) gives the following weighted fourth order Rellich inequality:

$$\int_B \frac{|\Delta u|^2}{|x|^{2m}} dx \geq H_{n,m} \int_B \frac{u^2}{|x|^{2m+4}} dx \quad u \in H^2(\Omega) \cap H_0^1(\Omega),$$

where

(7.39) $$H_{n,m} := \frac{(n+2m)^2 (n-4-2m)^2}{16}.$$

One can actually show that for $-\frac{n}{2} \leq m \leq \frac{n-4}{2}$,

(7.40) $$H_{n,m} = \inf_{u \in H^2(B) \cap H_0^1(B) \setminus \{0\}} \frac{\int_B \frac{|\Delta u|^2}{|x|^{2m}}}{\int_B \frac{u^2}{|x|^{2m+4}}} = \beta_{n,m} := \inf_{u \in H_0^2(B) \setminus \{0\}} \frac{\int_B \frac{|\Delta u|^2}{|x|^{2m}}}{\int_B \frac{u^2}{|x|^{2m+4}}},$$

which was computed in the last chapter.

The following theorem includes a large class of improved Hardy-Rellich inequalities as special cases.

THEOREM 7.2.4. Let $-\frac{n}{2} \leq m \leq \frac{n-4}{2}$ and let $P(x)$ be a HI-potential on a ball B of radius R in $\mathbb{R}^n$. Assume $\frac{P(r)}{P_r(r)} = -\frac{\lambda}{r} + f(r)$, where $f(r) \geq 0$ and $\lim_{r \to 0} r f(r) = 0$. If $\lambda \leq \frac{n}{2} + m$, then the following inequality holds for all $u \in H^2(B) \cap H_0^1(B)$

$$\int_B \frac{|\Delta u|^2}{|x|^{2m}} dx \geq H_{n,m} \int_B \frac{u^2}{|x|^{2m+4}} dx$$

(7.41) $$+ \beta(P;R) \left[\frac{(n+2m)^2}{4} + \frac{(n-2m-\lambda-2)^2}{4} \right] \int_B \frac{P(x)}{|x|^{2m+2}} u^2 dx.$$

7.2. HARDY-RELLICH INEQUALITIES ON $H^2(\Omega) \cap H^1_0(\Omega)$

Moreover, both constants $H_{n,m}$ and $\beta(P,R)$ are optimal.

Proof: Again we will use repeatedly inequality (7.18). Decomposing $u \in C^\infty(\bar{B}_R)$ into spherical harmonics $\Sigma_{k=0}^\infty u_k$, where $u_k(x) = f_k(|x|)\varphi_k(x)$, we can write

$$\frac{1}{n\omega_n}\int_{\mathbb{R}^n}\frac{|\Delta u_k|^2}{|x|^{2m}}dx$$
$$= \int_0^R r^{n-2m-1}f_k''(r)^2 dr + [(n-1)(2m+1)+2c_k]\int_0^R r^{n-2m-3}(f_k')^2 dr$$
$$+ c_k[c_k + (n-2m-4)(2m+2)]\int_0^R r^{n-2m-5}f_k(r)^2 dr$$
$$+ (n-1)f_k'(R)^2 R^{n-2m-2}$$
$$\geq (\frac{n+2m}{2})^2\int_0^R r^{n-2m-3}(f_k')^2 dr + \beta(P;R)\int_0^R r^{n-2m-1}P(x)(f_k')^2 dr$$
$$+ c_k[c_k + 2(\frac{n-\lambda-4}{2})^2 + (n-2m-4)(2m+2)]\int_0^R r^{n-2m-5}f_k(r)^2 dr$$
$$+ (n-1)f_k'(R)^2 R^{n-2m-2},$$

where we have used the fact that $c_k \geq 0$ to get the above inequality. We have by Theorem 4.4.2

$$\frac{1}{n\omega_n}\int_{\mathbb{R}^n}\frac{|\Delta u_k|^2}{|x|^{2m}}dx$$
$$\geq \beta_{n,m}\int_0^R r^{n-2m-5}f_k^2 dr$$
$$+ \beta(P;R)\frac{(n+2m)^2}{4}\int_0^R r^{n-2m-3}P(x)f_k^2 dr$$
$$+ \beta(P;R)\int_0^R r^{n-2m-1}P(x)(f_k')^2 dr$$
$$\geq \beta_{n,m}\int_0^R r^{n-2m-5}f_k^2 dr$$
$$+ \beta(P;R)(\frac{(n+2m)^2}{4} + \frac{(n-2m-\lambda-2)^2}{4})\int_0^R r^{n-2m-3}P(x)f_k^2 dr$$
$$\geq \frac{\beta_{n,m}}{n\omega_n}\int_B \frac{u_k^2}{|x|^{2m+4}}dx$$
$$+ \frac{\beta(P;R)}{n\omega_n}(\frac{(n+2m)^2}{4} + \frac{(n-2m-\lambda-2)^2}{4})\int_B \frac{P(x)}{|x|^{2m+2}}u_k^2 dx.$$

Hence, (7.41) holds and the proof is complete. $\square$

The following is immediate from Theorem 7.2.4 and from the fact that $\lambda = 2$ for the HI-potential under consideration.

COROLLARY 7.2.3. *Let Ω be a smooth bounded domain in $\mathbb{R}^n$, $n \geq 4$ and $R = \sup_{x \in \Omega} |x|$. Then the following holds for all $u \in H^2(\Omega) \cap H^1_0(\Omega)$*

(1) Choose $\rho \geq R(e^{e^{e^{\cdot^{\cdot^{\cdot^{e}}}}}}$ $e(k-times)$) so that $R\frac{\varphi'(R)}{\varphi(R)} \geq -\frac{n}{2}$. Then

$$\int_\Omega |\Delta u(x)|^2 dx \geq \frac{n^2(n-4)^2}{16} \int_\Omega \frac{u^2}{|x|^4} dx$$

$$+ (1 + \frac{n(n-4)}{8}) \sum_{j=1}^{k} \int_\Omega \frac{u^2}{|x|^4} (\prod_{i=1}^{j} \log^{(i)} \frac{\rho}{|x|})^{-2} dx.$$

(2) Let X_i be defined as in Chapter 1, then

$$\int_\Omega |\Delta u(x)|^2 dx \geq \frac{n^2(n-4)^2}{16} \int_\Omega \frac{u^2}{|x|^4} dx$$

$$+ (1 + \frac{n(n-4)}{8}) \sum_{i=1}^{n} \int_\Omega \frac{u^2}{|x|^4} X_1^2(\frac{|x|}{R}) X_2^2(\frac{|x|}{R}) ... X_i^2(\frac{|x|}{R}) dx.$$

Moreover, all constants in the above inequalities are best constants.

THEOREM 7.2.5. Assume $n \geq 4$ and let P be a HI-potential on a ball B of radius R in $\mathbb{R}^n$. Then the following holds for all $u \in H^2(B) \cap H_0^1(B)$:

(7.42) $$\int_B |\Delta u|^2 dx \geq \frac{n^2(n-4)^2}{16} \int_B \frac{u^2}{|x|^4} dx$$

$$+ \beta(P, R) \frac{n^2}{4} \int_B \frac{P(x)}{|x|^2} u^2 dx + \frac{\mu_n^2}{R^2} \|u\|_{H_0^1},$$

where μ_n is defined in Corollary 7.1.1.

Proof: Decomposing again $u \in C^\infty(\bar{B}_R)$ into its spherical harmonics $\Sigma_{k=0}^\infty u_k$ where $u_k(x) = f_k(|x|)\varphi_k(x)$, we calculate

$$\frac{1}{n\omega_n} \int_{\mathbb{R}^n} |\Delta u_k|^2 dx = \int_0^R r^{n-1} f_k''(r)^2 dr + [n-1+2c_k] \int_0^R r^{n-3} (f_k')^2 dr$$

$$+ c_k[c_k + n - 4] \int_0^R r^{n-5} f_k(r)^2 dr$$

$$+ (n-1) f_k'(R)^2 R^{n-2m-2}$$

$$\geq \frac{n^2}{4} \int_0^R r^{n-3} (f_k')^2 dr + \frac{\mu_n^2}{R^2} \int_0^R r^{n-1} (f_k')^2 dr$$

$$+ c_k \int_0^R r^{n-3} (f_k')^2 dr$$

$$\geq \frac{n^2(n-4)^2}{16} \int_0^R r^{n-5} f_k^2 dr$$

$$+ \beta(P; R) \frac{n^2}{4} \int_0^R P(r) r^{n-3} f_k^2 dr$$

$$+ \frac{\mu_n^2}{R^2} \int_0^R r^{n-1} (f_k')^2 dr + c_k \frac{\mu^2}{R^2} \int_0^R r^{n-3} f_k^2 dr$$

$$= \frac{n^2(n-4)^2}{16 n\omega_n} \int_{\mathbb{R}^n} \frac{u_k^2}{|x|^{2m+4}} dx$$

$$+ \frac{\beta(P; R)}{n\omega_n} \frac{n^2}{4} \int_{\mathbb{R}^n} \frac{P(x)}{|x|^2} u_k^2 dx + \frac{\mu_n^2}{n\omega_n R^2} \|u_k\|_{P_0^{1,2}}.$$

Hence (7.42) holds. □

7.3. Further comments

The classical Hardy-Rellich inequality (6.48), which was first proved by Rellich [**242**] for functions u in $H_0^2(\Omega)$ was extended to functions in $H^2(\Omega) \cap H_0^1(\Omega)$ by J.W. Dold, V.A. Galaktionov, A.A. Lacey and J.L. Vazquez in [**121**]. Motivated by 4th-order nonlinear eigenvalue problems with Navier boundary conditions (see next chapter), Moradifam extended in [**223**] the approach of Ghoussoub-Moradifam in [**163**] in the case of $H_0^2(\Omega)$ to provide a unified approach and obtain various weighted and improved Hardy-Rellich inequalities for $H^2(\Omega) \cap H_0^1(\Omega)$. Bessel pairs are again at the heart of these developments.

CHAPTER 8

Critical Dimensions for 4th Order Nonlinear Eigenvalue Problems

This chapter contains applications of various improved and weighted Hardy-Rellich inequalities to fourth order nonlinear elliptic eigenvalue problems of the form $\Delta^2 u = \lambda f(u)$ on Ω, under the Navier boundary condition $u = \Delta u = 0$ on $\partial\Omega$, or its Dirichlet counterpart $u = \partial_\nu u = 0$ on $\partial\Omega$. Here again $\lambda \geq 0$ is a parameter, Ω is a bounded domain in $\mathbb{R}^n$, $n \geq 2$, and ∂_ν denotes the normal derivative on $\partial\Omega$. The nonlinearity f can either be as in the Gelfand problem $f(u) = e^u$ or as in the MEMS model $f(u) = (1-u)^{-p}$ with $p > 0$. Recent results –not included here– show that the extremal solutions for these equations are regular as long as we are in low dimensions. New weighted Hardy-Rellich inequalities both on H_0^2 and on $H^2 \cap H_0^1$ are used to show that the extremals are however singular in high dimensions.

8.1. Fourth order nonlinear eigenvalue problems

There are two fourth order extensions of the nonlinear eigenvalue problem (Q_λ) studied in Chapter 4, namely,

$$(N_\lambda) \quad \begin{cases} \Delta^2 u = \lambda f(u) & \text{in } \Omega \\ u = \Delta u = 0 & \text{on } \partial\Omega, \end{cases}$$

and its Dirichlet counterpart

$$(D_\lambda) \quad \begin{cases} \Delta^2 u = \lambda f(u) & \text{in } \Omega \\ u = \partial_\nu u = 0 & \text{on } \partial\Omega, \end{cases}$$

where again $\lambda \geq 0$ is a parameter, Ω is a bounded domain in $\mathbb{R}^n$, $n \geq 2$, and where ∂_ν denote the normal derivative on $\partial\Omega$. The nonlinearity f may again satisfy one of the conditions (R) or (S) considered in Chapter 4.

Problem (Q_λ) is heavily dependent on the maximum principle, which exposes a major hurdle in the study of (D_λ) since for general domains there is no maximum principle for Δ^2 with Dirichlet boundary conditions. But if we restrict our attention to the unit ball then one does have a weak form of the maximum principle, the Boggio principle, which will be sufficient for our purpose.

Problem (N_λ) with Navier boundary conditions does not suffer from the lack of a maximum principle regardless of the domain Ω, and the structure of the corresponding bifurcation diagram can be analyzed in general domains.

DEFINITION 8.1.1. Given a smooth solution u of (N_λ) (resp., (D_λ)), we say that u is a *semi-stable solution* if

$$(8.1) \quad \int_\Omega \lambda f'(u)\psi^2 dx \leq \int_\Omega (\Delta\psi)^2 dx \quad \forall \psi \in H^2(\Omega) \cap H_0^1(\Omega) \quad (\text{resp.} \forall \psi \in H_0^2(\Omega)).$$

DEFINITION 8.1.2. A smooth solution u of (N_λ) (resp., (D_λ)) is said to be *minimal* provided $u \leq v$ a.e. in Ω for any solution v of (N_λ) (resp., (D_λ)).

The *extremal parameter* is defined as

$$\lambda^* := \sup\{\lambda > 0 : \text{there exists a smooth solution of } (N_\lambda) \text{ (resp., } (D_\lambda))\}.$$

The following result has been proved in various levels of generality by several authors. For details we refer for example to [24, 54, 83, 112, 130, 174].

THEOREM 8.1.3. *Let Ω be any bounded smooth domain in $\mathbb{R}^n$ in the case of (N_λ), and assume Ω is the unit ball in the case of (D_λ). Assume the nonlinearity f satisfies either one of conditions (R) or (S). The following assertions then hold:*

(1) $0 < \lambda^* < \infty$.
(2) *For each $0 < \lambda < \lambda^*$, there exists a smooth minimal solution u_λ of (N_λ) (resp., (D_λ)). Moreover the minimal solution u_λ is semi-stable.*
(3) *For each $x \in \Omega$, $\lambda \mapsto u_\lambda(x)$ is strictly increasing on $(0, \lambda^*)$, and it therefore makes sense to define $u^*(x) := \lim_{\lambda \nearrow \lambda^*} u_\lambda(x)$, which is called the extremal solution.*
(4) *There are no solutions for $\lambda > \lambda^*$.*

Our main interest is in the regularity of the extremal solution u^* associated with (N_λ) (resp., (D_λ)). It is standard to show that u^* is a "weak solution" of (N_{λ^*}) (resp., (D_λ)) in a suitable sense that we shall not define here since it will not be needed in the sequel. One can then proceed to show that u^* it is the unique weak solution in a fairly broad class of solutions. Regularity results on u^* translate into regularity properties for any weak semi-stable solution. Indeed, weak semi-stable solution is either the classical solution u_λ or the extremal solution u^*. Our preference for not stating the results in this generality is to avoid the technical details of defining precisely what we mean by a suitable weak solution.

8.2. A Dirichlet boundary value problem with an exponential nonlinearity

Consider the fourth order elliptic problem

(8.2) $$\begin{cases} \Delta^2 u = \lambda e^u & \text{in } B \\ u = \partial_\nu u = 0 & \text{on } \partial B, \end{cases}$$

where B is the unit ball in $\mathbb{R}^n$, $n \geq 1$, ν is the exterior unit normal vector and $\lambda \geq 0$ is a parameter. We shall consider the question whether u^* is regular (i.e., $u^* \in L^\infty(B)$) or singular (i.e., $u^* \notin L^\infty(B)$).

It was established by Davila et al. [112] that the extremal solution u^* is regular provided the dimension $n \leq 12$. Our goal here is to use improved Hardy-Rellich type inequalities to show that u^* is singular for $n \geq 13$. Now as in the second order case one can establish that the only singular, semi-stable weak solution for (N_λ) must occur at $\lambda = \lambda^*$ and must be the extremal solution u^*. One is therefore tempted to consider the natural candidate $u = -4\log(|x|)$, which is clearly a singular weak solution for $\lambda = \lambda(n) := 8(n-2)(n-4)$. By the Hardy-Rellich inequality, u is semi-stable provided $n \geq 13$, since then we have for all $\psi \in H_0^2(\Omega)$,

(8.3) $$\lambda(n) \int_\Omega e^u \psi^2 \, dx \leq \frac{n^2(n-4)^2}{16} \int_\Omega \frac{\psi^2}{|x|^4} \, dx \leq \int_\Omega (\Delta \psi)^2 \, dx.$$

The problem here is that $u = -4\log(|x|)$ does not satisfy the boundary conditions, and one would like to perturb it in such a way that it does, by considering functions of the form,

$$u(x) = -4\log(|x|) - \frac{4}{m} + \frac{4}{m}|x|^m, \quad (8.4)$$

where $m > 0$. But now one should settle for testing stability on sub-solutions instead. Moreover, the Hardy-Rellich (8.3) will have to be replaced by a more subtle one. The following lemma is crucial to prove singularity of the extremal solution of the equation (8.2), a proof of which may be found in [**222**].

LEMMA 8.2.1. *Suppose there exists a radial function $u \in H^2(B) \cap W^{4,\infty}_{loc}(B\setminus\{0\})$ with $u \notin L^\infty(B)$, such that for some $\beta > \lambda' > 0$, we have*

$$(8.5) \quad \begin{cases} \Delta^2 u \leq \lambda' e^u & \text{for } 0 < r < 1, \\ u(1) = 0, \ u'(1) = 0, \end{cases}$$

and

$$(8.6) \quad \beta \int_B e^u \varphi^2 \, dx \leq \int_B (\Delta \varphi)^2 \, dx \quad \text{for all } \varphi \in C_0^\infty(B).$$

Then the corresponding extremal solution u^* is singular.

In order to apply the above lemma we will need the following improvement of the Hardy-Rellich inequality.

THEOREM 8.2.2. *Let $n \geq 5$ and let B denote the unit ball in $\mathbb{R}^n$. Then the following improved Hardy-Rellich inequality holds for all $u \in C_0^\infty(B)$,*

$$(8.7) \quad \int_B |\Delta u|^2 \, dx \geq \frac{(n-2)^2(n-4)^2}{16} \int_B \frac{u^2 \, dx}{(|x|^2 - \frac{9}{10}|x|^{\frac{n}{2}+1})(|x|^2 - |x|^{\frac{n}{2}})}$$

$$+ \frac{(n-1)(n-4)^2}{4} \int_B \frac{u^2 \, dx}{|x|^2(|x|^2 - |x|^{\frac{n}{2}})}.$$

In particular, the following inequality holds:

$$(8.8) \quad \int_B |\Delta u|^2 \, dx \geq \frac{n^2(n-4)^2}{16} \int_B \frac{u^2 \, dx}{|x|^2(|x|^2 - |x|^{\frac{n}{2}})}.$$

Proof: Let $\varphi := r^{-\frac{n}{2}+1} - \frac{9}{10}$. Since

$$-\frac{\varphi'' + \frac{(n-1)}{r}\varphi'}{\varphi} = \frac{(n-2)^2}{4} \cdot \frac{1}{r^2 - \frac{9}{10}r^{\frac{n}{2}+1}},$$

The couple $(1, \frac{(n-2)^2}{4} \frac{1}{r^2 - \frac{9}{10}r^{\frac{n}{2}+1}})$ is therefore a Bessel pair on $(0,1)$. By Theorem 6.1.1 the following inequality holds for all $u \in C_0^\infty(B)$.

$$(8.9) \quad \int_B |\Delta u|^2 dx \geq \frac{(n-2)^2}{4} \int_B \frac{|\nabla u|^2 \, dx}{|x|^2 - \frac{9}{10}|x|^{\frac{n}{2}+1}} + (n-1) \int_B \frac{|\nabla u|^2 \, dx}{|x|^2}.$$

Let $V(r) := \frac{1}{r^2 - \frac{9}{10}r^{\frac{n}{2}+1}}$, and note that

$$(8.10) \quad \frac{V_r}{V} = -\frac{2}{r} + \frac{9}{10}(\frac{n-2}{2})\frac{r^{\frac{n}{2}-2}}{1 - \frac{9}{10}r^{\frac{n}{2}-1}} \geq -\frac{2}{r},$$

which yields that $\psi(r) = r^{-\frac{n}{2}+2} - 1$ is a positive super-solution for the ODE

$$(8.11) \quad y'' + (\frac{n-1}{r} + \frac{V_r}{V})y'(r) + \frac{W_1(r)}{V(r)}y = 0,$$

where
$$W_1(r) = \frac{(n-4)^2}{4(r^2 - r^{\frac{n}{2}})(r^2 - \frac{9}{10}r^{\frac{n}{2}+1})}.$$

Hence the ODE (8.11) has a positive solution, which means that (V, W_1) is a Bessel pair on $(0, 1)$. It then follows from Theorem 6.1.1 that

(8.12) $$\int_B \frac{|\nabla u|^2 \, dx}{|x|^2 - \frac{9}{10}|x|^{\frac{n}{2}+1}} \geq (\frac{n-4}{2})^2 \int_B \frac{u^2 \, dx}{(|x|^2 - \frac{9}{10}|x|^{\frac{n}{2}+1})(|x|^2 - |x|^{\frac{n}{2}})}.$$

Similarly,

(8.13) $$\int_B \frac{|\nabla u|^2 \, dx}{|x|^2} \geq (\frac{n-4}{2})^2 \int_B \frac{u^2 \, dx}{|x|^2(|x|^2 - |x|^{\frac{n}{2}})}.$$

Combine the above two inequalities with (8.9) to get (8.7). □

THEOREM 8.2.3. *The following estimates hold:*
  (1) *If $n \geq 32$, then Lemma 8.2.1 holds with $u := w_2$, $\lambda'_n = 8(n-2)(n-4)e^2$ and $\beta = H_n := \frac{n^2(n-4)^2}{16} > \lambda'_n$.*
  (2) *If $13 \leq n \leq 31$, then Lemma 8.2.1 holds with $u := w_{3.5}$ and $\lambda'_n < \beta_n$ given in Table 1.*

The extremal solution is therefore singular for dimensions $n \geq 13$.

Proof: 1) Assume first that $n \geq 32$, then
$$8(n-2)(n-4)e^2 < \frac{n^2(n-4)^2}{16},$$
and
$$\Delta^2 w_2 = 8(n-2)(n-4)\frac{1}{r^4} \leq 8(n-2)(n-4)e^2 e^{w_2}.$$

Moreover,
$$8(n-2)(n-4)e^2 \int_B e^{w_2} \varphi^2 \leq H_n \int_B e^{-4\log(|x|)} \varphi^2 = H_n \int_B \frac{\varphi^2}{|x|^2} \leq \int_B |\Delta \varphi|^2.$$

Thus it follows from Lemma 8.2.1 that u^* is singular and $\lambda^* \leq 8(n-2)(n-4)e^2$.

2) Assume $13 \leq n \leq 31$. We shall show that $u = w_{3.5}$ satisfies the assumptions of Lemma 8.2.1 for each dimension $13 \leq n \leq 31$. Using Maple, one can verify that for each dimension $13 \leq n \leq 31$, inequality (8.5) holds for λ'_n given by Table 1. Then, by using Maple again, one easily sees that there exists $\beta_n > \lambda'_n$ such that
$$\frac{(n-2)^2(n-4)^2}{16} \frac{1}{(|x|^2 - 0.9|x|^{\frac{n}{2}+1})(|x|^2 - |x|^{\frac{n}{2}})}$$
$$+ \frac{(n-1)(n-4)^2}{4} \frac{1}{|x|^2(|x|^2 - |x|^{\frac{n}{2}})}$$
$$\geq \beta_n e^{w_{3.5}}.$$

The above inequality and the improved Hardy-Rellich inequality (8.7) guarantee that the stability condition (8.6) holds for $\beta_n > \lambda'$. Hence by Lemma 8.2.1 the extremal solution is singular for $13 \leq n \leq 31$. The values of λ_n and β_n are shown in Table 1.

REMARK 8.2.4. Note that the classical Hardy-Rellich inequality applied with $u := w_{3.5}$ yields –via Lemma 8.2.1– that u^* is singular in dimensions $n \geq 22$. For the remaining range $13 \leq n \leq 21$, one needs the improved Hardy-Rellich inequality (8.7). Note also that the values of λ'_n and β_n in Table 1 are not optimal.

Table 1. Summary

n	λ'_n	β_n
$n \geq 32$	$8(n-2)(n-4)e^2$	H_n
31	20000	86900
30	18500	76500
29	17000	67100
28	16000	58500
27	14500	50800
26	13500	43870
25	12200	37630
24	11100	32050
23	10100	27100
22	9050	22730
21	8150	18890
20	7250	15540
19	6400	12645
18	5650	10155
17	4900	8035
16	4230	6250
15	3610	4765
14	3050	3545
13	2525	2560

8.3. A Dirichlet boundary value problem with a MEMS nonlinearity

Consider now the problem

(8.14)
$$\begin{cases} \Delta^2 u = \frac{\lambda}{(1-u)^2} & \text{in } B \subseteq \mathbb{R}^n, \\ 0 < u < 1 & \text{in } B, \\ u = \partial_\nu u = 0 & \text{on } \partial B. \end{cases}$$

This equation models a simple Micro-ElectroMechanical System (MEMS) (see for example [130]). It was proved in [99] that the extremal u^* is regular provided $n \leq 8$. We are interested here in showing that u^* is singular (i.e., $\sup_B u^* = 1$) for dimensions $n \geq 9$.

One can again establish that the only singular, semi-stable weak solution for (N_λ) must occur at $\lambda = \lambda^*$ and must be the extremal solution u^*. The natural candidate $u(x) = 1 - |x|^{\frac{4}{3}}$ is a singular weak solution that corresponds to the voltage $\lambda(n) := \frac{8}{9}(n - \frac{2}{3})(n - \frac{8}{3})$. It is also semi-stable when $n \geq 9$, since then $2\lambda(n) \leq \frac{n^2(n-4)^2}{16}$ and by the Hardy-Rellich inequality, we would have for all $\psi \in H_0^2(\Omega)$,

$$\int_B |\Delta \varphi|^2 \, dx \geq H_n \int_B \frac{\varphi^2}{|x|^4} \, dx \geq \int_B \frac{2\lambda(n)}{(1-u)^3} \varphi^2 \, dx.$$

Again, the problem here is that $u(x) = 1 - |x|^{\frac{4}{3}}$ does not satisfy the boundary conditions, and one would like to perturb it in such a way that it does, by considering for $m > 0$,

(8.15) $$w_m(x) := 1 - \frac{3m}{3m-4}|x|^{4/3} + \frac{4}{3m-4}|x|^m,$$

which satisfies the right boundary conditions: $w_m(1) = w'_m(1) = 0$. Analogous to the exponential case, we have the following lemma, which allows us to consider sub-solutions and to use improved Hardy-Rellich inequalities. A proof can be found in [99] or [130].

LEMMA 8.3.1. *Suppose there exists a singular radial function* $w \in H^2(B)$ *with* $\frac{1}{1-w} \in L^\infty_{loc}(\bar{B} \setminus \{0\})$ *such that for some* $\beta > \lambda' > 0$, *we have*

(8.16) $$\begin{cases} \Delta^2 w \leq \frac{\lambda'}{(1-w)^2} & \text{for } 0 < r < 1, \\ w(1) = 0, \ w'(1) = 0, \end{cases}$$

and

(8.17) $$2\beta \int_B \frac{\varphi^2}{(1-w)^3} dx \leq \int_B (\Delta\varphi)^2 \, dx \quad \text{for all } \varphi \in H^2_0(B).$$

Then, the corresponding extremal solution u^* *is necessarily singular.*

In order to show that the extremal solution is singular for $n \geq 9$, we need to distinguish between three different ranges for the dimension. For each range, we will need a suitable Hardy-Rellich type inequality. The range where $n \geq 17$ will only require the classical inequality, while the cases $10 \leq n \leq 16$ and $n = 9$ will require the two improved ones that we now establish by using Theorem 6.1.1.

COROLLARY 8.3.1. *Let* $n \geq 5$ *and* B *be the unit ball in* $\mathbb{R}^n$. *Then the following improved Hardy-Rellich inequality holds for all* $\varphi \in C_0^\infty(B)$:

(8.18) $$\int_B (\Delta\varphi)^2 \, dx \geq \frac{(n-2)^2(n-4)^2}{16} \int_B \frac{\varphi^2}{(|x|^2 - |x|^{\frac{n}{2}+1})(|x|^2 - |x|^{\frac{n}{2}})} dx$$
$$+ \frac{(n-1)(n-4)^2}{4} \int_B \frac{\varphi^2}{|x|^2(|x|^2 - |x|^{\frac{n}{2}})} dx.$$

Proof: Let $0 < \alpha < 1$ and define $y(r) := r^{-\frac{n}{2}+1} - \alpha$. Since

$$-\frac{y'' + \frac{(n-1)}{r}y'}{y} = \frac{(n-2)^2}{4} \frac{1}{r^2 - \alpha r^{\frac{n}{2}+1}},$$

the couple $\left(1, \frac{(n-2)^2}{4} \frac{1}{r^2 - \alpha r^{\frac{n}{2}+1}}\right)$ is a Bessel pair on $(0, 1)$. By Theorem 6.1.1(4) the following inequality then holds for all $\varphi \in C_0^\infty(B)$,

(8.19) $$\int_B (\Delta\varphi)^2 dx \geq \frac{(n-2)^2}{4} \int_B \frac{|\nabla\varphi|^2}{|x|^2 - \alpha|x|^{\frac{n}{2}+1}} dx + (n-1) \int_B \frac{|\nabla\varphi|^2}{|x|^2} dx.$$

Set $V(r) := \frac{1}{r^2 - \alpha r^{\frac{n}{2}+1}}$ and note that

$$\frac{V_r}{V} = -\frac{2}{r} + \frac{\alpha(n-2)}{2} \frac{r^{\frac{n}{2}-2}}{1 - \alpha r^{\frac{n}{2}-1}} \geq -\frac{2}{r}.$$

The function $y(r) = r^{-\frac{n}{2}+2} - 1$ is decreasing and is then a positive super-solution on $(0, 1)$ for the ODE

$$y'' + \left(\frac{n-1}{r} + \frac{V_r}{V}\right)y'(r) + \frac{W_1(r)}{V(r)} y = 0,$$

8.3. A DIRICHLET BOUNDARY VALUE PROBLEM WITH A MEMS NONLINEARITY

where
$$W_1(r) = \frac{(n-4)^2}{4(r^2 - r^{\frac{n}{2}})(r^2 - \alpha r^{\frac{n}{2}+1})}.$$

Hence, by Theorem 6.1.1(2) we deduce that for all $\varphi \in C_0^\infty(B)$,

$$\int_B \frac{|\nabla \varphi|^2}{|x|^2 - \alpha|x|^{\frac{n}{2}+1}} \, dx \geq \left(\frac{n-4}{2}\right)^2 \int_B \frac{\varphi^2}{(|x|^2 - \alpha|x|^{\frac{n}{2}+1})(|x|^2 - |x|^{\frac{n}{2}})} \, dx.$$

Similarly, for $V(r) = \frac{1}{r^2}$ we have that for all $\varphi \in C_0^\infty(B)$,

$$\int_B \frac{|\nabla \varphi|^2}{|x|^2} \, dx \geq \left(\frac{n-4}{2}\right)^2 \int_B \frac{\varphi^2}{|x|^2(|x|^2 - |x|^{\frac{n}{2}})} \, dx.$$

Combining the above two inequalities with (8.19) and letting $\alpha \to 1$ we obtain (8.18). □

COROLLARY 8.3.2. *Let $n = 9$ and consider $\varphi(r) := r^{-\frac{n}{2}+1} + r - 1.9$ and $\psi(r) := r^{-\frac{n}{2}+2} + 20r^{-1.69} + 10r^{-1} + 10r + 7r^2 - 48$. Then, the following inequality holds for all $\varphi \in C_0^\infty(B)$:*

(8.20) $$\int_B (\Delta \varphi)^2 \, dx \geq \int_B Q(|x|) \left(P(|x|) + \frac{n-1}{|x|^2} \right) \varphi^2 \, dx,$$

where

$$P(r) := -\frac{\varphi''(r) + \frac{n-1}{r}\varphi'(r)}{\varphi(r)} \quad \text{and} \quad Q(r) := -\frac{\psi''(r) + \frac{n-3}{r}\psi'(r)}{\psi(r)}.$$

Proof: By definition the couple of functions $(1, P(r))$ is a Bessel pair on $(0,1)$. One can easily see that $P(r) \geq \frac{2}{r^2}$. Hence, by Theorem 6.1.1(4) the following inequality holds for all $\varphi \in C_0^\infty(B)$,

(8.21) $$\int_B (\Delta \varphi)^2 dx \geq \int_B P(|x|)|\nabla \varphi|^2 + (n-1) \int_B \frac{|\nabla \varphi|^2}{|x|^2} \, dx.$$

Using Maple, it is easy to see that
$$\frac{P_r}{P} \geq -\frac{2}{r} \quad \text{in} \quad (0,1),$$

and therefore $\psi(r)$ is a positive super-solution for the ODE
$$y'' + \left(\frac{n-1}{r} + \frac{P_r(r)}{P(r)}\right) y'(r) + \frac{P(r)Q(r)}{P(r)} y = 0 \text{ on } (0,1).$$

Hence, by Theorem 6.1.1(2) we have for all $\varphi \in C_0^\infty(B)$
$$\int_B P(|x|)|\nabla \varphi|^2 \, dx \geq \int_B P(|x|)Q(|x|)\varphi^2 \, dx,$$

and similarly
$$\int_B \frac{|\nabla \varphi|^2}{|x|^2} \, dx \geq \int_B \frac{Q(|x|)}{|x|^2} \varphi^2 \, dx,$$

since $\psi(r)$ is a positive solution for the ODE
$$y'' + \frac{n-3}{r} y'(r) + Q(r)y = 0.$$

Combining the above two inequalities with (8.21) we get (8.20). □

THEOREM 8.3.1. *The following statements hold:*
(1) *If $n \geq 31$, then Lemma 8.3.1 holds with $w := w_2$, $\lambda' := 27\bar{\lambda}_n < \beta := \frac{H_n}{2}$.*

(2) If $17 \leq n \leq 30$, then Lemma 8.3.1 holds with $w := w_3$, $\lambda' := \frac{H_n}{2} - 1 < \beta := \frac{H_n}{2}$.
(3) If $10 \leq n \leq 16$, then Lemma 8.3.1 holds with $w := w_3$, $\lambda'_n < \beta_n$ given in Table 2.
(4) If $n = 9$, then Lemma 8.3.1 holds with $w := w_{2.8}$, $\lambda'_9 := 366 < \beta_9 := 368.5$.

The extremal solution is therefore singular for dimension $n \geq 9$.

TABLE 2. Summary

n	w	λ'_n	β_n
9	$w_{2.8}$	366	366.5
10	w_3	450	487
11	w_3	560	739
12	w_3	680	1071
13	w_3	802	1495
14	w_3	940	2026
15	w_3	1100	2678
16	w_3	1260	3469
$17 \leq n \leq 30$	w_3	$H_n/2 - 1$	$H_n/2$
$n \geq 31$	w_2	$27\bar{\lambda}_n$	$H_n/2$

Proof: 1) Assume first that $n \geq 31$, then $27\bar{\lambda} \leq \frac{H_n}{2}$. We shall show that w_2 is a singular $H^2(B)$–weak sub-solution of $(P)_{27\bar{\lambda}}$ so that (8.17) holds with $\beta = \frac{H_n}{2}$. Indeed, write

$$w_2 := 1 - |x|^{\frac{4}{3}} - 2(|x|^{\frac{4}{3}} - |x|^2) = \bar{u} - \varphi_0,$$

where $\varphi_0 := 2(|x|^{\frac{4}{3}} - |x|^2)$, and note that $w_2 \in H_0^2(B)$, $\frac{1}{1-w_2} \in L^3(B)$, $0 \leq w_2 \leq 1$ in B, and

$$\Delta^2 w_2 = \frac{3\bar{\lambda}}{r^{\frac{8}{3}}} \leq \frac{27\bar{\lambda}}{(1-w_2)^2} \quad \text{in } B \setminus \{0\}.$$

So w_2 is $H^2(B)$–weak sub-solution of $(P)_{27\bar{\lambda}}$. Moreover, since $\varphi_0 \geq 0$, we get from the standard Hardy-Rellich inequality that

$$H_n \int_B \frac{\varphi^2}{(1-w_2)^3}\,dx = H_n \int_B \frac{\varphi^2}{(|x|^{\frac{4}{3}} + \varphi_0)^3}\,dx \leq H_n \int_B \frac{\varphi^2}{|x|^4}\,dx \leq \int_B (\Delta\varphi)^2\,dx$$

for all $\varphi \in H_0^2(B)$. It follows from Lemma 8.3.1 that u^* is singular.

2) Assume $17 \leq n \leq 30$ and consider now the function

$$w_3 := 1 - \frac{9}{5}r^{\frac{4}{3}} + \frac{4}{5}r^3.$$

8.3. A DIRICHLET BOUNDARY VALUE PROBLEM WITH A MEMS NONLINEARITY

Note that $0 \leq w_3 \leq 1$ in B, $w_3 \in H_0^2(B)$ and $\frac{1}{1-w_3} \in L^3(B)$. Moreover, by the standard Hardy-Rellich inequality, we have for any $\varphi \in H_0^2(B)$,

$$H_n \int_B \frac{\varphi^2}{(1-w_3)^3} dx = 125 H_n \int_B \frac{\varphi^2}{(9r^{\frac{4}{3}} - 4r^3)^3} dr \leq 125 H_n \sup_{0<r<1} \frac{1}{(9-4r^{\frac{5}{3}})^3}$$

$$\times \int_B \frac{\varphi^2 \, dx}{r^4} dx$$

$$= H_n \int_B \frac{\varphi^2}{r^4} \leq \int_B (\Delta \varphi)^2 \, dx.$$

An easy computation shows that

$$\frac{H_n - 2}{2(1-w_3)^2} - \Delta^2 w_3 = \frac{25(H_n - 2)}{2(9r^{\frac{4}{3}} - 4r^3)^2} - \frac{9\bar{\lambda}}{5r^{\frac{8}{3}}} - \frac{12}{5}\frac{n^2 - 1}{r}$$

$$= \frac{25[n^2(n-4)^2 - 32]}{32(9r^{\frac{4}{3}} - 4r^3)^2} - \frac{8(n-\frac{2}{3})(n-\frac{8}{3})}{5r^{\frac{8}{3}}} - \frac{12}{5}\frac{n^2 - 1}{r}.$$

By using Maple one can verify that this final quantity is nonnegative on $(0, 1)$ whenever $17 \leq n \leq 30$, and hence w_3 is a $H^2(B)$−weak sub-solution of $(P_{\frac{H_n}{2}-1})$. It follows from Lemma 8.3.1 that u^* is singular.

3) Assume $10 \leq n \leq 30$. We shall prove that again $w := w_3$ satisfies the assumptions of Lemma 8.3.1. Indeed, using Maple, we show that for each dimension $10 \leq n \leq 16$, inequality (8.16) holds with λ'_n given by Table 2. Then, by using Maple again, we show that for each dimension $10 \leq n \leq 16$, the following inequality holds

$$\frac{(n-2)^2(n-4)^2}{16} \frac{1}{(|x|^2 - |x|^{\frac{n}{2}+1})(|x|^2 - |x|^{\frac{n}{2}})} + \frac{(n-1)(n-4)^2}{4} \frac{1}{|x|^2(|x|^2 - |x|^{\frac{n}{2}})}$$

$$\geq \frac{2\beta_n}{(1-w_3)^3}.$$

where β_n is again given by Table 1. The above inequality and the Hardy-Rellich inequality (8.8) guarantee that the stability condition (8.17) holds with $\beta := \beta_n$. Since $\beta_n > \lambda'_n$, we deduce from Lemma 8.3.1 that the extremal solution is singular for $10 \leq n \leq 16$.

4) Suppose now $n = 9$ and consider $w := w_{2.8}$. Using Maple on can see that

$$\Delta^2 w \leq \frac{366}{(1-w)^2} \quad \text{in } B,$$

and

$$\frac{723}{(1-w)^3} \leq Q(r)\left(P(r) + \frac{n-1}{r^2}\right) \quad \text{for all } r \in (0,1),$$

where P and Q are given in (8.27). Since $723 > 2 \times 366$, Lemma 8.3.1 yields that the extremal solution u^* is singular in dimension $n = 9$. $\square$

8.4. A Navier boundary value problem with a MEMS nonlinearity

Consider the problem

(8.22) $$\begin{cases} \Delta^2 u = \frac{\lambda}{(1-u)^2} & \text{in } B \subset \mathbb{R}^n, \\ 0 < u < 1 & \text{in } \Omega, \\ u = \Delta u = 0 & \text{on } \partial B, \end{cases}$$

where B is the unit ball and $\lambda > 0$ is a parameter. Similar to the corresponding problem with Dirichlet boundary condition, we shall show that the extremal solution is singular for $n \geq 9$. Again, the problem here is that $u(x) = 1 - |x|^{\frac{4}{3}}$ does not satisfy the boundary conditions, and one would like to perturb it in such a way that it does, by considering for $m > \frac{4}{3}$, the functions

(8.23) $$u_m := 1 - a_{n,m} r^{\frac{4}{3}} + b_{n,m} r^m,$$

where

$$a_{n,m} := \frac{m(n+m-2)}{m(n+m-2) - \frac{4}{3}(n-2/3)} \quad \text{and} \quad b_{n,m} := \frac{\frac{4}{3}(n-2/3)}{m(n+m-2) - \frac{4}{3}(n-2/3)}.$$

Note that u_m satisfies the right boundary conditions: $u_m(1) = \Delta u_m(1) = 0$. Analogous to the case with the Dirichlet boundary condition, one can prove the following lemma (see [**221**] and [**130**]), which allows us to consider sub-solutions and to use improved Hardy-Rellich inequalities.

LEMMA 8.4.1. *Suppose there exists a singular radial function $u \in H^2(B) \cap W_{loc}^{4,\infty}(B \setminus \{0\})$ with $\frac{1}{1-u} \in L_{loc}^\infty(\bar{B} \setminus \{0\})$ such that for some $\beta > \lambda' > 0$, we have*

(8.24) $$\begin{cases} \Delta^2 u \leq \frac{\lambda'}{(1-u)^2} & \text{for } 0 < r < 1, \\ u(1) = 0, \ \Delta u|_{r=1} = 0, \end{cases}$$

and

(8.25) $$2\beta \int_B \frac{\varphi^2}{(1-u)^3} dx \leq \int_B (\Delta \varphi)^2 dx \quad \text{for all } \varphi \in H^2(B) \cap H_0^1(B).$$

Then the corresponding extremal solution u^* is necessarily singular.

We have to distinguish again between three different ranges for the dimension. For each range, we will need a suitable Hardy-Rellich type inequality on $H^2 \cap H_0^1$.

COROLLARY 8.4.1. *Let $n \geq 5$ and B be the unit ball in $\mathbb{R}^n$. Then the following improved Hardy-Rellich inequality holds for all $\varphi \in H^2(B) \cap H_0^1(B)$:*

(8.26) $$\int_B (\Delta \varphi)^2 dx \geq \frac{(n-2)^2(n-4)^2}{16} \int_B \frac{\varphi^2 \, dx}{(|x|^2 - \frac{n}{2(n-1)}|x|^{\frac{n}{2}+1})(|x|^2 - |x|^{\frac{n}{2}})} \\ + \frac{(n-1)(n-4)^2}{4} \int_B \frac{\varphi^2 \, dx}{|x|^2(|x|^2 - |x|^{\frac{n}{2}})}.$$

Proof: Let $\alpha := \frac{n}{2(n-1)}$ and $V(r) := \frac{1}{r^2 - \alpha r^{\frac{n}{2}+1}}$ and note that

$$\frac{V_r}{V} = -\frac{2}{r} + \frac{\alpha(n-2)}{2} \frac{r^{\frac{n}{2}-2}}{1 - \alpha r^{\frac{n}{2}-1}} \geq -\frac{2}{r}.$$

8.4. A NAVIER BOUNDARY VALUE PROBLEM WITH A MEMS NONLINEARITY

The function $y(r) = r^{-\frac{n}{2}+2} - 1$ is decreasing and is then a positive super-solution on $(0,1)$ for the ODE

$$y'' + \left(\frac{n-1}{r} + \frac{V_r}{V}\right)y'(r) + \frac{W_1(r)}{V(r)}y = 0,$$

where

$$W_1(r) = \frac{(n-4)^2}{4(r^2 - r^{\frac{n}{2}})(r^2 - \alpha r^{\frac{n}{2}+1})}.$$

Hence, by Theorem 6.1.1 we deduce

$$\int_B \frac{|\nabla \varphi|^2 \, dx}{|x|^2 - \alpha |x|^{\frac{n}{2}+1}} \geq \left(\frac{n-4}{2}\right)^2 \int_B \frac{\varphi^2 \, dx}{(|x|^2 - \alpha |x|^{\frac{n}{2}+1})(|x|^2 - |x|^{\frac{n}{2}})}$$

for all $\varphi \in H^2(B) \cap H_0^1(B)$. Similarly, for $V(r) = \frac{1}{r^2}$ we have that

$$\int_B \frac{|\nabla \varphi|^2 \, dx}{|x|^2} \geq \left(\frac{n-4}{2}\right)^2 \int_B \frac{\varphi^2 \, dx}{|x|^2(|x|^2 - |x|^{\frac{n}{2}})}$$

for all $\varphi \in H^2(B) \cap H_0^1(B)$. Combining the above two inequalities with (7.24) we get (8.26).

COROLLARY 8.4.2. *Let $n \geq 7$ and B be the unit ball in $\mathbb{R}^n$. Then the following improved Hardy-Rellich inequality holds for all $\varphi \in H^2(B) \cap H_0^1(B)$:*

(8.27) $$\int_B |\Delta u|^2 \, dx \geq \int_B W(|x|) u^2 \, dx,$$

where

(8.28) $$W(r) = K(r)\left(\frac{(n-2)^2}{4(r^2 - \frac{n}{2(n-1)} r^{\frac{n}{2}+1})} + \frac{(n-1)}{r^2}\right),$$

$$K(r) = -\frac{\varphi''(r) + \frac{(n-3)}{r}\varphi'(r)}{\varphi(r)},$$

and

$$\varphi(r) = r^{-\frac{n}{2}+2} + 9r^{-2} + 10r - 20.$$

Proof: Let $\alpha := \frac{n}{2(n-1)}$ and $V(r) := \frac{1}{r^2 - \alpha r^{\frac{n}{2}+1}}$. Then φ is a sub-solution for the ODE

$$y'' + \left(\frac{n-1}{r} + \frac{V_r}{V}\right)y'(r) + \frac{W_2(r)}{V(r)}y = 0,$$

where

$$W_2(r) = \frac{K(r)}{r^2 - \alpha r^{\frac{n}{2}+1}}.$$

Hence, by Theorem 6.1.1 we have

(8.29) $$\int_B \frac{|\nabla u|^2 \, dx}{|x|^2 - \alpha |x|^{\frac{n}{2}+1}} \geq \int_B W_2(|x|) u^2 \, dx.$$

Similarly

(8.30) $$\int_B \frac{|\nabla u|^2 \, dx}{|x|^2} \geq \int_B W_3(|x|) u^2 \, dx,$$

where

$$W_3(r) = \frac{K(r)}{r^2}.$$

Combining the above two inequalities with (7.24), we get the improved Hardy-Rellich inequality (8.27).

THEOREM 8.4.2. *The following upper bounds on λ^* hold in large dimensions.*
(1) *If $n \geq 31$, then Lemma 8.4.1 holds with $u := w_2$, and $\lambda'_n := 27\bar{\lambda} < \beta = \frac{H_n}{2}$.*
(2) *If $16 \leq n \leq 30$, then Lemma 8.4.1 holds with $u := w_3$, and $\lambda'_n := \frac{H_n}{2} - 1 < \beta_n := \frac{H_n}{2}$.*
(3) *If $10 \leq n \leq 15$, then Lemma 8.4.1 holds with $u := w_3$, $\lambda'_n < \beta_n$ given in Table 3.*
(4) *If $n = 9$, then Lemma 8.4.1 holds with $u := w_{2.8}$, $\lambda'_9 := 249 < \beta_9 := 251$.*

The extremal solution is therefore singular for dimensions $n \geq 9$.

Proof: 1) Assume first that $n \geq 31$, then it is easy to see that $a_{n,2} < 3$ and $a_{n,2}^3 \bar{\lambda} \leq 27\bar{\lambda} < \frac{H_n}{2}$. We shall show that w_2 is a singular weak sub-solution of $(P_{a_{n,2}^3,\bar{\lambda}})$ which is stable. Note that $w_2 \in H^2(B)$, $\frac{1}{1-w_2} \in L^3(B)$, $0 \leq w_2 \leq 1$ in B, and
$$\Delta^2 w_2 \leq \frac{a_{n,2}^3 \bar{\lambda}}{(1-w_2)^2} \quad \text{in } B \setminus \{0\}.$$
So w_2 is a weak sub-solution of $(P_{27\bar{\lambda}})$. Moreover,
$$w_2 = 1 - |x|^{\frac{4}{3}} + (a_{n,2} - 1)(|x|^{\frac{4}{3}} - |x|^2) \leq 1 - |x|^{\frac{4}{3}}.$$
Since $27\bar{\lambda} \leq \frac{H_n}{2}$, we get that
$$54\bar{\lambda} \int_B \frac{\varphi^2 \, dx}{(1-w_2)^3} \leq H_n \int_B \frac{\varphi^2 \, dx}{(1-w_2)^3} \leq H_n \int_B \frac{\varphi^2 \, dx}{|x|^4} \leq \int_B (\Delta \varphi)^2 \, dx$$
for all $\varphi \in C_0^\infty(B)$. Hence, w_2 is stable. Thus it follows from Lemma 8.4.1 that u^* is singular.

2) Assume $16 \leq n \leq 30$ and consider
$$w_3 := 1 - a_{n,3} r^{\frac{4}{3}} + b_{n,3} r^3.$$
We show that it is a singular weak sub-solution of $(P_{\frac{H_n}{2}-1})$ which is stable. Indeed, we clearly have $0 \leq w_3 \leq 1$ a.e. in B, $w_3 \in H^2(B)$ and $\frac{1}{1-w_3} \in L^3(B)$. Note that
$$H_n \int_B \frac{\varphi^2 \, dx}{(1-w_3)^3} = H_n \int_B \frac{\varphi^2 \, dx}{(a_{n,m} r^{\frac{4}{3}} - b_{n,m} r^m)^3}$$
$$\leq \sup_{0<r<1} \frac{H_n}{(a_{n,m} - b_{n,m} r^{m-\frac{4}{3}})^3} \int_B \frac{\varphi^2 \, dx}{r^4}$$
$$= H_n \int_B \frac{\varphi^2 \, dx}{r^4} \leq \int_B (\Delta \varphi)^2 \, dx.$$

Using maple one can verify that for $16 \leq n \leq 31$
$$\Delta^2 w_3 \leq \frac{\frac{H_n}{2} - 1}{(1-w_3)^2} \quad \text{on } (0,1).$$
Hence w_3 is a sub-solution of $(P_{\frac{H_n}{2}-1})$. By Lemma 8.4.1 u^* is singular.

3) Assume $10 \leq n \leq 15$. We shall show that w_3 satisfies the assumptions of Lemma 8.4.1 for each dimension $10 \leq n \leq 15$. Using Maple, for each dimension

$10 \leq n \leq 15$, one can verify that inequality (19.3) holds for λ'_n given by Table 3. Then, by using Maple again, we show that there exists $\beta_n > \lambda'_n$ such that

$$(8.31) \quad \frac{2\beta_n}{(1-w_3)^3} \leq \frac{(n-2)^2(n-4)^2}{16} \frac{1}{(|x|^2 - \frac{n}{2(n-1)}|x|^{\frac{n}{2}+1})(|x|^2 - |x|^{\frac{n}{2}})}$$
$$+ \frac{(n-1)(n-4)^2}{4} \frac{1}{|x|^2(|x|^2 - |x|^{\frac{n}{2}})}.$$

The above inequality and improved Hardy-Rellich inequality (8.26) guarantee that the stability condition (8.6) holds for $\beta_n > \lambda'$. Hence by Lemma 8.4.1 the extremal solution is singular for $10 \leq n \leq 15$. The values of λ_n and β_n are shown in Table 3.

TABLE 3. Summary

n	λ'_n	β_n
9	249	251
10	320	367
11	405	574
12	502	851
13	610	1211
14	730	1668
15	860	2235
$16 \leq n \leq 30$	$\frac{H_n}{2} - 1$	$\frac{H_n}{2}$
$n \geq 31$	$27\bar{\lambda}$	$\frac{H_n}{2}$

4) Let u:=$w_{2.8}$. Using Maple on can see that

$$\Delta^2 u \leq \frac{249}{(1-u)^2} \quad \text{in} \quad B,$$

and

$$\frac{502}{(1-u(r))^3} \leq W(r) \quad \text{for all} \quad r \in (0,1),$$

where W is given by (8.28). Since, $502 > 2 \times 249$, Lemma 8.4.1 yields that the extremal solution u^* is singular in dimension $n = 9$.

8.5. Further comments

The Boggio principle was first established in [55]. The first (truly supercritical) results concerning the boundedness of the extremal solution in a fourth order problem are due to J. Davila, L. Dupaigne, I. Guerra and M. Montenegro [112], where they examined the problem (D_λ) on the unit ball in $\mathbb{R}^n$ with $f(t) = e^t$. They showed that the extremal solution u^* is bounded if and only if $n \leq 12$. Their approach is heavily dependent on the fact that Ω is the unit ball. Even in this situation there are two main hurdles, the first being that the standard energy estimate approach, which was so successful in the second order case, does not appear to work in the fourth order case. The second is the fact that it is quite hard to construct explicit solutions of (D_λ) on the unit ball that satisfy both boundary conditions, which is needed to show that the extremal solution is unbounded for $n \geq 13$. So one needs to find an explicit singular, semi-stable solution which satisfies the first boundary condition, and then to perturb it enough to satisfy the second boundary condition but not too much so as not to lose the semi-stability. Davila et al. [**112**] succeeded in doing so for $n \geq 32$, but they were forced to use a computer assisted proof to show that the extremal solution is unbounded for the intermediate dimensions $13 \leq n \leq 31$. Using various improved Hardy-Rellich inequalities from Ghoussoub-Moradifam [**163**] the need for the computer assisted proof was removed in Moradifam [**222**]. The case where $f(t) = (1-t)^{-2}$ was settled at the same time in Cowan-Esposito-Ghoussoub-Moradifam [**99**], who showed that the extremal solution associated with (D_λ) is a classical solution if and only if $n \leq 8$. This problem is also studied by Moradifam [**221**] under the Navier boundary condition.

Cowan-Esposito-Ghoussoub tackled in [**98**] the regularity of the extremal solution of the nonlinear eigenvalue problem $\Delta^2 u = \lambda f(u)$ on a general bounded domain Ω in $\mathbb{R}^n$ with the Navier boundary condition $u = \Delta u = 0$ on $\partial \Omega$. They establish –among other things– that the extremal solution u^* is smooth if $f(t) = e^t$ and $n \leq 8$, or if $f(t) = (1+t)^p$ and $n < \frac{8p}{p-1}$. They also show that if $f(t) = (1-t)^{-p}$, $p > 1$ and $p \neq 3$, then u^* is smooth for $n \leq \frac{8p}{p+1}$. More recently, Cowan-Ghoussoub [**102**] extended these results by showing that u^* is smooth provided

$$n < 2 + 4\sqrt{2} + 4\sqrt{2-\sqrt{2}} = 10.718... \quad \text{when } f(u) = e^u,$$

and

$$n < \frac{4p}{p-1} + \frac{4(p+1)}{p-1}\left(\sqrt{\frac{2p}{p+1}} + \sqrt{\frac{2p}{p+1} - \sqrt{\frac{2p}{p+1}}} - \frac{1}{2}\right) \quad \text{when } f(u) = (u+1)^p.$$

The expected optimal results are those obtained above in the case of radial domains, e.g., u^* is smooth for $n \leq 12$ when $f(t) = e^t$ and for $n \leq 8$ when $f(t) = (1-t)^{-2}$. It is worth noting that this latest improvement is based on the following observation, which is not unrelated to the relationship between Hardy and Hardy-Rellich inequalities: If u is a semi-stable solution of (N_1), that is if

(8.32) $\qquad \int_\Omega f'(u)\psi^2 dx \leq \int_\Omega (\Delta \psi)^2 \, dx \quad \text{for all } \psi \in H^2(\Omega) \cap H_0^1(\Omega),$

then

(8.33) $\qquad \int_\Omega \sqrt{f'(u_\lambda)}\varphi^2 \leq \int_\Omega |\nabla \varphi|^2 \, dx \quad \text{for all } \varphi \in H_0^1(\Omega).$

The lack of a maximum principle for the bi-Laplacian under Dirichlet boundary conditions contributes to keeping the following problem elusive.

Open problem (8): Develop the Dirichlet counterpart (D_λ) of (N_λ) for general bounded domains Ω in $\mathbb{R}^n$.

Part 3

Hardy Inequalities for General Elliptic Operators

CHAPTER 9

General Hardy Inequalities

We consider general Hardy inequalities of the form
$$\int_\Omega |\nabla u|_A^2 dx \geq \frac{1}{4} \int_\Omega \frac{|\nabla E|_A^2}{E^2} u^2 dx, \quad u \in H_0^1(\Omega)$$
where E is a positive function defined in Ω, and $A(x)$ is a $n \times n$ symmetric, uniformly positive definite matrix defined in Ω with $|\xi|_A^2 := \langle A(x)\xi, \xi \rangle$ for $\xi \in \mathbb{R}^n$. Our basic assumption will be that $-\text{div}(A\nabla E)\,dx$ is a nonnegative nonzero finite measure on Ω, which we shall denote by $\mu := \mu_{A,E}$.

It is shown that the above inequality is optimal in either one of the following two cases:
- E is an *interior weight*, that is $E = +\infty$ on the support of μ, or
- E is a *boundary weight*, meaning that $E = 0$ on $\partial\Omega$.

The best constant is not attained in either situation, and in the latter case, the following improvement
$$\int_\Omega |\nabla u|_A^2 dx \geq \frac{1}{4} \int_\Omega \frac{|\nabla E|_A^2}{E^2} u^2 dx + \frac{1}{2} \int_\Omega \frac{u^2}{E} d\mu, \quad u \in H_0^1(\Omega)$$
is shown to be optimal, yet still not attained. Optimal weighted versions of these inequalities are also established. Optimal analogous versions of the above inequalities are established for $p \neq 2$. Many of the Hardy inequalities obtained in the previous sections can be obtained, via the above approach, by using suitable choices for the function E and the matrix $A(x)$.

9.1. A general inequality involving interior and boundary weights

Throughout this chapter, we shall assume that Ω is a bounded connected domain in $\mathbb{R}^n$ (unless otherwise mentioned) with smooth boundary and $A(x) = (a^{i,j}(x))$ is a $n \times n$ symmetric, uniformly positive definite matrix with $a^{i,j} \in C^\infty(\overline{\Omega})$. For $\xi \in \mathbb{R}^n$, we define $|\xi|_A^2 := |\xi|_{A(x)}^2 := A(x)\xi \cdot \xi$. We start by introducing and justifying the needed concepts.

If E is a given positive C^1-function on Ω, we consider for any $u \in C_c^\infty(\Omega)$ the function $v := E^{\frac{-1}{2}}u$. A formal calculation shows the following identity
$$|\nabla u|_A^2 - \frac{|\nabla E|_A^2}{4E^2} u^2 = E|\nabla v|_A^2 + \frac{A\nabla E \cdot \nabla(v^2)}{2} \quad \text{on } \Omega,$$
which once integrated over Ω, yields
$$(9.1) \quad \int_\Omega |\nabla u|_A^2 dx - \frac{1}{4}\int_\Omega \frac{|\nabla E|_A^2}{E^2} u^2 dx = \int_\Omega E|\nabla v|_A^2 dx - \frac{1}{2}\int_\Omega \frac{u^2}{E} \text{div}(A\nabla E)dx.$$
Take now $\beta \neq 1$ and apply (9.1) with $E_0 := E^\beta$. After collecting like terms, we obtain the following basic inequality.

THEOREM 9.1.1. *Let $A(x)$ denote a uniformly positive definite $N \times N$ matrix with smooth coefficients defined on Ω. Suppose E is a smooth positive function on Ω and fix a constant β with $1 \leq \beta \leq 2$. Then, for all $\psi \in H_0^1(\Omega)$ we have*
$$(9.2) \quad \int_\Omega |\nabla \psi|_A^2 \geq \frac{\beta(2-\beta)}{4}\int_\Omega \frac{|\nabla E|_A^2}{E^2}\psi^2 + \frac{\beta}{2}\int_\Omega \frac{-\text{div}(A\nabla E)}{E}\psi^2,$$

where $\int_\Omega |\nabla \psi|_A^2 = \int_\Omega A(x) \nabla \psi \cdot \nabla \psi \, dx$.

Let us now write $\mathcal{L}_A(E) := -\text{div}(A\nabla E)$ whenever the latter is defined on a function E that is positive in Ω. If we further assume that $\mathcal{L}_A(E) := -\text{div}(A\nabla E) \geq 0$ in Ω, then by taking $\beta = 1$, we get for all $u \in H_0^1(\Omega)$,

$$(9.3) \qquad \int_\Omega |\nabla u|_A^2 \, dx \geq \frac{1}{4} \int_\Omega \frac{|\nabla E|_A^2}{E^2} u^2 \, dx.$$

From this we see that the optimal constant

$$C(E) := \inf \left\{ \frac{\int_\Omega |\nabla u|_A^2}{\int_\Omega \frac{|\nabla E|_A^2}{E^2} u^2} dx \, : \, u \in H_0^1(\Omega) \backslash \{0\} \right\} \geq \frac{1}{4}.$$

It is possible to show that for all non-zero $u \in H_0^1(\Omega)$ we have

$$\int_\Omega E |\nabla v|_A^2 \, dx > 0,$$

where v is defined as above. In view of (9.1), this means that if $C(E)$ is attained then necessarily $C(E) > \frac{1}{4}$.

We are interested here in the case where $C(E) = \frac{1}{4}$, which means $C(E)$ cannot be attained and consequently $\frac{|\nabla E|_A^2}{E^2}$ needs to be singular. Indeed, if $\frac{|\nabla E|_A^2}{E^2} \in L^p(\Omega)$ for some $p > \frac{n}{2}$, then $H_0^1(\Omega)$ is compactly embedded in $L^2(\Omega, \frac{|\nabla E|_A^2}{E^2} dx)$ and one could then apply standard compactness arguments to show that $C(E)$ is attained.

There are two obvious ways to ensure that $\frac{|\nabla E|_A^2}{E^2}$ is singular, which naturally lead to considering the following two classes of functions E (weights). But first we recall the following notion.

DEFINITION 9.1.2. The *box-counting dimension (or entropy dimension)* of a compact subset K of $\mathbb{R}^n$ is defined as

$$dim_{box}(K) := n - \lim_{r \searrow 0} \frac{\log(\mathcal{H}^n(K_r))}{\log(r)},$$

provided this limit exists. Here $K_r := \{x \in \Omega : \text{dist}(x, K) < r\}$ and $\mathcal{H}^\alpha$ denotes the α-dimensional Hausdorff measure.

DEFINITION 9.1.3. Suppose $E > 0$ in Ω and assume that $\mu := \mathcal{L}_A(E) \, dx$ is a nonnegative nonzero finite measure on $\mathbb{R}^n$ with support $K \subset \Omega$. Say that
 (1) E is a *boundary weight* on Ω, if $E \in H_0^1(\Omega)$.
 (2) E is an *interior weight* on Ω, if $E \in C^\infty(\overline{\Omega} \backslash K)$, $E = +\infty$ on K and $dim_{box}(K) < n - 2$.

Let Ω be a domain in $\mathbb{R}^n$. We leave it as an exercise to verify that:
 - $C_c^{0,1}(\Omega \backslash K)$ is dense in $W_0^{1,p}(\Omega)$ provided K is compact and $dim_{box}(K) < n - p$ (Hint: Use appropriate Lipschitz cut-off functions).
 - if E is either an interior weight or a boundary weight on Ω, then it is bounded away from zero on compact subsets of Ω (Hint: Use the maximum principle).

The following theorem contains the main inequalities.

THEOREM 9.1.4. *Consider Ω and A as defined above.*

9.1. A GENERAL INEQUALITY INVOLVING INTERIOR AND BOUNDARY WEIGHTS 127

(1) If E is either an interior or a boundary weight on Ω, then for all $u \in H_0^1(\Omega)$,

(9.4) $$\int_\Omega |\nabla u|_A^2 dx - \frac{1}{4}\int_\Omega \frac{|\nabla E|_A^2}{E^2} u^2 dx \geq 0.$$

Moreover $\frac{1}{4}$ is the optimal constant, and is not attained in $H_0^1(\Omega)$.

(2) If E is a boundary weight on Ω, then for all $u \in H_0^1(\Omega)$,

(9.5) $$\int_\Omega |\nabla u|_A^2 dx - \frac{1}{4}\int_\Omega \frac{|\nabla E|_A^2}{E^2} u^2 dx \geq \frac{1}{2}\int_\Omega \frac{u^2}{E} d\mu.$$

Moreover $\frac{1}{2}$ is optimal (once one fixes $\frac{1}{4}$) and is not attained in $H_0^1(\Omega)$.

We start the proof of Theorem 9.1.4 by essentially justifying the computation in (9.1).

LEMMA 9.1.5. *(i) Suppose E is an interior weight on Ω, then for all $u \in C_c^{0,1}(\Omega \backslash K)$,*

(9.6) $$\int_\Omega |\nabla u|_A^2 dx - \frac{1}{4}\int_\Omega \frac{|\nabla E|_A^2}{E^2} u^2 dx \geq \int_\Omega E|\nabla v|_A^2 dx,$$

where $v := E^{\frac{-1}{2}} u$.

(ii) Suppose E is a boundary weight on Ω, then for all $u \in H_0^1(\Omega)$,

(9.7) $$\int_\Omega |\nabla u|_A^2 dx - \frac{1}{4}\int_\Omega \frac{|\nabla E|_A^2}{E^2} u^2 dx \geq \int_\Omega E|\nabla v|_A^2 dx + \frac{1}{2}\int_\Omega \frac{u^2}{E} d\mu.$$

Proof: (i) Since E is smooth away from K and noting the supports of both u and v, the integration by parts used in obtaining (9.1) is valid.

(ii) Now suppose E is a boundary weight, and extend it to all of $\mathbb{R}^n$ by setting $E = 0$ outside of $\overline{\Omega}$ and let E_ε denote the ε-mollification of E. Let $u \in C_c^\infty(\Omega)$, $v_\varepsilon := E_\varepsilon^{\frac{-1}{2}} u$ and define $F_\varepsilon := \mathcal{L}_A(E_\varepsilon)$. Now one easily obtains (9.1) but with E and v replaced with $E_\varepsilon, v_\varepsilon$. Standard arguments show that $uE_\varepsilon^{-1} \to uE^{-1}$ in $H_0^1(\Omega)$, $|\nabla E_\varepsilon|_A^2 E_\varepsilon^{-2} \to |\nabla E|_A^2 E^{-2}$, $E_\varepsilon|\nabla v_\varepsilon|_A^2 \to E|\nabla v|_A^2$ a.e. in Ω and $uF_\varepsilon \rightharpoonup u\mu$ in $H^{-1}(\Omega)$. Using these results along with Fatou's lemma allows us to pass to the limit. □

The next lemma provides a supply of test functions needed to evaluate the best constants.

LEMMA 9.1.6. *Suppose E is an interior weight on Ω with $0 < \gamma := \min_{\partial\Omega} E$, and let g be a solution to $\mathcal{L}_A(g) = 0$ in Ω with $g = E$ on $\partial\Omega$. Then,*

(1) $u_t := E^t - g^t \in H_0^1(\Omega)$ for $0 < t < \frac{1}{2}$.
(2) The function $I(t) := \int_\Omega |\nabla E|_A^2 E^{2t-2} dx$ is finite for $t < \frac{1}{2}$ and $I(t) \to +\infty$ as $t \nearrow \frac{1}{2}$.
(3) If $E = \gamma > 0$ on $\partial\Omega$, then for each $0 < t < \frac{1}{2}$ and $\tau > \frac{1}{2}$, the function $v_{t,\tau} := E^t \log^\tau(\gamma^{-1} E)$ belongs to $H_0^1(\Omega)$.
(4) For each $0 < t < \frac{1}{2}$, the function $J_t(\tau) := \int_\Omega E^{2t-2}|\nabla E|_A^2 \log^{2\tau-2}(\gamma^{-1}E) dx \to +\infty$ as $\tau \searrow \frac{1}{2}$.

Proof: We prove the results up to some integration by parts, which can be justified by regularizing the measure, integrating by parts and passing to limits. For (1) and (2), we fix $0 < t < \frac{1}{2}$ and note that $|\nabla u_t|^2 \leq CE^{2t-2}|\nabla E|_A^2 +$

$Cg^{2t-2}|\nabla g|_A^2$ where C is some uniform constant. Now multiply $\mathcal{L}_A(E) = \mu$ by E^{2t-1} and integrate over Ω to obtain

$$\begin{aligned}
(1-2t)\int_\Omega E^{2t-2}|\nabla E|_A^2 dx &= -\int_{\partial\Omega} g^{2t-1}(A\nabla E)\cdot\nu d\sigma \\
&= \varepsilon(t) - \int_{\partial\Omega}(A\nabla E)\cdot\nu d\sigma \\
&= \varepsilon(t) - \int_\Omega \operatorname{div}(A\nabla E)dx \\
&= \varepsilon(t) + \mu(\Omega),
\end{aligned}$$

where $\varepsilon(t) \to 0$ as $t \nearrow \frac{1}{2}$. Note $\int_\Omega E^{2t-1}d\mu = 0$ since $t < \frac{1}{2}$ and $E = \infty$ on K. From this we see that $I(t) = \int_\Omega |\nabla E|_A^2 E^{2t-2}dx < \infty$ and so $u_t \in H_0^1(\Omega)$. We also see that $\lim_{t \nearrow \frac{1}{2}} I(t) = \infty$.

For (3) take $0 < t < \frac{1}{2}$, $\tau > \frac{1}{2}$ and $v_{t,\tau}$ defined as above. One easily sees that $v_{t,\tau}$ is continuous near $\partial\Omega$ and vanishes on $\partial\Omega$. So to show $v_{t,\tau} \in H_0^1(\Omega)$ it is sufficient to establish that

$$w_1 := E^{2t-2}|\nabla E|^2 \log^{2\tau}(\gamma^{-1}E), \quad w_2 := E^{2t-2}|\nabla E|^2 \log^{2\tau-2}(\gamma^{-1}E) \in L^1(\Omega).$$

These functions are only singular near K and $\partial\Omega$. Now set $W_\tau := E^{2t-2}|\nabla E|^2 \log^{2\tau-2}(\gamma^{-1}E)$ and so $w_2 = W_\tau$ and $w_1 = W_{\tau+1}$. Now suppose $t' \in (t, \frac{1}{2})$. We have

$$W_{\tau+1} = E^{2t'-2}|\nabla E|^2 \frac{\log^{2\tau}(\gamma^{-1}E)}{E^{2t'-2t}} \leq CE^{2t'-2}|\nabla E|^2 \quad \text{near } K,$$

and so $w_1 = W_{\tau+1} \in L^1(K_\varepsilon)$ where K_ε is a small neighborhood of K. Now note that w_2 is better behaved than w_1 near K and so we also have $w_2 \in L^1(K_\varepsilon)$. Define $\Omega_\varepsilon := \{x \in \Omega : E(x) < \gamma + \varepsilon\}$ and take $\varepsilon > 0$ sufficiently small such that $K \subset \Omega\backslash\Omega_{2\varepsilon}$. Using the co-area formula, we have

$$\begin{aligned}
\int_{\Omega_\varepsilon} E^{2t-2}|\nabla E|^2 \log^{2\tau-2}(\gamma^{-1}E)dx &\leq \sup_{\Omega_\varepsilon}|\nabla E| \int_{\Omega_\varepsilon} E^{2t-2}\log^{2\tau-2}(\gamma^{-1}E)|\nabla E|dx \\
&\leq C\int_1^{1+\frac{\varepsilon}{\gamma}} s^{2t-2}\log^{2\tau-2}(s)ds,
\end{aligned}$$

which is finite for $\tau > \frac{1}{2}$. So we see that $w_2 \in L^1(\Omega_\varepsilon)$ for sufficiently small $\varepsilon > 0$ and noting that w_1 is better behaved near $\partial\Omega$ than w_2 we have the same for w_1. Combining these results we see that $v_{t,\tau} \in H_0^1(\Omega)$.

Fix $0 < t < \frac{1}{2}$ and $\tau > \frac{1}{2}$. By Hopf's lemma we have that $|\nabla E|$ is bounded away from zero on Ω_ε for $\varepsilon > 0$ sufficiently small. By fixing such an $\varepsilon > 0$ we get

$$\begin{aligned}
J_t(\tau) &\geq C\int_{\Omega_\varepsilon} E^{2t-2}\log^{2\tau-2}(\gamma^{-1}E)|\nabla E|dx \\
&\geq \tilde{C}\int_1^{1+\frac{\varepsilon}{\gamma}} s^{2t-2}\log^{2\tau-2}(s)ds,
\end{aligned}$$

and a simple computation shows that the last integral becomes unbounded as $\tau \searrow \frac{1}{2}$. $\square$

Proof of Theorem 9.1.4: (1) Using Lemma 9.1.5 and, in the case where E is a interior weight on Ω, the fact that $C_c^{0,1}(\Omega\backslash K)$ is dense in $H_0^1(\Omega)$, one obtains (9.4). We now show that the constant is optimal. Suppose E is an interior weight on Ω

and define $E_\varepsilon := \varepsilon + E$, $g_\varepsilon := \varepsilon + g$ where $\varepsilon > 0$. Define $I_\varepsilon(t) := \int_\Omega |\nabla E_\varepsilon|_A^2 E_\varepsilon^{2t-2} dx$. As in the proof of Lemma 9.1.6 one can show that for each $\varepsilon > 0$, $\lim_{t \nearrow \frac{1}{2}} I_\varepsilon(t) = \infty$. We use $u_{t,\varepsilon} := E_\varepsilon^t - g_\varepsilon^t$ as test functions. Let $0 < t < \frac{1}{2}$ and $\varepsilon > 0$. Then

$$Q_{t,\varepsilon} := \frac{\int_\Omega |\nabla u_{t,\varepsilon}|_A^2 dx}{\int_\Omega \frac{|\nabla E_\varepsilon|_A^2}{E_\varepsilon^2} u_{t,\varepsilon}^2 dx} \le \frac{t^2 I_\varepsilon(t) + C_0 + C_1 \sqrt{I_\varepsilon(t)}}{I_\varepsilon(t) - C_2 I_\varepsilon(\frac{t}{2}) - C_3 I_\varepsilon(0)},$$

where the constants C_k possibly depend on ε. From this, we see that $\lim_{t \nearrow \frac{1}{2}} Q_{t,\varepsilon} = \frac{1}{4}$ after recalling that $Q_{t,\varepsilon} \ge \frac{1}{4}$. Now fix $\varepsilon > 0$ and let $u \in C_c^\infty(\Omega)$ be non-zero. A simple computation shows that

$$\frac{\int_\Omega |\nabla u|_A^2 dx}{\int_\Omega \frac{|\nabla E|_A^2}{E^2} u^2 dx} \le \frac{\int_\Omega |\nabla u|_A^2 dx}{\int_\Omega \frac{|\nabla E_\varepsilon|_A^2}{E_\varepsilon^2} u^2 dx},$$

which, when combined with the above facts, gives the desired best constant result. One can then use (9.6) to see that $\frac{1}{4}$ is not attained in $H_0^1(\Omega)$.

Now suppose E is a boundary weight on Ω, $\varepsilon > 0$ and $t > \frac{1}{2}$. Define $f_\varepsilon(z) := z^{2t-1} - \varepsilon^{2t-1}$ for $z > \varepsilon$ and 0 otherwise. Using $f_\varepsilon(E) \in H_0^1(\Omega)$ as a test function in the PDE associated with E one obtains, after sending $\varepsilon \searrow 0$,

$$(9.8) \qquad (2t-1) \int_\Omega E^{2t-2} |\nabla E|_A^2 dx = \int_\Omega E^{2t-1} d\mu,$$

which shows that $E^t \in H_0^1(\Omega)$ for $\frac{1}{2} < t \le 1$. To see that $\frac{1}{4}$ is optimal in (9.4), use that E^t is a minimizing sequence as $t \searrow \frac{1}{2}$.

(2) Suppose E is a boundary weight on Ω, and let $\frac{1}{2} < t < 1$ so that $E^t \in H_0^1(\Omega)$. Using (9.8) we have

$$\frac{\int_\Omega |\nabla E^t|_A^2 dx - \frac{1}{4} \int_\Omega \frac{|\nabla E|_A^2}{E^2} (E^t)^2}{\int_\Omega \frac{(E^t)^2}{E} d\mu} = \frac{t}{2} + \frac{1}{4},$$

which shows that $\frac{1}{2}$ is optimal. $\square$

One can also establish the following refinement of Theorem 9.1.4 (1).

COROLLARY 9.1.1. *Let $E \in C^\infty(\overline{\Omega})$ with $E > 0$, $\mathcal{L}_A(E) \ge 0$ in Ω and suppose that $\Gamma := \{x \in \partial\Omega : E(x) = 0\}$ contains $B(x_0, r) \cap \partial\Omega$ for some $x_0 \in \partial\Omega$ and $r > 0$. Then (9.4) is optimal.*

Proof: The only issue is whether $\frac{1}{4}$ is optimal. Without loss of generality assume that $0 \in \partial\Omega$ and $B(0, 2R) \cap \partial\Omega \subset \Gamma$. Suppose $0 < r < R$ and define

$$\varphi(x) := \begin{cases} 1 & x \in \Omega(r) \\ \frac{R-|x|}{R-r} & x \in \Omega(R) \backslash \Omega(r) \\ 0 & x \in \Omega \backslash \Omega(R), \end{cases}$$

where $\Omega(r) := B(0,r) \cap \Omega$. Define $u_t := E^t \varphi$ which can be shown to be an element of $H_0^1(\Omega)$ for $t > \frac{1}{2}$. One uses u_t as $t \searrow \frac{1}{2}$ as a minimizing sequence along with arguments similar to the above to show that $\frac{1}{4}$ is optimal. $\square$

We now look at various examples of Hardy inequalities which can be obtained after making suitable choices of weights E and matrices A. In most of the examples we will take A to be the identity matrix.

COROLLARY 9.1.2. *(Classical Hardy's inequality)* Let Ω denote a domain in $\mathbb{R}^n$ which contains the origin.

(1) If $n \geq 3$, set $E(x) := |x|^{2-n}$. Then $-\Delta E = c\delta_0$ where $c > 0$ and δ_0 is the Dirac mass at 0. Also $\frac{|\nabla E|^2}{4E^2} = \left(\frac{n-2}{2}\right)^2 \frac{1}{|x|^2}$ and so *(9.4)* gives the classical Hardy inequality:

(9.9) $$\int_\Omega |\nabla u|^2 dx \geq \frac{1}{4} \int_\Omega \frac{u^2}{|x|^2} dx \quad \text{for all } u \in C_c^\infty(\Omega).$$

(2) If $n = 2$ and Ω is bounded, set $E(x) := -\log(R^{-1}|x|)$ where $R := \sup_\Omega |x|$. Then $-\Delta E = c\delta_0$ where $c > 0$ and *(9.4)* then gives

(9.10) $$\int_\Omega |\nabla u|^2 dx \geq \frac{1}{4} \int_\Omega \frac{u^2}{|x|^2 \log^2(R^{-1}|x|)} dx \quad \text{for all } u \in C_c^\infty(\Omega).$$

COROLLARY 9.1.3. *(Multipolar Hardy inequality)* Let $x_1, ..., x_k$ be k points in Ω. Then there exists a potential V, which behaves like $\frac{1}{|x-x_i|^2}$ near each singularity x_i such that

(9.11) $$\int_\Omega |\nabla u|^2 dx \geq \frac{1}{(n-2)^2} \int_\Omega V(x) u^2 dx \quad \text{for all } u \in H_0^1(\Omega).$$

Proof: Apply *(9.4)* with $E(x) = \sum_{i=1}^k \frac{1}{|x-x_i|^{n-2}}$ and note that the potential

$$V(x) = \left| \frac{\sum_{i=1}^k \frac{x-x_i}{|x-x_i|^n}}{\sum_{i=1}^k \frac{1}{|x-x_i|^{n-2}}} \right|^2$$

satisfies $V(x) = \frac{1}{(n-2)^2} \left|\frac{\nabla E}{E}\right|^2$. We then obtain that

(9.12) $$\int_\Omega |\nabla u|^2 dx - \frac{1}{(n-2)^2} \int_\Omega V(x) u^2 dx \geq \int_\Omega E|\nabla v|^2 dx,$$

where $v := E^{\frac{-1}{2}} u$. Note now that for ϵ small enough, we have for each $x \in B(x_j, \epsilon)$, that

$$V(x) = \frac{1}{|x-x_j|^2} + O(|x-x_j|^{n-4}).$$

□

COROLLARY 9.1.4. *(Hardy's boundary inequalities)* Let Ω be a domain in $\mathbb{R}^n$ and set $\delta(x) := \text{dist}(x, \partial\Omega)$.

(1) If Ω is bounded and convex, then δ is concave, $-\Delta\delta \geq 0$ in Ω, and *(9.5)* applied with $E(x) := \delta(x)$ gives –an improved version of– the following inequality

(9.13) $$\int_\Omega |\nabla u|^2 dx \geq \frac{1}{4} \int_\Omega \frac{u^2}{\delta^2} dx \quad \text{for all } u \in H_0^1(\Omega).$$

Moreover the constant $\frac{1}{4}$ is optimal and not attained.
In particular, if $\Omega = B$ the unit ball in $\mathbb{R}^n$, then *(9.5)* applied with $E(x) := 1 - |x|$ gives

(9.14) $$\int_B |\nabla u|^2 dx \geq \frac{1}{4} \int_B \frac{u^2}{(1-|x|)^2} dx + \frac{n-1}{2} \int_B \frac{u^2}{|x|(1-|x|)} dx \quad \text{for all } u \in C_c^\infty(B).$$

(2) If $\Omega := (0,\infty) \times (0,\infty)$, then $E(x) := \mathrm{dist}(x, \partial\Omega) = \min\{x_1, x_2\}$, and $-\Delta E = \sqrt{2}\sigma$ where σ is the measure associated with the line $\Gamma := \{x : x_2 = x_1\}$. (9.1) then gives, for all $u \in C_c^\infty(\Omega)$,

$$(9.15) \quad \int_\Omega |\nabla u|^2 dx \geq \frac{1}{4} \int_\Omega \frac{u^2}{(\min\{x_1, x_2\})^2} dx + \frac{1}{\sqrt{2}} \int_\Gamma \frac{u^2}{\min\{x_1, x_2\}} d\sigma.$$

$\square$

COROLLARY 9.1.5. **(Hardy inequalities involving general distance functions)** Let Ω be a domain in $\mathbb{R}^n$ and suppose M is a piecewise smooth surface of codimension $k \leq n$, $k \neq 2$. If $d(x) := \mathrm{dist}(x, M)$ is such that $-\Delta d^{2-k} \geq 0$ in $\Omega \setminus M$, then (9.1) applied with $E(x) := d(x)^{2-k}$ yields the following inequality:

$(9.16) \quad \int_\Omega |\nabla u|^2 dx \geq \frac{(k-2)^2}{4} \int_\Omega \frac{u^2}{d^2} dx$ for all $u \in H_0^1(\Omega \setminus M)$.

$\square$

In general the Poincaré inequality

$$\int_\Omega |\nabla u|^2 dx \geq C \int_\Omega u^2 dx \quad \text{for } u \in C_c^\infty(\Omega),$$

does not hold for unbounded domains. It is however the case that for some of them the inequality does in fact hold, as in the case of the unbounded slab $\Omega := \{x \in \mathbb{R}^n : 0 < x_n < \pi\}$. We can actually use (9.1) to show a slightly stronger result. Indeed, set $E(x) := \sin(x_n)$ into (9.4) and drop a term to arrive at the following result.

COROLLARY 9.1.6. **(Poincaré's inequality in an unbounded slab)** Consider the domain $\Omega := \{x \in \mathbb{R}^n : 0 < x_n < \pi\}$, then

$(9.17) \quad \int_\Omega |\nabla u|^2 dx \geq \frac{1}{4} \int_\Omega \frac{u(x)^2}{\tan^2(x_n)} dx + \frac{1}{2} \int_\Omega u^2 dx \quad$ for all $u \in C_c^\infty(\Omega)$.

$\square$

One can also prove the following classical result.

COROLLARY 9.1.7. **(Trace theorem)** Let Ω be a domain in $\mathbb{R}^n$, $n \geq 3$ such that $B \subset\subset \Omega$ (where B is the unit ball). Then,

$(9.18) \quad \int_\Omega |\nabla u|^2 dx \geq \frac{c}{2} \int_{\partial B} u^2 d\sigma \quad$ for all $u \in C_c^\infty(\Omega)$.

Proof: Define

$$E(x) := \begin{cases} 1 & |x| < 1 \\ \frac{1}{|x|^{n-2}} & |x| > 1. \end{cases}$$

A computation shows that $-\Delta E = c\sigma$ where $c > 0$ and where σ is the surface measure associated with ∂B. Putting this E into (9.1) and dropping a couple of terms gives the result.

$\square$

REMARK 9.1.7. All examples above dealt with the case where A is the identity matrix. Here is an example of a different nature. Suppose Ω is an open subset of $\mathbb{R}^N = \mathbb{R}^n \times \mathbb{R}^k$ and use the notation $\xi = (x, y)$ for a given $\xi \in \Omega$. For $\gamma > 0$, define the vector field $\nabla_\gamma := (\nabla_x, |x|^\gamma \nabla_y)$ and the *Baouendi-Grushin operator* $\mathcal{L}_A := -\Delta_x - |x|^{2\gamma} \Delta_y$. Take

$$A(\xi) := \begin{pmatrix} I_n & 0 \\ 0 & |x|^{2\gamma} I_k \end{pmatrix}$$

where I_n, I_k are the identity matrices of size n and k. Then $|\nabla_\gamma E|^2 = |\nabla E|_A^2$ and $-\mathrm{div}(A\nabla E) = \mathcal{L}_A(E)$.

$\square$

9.2. Best pair of constants and eigenvalue estimates

We fix a weight $E > 0$ in Ω, and consider functions $\tilde{E} := f(E)$ of E, where $f : (0, \infty) \to (0, \infty)$ is a smooth function. By using now $\tilde{E}$ in (9.1) instead of E, we get the following useful result.

COROLLARY 9.2.1. *For all $u \in C_c^\infty(\Omega)$, we have*
$$\int_\Omega |\nabla u|_A^2 dx \geq \int_\Omega |\nabla E|_A^2 \left(\frac{f'(E)^2}{4f(E)^2} - \frac{f''(E)}{2f(E)} \right) u^2 dx + \frac{1}{2} \int_\Omega \frac{f'(E)\mathcal{L}_A(E)}{f(E)} u^2 dx. \tag{9.19}$$

An important example is the case $f(E) := E^t$ where $0 < t < 1$. Actually, one can use $E(x) := \delta(x)^t$ with $\delta(x) := \text{dist}(x, \partial\Omega)$ to show that if one drops the requirement that μ is a *finite* measure (and just assumes μ a locally finite measure), then (9.4) need not be optimal. Indeed, we have the following example.

EXAMPLE 9.2.1. Take Ω a bounded convex domain in $\mathbb{R}^n$ and set $\delta(x) := \text{dist}(x, \partial\Omega)$. Fix $\frac{1}{2} < t < 1$ and set $E := \delta^t \in H_0^1(\Omega)$. Then
$$\frac{|\nabla E|^2}{E^2} = \frac{t^2}{\delta^2}, \qquad \mu := -\Delta E = t(1-t)\delta^{t-2} + t\delta^{t-1}(-\Delta\delta) \geq 0 \quad \text{in } \Omega,$$
and so putting E into (9.4) gives
$$\int_\Omega |\nabla u|^2 dx \geq \frac{t^2}{4} \int_\Omega \frac{u^2}{\delta^2} dx,$$
for $u \in H_0^1(\Omega)$. This shows that (9.4) was not optimal. This apparent failure of theorem 9.1.4 is due to the fact that μ is not a finite measure. Use the co-area formula to show that $\delta^{t-2} \notin L^1(\Omega)$.

COROLLARY 9.2.2. *Suppose E is an interior weight on Ω with $E = 1$ on $\partial\Omega$.*

(1) *By using $f(E) := (\log(E))^{\frac{1}{2}}$ one obtains the inequality*
$$\int_\Omega |\nabla u|_A^2 dx \geq \frac{1}{16} \int_\Omega \frac{3 + 4\log(E)}{E^2 \log^2(E)} |\nabla E|_A^2 u^2 dx, \quad u \in H_0^1(\Omega). \tag{9.20}$$

(2) *Taking instead $f(E) := E\log(E)$ yields*
$$\int_\Omega |\nabla u|_A^2 dx \geq \frac{1}{4} \int_\Omega \frac{\log^2(E) + 1}{E^2 \log^2(E)} u^2 dx, \quad u \in H_0^1(\Omega). \tag{9.21}$$

We now give an alternate way to view best constants in (9.5) where boundary weights are considered. For that set
$$\mathcal{C} = \{(\beta, \alpha) \in \mathbb{R}^2 \text{ such that (9.22) below is satisfied}\}$$

$$\int_\Omega |\nabla u|_A^2 dx \geq \alpha \int_\Omega \frac{|\nabla E|_A^2}{E^2} u^2 dx + \beta \int_\Omega \frac{u^2}{E} d\mu, \quad u \in H_0^1(\Omega). \tag{9.22}$$

We now identify the set $\mathcal{C}$.

THEOREM 9.2.2. *Suppose $E \in L^\infty(\Omega)$ is a boundary weight on Ω. Then*
$$\mathcal{C} = \left\{ (\beta, \alpha) : \beta > \frac{1}{2}, \alpha \leq \beta - \beta^2 \right\} \cup \left(-\infty, \frac{1}{2}\right] \times \left(-\infty, \frac{1}{4}\right].$$

Moreover, the inequality (9.22) does attain on $\Gamma := \{(\tau, \tau - \tau^2) : \tau > \frac{1}{2}\} \subset \partial\mathcal{C}$ and does not attain on $\partial\mathcal{C} \backslash \Gamma$.

9.2. BEST PAIR OF CONSTANTS AND EIGENVALUE ESTIMATES

Proof: Denote $\mathcal{C}' := \{(\beta, \alpha) : \beta > \frac{1}{2}, \alpha \leq \beta - \beta^2\} \cup \left(-\infty, \frac{1}{2}\right] \times \left(-\infty, \frac{1}{4}\right]$. Using similar arguments to the above one can show that $E^t \in H_0^1(\Omega)$ for all $t > \frac{1}{2}$. Suppose now $(\beta, \alpha) \in \mathcal{C}$. If $\beta > \frac{1}{2}$, then testing (9.22) on $u := E^\beta$ shows that $\alpha \leq \beta - \beta^2$. If $\beta \leq \frac{1}{2}$ then testing (9.22) on $u := E^t$ and sending $t \searrow \frac{1}{2}$ shows that $\alpha \leq \frac{1}{4}$. It follows that $\mathcal{C} \subset \mathcal{C}'$.

Now for the other inclusion, we fix $t \geq 1$ and put $E_2 := E^t$. Then we have

$$\frac{|\nabla E_2|_A^2}{E_2^2} = t^2 \frac{|\nabla E|_A^2}{E^2}, \qquad \frac{\mathcal{L}_A(E_2)}{E_2} = t(1-t)\frac{|\nabla E|_A^2}{E^2} + t\frac{\mathcal{L}_A(E)}{E}.$$

Putting $E = E_2$ into (9.1) we obtain

$$(9.23) \qquad \int_\Omega |\nabla u|_A^2 dx \geq \left(\frac{t}{2} - \frac{t^2}{4}\right) \int_\Omega \frac{|\nabla E|_A^2}{E^2} u^2 dx + \frac{t}{2} \int_\Omega \frac{u^2}{E} d\mu,$$

and so we see that $(\frac{t}{2}, \frac{t}{2} - \frac{t^2}{4}) \in \mathcal{C}$ for all $t \geq 1$. From this we infer that the curve $\alpha = \beta - \beta^2$ for $\beta \geq \frac{1}{2}$ is contained in $\mathcal{C}$. It is straightforward to check that the remaining portion of $\partial \mathcal{C}'$ is contained in $\mathcal{C}$.

In order to identify for which couples (α, β) the inequality is attained, we note first that while proving (9.1), one drops the term $\int_\Omega E |\nabla \left(\frac{u}{\sqrt{E}}\right)|_A^2 dx$, which is positive for non-zero u provided u is not a multiple of $\sqrt{E}$. Since $\sqrt{E} \notin H_0^1(\Omega)$ this was never an issue. However, the situation is different when we consider E^t for $t > 1$. Now to see that the inequality does not attain when $(\beta, \alpha) \in \partial \mathcal{C} \setminus \Gamma$, use the fact that (9.4) does not attain in H_0^1 and the fact that $\mu \geq 0$. On the other hand, to see that the inequality does attain on the remaining portion of $\partial \mathcal{C}$ note that (9.23) attains at $u := E^{\frac{t}{2}} \in H_0^1(\Omega)$ for $t > 1$. □

We now consider some implications of the above inequality to the problem of evaluating the first eigenvalue of $\mathcal{L}_A$ in $H_0^1(\Omega)$, which we denote by $\lambda_A(\Omega)$.

PROPOSITION 9.2.1. *Let Ω be a bounded subset of $\mathbb{R}^n$, then*

$$(9.24) \qquad \lambda_A(\Omega) \geq \sup\left\{\frac{\pi^2}{4\|E\|_\infty^2}; E > 0, \mathcal{L}_A(E) \geq 0 \text{ and } |\nabla E|_A^2 = 1 \text{ a.e. in } \Omega\right\}.$$

Proof: To show this, it suffices to consider $f(z) := \sin^2\left(\frac{\pi z}{2\|E\|_{L^\infty}}\right)$ in the above result and to drop the term involving the measure. □

We now give a result relating $\lambda_A(\Omega)$ to the first eigenvalue $\lambda_A(B)$ of $\mathcal{L}_A$ on subdomains B of Ω, in the case where $E > 0$ is the first eigenfunction of $\mathcal{L}_A$ on Ω.

COROLLARY 9.2.3. *Suppose $(E, \lambda_A(\Omega))$ is the first eigenpair (with $E > 0$) of $\mathcal{L}_A$ on $H_0^1(\Omega)$. For $B \subset \Omega$ we set*

$$\underline{\alpha}(B) := \inf_B \frac{|\nabla E|_A^2}{E^2}, \qquad \overline{\alpha}(B) := \sup_B \frac{|\nabla E|_A^2}{E^2}.$$

(i) *If $\underline{\alpha}(B) > \lambda_A(\Omega)$ then*

$$4\lambda_A(B) \geq \frac{(\underline{\alpha}(B) + \lambda_A(\Omega))^2}{\underline{\alpha}(B)}.$$

(ii) *If $\lambda_A(\Omega) > \overline{\alpha}(B)$ then*

$$4\lambda_A(B) \geq \frac{(\overline{\alpha}(B) + \lambda_A(\Omega))^2}{\overline{\alpha}(B)}.$$

Proof: Let $B \subset \Omega$ and let $u \in C_c^\infty(B)$ with $\int_B u^2 = 1$. Using (9.23), one gets

$$2 \int_B |\nabla u|_A^2 dx \geq (t - \frac{t^2}{2}) \inf_B \frac{|\nabla E|_A^2}{E^2} + \lambda_A(\Omega) t,$$

for $0 < t < 2$. If $t > 2$ then we get the same expression but with the infimum replaced with supremum. Now take the infimum over u and in case (i) set $t := 1 + \frac{\lambda_A(\Omega)}{\underline{\alpha}(B)} < 2$ and in case (ii) set $t := 1 + \frac{\lambda_A(\Omega)}{\overline{\alpha}(B)} > 2$ to see the result. □

9.3. Weighted Hardy inequalities for general elliptic operators

We now examine weighted versions of the above inequalities. We start with general analogs of Caffarelli-Kohn-Nirenberg inequalities. For that we consider for each $t \in \mathbb{R}$ the norm

$$\|u\|_t^2 := \int_\Omega E^{2t} |\nabla u|_A^2 dx.$$

- If E is an interior weight on Ω, we define $X_t := X_t(A, E)$ to be the completion of $C_c^{0,1}(\Omega \setminus K)$ for the norm $\|\cdot\|_t$.
- If E is a boundary weight on Ω, X_t is then taken to be the completion of $C_c^{0,1}(\Omega)$ for that norm.

REMARK 9.3.1. One should note that if E is an interior weight on Ω and $t > \frac{1}{2}$, then X_t does not contain $C_c^\infty(\Omega)$. To see this, use (9.26) to note that if $C_c^\infty(\Omega) \subset X_t$ then $E^t \in H_{loc}^1(\Omega)$, which we know to be false. For $t < \frac{1}{2}$ we do have $C_c^\infty(\Omega) \subset X_t$.

THEOREM 9.3.2. Suppose $0 \neq t < \frac{1}{2}$ and E an interior weight on Ω. Then for all $u \in X_t$,

(9.25)
$$\int_\Omega E^{2t} |\nabla u|_A^2 dx \geq (t - \frac{1}{2})^2 \int_\Omega |\nabla E|_A^2 E^{2t-2} u^2 dx$$
$$+ (\frac{1}{2} - t) \int_\Omega -\operatorname{div}(A \nabla E) E^{2t-1} u^2 dx.$$

In particular, we have for all $u \in X_t$,

(9.26)
$$\int_\Omega E^{2t} |\nabla u|_A^2 dx \geq (t - \frac{1}{2})^2 \int_\Omega |\nabla E|_A^2 E^{2t-2} u^2 dx.$$

Moreover the constant $(t - \frac{1}{2})^2$ is optimal and is not attained in X_t.

Proof: Let $t \neq 0, \frac{1}{2}$, $u \in C_c^{0,1}(\Omega \setminus K)$ and define $w := E^t u \in C_c^{0,1}(\Omega \setminus K)$. Put w into inequality (9.1.1), that is

$$\int_\Omega |\nabla w|_A^2 dx \geq \frac{1}{4} \int_\Omega \frac{|\nabla E|_A^2}{E^2} w^2 dx - \frac{1}{2} \int_\Omega \frac{w^2}{E} \operatorname{div}(A \nabla E) dx,$$

and re-group to obtain (9.25). We now show the constant in (9.26) is optimal. Let $v_m \in C_c^{0,1}(\Omega \setminus K)$ be such that

$$D_m := \frac{\int_\Omega |\nabla v_m|_A^2 dx}{\int_\Omega \frac{|\nabla E|_A^2}{E^2} v_m^2 dx} \to \frac{1}{4}.$$

Define $u_m := E^{-t} v_m \in X_t$. A computation shows that

$$\frac{\int_\Omega E^{2t} |\nabla u_m|_A^2 dx}{\int_\Omega |\nabla E|_A^2 E^{2t-2} u_m^2 dx} = D_m + t^2 - t,$$

and since $D_m \to \frac{1}{4}$ we see that $(t-\frac{1}{2})^2$ is optimal.
For the case $\gamma := \min_{\partial\Omega} E > 0$, we can show the constant is not attained by using later results on improvements. If $\gamma = 0$ we then substitute w into (9.1) instead of (9.4) and hold onto the extra term $\int_\Omega E|\nabla(E^{t-\frac{1}{2}}u)|_A^2 dx$ to see that the optimal constant is not attained. $\square$

THEOREM 9.3.3. *Under the same conditions on Ω.*
(1) *If $0 \neq t < \frac{1}{2}$ and E is a boundary weight on Ω, then for all $u \in X_t$,*

$$\text{(9.27)} \qquad \int_\Omega E^{2t}|\nabla u|_A^2 dx - (t-\frac{1}{2})^2 \int_\Omega |\nabla E|_A^2 E^{2t-2} u^2 dx \geq 0.$$

Moreover the constant is optimal and not attained.
(2) *If $0 \neq t < \frac{1}{2}$ and E is a boundary weight on Ω, then for all $u \in X_t$,*

$$\text{(9.28)} \qquad \int_\Omega E^{2t}|\nabla u|_A^2 dx - (t-\frac{1}{2})^2 \int_\Omega |\nabla E|_A^2 E^{2t-2} u^2 dx \geq (\frac{1}{2}-t) \int_\Omega E^{2t-1} u^2 d\mu.$$

Again, the constant on the right is optimal and not attained.
(3) *If $t > \frac{1}{2}$ and $E \in L^\infty(\Omega)$ is a boundary weight on Ω, then*

$$\inf\left\{ \frac{\int_\Omega E^{2t}|\nabla u|_A^2 dx}{\int_\Omega |\nabla E|_A^2 E^{2t-2} u^2 dx} \ : \ u \in X_t \backslash \{0\} \right\} = 0.$$

Proof: We first prove (9.28) for $u \in C_c^{0,1}(\Omega)$, which then gives (9.27) for the same class of u's. Suppose $0 \neq t < \frac{1}{2}$ and E is a boundary weight on Ω. We use the notation introduced in the proof of Lemma 9.1.5, namely E_ε is the standard mollification of E and $F_\varepsilon := \mathcal{L}_A(E_\varepsilon)$. Recall that for any $u \in C_c^{0,1}(\Omega)$ we have $uF_\varepsilon \to u\mu$ in $H^{-1}(\Omega)$ and that

$$\int_\Omega |\nabla v|_A^2 dx \geq \frac{1}{4} \int_\Omega \frac{|\nabla E_\varepsilon|_A^2}{E_\varepsilon^2} v^2 dx + \frac{1}{2} \int_\Omega \frac{v^2}{E_\varepsilon} F_\varepsilon dx,$$

for all $v \in H_0^1(\Omega)$. Now let $u \in C_c^{0,1}(\Omega)$ and set $v := E_\varepsilon^t u \in C_c^{0,1}(\Omega)$. Putting v into the above gives

$$\text{(9.29)} \quad \int_\Omega E_\varepsilon^{2t}|\nabla u|_A^2 dx \geq (t-\frac{1}{2})^2 \int_\Omega |\nabla E_\varepsilon|_A^2 E_\varepsilon^{2t-2} u^2 dx + (\frac{1}{2}-t) \int_\Omega E_\varepsilon^{2t-1} u^2 F_\varepsilon dx.$$

Since $E_\varepsilon^{2t} \to E^{2t}$ in $L_{loc}^1(\Omega)$, we have

$$\int_\Omega E_\varepsilon^{2t}|\nabla u|_A^2 dx \to \int_\Omega E^{2t}|\nabla u|_A^2 dx,$$

and using similar ideas from the proof of Lemma 9.1.5 one can show that

$$\int_\Omega E_\varepsilon^{2t-1} u^2 F_\varepsilon dx \to \int_\Omega E^{2t-1} u^2 d\mu.$$

Using these results, sending $\varepsilon \searrow 0$ in (9.29) and after an application of Fatou's lemma we arrive at (9.28) for $u \in C_c^{0,1}(\Omega)$.
Now we show the constants are optimal. Recalling the proof of Theorem 9.1.4 there exists $v_m \in C_c^\infty(\Omega)$ such that

$$D_m := \frac{\int_\Omega |\nabla v_m|_A^2 dx}{\int_\Omega \frac{|\nabla E|_A^2}{E^2} v_m^2 dx} \to \frac{1}{4} \quad \text{and} \quad F_m := \frac{\int_\Omega |\nabla v_m|_A^2 dx - \frac{1}{4} \int_\Omega \frac{|\nabla E|_A^2}{E^2} v_m^2 dx}{\int_\Omega \frac{v_m^2}{E} d\mu} \to \frac{1}{2}.$$

Define $u_m := E^{-t}v_m$ which one easily sees is an element of X_t. Then

$$\Phi_m := \frac{\int_\Omega E^{2t}|\nabla u_m|_A^2 dx}{\int_\Omega |\nabla E|_A^2 E^{2t-2} u_m^2 dx} = D_m + t^2 - 2t \frac{\int_\Omega E^{-1} v_m \nabla v_m \cdot A\nabla E dx}{\int_\Omega \frac{|\nabla E|_A^2}{E^2} v_m^2 dx},$$

and

$$\Psi_m := \frac{\int_\Omega E^{2t}|\nabla u_m|_A^2 dx - (t-\tfrac{1}{2})^2 \int_\Omega |\nabla E|_A^2 E^{2t-2} u_m^2 dx}{\int_\Omega E^{2t-1} u_m^2 d\mu}$$

$$= F_m + \frac{t \int_\Omega \frac{|\nabla E|_A^2}{E^2} v_m^2 dx - 2t \int_\Omega E^{-1} v_m \nabla v_m \cdot A\nabla E dx}{\int_\Omega \frac{v_m^2}{E} d\mu}.$$

Using $E_\varepsilon, F_\varepsilon$ as defined above one can show as above that

(9.30) $$2 \int_\Omega E^{-1} v_m \nabla v_m \cdot A \nabla E dx = \int_\Omega \frac{v_m^2}{E} d\mu + \int_\Omega \frac{|\nabla E|_A^2}{E^2} v_m^2 dx.$$

From this we see that

$$\Phi_m = D_m + t^2 - t - t \frac{\int_\Omega \frac{v_m^2}{E} d\mu}{\int_\Omega \frac{|\nabla E|_A^2}{E^2} v_m^2 dx},$$

and noting that

$$\frac{\int_\Omega \frac{v_m^2}{E} d\mu}{\int_\Omega \frac{|\nabla E|_A^2}{E^2} v_m^2 dx} = \frac{D_m - \tfrac{1}{4}}{F_m} \to 0,$$

we get that (9.27) is optimal. Similarly one sees using (9.30) that $\Psi_m = F_m - t$ and hence (9.28) is optimal.

To show the constants are not attained, we hold on as usual to the extra term that we dropped in the above calculations. Since $\int_\Omega E^{-1} |\nabla E|_A^2 dx = \infty$ one can show this extra term is positive for $u \in X_t \backslash \{0\}$.

(3) Take $t > \tfrac{1}{2}$ and let E be a boundary weight on Ω. For $\varepsilon, \tau > 0$ small, define

$$u_{\varepsilon,\tau}(x) := \begin{cases} 0 & E < \varepsilon \\ E^\tau - \varepsilon^\tau & E > \varepsilon. \end{cases}$$

Then $u_{\varepsilon,\tau} \in X_t$. Now use the sequence u_m, where $u_m := u_{\varepsilon_m, \tau_m}$, $\varepsilon_m := m^{-m}$, and $\tau_m := m^{-1}$ to get the desired result. $\square$

We now investigate the possibility of more general weighted inequalities of the form

$$\int_\Omega W(x) |\nabla u|_A^2 dx \geq \int_\Omega U(x) u^2 dx, \qquad u \in C_c^{0,1}(\Omega \backslash K).$$

THEOREM 9.3.4. *Suppose E is an interior weight on Ω with $\gamma := \min_{\partial \Omega} E$ and $0 < f \in C^\infty(\gamma, \infty)$. Then, for all $u \in C_c^{0,1}(\Omega \backslash K)$ we have*

(9.31) $$\int_\Omega f(E)^2 |\nabla u|_A^2 dx \geq \int_\Omega |\nabla E|_A^2 \left(\frac{f(E)^2}{4E^2} + f(E) f''(E) \right) u^2 dx.$$

Moreover, this is optimal in the sense that the optimal constant is equal to 1, provided either $\liminf_{z \to \infty} f''(z) > 0$ or $\lim_{z \to \infty} \frac{z^2 f''(z)}{f(z)} = 0$.

Proof: Let $u \in C_c^{0,1}(\Omega \backslash K)$ and define $w := f(E)u \in C_c^{0,1}(\Omega \backslash K)$. Putting w into (9.4), integrating by parts and re-grouping gives (9.31). Let $v_m \in C_c^{0,1}(\Omega \backslash K)$ be such that

$$D_m := \frac{\int_\Omega |\nabla v_m|^2 dx}{\int_\Omega \frac{|\nabla E|_A^2}{E^2} v_m^2 dx} \to \frac{1}{4}.$$

Without loss of generality we can assume the supports of v_m concentrate on K. Define $u_m := \frac{v_m}{f(E)} \in C_c^{0,1}(\Omega \backslash K)$. Then a computation shows that

$$Q_m := \frac{\int_\Omega f(E)^2 |\nabla u_m|_A^2 dx}{\int_\Omega |\nabla E|_A^2 \left(\frac{f(E)^2}{4E^2} + f(E)f''(E)\right) u_m^2 dx}$$

$$= \frac{\int_\Omega |\nabla v_m|_A^2 dx + \int_\Omega \frac{|\nabla E|_A^2 f''(E)}{f(E)^2} v_m^2 dx}{\int_\Omega \frac{|\nabla E|_A^2}{4E^2} v_m^2 dx + \int_\Omega \frac{|\nabla E|_A^2 f''(E)}{f(E)^2} v_m^2 dx}.$$

Now suppose $\liminf_{z \to \infty} f''(z) > 0$. The monotonicity of $x \mapsto \frac{\alpha+x}{\beta+x}$, where α and β are positive constants, shows $Q_m \to 1$. Now suppose $\lim_{z \to \infty} \frac{z^2 f''(z)}{f(z)} = 0$. Using this and the fact that the v_m's support concentrates on K one easily sees that

$$\frac{\int_\Omega \frac{|\nabla E|_A^2 f''(E)}{f(E)^2} v_m^2 dx}{\int_\Omega \frac{|\nabla E|_A^2}{4E^2} v_m^2 dx} \to 0,$$

from which follows that $Q_m \to 1$. □

9.4. Non-quadratic general Hardy inequalities for elliptic operators

For $1 < p \le n$, we define $\mathcal{L}_{A,p}(E) := -\text{div}(|\nabla E|_A^{p-2} A \nabla E)$.

Interior case. Suppose μ is a nonnegative nonzero finite measure supported on $K \subset \Omega$, $dim_{box}(K) < n - p$ (hence $C_c^{0,1}(\Omega \backslash K)$ is dense in $W_0^{1,p}(\Omega)$) and that $0 < E$ is a solution of

(9.32) $$\mathcal{L}_{A,p}(E) = \mu \quad \text{in } \Omega.$$

By regularity theory (see [D], [T]) there is some $0 < \sigma < 1$ such that $E \in C^{1,\sigma}(\Omega \backslash K)$ and by the maximum principle (see [V]) $E > 0$ in $\Omega \backslash K$. Moreover, if $\mu = \delta_0$, then one can show that $E(0) = +\infty$.

THEOREM 9.4.1. *Suppose E is as above.*
(i) *Then for all $u \in W_0^{1,p}(\Omega)$,*

(9.33) $$\int_\Omega |\nabla u|_A^2 dx \ge \left(\frac{p-1}{p}\right)^p \int_\Omega \frac{|\nabla E|_A^p}{E^p} |u|^p dx.$$

(ii) *Suppose $E = \infty$ on K and $E = \gamma$ on $\partial\Omega$, where γ is a non-negative constant. Then the constant $(\frac{p-1}{p})^p$ in (9.33) is optimal.*

9. GENERAL HARDY INEQUALITIES

Proof: (i) Let $u \in C_c^{0,1}(\Omega \backslash K)$. Then $\nabla E^{1-p} = (1-p)E^{-p}\nabla E$, and dotting both sides with $|\nabla E|_A^{p-2} A\nabla E|u|^p$ and integrating over Ω gives

$$
\begin{aligned}
(1-p) \int_\Omega \frac{|\nabla E|_A^p}{E^p} |u|^p dx &= \int_\Omega \nabla E^{1-p} \cdot \left(|\nabla E|_A^{p-2} A \nabla E |u|^p\right) dx \\
&= \int_\Omega E^{1-p} |u|^p d\mu \\
&\quad - \int_\Omega E^{1-p} |\nabla E|_A^{p-2} A\nabla E \cdot p|u|^{p-2} u \nabla u\, dx \\
&= -\int_\Omega E^{1-p} |\nabla E|_A^{p-2} A\nabla E \cdot p|u|^{p-2} u \nabla u\, dx,
\end{aligned}
$$

where we have used the divergence theorem and the fact that $u = 0$ on K. Now using the Cauchy-Schwarz inequality on the inner product induced by $A(x)$ we see that

$$
\frac{p-1}{p} \int_\Omega \frac{|\nabla E|_A^p}{E^p} |u|^p dx \le \int_\Omega \frac{|\nabla E|_A^{p-1} |u|^{p-1}}{E^{p-1}} |\nabla u|_A\, dx.
$$

We now apply Hölder's inequality on the right after recalling that $(p-1)p' = p$ where p' is the conjugate of p. Now use density to extend to all of $W_0^{1,p}(\Omega)$.

(ii) We first consider the case $\gamma > 0$. We begin by showing that $u_t := E^t - \gamma^t \in W_0^{1,p}(\Omega)$ for $0 < t < \frac{p-1}{p}$. Fix $0 < t < \frac{p}{p-1}$ and multiply (9.32) by E^{tp-p+1} and integrate over Ω to get

$$
\begin{aligned}
0 &= \int_\Omega E^{tp-p+1} d\mu \\
&= (tp-p+1) \int_\Omega |\nabla E|_A^2 E^{tp-p} dx - \gamma^{tp-p+1} \int_{\partial \Omega} |\nabla E|_A^{p-2} A\nabla E \cdot \nu\, d\mathcal{H}^{n-1} \\
&= (tp-p+1) \int_\Omega |\nabla E|_A^2 E^{tp-p} dx - \gamma^{tp-p+1} \int_\Omega \mathrm{div}(|\nabla E|_A^{p-2} A\nabla E) dx \\
&= (tp-p+1) \int_\Omega |\nabla E|_A^2 E^{tp-p} dx + \gamma^{tp-p+1} \mu(\Omega),
\end{aligned}
$$

where the first integral is zero since $E = \infty$ on K and $tp - p + 1 < 0$. Re-arranging this we arrive at

$$
\int_\Omega |\nabla E^t|_A^p dx = \frac{\mu(\Omega) \gamma^{tp-p+1} t^p}{p - tp - 1},
$$

from which we see that $E^t \in W^{1,p}(\Omega)$ for $0 < t < \frac{p-1}{p}$ and that

$$
\lim_{t \nearrow \frac{p-1}{p}} \int_\Omega |\nabla E|_A^p E^{tp-p} dx = \infty.
$$

Put t as above and set $u_t := E^t - \gamma^t \in W_0^{1,p}(\Omega)$. By the binomial theorem we have

$$
(1+x)^p = \sum_{m=0}^\infty (p,m) x^m,
$$

for all $|x| \le 1$ where (p,m) are the binomial coefficients. One should note that (p,m) is eventually alternating and since we have convergence at $x = -1$ we see that $\sum_m (p,m)(-1)^m$ converges. Now we have

$$
|u_t|^p = E^{tp} \left|1 - \frac{\gamma^t}{E^t}\right| = E^{tp} \sum_{m=0}^\infty (p,m) \frac{(-1)^m \gamma^{tm}}{E^{tm}},
$$

and defining
$$Q_t := \frac{\int_\Omega \frac{|\nabla E|_A^p}{E^p}|u_t|^p dx}{\int_\Omega |\nabla u_t|_A^p dx},$$

we have
$$Q_t - \frac{1}{t^p} = \frac{\int_\Omega |\nabla E|_A^p E^{tp-p}\left(\sum_{m=1}^\infty (p,m)(-1)^m \frac{\gamma^{tm}}{E^{tm}}\right) dx}{t^p \int_\Omega |\nabla E|_A^p E^{tp-p} dx},$$

and so
$$\begin{aligned}
\left|Q_t - \frac{1}{t^p}\right| &\leq \frac{1}{t^p}\sum_{m=1}^\infty |(p,m)|\frac{\int_\Omega |\nabla E|_A^p E^{tp-p}\gamma^{tm} E^{-tm} dx}{\int_\Omega |\nabla E|_A^p E^{tp-p} dx} \\
&= \frac{1}{t^p}\sum_{m=1}^\infty |(p,m)|\frac{p-tp-1}{p-tp-1+tm} \\
&\leq \frac{p-tp-1}{t^{p+1}}\sum_{m=1}^\infty \frac{|(p,m)|}{m} \\
&=: \frac{p-tp-1}{t^{p+1}}C_p,
\end{aligned}$$

in such a way that
$$\lim_{t \nearrow \frac{p-1}{p}} \left|Q_t - \frac{1}{t^p}\right| = 0,$$

which shows that the constant in (9.33) is optimal.

Now we handle the case $\gamma = 0$. Let $\mathcal{L}_{A,p}(E) = \mu$ in Ω and $E = 0$ on $\partial\Omega$ and define $E_\varepsilon := \varepsilon + E$ where $\varepsilon > 0$. Then $\mathcal{L}_{A,p}(E_\varepsilon) = \mu$ in Ω and $E_\varepsilon = \varepsilon$ on $\partial\Omega$. For $u \in C_c^\infty(\Omega)$ non-zero we have, after some simple algebra,

$$\frac{\int_\Omega |\nabla u|_A^p dx}{\int_\Omega \frac{|\nabla E|_A^p}{E^p}|u|^p dx} \leq \frac{\int_\Omega |\nabla u|_A^p dx}{\int_\Omega \frac{|\nabla E_\varepsilon|_A^p}{E_\varepsilon^p}|u|^p dx},$$

which shows the constant is also optimal in the case of $\gamma = 0$. $\square$

Boundary case. Analogously to the quadratic case we will be interested in the validity of (9.33) when E is a solution to
$$\begin{aligned}
\mathcal{L}_{A,p}(E) &= \mu & \text{in } \Omega, \\
E &= 0 & \text{on } \partial\Omega
\end{aligned}$$

where μ is a nonnegative nonzero finite measure and where we impose some added regularity restrictions to E or μ. Recall in the quadratic case we added the condition that $E \in H_0^1(\Omega)$. For simplicity we will assume that μ is smooth, say $d\mu = f dx$ where $0 \leq f \in C^\infty(\overline{\Omega})$ is non-zero. One can show that $E \in C^{1,\sigma}(\overline{\Omega})$ for some $0 < \sigma < 1$.

THEOREM 9.4.2. *Suppose E is a positive solution to $\mathcal{L}_{A,p}(E) = \mu$ in Ω where μ is as above. Then the following hold:*

(1) *For all $u \in W_0^{1,p}(\Omega)$,*

(9.34) $$\int_\Omega |\nabla u|_A^p dx \geq \left(\frac{p-1}{p}\right)^p \int_\Omega \frac{|\nabla E|_A^p}{E^p}|u|^p dx + \left(\frac{p-1}{p}\right)^{p-1}\int_\Omega \frac{|u|^p}{E^{p-1}} d\mu,$$

and in particular,

$$\int_\Omega |\nabla u|_A^p dx \geq \left(\frac{p-1}{p}\right)^p \int_\Omega \frac{|\nabla E|_A^p}{E^p}|u|^p dx. \tag{9.35}$$

(2) If $E = 0$ on $\partial\Omega$, then (9.35) is optimal.
(3) If $E = 0$ on $\partial\Omega$, then the constant is also optimal in (9.34) i.e.,

$$\inf\left\{\frac{\int_\Omega |\nabla u|_A^p dx - \left(\frac{p-1}{p}\right)^p \int_\Omega \frac{|\nabla E|_A^p}{E^p}|u|^p dx}{\int_\Omega \frac{|u|^p}{E^{p-1}}d\mu} : u \in W_0^{1,p}(\Omega), u \neq 0\right\} = \left(\frac{p-1}{p}\right)^{p-1}.$$

Proof: (1) Suppose E is a positive solution to $\mathcal{L}_{A,p}(E) = \mu$ in Ω and let $u \in C_c^\infty(\Omega)$. From the proof of Theorem 9.4.1 we have

$$(p-1)\int_\Omega \frac{|\nabla E|_A^p}{E^p}|u|^p dx + \int_\Omega \frac{|u|^p}{E^{p-1}}d\mu = p\int_\Omega \frac{|\nabla E|_A^{p-2}}{E^{p-1}} A\nabla E \cdot \nabla u |u|^{p-2} u \, dx$$

$$\leq p\left(\int_\Omega \frac{|\nabla E|_A^p}{E^p}|u|^p dx\right)^{\frac{1}{p'}}\left(\int_\Omega |\nabla u|_A^p dx\right)^{\frac{1}{p}}.$$

Now let q denote p' and

$$B := \int_\Omega \frac{|\nabla E|_A^p}{E^p}|u|^p dx, \quad C := \int_\Omega \frac{|u|^p}{E^{p-1}}d\mu, \quad D := \int_\Omega |\nabla u|_A^p dx.$$

Using Young's inequality with $t > 0$ we arrive at

$$\frac{(p-1)}{p}B + \frac{C}{p} \leq B^{\frac{1}{q}}D^{\frac{1}{p}} \leq tB + C(t)D,$$

where $C(t) := p^{-1}q^{\frac{-p}{q}}t^{\frac{-p}{q}}$, and so

$$\frac{1}{C(t)}\left(\frac{p-1}{p} - t\right)B + \frac{1}{pC(t)}C \leq D,$$

for all $t > 0$. Picking $t = q^2$ gives the desired result.
(2) Let $t > \frac{p-1}{p}$, multiply $\mathcal{L}_{A,p}(E) = \mu$ by E^{tp-p+1} and integrate over Ω to obtain

$$\int_\Omega E^{tp-p+1}d\mu = (tp - p + 1)\int_\Omega |\nabla E|_A^p E^{tp-p}dx, \tag{9.36}$$

which shows that $E^t \in W_0^{1,p}(\Omega)$ for $p > \frac{p-1}{p}$. Consider $u_t := E^t$ as a minimizing sequence and send $t \searrow \frac{p-1}{p}$ to see that (9.35) is optimal.
(3) Again use $u_t := E^t$, send $t \searrow \frac{p}{p-1}$, and use (9.36) to get the result. □

An important example is when $A(x)$ is the identity matrix and $E(x) = \delta(x) :=$ dist$(x, \partial\Omega)$ so that $|\nabla \delta| = 1$ a.e.. Then $\mathcal{L}_{A,p}(\delta) = -\text{div}(|\nabla \delta|^{p-2}\delta) = -\Delta\delta =: \mu$ which is non-negative if we further assume that Ω is convex. In this case we have the L^p analog of (9.13).

COROLLARY 9.4.1. *Suppose Ω is convex and $\delta(x) := \text{dist}(x, \partial\Omega)$. Then for $1 < p < \infty$ and $u \in W_0^{1,p}(\Omega)$ we have*

$$\int_\Omega |\nabla u|^p dx - \left(\frac{p-1}{p}\right)^p \int_\Omega \frac{|u|^p}{\delta^p}dx \geq \left(\frac{p-1}{p}\right)^{p-1}\int_\Omega \frac{|u|^p}{\delta^{p-1}}d\mu,$$

where $d\mu := -\Delta\delta \, dx$. Moreover all constants are optimal.

9.5. Further comments

Adimurthi and Sekar [9] may have been the first to study the role of the fundamental solution while studying Hardy inequalities for general elliptic operators, though some say that the ideas go back at least to Jacobi's 1837 paper [190]. They established the inequality

$$(9.37) \qquad \int_\Omega |\nabla u|_A^p dx - \left(\frac{p-1}{p}\right)^p \int_\Omega \frac{|\nabla E|_A^p}{E} |u|^p dx \geq 0,$$

where $u \in W_0^{1,p}(\Omega)$. Their approach was to look at functions E which solve

$$\begin{aligned} \mathcal{L}_{A,p}(E) &= \delta_0 &&\text{in } \Omega \\ E &= 0 &&\text{on } \partial\Omega, \end{aligned}$$

where $0 \in \Omega$ and where δ_0 is the Dirac mass at 0. They posed the question as to whether $(\frac{p-1}{p})^p$ is optimal in (9.37)? That was the starting point of C. Cowan [95], who not only showed that this is indeed the case in many more situations, whenever $1 < p < n$, but also initiated a penetrating analysis of general Hardy inequalities, their best constants, their improvements as well as their weighted versions. Cowan's work also included boundary weights initiated by Brezis-Marcus [61], and Brezis-Marcus-Shafrir [62] in the case of the distance to the boundary (Corollary 9.1.4). All results of this chapter and the following one originate in his paper [95].

Important work on multipolar Hardy inequalities was done by Felli-Terracini [138] and Bosi-Dolbeault-Esteban [57] who established that for $\mu \in (0, \frac{(n-2)^2}{4}]$, there exists $0 < C_k < \pi^2$ such that for every $u \in C_c^\infty(\mathbb{R}^n)$,

$$(9.38) \qquad \int_{\mathbb{R}^n} |\nabla u|^2 \, dx \geq \mu \sum_{i=1}^k \int_{\mathbb{R}^n} \frac{u^2}{|x-x_i|^2} \, dx - \frac{C_k + (k+1)\mu}{d^2} \int_{\mathbb{R}^n} u^2 \, dx,$$

where $d = \frac{1}{2}\min\{|x_i - x_j|; i \neq j\}$. Certain improved multipolar Hardy inequalities were recently established by Cacazu-Zuazua [108]. As mentioned above, the following question is still unresolved.

Open problem (9): Find the optimal constant $C > 0$ in the following multipolar Hardy inequality when the number of poles k is larger than 2.

$$(9.39) \qquad \int_\Omega |\nabla u|^2 \, dx \geq C \sum_{i=1}^k \int_\Omega \frac{u^2}{|x-x_i|^2} \, dx \quad \text{for every } u \in H_0^1(\Omega).$$

It is easy to see that the best constant C^* in (9.39) satisfies $\frac{(n-2)^2}{4k} \leq C^* \leq \frac{(n-2)^2}{4}$ and that $C^* = (\frac{n-2}{2})^2$ if and only if $k = 1$.

CHAPTER 10

Improved Hardy Inequalities For General Elliptic Operators

This chapter addresses the possibility of improving the Hardy inequality for general elliptic operators (9.4) in the spirit of Chapters 2 and 4, namely by providing suitable conditions on non necessarily radial potentials V that will yield inequalities of the following type:

$$\int_\Omega |\nabla u|_A^2 dx - \frac{1}{4}\int_\Omega \frac{|\nabla E|_A^2}{E^2} u^2 dx \geq \int_\Omega V(x) u^2 dx.$$

Necessary and sufficient conditions on V are obtained (now in terms of the solvability of a linear PDE) for the above inequality to hold. Analogous results involving improvements are obtained for the weighted versions. Optimal inequalities and improvements are also established for functions in $H^1(\Omega)$.

10.1. General Hardy inequalities with improvements

DEFINITION 10.1.1. Suppose E is an interior weight on Ω with K being the support of μ_E. A non-negative function $V \in C^\infty(\Omega\setminus K)$ is said to be *an improving potential for E* if for all $u \in H_0^1(\Omega)$,

(10.1) $$\int_\Omega |\nabla u|_A^2 dx - \frac{1}{4}\int_\Omega \frac{|\nabla E|_A^2}{E^2} u^2 dx \geq \int_\Omega V(x) u^2 dx.$$

If E is a boundary weight, then V is said to be *an improving potential for E* provided the same condition holds except that it is now supposed to satisfy $V \in C^\infty(\Omega)$.

The next theorem gives necessary and sufficient conditions for V to be an improving potential of E in terms of the solvability of a singular linear equation. For the necessary direction we will need to assume one of these two conditions on Ω.

(B1) If E is an interior weight on Ω with K being the set on which E is infinite, this condition assumes that that there exists a sequence $(\Omega_m)_m$ of non-empty subdomains of Ω which are connected, have a smooth boundary, and satisfy $\Omega_m \subset\subset \Omega\setminus K$, $\Omega_m \subset\subset \Omega_{m+1}$ and $\Omega\setminus K = \cup_m \Omega_m$.

(B2) If E is a boundary weight on Ω, this condition assumes that there exists a sequence $(\Omega_m)_m$ of non-empty subdomains of Ω which are connected, have a smooth boundary, satisfy $\Omega_m \subset\subset \Omega_{m+1}$ and $\Omega = \cup_m \Omega_m$.

THEOREM 10.1.2. *(Improvements for interior weights)* Suppose E is an interior weight on Ω and $0 \leq V \in C^\infty(\Omega\setminus K)$.

(1) If there exists $0 < \theta \in C^2(\Omega\setminus K)$ such that

(10.2) $$\frac{-\mathcal{L}_A(\theta)}{\theta} + \frac{|\nabla E|_A^2}{4E^2} + V \leq 0 \quad \text{in } \Omega\setminus K,$$

then, V is an improving potential for E.

(2) *Conversely, if V is an improving potential for E and that Ω satisfies (B1), then there exists $0 < \theta \in C^\infty(\Omega\backslash K)$ which satisfies (10.2).*

PROOF. (1) Suppose $V \in C^\infty(\Omega\backslash K)$ is non-negative and that θ is as in (10.2). Then $0 < \varphi := E^{-\frac{1}{2}} \in C^2(\Omega\backslash K)$ and satisfies

$$(10.3) \qquad -\mathcal{L}_A(\varphi) + \frac{A\nabla E \cdot \nabla \varphi}{E} + V\varphi \leq 0 \qquad \text{in } \Omega\backslash K.$$

For $u \in C_c^{0,1}(\Omega\backslash K)$, define $v := E^{-\frac{1}{2}}u$ and apply Lemma 9.1.5 to obtain

$$\int_\Omega |\nabla u|_A^2 dx - \frac{1}{4}\int_\Omega \frac{|\nabla E|_A^2}{E^2} u^2 dx = \int_\Omega E|\nabla v|_A^2 dx.$$

Now define $\psi \in C_c^{0,1}(\Omega\backslash K)$ by $v := \varphi\psi$. A calculation shows that

$$(10.4) \qquad E|\nabla v|_A^2 = E\psi^2|\nabla\varphi|_A^2 + E\varphi^2|\nabla\psi|_A^2 + 2E\varphi\psi A\nabla\varphi \cdot \nabla\psi,$$

and integrating by parts the last term over Ω, we obtain

$$\int_\Omega \psi^2 E|\nabla\varphi|_A^2 dx + 2\int_\Omega \varphi\psi E A\nabla\varphi \cdot \nabla\psi dx = \int_\Omega \psi^2\left(\mathcal{L}_A(\varphi)\varphi E - \varphi A\nabla E \cdot \nabla\varphi\right) dx$$
$$= \int_\Omega u^2 \left(\frac{\mathcal{L}_A(\varphi) - \frac{A\nabla E \cdot \nabla\varphi}{E}}{\varphi}\right) dx$$
$$=: Q.$$

By (10.3), we have $Q \geq \int_\Omega V(x)u^2 dx$ and so we see that

$$\int_\Omega |\nabla u|_A^2 dx - \frac{1}{4}\int_\Omega \frac{|\nabla E|_A^2}{E^2} u^2 dx \geq \int_\Omega E\varphi^2 |\nabla\psi|_A^2 dx + \int_\Omega V u^2 dx,$$

for all $u \in C_c^{0,1}(\Omega\backslash K)$. Since $C_c^{0,1}(\Omega\backslash K)$ is dense in $H_0^1(\Omega)$ and using Fatou's lemma, one can show (10.1) holds for all $u \in H_0^1(\Omega)$.

(2) Now suppose $V \in C^\infty(\Omega\backslash K)$ is an improving potential for E and that $(\Omega_m)_m$ is a sequence of subdomains satisfying assumption (B1). Define the elliptic operator P by

$$P(u) := \mathcal{L}_A(u) - \frac{|\nabla E|_A^2}{4E^2}u - Vu.$$

By a standard minimization argument along with the strong maximum principle, there exists $0 < \theta_m \in H_0^1(\Omega_m)$ such that

$$(10.5) \qquad \begin{aligned} P(\theta_m) &= \lambda_m \theta_m &&\text{in } \Omega_m \\ \theta_m &= 0 &&\text{on } \partial\Omega_m, \end{aligned}$$

where $0 \leq \lambda_m$, i.e., (θ_m, λ_m) is the first eigenpair of P in $H_0^1(\Omega_m)$. Since $H_0^1(\Omega_m) \subset H_0^1(\Omega_{m+1})$, we see that λ_m is decreasing and hence there exists $0 \leq \lambda$ such that $\lambda_m \searrow \lambda$. Let $x_0 \in \cap_m \Omega_m$ and suitably scale θ_m such that $\theta_m(x_0) = 1$ for all m. Now fix k and let $m > k+1$. Then

$$P(\theta_m) - \lambda_m \theta_m = 0 \qquad \text{in } \Omega_{k+1}.$$

Apply Harnack's inequality to the operator $P - \lambda_m$ to see that there exists some C_k such that

$$\sup_{\Omega_k}(\theta_m) \leq C_k \inf_{\Omega_k}(\theta_m) \leq C_k.$$

In other words, (θ_m) is bounded in $L^\infty_{loc}(\Omega\backslash K)$. By applying elliptic regularity theory and a bootstrap argument, one sees that $(\theta_m)_{m>k+1}$ is bounded in $C^{1,\alpha}(\Omega_k)$ for $\alpha < 1$, and after applying a diagonal argument one gets that there exists some non-zero $0 \leq \theta \in C^{1,\alpha}(\Omega\backslash K)$ such that $\theta_m \to \theta$ in $C^{1,\alpha}(\Omega_k)$ for all k. Using this convergence, one can pass to the limit in (10.5) to see that $P(\theta) = \lambda\theta$ in $\Omega\backslash K$ and after applying the strong maximum principle on Ω_m, we finally get that $\theta > 0$ in $\Omega\backslash K$. Regularity theory now implies that $\theta \in C^\infty(\Omega\backslash K)$. □

Essentially the same proof applies to the case of boundary weights – except now the measure μ does not drop out – and we get the following.

THEOREM 10.1.3. *(Improvements for boundary weights)* Suppose E is a boundary weight on Ω and $0 \leq V \in C^\infty(\Omega)$.

(1) Suppose $E \in C_0^1(\overline{\Omega})$ and that V is an improving potential for E. If Ω satisfies (B2), then there exists some $0 < \theta \in C^{1,\alpha}(\Omega)$ for all $\alpha < 1$ such that

$$(10.6) \qquad \frac{-\mathcal{L}_A(\theta)}{\theta} + \frac{|\nabla E|_A^2}{4E^2} + V \leq 0 \quad \text{in } \Omega.$$

(2) Suppose there exists some $0 < \varphi \in C^2(\Omega)$ such that

$$(10.7) \qquad \frac{-\mathcal{L}_A(\varphi)}{\varphi} + \frac{A\nabla E \cdot \nabla\varphi}{E\varphi} - \frac{\mu}{2E} + V \leq 0 \quad \text{in } \Omega.$$

Then V is an improving potential for E.

REMARK 10.1.4. Note that putting $\theta := E^{\frac{1}{2}}\varphi$ into (10.7) gives, at least formally, (10.6). Also one can replace μ by the absolutely continuous part of μ in (10.7).

The above theorems can be used in principle for computing best constants without the need for constructing appropriate minimizing sequences. Indeed, if $0 \leq V$ is an improving potential for the interior weight E, let $C(V) > 0$ denote the associated best constant, i.e.,

$$C(V) := \inf\left\{ \frac{\int_\Omega |\nabla u|_A^2 dx - \frac{1}{4}\int_\Omega \frac{|\nabla E|_A^2}{E^2} u^2 dx}{\int_\Omega V u^2 dx} \; : \; u \in H_0^1(\Omega)\backslash\{0\} \right\}.$$

Then one sees that

$$C(V) = \sup\left\{ c > 0 : \exists\, 0 < \theta \in C^2(\Omega\backslash K) \text{ s.t. } \frac{-\mathcal{L}_A(\theta)}{\theta} + \frac{|\nabla E|_A^2}{4E^2} + cV \leq 0 \text{ in } \Omega\backslash K \right\}.$$

Theorem 10.2.1 below will help in phrasing (in some cases) the above condition in terms of solvability of a linear ODE, which should again help in identifying best constants. Towards this end, we establish the following theorem that will lead to explicit examples of improving potentials.

THEOREM 10.1.5. (1) Suppose E is an interior weight on Ω, $0 < \gamma := \min_{\partial\Omega} E$ and $0 < f \in C^2((\gamma, \infty))$. Then for all $u \in C_c^{0,1}(\Omega\backslash K)$

$$\int_\Omega |\nabla u|_A^2 dx - \frac{1}{4}\int_\Omega \frac{|\nabla E|_A^2}{E^2} u^2 dx \geq \int_\Omega \frac{|\nabla E|_A^2}{f(E)}\left(-f''(E) - \frac{f'(E)}{E}\right) u^2 dx.$$

In particular by taking $f(E) := \sqrt{\log(\gamma^{-1}E)}$, we have for all $u \in H_0^1(\Omega)$,

$$(10.8) \qquad \int_\Omega |\nabla u|_A^2 dx - \frac{1}{4}\int_\Omega \frac{|\nabla E|_A^2}{E^2} u^2 dx \geq \frac{1}{4}\int_\Omega \frac{|\nabla E|_A^2}{E^2 \log^2(\gamma^{-1}E)} u^2 dx.$$

If $0 < \gamma = E$ on $\partial\Omega$, then $\frac{1}{4}$ (on the right hand side of (10.8)) is optimal.
(2) Suppose $E \in L^\infty(\Omega)$ is a boundary weight. Then, for all $u \in H_0^1(\Omega)$,

$$(10.9) \quad \int_\Omega |\nabla u|_A^2 dx - \frac{1}{4}\int_\Omega \frac{|\nabla E|_A^2}{E^2} u^2 dx \geq \frac{1}{4}\int_\Omega \frac{|\nabla E|_A^2}{E^2 \log^2\left(\frac{E}{e\|E\|_{L^\infty}}\right)} u^2 dx.$$

Proof: (1) Let E be an interior weight on Ω, $\gamma := \min_{\partial\Omega} E > 0$ and suppose $0 < f \in C^2((\gamma, \infty))$. Put $\varphi := f(E)$ into (10.3) to obtain the result.
Now take $f(E) := \sqrt{\log(\gamma^{-1}E)}$ to obtain (10.8) for all $u \in C_c^{0,1}(\Omega \setminus K)$ and extend to all of $H_0^1(\Omega)$ by density and by Fatou's lemma. We now show $\frac{1}{4}$ is optimal. Fix $0 < t < \frac{1}{2}$ and for $\tau > \frac{1}{2}$ define $u_\tau := E^t \log^\tau(\gamma^{-1}E)$. By Lemma 9.1.6, $u_\tau \in H_0^1(\Omega)$. A computation shows that

$$\frac{\int_\Omega |\nabla u_\tau|_A^2 dx - \frac{1}{4}\int_\Omega \frac{|\nabla E|_A^2}{E^2} u_\tau^2 dx}{\int_\Omega \frac{|\nabla E|_A^2}{E^2 \log^2(E\gamma^{-1})} u_\tau^2 dx} = (t^2 - \frac{1}{4})\frac{\int_\Omega E^{2t-2}|\nabla E|_A^2 \log^{2\tau}(E\gamma^{-1}) dx}{\int_\Omega E^{2t-2}|\nabla E|_A^2 \log^{2\tau-2}(E\gamma^{-1}) dx}$$

$$+\tau^2 + 2t\tau \frac{\int_\Omega E^{2t-2}|\nabla E|_A^2 \log^{2\tau-1}(E\gamma^{-1}) dx}{\int_\Omega E^{2t-2}|\nabla E|_A^2 \log^{2\tau-2}(E\gamma^{-1}) dx}$$

$$= (t^2 - \frac{1}{4})\frac{J_t(\tau+1)}{J_t(\tau)} + \tau^2 + 2t\tau \frac{J_t(\tau+1/2)}{J_t(\tau)},$$

where $J_t(\tau)$ is defined in Lemma 9.1.6. Sending $\tau \searrow \frac{1}{2}$ and using results from Lemma 9.1.6, we see that $\frac{1}{4}$ is optimal.

(2) Suppose $E \in L^\infty(\Omega)$ is a boundary weight on Ω. Here we use the notation from the proof of Lemma 9.1.5; $E_\varepsilon := \eta_\varepsilon * E$, $F_\varepsilon := \mathcal{L}_A(E_\varepsilon)$. Let $0 < f \in C^2((0, \|E\|_{L^\infty}])$. Starting at (9.7) for E_ε and decomposing v as usual, one arrives at

$$(10.10) \quad \int_\Omega |\nabla u|_A^2 dx - \frac{1}{4}\int_\Omega \frac{|\nabla E_\varepsilon|_A^2}{E_\varepsilon^2} u^2 dx \geq \int_\Omega \frac{|\nabla E_\varepsilon|_A^2}{f(E_\varepsilon)}\left(-f''(E_\varepsilon) - \frac{f'(E_\varepsilon)}{E_\varepsilon}\right) u^2 dx$$

$$+ \int_\Omega \left(\frac{f'(E_\varepsilon)}{f(E_\varepsilon)} + \frac{1}{2E_\varepsilon}\right) u^2 F_\varepsilon dx,$$

for all $u \in C_c^\infty(\Omega)$ by using methods similar to the proof of (i). Now take $f(z) := \sqrt{-\log(\frac{z}{e\|E\|_{L^\infty}})}$ and let $u \in C_c^\infty(\Omega)$. Then,

$$\int_\Omega |\nabla u|_A^2 dx - \frac{1}{4}\int_\Omega \frac{|\nabla E_\varepsilon|_A^2}{E_\varepsilon^2} u^2 dx \geq \frac{1}{4}\int_\Omega \frac{|\nabla E_\varepsilon|_A^2}{E_\varepsilon^2 \log^2(\frac{E_\varepsilon}{e\|E\|_{L^\infty}})} u^2 dx + I_\varepsilon,$$

where

$$I_\varepsilon := \frac{1}{2}\int_\Omega \frac{u^2}{E_\varepsilon^2}\left(1 + \frac{1}{\log(\frac{E_\varepsilon}{e\|E\|_{L^\infty}})}\right) F_\varepsilon dx.$$

Using methods similar to ones used in the proof of Lemma 9.1.5, one easily sees that $\lim_{\varepsilon \searrow 0} I_\varepsilon \geq 0$. This, standard results on convolutions, and Fatou's lemma yield the desired inequality for $u \in C_c^\infty(\Omega)$ and eventually to all of $H_0^1(\Omega)$. □

The next theorem allows us to transfer our knowledge of improvements from the non-weighted case to the weighted case, at least in the case where E is an interior weight.

THEOREM 10.1.6. (**Weighted interior improvements**) *Suppose E is an interior weight on Ω and $0 \leq V \in C^\infty(\Omega \backslash K)$. Then the following statements are equivalent:*

(1) *For all $u \in H_0^1(\Omega)$, we have*

$$\int_\Omega |\nabla u|_A^2 dx \geq \frac{1}{4} \int_\Omega \frac{|\nabla E|_A^2}{E^2} u^2 dx + \int_\Omega V u^2 dx. \tag{10.11}$$

(2) *For all $t \neq \frac{1}{2}$ and $u \in X_t$, we have*

$$\int_\Omega E^{2t} |\nabla u|_A^2 dx \geq (t - \frac{1}{2})^2 \int_\Omega |\nabla E|_A^2 E^{2t-2} u^2 dx + \int_\Omega V E^{2t} u^2 dx. \tag{10.12}$$

(3) *For all $u \in X_{\frac{1}{2}}$, we have*

$$\int_\Omega E |\nabla u|_A^2 dx \geq \int_\Omega V E u^2 dx. \tag{10.13}$$

Proof: (1) $\Rightarrow$ (2) If $t \neq \frac{1}{2}$ and $u \in C_c^{0,1}(\Omega \backslash K)$, define $v := E^t u \in C_c^{0,1}(\Omega \backslash K)$. Putting v into (10.11) and performing an integration by parts easily yield (10.12). (2) $\Rightarrow$ (3) Let $u \in C_c^{0,1}(\Omega \backslash K)$ which is an element of X_t for all t. Use (10.12) for u and sending $t \nearrow \frac{1}{2}$ gives (10.13).
(3) $\Rightarrow$ (1) Take $u \in C_c^{0,1}(\Omega \backslash K)$ and $v := E^{\frac{-1}{2}} u \in C_c^{0,1}(\Omega \backslash K)$. Putting v into (10.13) and integrating by parts gives (10.11) for such $u \in C_c^{0,1}(\Omega \backslash K)$. $\square$

Using similar arguments one can obtain a version of Theorem 10.1.6 for the case when E is a boundary weight on Ω.

10.2. Characterization of improving potentials via ODE methods

We now consider a special class of improving potentials V, namely those of the form $V(x) = f(E(x))|\nabla E(x)|_A^2$, where f is a one-dimensional function. In this case, we can use ODE methods to give a necessary and sufficient condition on V (actually) f to be an improving potential for E, at least in the case where E is an interior weight on Ω and $E = \gamma \geq 0$ on $\partial \Omega$. As in Theorem 10.1.2 we assume some geometrical properties of Ω.

THEOREM 10.2.1. (**Interior improvements using ODE methods**) *Suppose E is an interior weight on Ω such that $E = \gamma \geq 0$ on $\partial \Omega$, and that $\Omega_t := \{x \in \Omega : \gamma + \frac{1}{t} < E(x) < t\}$ is connected for sufficiently large t. For a function $0 \leq f \in C^\infty(\gamma, \infty)$, the following statements are equivalent:*

(1) *For all $u \in H_0^1(\Omega)$*

$$\int_\Omega |\nabla u|_A^2 dx - \frac{1}{4} \int_\Omega \frac{|\nabla E|_A^2}{E^2} u^2 dx \geq \int_\Omega f(E) |\nabla E|_A^2 u^2 dx. \tag{10.14}$$

(2) *There exists some $0 < h \in C^2(\gamma, \infty)$ such that*

$$h''(t) + \left(f(t) + \frac{1}{4t^2}\right) h(t) \leq 0, \quad \text{in } (\gamma, \infty). \tag{10.15}$$

Proof: Let E be an interior weight on Ω, $E = \gamma \geq 0$ on $\partial \Omega$ and $0 \leq f \in C^\infty(\gamma, \infty)$. To show that (2) $\Rightarrow$ (1) set $\theta := h(E)$ and use (2) along with Theorem 10.1.2 to get (1).

To prove the reverse direction, let
$$\Omega_m := \{x \in \Omega : \gamma + \frac{1}{t_m} < E(x) < t_m\}.$$
By hypothesis we can take Ω_m to be connected and non-empty for each m. Set
$$X = H^1_{0,E}(\Omega_m) := \{\varphi \in H^1_0(\Omega_m) : \varphi \text{ constant on the level sets of } E\},$$
and note that X is closed in $H^1_0(\Omega_m)$. Consider $X^\perp$ to be the space orthogonal to X in $H^1_0(\Omega_m)$, that is
$$\tilde{\varphi} \in X^\perp \quad \text{if and only if} \quad \int_{\Omega_m} \nabla\varphi \cdot \nabla\tilde{\varphi} = 0 \quad \forall \varphi \in X.$$
Define
$$Y := \{g(E); g \in H^1(\gamma + t_m^{-1}, t_m), \int_{\gamma+t_m^{-1}}^{t_m} g'(\tau)^2 d\tau < \infty \ \& \ g(\gamma + t_m^{-1}) = g(t_m) = 0\},$$
and note that the pointwise boundary condition on g makes sense since $g \in H^1(\gamma + t_m^{-1}, t_m) \subset C^{0,\frac{1}{2}}(\gamma + t_m^{-1}, t_m)$. We start by proving the following

Claim 1: $Y = X$.

Indeed, if $\varphi \in X$, it is constant along level sets of E and hence only depends on the value of E and so we can write $\varphi = g(E)$ for some function g. Using the co-area formula, we have

$$\int_{\Omega_m} |\nabla\varphi|^2 \, dx = \int_\Omega g'(E)^2 |\nabla E| |\nabla E| \, dx = \int_{\gamma+t_m^{-1}}^{t_m} \left(\int_{\{E=\tau\}} g'(\tau)^2 |\nabla E| dS(x) \right) d\tau$$
$$= \int_{\gamma+t_m^{-1}}^{t_m} g'(\tau)^2 \left(\int_{\partial\{E>\tau\}} |\nabla E| dS(x) \right) d\tau$$
$$= \int_{\gamma+t_m^{-1}}^{t_m} g'(\tau)^2 \left(\int_{\{E>\tau\}} -\Delta E dx \right) d\tau.$$

But recall that E is an interior weight with $-\Delta E = \mu$ where μ is nonnegative nonzero finite measure supported on K. So we have that $\int_{\{E>\tau\}} -\Delta E \, dx = \mu(K) > 0$, and hence we have
$$\int_{\Omega_m} |\nabla\varphi|^2 \, dx = \mu(K) \int_{\gamma+t_m^{-1}}^{t_m} g'(\tau)^2 d\tau.$$

Now we show the following:

Claim 2: For each $m \geq 1$, there exists $0 < \varphi_m \in X$ such that

(10.16) $\qquad -\Delta\varphi_m = |\nabla E|^2 (f(E) + \frac{1}{4E^2})\varphi_m + \lambda_m |\nabla E|^2 \varphi_m$ on Ω_m.

For that, define
$$F_m(\varphi) := \frac{1}{2} \int_{\Omega_m} |\nabla\varphi|^2 \, dx - \frac{1}{2} \int_{\Omega_m} |\nabla E|^2 (f(E) + \frac{1}{4E^2})\varphi^2 \, dx,$$
and set
$$M_m := \{\varphi \in X : J(\varphi) := \frac{1}{2} \int_{\Omega_m} |\nabla E|^2 \varphi^2 \, dx = 1\}.$$

Since $F(\varphi) \geq 0$, then by a standard minimization argument, there exists some $0 \leq \varphi_m \in M$ (hence non-zero), which minimizes F_m over M_m. So there is some $\lambda_m > 0$ such that $F'(\varphi_m) = \lambda_m J'(\varphi_m)$ in X, i.e.,

$$\int_{\Omega_m} \nabla\varphi_m \cdot \nabla\varphi\, dx = \int_{\Omega_m} |\nabla E|^2 \left(f(E) + \frac{1}{4E^2} + \lambda_m \right) \varphi_m \varphi\, dx \qquad \forall \varphi \in X.$$

Let $\psi \in C_c^\infty(\Omega_m)$ and write $\psi = \varphi + \tilde{\varphi} \in X + X^\perp$. Using the fact that $\varphi_m \in X$ we easily see that

$$\int_{\Omega_m} \nabla\varphi_m \cdot \nabla\psi\, dx = \int_{\Omega_m} |\nabla E|^2 \left(f(E) + \frac{1}{4E^2} + \lambda_m \right) \varphi_m \psi\, dx - I,$$

where

$$I := \int_{\Omega_m} |\nabla E|^2 \left(f(E) + \frac{1}{4E^2} + \lambda_m \right) \varphi_m \tilde{\varphi}\, dx.$$

We now show that $I = 0$. Since $\tilde{\varphi} \in X^\perp$, we have for all $\varphi \in X$, and hence for any $g \in C^2$ such that $\varphi = g(E)$,

$$\begin{aligned}
0 &= \int_{\Omega_m} \nabla\varphi \cdot \nabla\tilde{\varphi}\, dx \\
&= \int_{\Omega_m} g'(E) \nabla E \cdot \nabla\tilde{\varphi}\, dx \\
&= -\int_{\Omega_m} g''(E)|\nabla E|^2 \tilde{\varphi}\, dx + \int_{\Omega_m} g'(E)(-\Delta E)\, dx + \int_{\partial\Omega_m} \tilde{\varphi} g'(E) \nabla E \cdot \nu \\
&= -\int_{\Omega_m} g''(E)|\nabla E|^2 \tilde{\varphi}\, dx,
\end{aligned}$$

since the boundary integral is zero and the integral involving $-\Delta E$ is also zero since $-\Delta E = \mu$ is supported on K which is in the complement of Ω_m.

Let now g_m be such that $\varphi_m = g_m(E)$ and solve the boundary value problem

$$g_0''(\tau) = \left(f(\tau) + \tfrac{1}{4\tau^2} + \lambda_m \right) g_m(\tau) \quad \text{for } \gamma + t_m^{-1} < \tau < t_m,$$

with boundary conditions $g(\gamma + t_m^{-1}) = g(t_m) = 0$. This has a solution g_0 in $C^{2,\frac{1}{2}}$, and therefore

$$0 = \int_{\Omega_m} g_0''(E)|\nabla E|^2 \tilde{\varphi}\, dx \qquad \forall \varphi \in X.$$

By using the boundary value problem, one gets

$$0 = \int_{\Omega_m} \left(f(E) + \frac{1}{4E^2} + \lambda_m \right) g_m(E)|\nabla E|^2 \tilde{\varphi}\, dx$$

or

$$0 = \int_{\Omega_m} |\nabla E|^2 \left(f(E) + \frac{1}{4E^2} + \lambda_m \right) \varphi_m \tilde{\varphi}\, dx,$$

which means that $I = 0$. It follows that $\varphi_m \in X$ and satisfies (10.16). Now the maximum principle insures that $\varphi_m > 0$ and Claim 2) is proved.

To finish the proof of 1) $\Rightarrow$ 2) in Theorem 10.2.1, we first note that even though Claims 1) and 2) were proved when $A = I$, they do hold for any uniformly positive definite $A(x)$ with smooth coefficients on Ω. In other words, we have a sequence $\varphi_m > 0$ with $\varphi_m = 0$ on $\partial\Omega_m$ such that

(10.17) $\qquad \mathcal{L}_A(\varphi_m) = |\nabla E|^2 (f(E) + \tfrac{1}{4E^2})\varphi_m + \lambda_m |\nabla E|^2 \varphi_m$ on Ω_m.

Since $H^1_{0,E}(\Omega_m) \subset H^1_{0,E}(\Omega_{m+1})$ one sees that λ_m is decreasing and from (10.14) that $\lambda_m \geq 1$ and hence there exists some $\lambda \geq 1$ such that $\lambda_m \searrow \lambda$. By suitably scaling φ_m as before and after an application of Harnack's inequality we can assume that $\varphi_m \to \varphi$ in $C^{1,\alpha}_{loc}(\Omega \backslash K)$ where $\varphi \geq 0$ is nonzero and constant on level sets of E. Passing to the limit shows that

(10.18) $$\mathcal{L}_A(\varphi) = |\nabla E|^2(f(E) + \tfrac{1}{4E^2})\varphi + \lambda |\nabla E|^2 \varphi \quad \text{on } \Omega \backslash K.$$

The strong maximum principle argument shows that $\varphi > 0$ in $\Omega \backslash K$. But φ is constant on level sets of E, hence $\varphi = h(E)$ for some $0 < h$ in (γ, ∞) and since φ is smooth on $\Omega \backslash K$ we see that h is smooth on (γ, ∞). Writing the equation for φ in terms of h gives

$$-h''(E)|\nabla E|^2_A = h(E)\left(f(E) + \frac{1}{4E^2} + \lambda\right)|\nabla E|^2_A \quad \text{in } \Omega \backslash K.$$

Using Hopf's lemma, we can cancel the gradients and we are done. □

Using the vast knowledge of ODE's one can use the above theorem to obtain various results concerning potentials of the form $V(x) = |\nabla E|^2_A f(E)$. We can establish, for example, the following negative result.

COROLLARY 10.2.1. *Suppose E is an interior weight on Ω and $E = 0$ on $\partial \Omega$. Then there is no $0 < f \in C(0, \infty)$ such that*

$$\int_\Omega |\nabla u|^2_A dx - \frac{1}{4}\int_\Omega \frac{|\nabla E|^2_A}{E^2} u^2 dx \geq \int_\Omega f(E) |\nabla E|^2_A u^2 dx, \quad \text{for all } u \in H^1_0(\Omega).$$

Proof: Suppose there is such a function f. Using the proof of Theorem 10.2.1 one sees that there is some $0 < h \in C^2(0, \infty)$ such that

$$h''(t) + \lambda\left(f(t) + \frac{1}{4t^2}\right) h(t) = 0,$$

in $(0, \infty)$ where $\lambda \geq 1$. Now set $h(t) = \sqrt{t}\, y(t)$ to see that

$$0 = y''(t) + \frac{y'(t)}{t} + y(t)\left(\lambda f(t) + \frac{\lambda - 1}{4t^2}\right),$$

in $(0, \infty)$ and $y(t) > 0$. But Lemma 1.1.2.(4) shows that this is impossible. □

Other than some regularity issues this ODE approach extends immediately to the case where E is a boundary weight in Ω, which we shall consider here in the context of an extension of an inequality by Avkhadiev-Wirths. For that suppose μ is a nonnegative nonzero locally finite measure in Ω (possibly unbounded) and $0 < E \in L^\infty(\Omega)$ is a solution to

$$\begin{aligned}
\mathcal{L}_A(E) &= \mu & \text{in } \Omega \\
|\nabla E|_A &= 1 & \text{a.e. in } \Omega \\
E &= 0 & \text{on } \partial\Omega.
\end{aligned}$$

THEOREM 10.2.2. *Suppose E is as above, and let $\lambda_0 = 0.940...$ denote the Lambs constant, that is the first positive zero of $J_0(t) - 2tJ_1(t)$ where J_n is the Bessel function of order n. Then for all $u \in C_c^\infty(\Omega)$,*

(10.19) $$\int_\Omega |\nabla u|^2_A dx \geq \frac{1}{4}\int_\Omega \frac{u^2}{E^2} dx + \frac{\lambda_0^2}{\|E\|^2_{L^\infty}} \int_\Omega u^2 dx.$$

Proof: Let E be as above and extend it to all of $\mathbb{R}^n$ by setting $E = 0$ on $\mathbb{R}^n \backslash \overline{\Omega}$. Let E_ε denote the ε-mollification of E and $F_\varepsilon := \mathcal{L}_A(E_\varepsilon)$. Returning to the proof of Theorem 10.1.5 (ii) we have

$$\int_\Omega |\nabla u|_A^2 dx - \frac{1}{4}\int_\Omega \frac{|\nabla E_\varepsilon|_A^2}{E_\varepsilon^2} u^2 dx \geq \int_\Omega \frac{|\nabla E_\varepsilon|_A^2}{f(E_\varepsilon)}\left(-f''(E_\varepsilon) - \frac{f'(E_\varepsilon)}{E_\varepsilon}\right) u^2 dx + I_\varepsilon,$$

where

$$I_\varepsilon := \int_\Omega \left(\frac{f'(E_\varepsilon)}{f(E_\varepsilon)} + \frac{1}{2E_\varepsilon}\right) u^2 F_\varepsilon dx,$$

for $u \in C_c^\infty(\Omega)$ and $0 < f \in C^2((0, \|E\|_{L^\infty}])$. Now set $\lambda := \frac{\lambda_0^2}{\|E\|_{L^\infty}^2}$ where λ_0 is Lambs constant and define $f(t) := J_0(\sqrt{\lambda} t)$. It is possible to show that

$$f(t) > 0, \qquad \frac{1}{f(t)}\left(-f''(t) - \frac{f'(t)}{t}\right) = \lambda, \qquad l(t) := \frac{f'(t)}{f(t)} + \frac{1}{2t} \geq 0$$

in $(0, \|E\|_{L^\infty})$. Fixing $u \in C_c^\infty(\Omega)$ and substituting this f into the above gives

$$\int_\Omega |\nabla u|_A^2 dx - \frac{1}{4}\int_\Omega \frac{|\nabla E_\varepsilon|_A^2}{E_\varepsilon^2} u^2 dx \geq \frac{\lambda_0^2}{\|E\|_{L^\infty}^2} \int_\Omega |\nabla E_\varepsilon|_A^2 u^2 dx + I_\varepsilon,$$

after noting that $\|E_\varepsilon\|_{L^\infty} \leq \|E\|_{L^\infty}$ and where $I_\varepsilon := \int_\Omega l(E_\varepsilon) u^2 F_\varepsilon dx$. It is possible to show that $l \in C^\infty((0, \|E\|_{L^\infty}])$. A standard argument shows that $l(E_\varepsilon) u \to l(E) u$ in $H_0^1(\Omega)$ and $u F_\varepsilon dx \rightharpoonup u \mu$ in $H^{-1}(\Omega)$ and hence one can conclude that $\liminf_{\varepsilon \searrow 0} I_\varepsilon \geq 0$. Passing to the limit (as $\varepsilon \searrow 0$) in the remaining integrals gives the desired result. $\square$

We can now deduce the following inequality of Avkhadiev and Wirths. Given a domain Ω in $\mathbb{R}^n$ we say it has *finite in-radius* if $\delta(x) := \mathrm{dist}(x, \partial\Omega)$ is bounded in Ω. Note that there are unbounded convex domains –such as a cylinder– with infinite diameter but with a finite in-radius.

THEOREM 10.2.3. **(Avkhadiev-Wirths)** *Suppose Ω is a convex domain in $\mathbb{R}^n$ with finite in-radius δ. Then, the following holds for all $u \in H_0^1(\Omega)$,*

(10.20) $\qquad \int_\Omega |\nabla u|^2 dx \geq \frac{1}{4}\int_\Omega \frac{u^2}{\delta(x)^2} dx + \frac{\lambda_0^2}{\|\delta\|_{L^\infty}^2} \int_\Omega u^2 dx,$

and the inequality is optimal.

10.3. Hardy inequalities on $H^1(\Omega)$

Let K be a compact subset of Ω with $\dim_{box}(K) < n-2$. We shall use that in this case $C_c^{0,1}(\overline{\Omega}\backslash K)$ is dense in $H^1(\Omega)$.

DEFINITION 10.3.1. We say that E is a Neumann interior weight on Ω provided there exists some compact $K \subset \Omega$, $\dim_{box}(K) < n-2$ and $E \in C^\infty(\overline{\Omega}\backslash K)$ such that $E = \infty$ on K, $\inf_\Omega E > 0$, $\mathcal{L}_A(E) + E$ is a nonnegative nonzero measure μ whose support is K, and $A\nabla E \cdot \nu = 0$ on $\partial\Omega$, where $\nu(x)$ denotes the outward normal vector at $x \in \partial\Omega$.

THEOREM 10.3.2. *Suppose E is a Neumann interior weight on Ω. Then,*

(1) *For $u \in C_c^{0,1}(\overline{\Omega}\backslash K)$ and $v := E^{\frac{-1}{2}} u$, we have*

(10.21) $\qquad \int_\Omega |\nabla u|_A^2 dx + \frac{1}{2}\int_\Omega u^2 dx \geq \frac{1}{4}\int_\Omega \frac{|\nabla E|_A^2}{E^2} u^2 dx + \int_\Omega E|\nabla v|_A^2 dx.$

(2) For all $u \in H^1(\Omega)$, we have

(10.22) $$\int_\Omega |\nabla u|_A^2 dx + \frac{1}{2}\int_\Omega u^2 dx \geq \frac{1}{4}\int_\Omega \frac{|\nabla E|_A^2}{E^2} u^2 dx.$$

Moreover $\frac{1}{4}$ and $\frac{1}{2}$ are optimal in the sense that if one fixes $\frac{1}{4}$ then $\frac{1}{2}$ is optimal and vice-versa. Also the inequality is not attained.

PROOF. (1) Let $u \in C_c^{0,1}(\overline{\Omega}\backslash K)$ and define $v := E^{\frac{-1}{2}} u$. Then

$$|\nabla u|_A^2 = E|\nabla v|_A^2 + \frac{|\nabla E|_A^2}{4E^2} u^2 + v\nabla v \cdot A\nabla E,$$

and integrating this over Ω gives

(10.23) $$\int_\Omega |\nabla u|_A^2 dx + \frac{1}{2}\int_\Omega u^2 dx = \frac{1}{4}\int_\Omega \frac{|\nabla E|_A^2}{E^2} u^2 dx + \int_\Omega E|\nabla v|_A^2 dx.$$

(2) Using (1) and the fact that $C_c^{0,1}(\overline{\Omega}\backslash K)$ is dense in $H^1(\Omega)$ one obtains (10.22) for all $u \in H^1(\Omega)$. We now show the constants are optimal.

We first show that $E^t \in H^1(\Omega)$ for $0 < t < \frac{1}{2}$. As in the proof of Lemma 9.1.6, one can proceed with the following formal calculations, which can be justified by first regularizing the measure, obtaining approximate solutions and then passing to the limit. Fix $0 < t < \frac{1}{2}$ and multiply $\mathcal{L}_A(E) + E = \mu$ by E^{2t-1} and integrate over Ω using integration by parts and the fact that $E = \infty$ on K along with the boundary conditions of E to see that

(10.24) $$\int_\Omega E^{2t} dx = (1-2t)\int_\Omega E^{2t-2}|\nabla E|_A^2 dx,$$

which shows that $E^t \in H^1(\Omega)$ for $0 < t < \frac{1}{2}$. To show the constants are optimal, use E^t as a minimizing sequence as $t \nearrow \frac{1}{2}$. A computation shows

$$\frac{\int_\Omega |\nabla E^t|_A^2 dx + \frac{1}{2}\int_\Omega E^{2t} dx}{\int_\Omega \frac{|\nabla E|_A^2}{E^2} E^{2t} dx} = t^2 + \frac{1}{2} - t,$$

and we see that $\frac{1}{4}$ is optimal. One similarly shows that $\frac{1}{2}$ is optimal.

To show that the inequality is not attained, we again hold on to the extra term that we dropped in the above calculations. This term is positive for non-zero $u \in H^1(\Omega)$ provided $E^{\frac{1}{2}} \notin H^1(\Omega)$ which is the case in view of (10.24). □

EXAMPLE 10.3.3. Let B denote the unit ball in $\mathbb{R}^3$ and set $E(x) := |x|^{-1} e^{|x|}$. A computation shows that

$$-\Delta E + E = 4\pi^2 \delta_0 \quad \text{in } B,$$

where $\partial_\nu E = 0$ on ∂B. Here δ_0 is the Dirac mass at 0. Putting E into (10.22) we see that

$$\int_B |\nabla u|^2 dx + \frac{1}{2}\int_B u^2 dx \geq \frac{1}{4}\int_B \frac{(1-|x|)^2}{|x|^2} u^2 dx, \quad u \in H^1(B).$$

Also the constants are optimal and are not attained.

10.3. HARDY INEQUALITIES ON $H^1(\Omega)$

We now examine weighted versions of (10.22). Suppose E is a Neumann interior weight on Ω and as usual we let K denote the support of μ. For $t \neq \frac{1}{2}$ and $u \in C_c^{0,1}(\overline{\Omega}\backslash K)$ we define

$$\|u\|_t^2 := \begin{cases} \int_\Omega E^{2t}|\nabla u|^2 dx + \int_\Omega E^{2t}u^2 dx & t < \frac{1}{2} \\ \int_\Omega E^{2t}|\nabla u|^2 dx & t > \frac{1}{2}, \end{cases}$$

and we let Y_t denote the completion of $C_c^{0,1}(\overline{\Omega}\backslash K)$ with respect to this norm.

THEOREM 10.3.4. *Suppose E is a Neumann interior weight on Ω and $t \neq \frac{1}{2}$. Then,*

$$\int_\Omega E^{2t}|\nabla u|_A^2 dx + \left(\frac{1}{2} - t\right)\int_\Omega E^{2t}u^2 dx \geq \left(t - \frac{1}{2}\right)^2 \int_\Omega E^{2t-2}|\nabla E|_A^2 u^2 dx,$$

for all $u \in Y_t$. Moreover the constants are optimal and not attained.

Note in particular that for $t > \frac{1}{2}$ one only has a gradient term on the left hand side and so we can conclude that $C^\infty(\overline{\Omega})$ is not contained in Y_t for $t > \frac{1}{2}$.

Proof: Suppose E is a Neumann interior weight on Ω, $t \neq \frac{1}{2}$ and let $u \in C_c^{0,1}(\overline{\Omega}\backslash K)$. Putting $E^t u$ into (10.21) gives

$$\int_\Omega E^{2t}|\nabla u|_A^2 dx + \left(\frac{1}{2} - t\right)\int_\Omega E^{2t}u^2 dx \geq \left(t - \frac{1}{2}\right)^2 \int_\Omega E^{2t-2}|\nabla E|_A^2 u^2 dx$$
$$+ \int_\Omega E|\nabla w|_A^2 dx,$$

where $w := E^{t-\frac{1}{2}}u$. To show the constants are optimal one takes the same approach as in Theorem 9.3.2. To show the optimal constants are not obtained otherwise, suppose we have equality for some nonzero $u \in Y_t$. It is then easily seen that $\sqrt{E} \in H^1(\Omega)$ which we know is not the case. $\square$

We now examine improvements of (10.22).

THEOREM 10.3.5. *Suppose E is a Neumann interior weight on Ω. Then,*

(1) *Suppose $V \in C^\infty(\Omega\backslash K)$ is such that for some $0 < \varphi \in C^2(\Omega\backslash K) \cap C^1(\overline{\Omega}\backslash K)$, we have*

(10.25) $\quad -\mathcal{L}_A(\varphi) + \frac{A\nabla E \cdot \nabla \varphi}{E} + V\varphi \leq 0$ *in $\Omega\backslash K$ and $A\nabla\varphi \cdot \nu \geq 0$ on $\partial\Omega$.*

Then, for all $u \in H^1(\Omega)$,

(10.26) $\quad \displaystyle\int_\Omega |\nabla u|_A^2 + \frac{1}{2}\int_\Omega u^2 dx - \frac{1}{4}\int_\Omega \frac{|\nabla E|_A^2}{E^2}u^2 dx \geq \int_\Omega V(x)u^2 dx.$

(2) *Conversely, suppose $0 \leq V \in C^\infty(\overline{\Omega}\backslash K)$ is such that (10.26) holds for all $u \in H^1(\Omega)$, and that $\{x \in \Omega : E(x) < t\}$ is connected for sufficiently large t, then there exists $0 < \theta \in C^\infty(\overline{\Omega}\backslash K)$ such that*

(10.27) $\quad -\mathcal{L}_A(\theta) - \frac{\theta}{2} + \frac{|\nabla E|_A^2}{4E^2}\theta + V\theta \leq 0$ *in $\Omega\backslash K$ and $A\nabla\theta \cdot \nu = 0$ on $\partial\Omega$.*

Note that one can go from (10.25) to (10.27) by using the change of variables $\theta = \varphi E^{\frac{1}{2}}$ in the case that $A\nabla\varphi \cdot \nu = 0$ on $\partial\Omega$.

PROOF. The proof is similar to the proof of Theorem 10.1.2. $\square$

REMARK 10.3.6. One can obtain an analogous version of Theorem 10.1.6 for the case where E is an interior weight on Ω satisfying a Neumann boundary condition.

10.4. Hardy inequalities for exterior and annular domains

In this section we obtain optimal Hardy inequalities which are valid on exterior and annular domains. These inequalities will be valid for functions u which are nonzero on various portions of the boundary. For simplicity we only consider the case where $A(x)$ is the identity matrix and hence $\mathcal{L}_A = -\Delta$. The results immediately generalize to the case where $A(x)$ is not the identity matrix. We first examine the exterior domain case.

Condition (Ext.): We suppose that $E > 0$ in $\mathbb{R}^n$, $-\Delta E$ is a nonnegative nonzero finite measure (which we denote by μ) with compact support K and we let Ω denote a connected exterior domain in $\mathbb{R}^n$ with $\text{dist}(K, \Omega) > 0$. In addition we assume that the compliment of Ω denoted by Ω^c is connected, that $\lim_{|x| \to \infty} E = 0$ and $\partial_\nu E \geq 0$ on $\partial \Omega$.

We will work in the following function space. Let $D^1(\Omega \cup \partial\Omega)$ denote the completion of $C_c^\infty(\Omega \cup \partial\Omega)$ with respect to the norm $\|\nabla u\|_{L^2(\Omega)}$. Note that we don't require u to be zero on the boundary of $\partial\Omega$. We then have the following theorem.

THEOREM 10.4.1. *Suppose E, μ, K, Ω are as in condition (Ext.). Then,*
(1) For all $u \in D^1(\Omega \cup \partial\Omega)$ we have

$$\int_\Omega |\nabla u|^2 dx \geq \frac{1}{4} \int_\Omega \frac{|\nabla E|^2}{E^2} u^2 dx. \tag{10.28}$$

Moreover the constant is optimal and not attained.
(2) For all $u \in D^1(\Omega \cup \partial\Omega)$ we have

$$\int_\Omega |\nabla u|^2 dx \geq \frac{1}{4} \int_\Omega \frac{|\nabla E|^2}{E^2} u^2 dx + \frac{1}{2} \int_{\partial\Omega} \frac{u^2 \partial_\nu E}{E} dS(x). \tag{10.29}$$

Proof: Let $u \in C_c^\infty(\Omega \cup \partial\Omega)$ and set $v := E^{\frac{-1}{2}} u$. As before, we have

$$|\nabla u|^2 - \frac{|\nabla E|^2 u^2}{4E^2} = E|\nabla v|^2 + v \nabla v \cdot \nabla E, \quad \text{in } \Omega. \tag{10.30}$$

Integrating the last term by parts gives

$$\int_\Omega v \nabla v \cdot \nabla E \, dx = \frac{1}{2} \int_{\partial\Omega} \frac{u^2 \partial_\nu E}{E} dS(x).$$

We obtain (10.29) by integrating (10.30) over Ω and since $\partial_\nu E \geq 0$ on $\partial\Omega$ we obtain (10.28). We now show the constant is optimal. For large R, we set $\Omega_R := \Omega \cap B_R$ where B_R is the ball centered at 0 with radius R. Let $\frac{1}{2} < t < 1$, multiply $-\Delta E = \mu$ by E^{2t-1} and integrate over Ω_R to obtain

$$(2t-1) \int_{\Omega_R} E^{2t-2} |\nabla E|^2 dx = \int_{\partial\Omega} \partial_\nu E E^{2t-1} dS(x) + \int_{\partial B_R} \partial_\nu E E^{2t-1} dS(x).$$

Using a Newtonian potential argument one can show that as $R \to \infty$ the surface integral over the ball B_R goes to zero. It follows that

$$(2t-1) \int_\Omega E^{2t-2} |\nabla E|^2 dx = \int_{\partial\Omega} \partial_\nu E E^{2t-1} dS(x), \tag{10.31}$$

and so $\int_\Omega |\nabla E^t|^2 dx < \infty$. Using this and a standard cut-off function argument, one sees that $E^t \in D^1(\Omega \cup \partial\Omega)$. Now one uses E^t as a minimizing sequence as $t \searrow \frac{1}{2}$ to show that $\frac{1}{4}$ is optimal. We now show the constant is not attained. Assume that $x_0 \in \partial\Omega$ is such that $E(x_0) = \min_{\partial\Omega} E$. Then by Hopf's lemma $\partial_\nu E(x_0) > 0$ and so using this along with continuity and (10.31) one sees that $E^{\frac{1}{2}} \notin D^1(\Omega \cup \partial\Omega)$. Now to finish the proof it will be sufficient to show that $\int_\Omega E|\nabla v|^2 dx > 0$ for all nonzero $u \in D^1(\Omega \cup \partial\Omega)$. But the only nonzero u for which this integral is zero are multiples of $E^{\frac{1}{2}}$ which are not in $D^1(\Omega \cup \partial\Omega)$. $\square$

EXAMPLE 10.4.2. Take Ω to be an exterior domain in $\mathbb{R}^n$ where $n \geq 3$, $0 \notin \overline{\Omega}$, and such that $\nu(x) \cdot x \leq 0$ on $\partial\Omega$, where $\nu(x)$ is the outward pointing normal. Define $E(x) := |x|^{2-n}$ and use Theorem 10.4.1 to see that

$$(10.32) \qquad \int_\Omega |\nabla u|^2 dx \geq \left(\frac{n-2}{2}\right)^2 \int_\Omega \frac{u^2}{|x|^2} dx,$$

for all $u \in D^1(\Omega \cup \partial\Omega)$. Moreover the constant is optimal and not attained. In fact using (ii) from the same theorem shows that we can add the following nonnegative term to the right hand side of (10.32):

$$\frac{(n-2)}{2} \int_{\partial\Omega} \frac{u^2(-x \cdot \nu)}{|x|^2} dS(x).$$

We now examine the annular domain case.

Condition (Annul.): We assume that $\Omega_1 \subset\subset \Omega_2$ are two bounded connected domains in $\mathbb{R}^n$ with smooth boundaries and $\Omega := \Omega_2 \setminus \overline{\Omega_1}$ is connected. In addition, we assume that $E > 0$ in Ω_2 with $-\Delta E = \mu$ in Ω_2 where μ is a nonnegative nonzero finite measure supported on $K \subset \Omega_1$. We also suppose that $\partial_\nu E \leq 0$ on $\partial\Omega_1$.

THEOREM 10.4.3. Suppose Ω, K, E are as in condition (Annul.). Then,

(1) For all $u \in H^1(\Omega)$ with $u = 0$ on $\partial\Omega_2$, we have

$$(10.33) \qquad \int_\Omega |\nabla u|^2 dx \geq \frac{1}{4} \int_\Omega \frac{|\nabla E|^2}{E^2} u^2 dx.$$

Moreover the constant is optimal and not attained if we assume that $E = 0$ on $\partial\Omega_2$.

(2) For all $u \in H^1(\Omega)$ with $u = 0$ on $\partial\Omega_2$, we have

$$(10.34) \qquad \int_\Omega |\nabla u|^2 dx \geq \frac{1}{4} \int_\Omega \frac{|\nabla E|^2}{E^2} u^2 dx + \frac{1}{2} \int_{\partial\Omega} \frac{u^2 \partial_\nu E}{E} dS(x).$$

Proof: The proofs of (10.33) and (10.34) are similar to the previous theorem so we omit the details. We now show the constant is optimal. Let $H_0^1(\Omega \cup \partial\Omega_1)$ denote $\{u \in H^1(\Omega) : u = 0$ on $\partial\Omega_2\}$. Again we multiply $-\Delta E = \mu$ by E^{2t-1} for $\frac{1}{2} < t < 1$ and integrate over Ω to obtain

$$(2t-1) \int_\Omega E^{2t-2} |\nabla E|^2 dx = -\int_{\partial\Omega_1} \partial_\nu E E^{2t-1} dS(x),$$

which shows that $E^t \in H_0^1(\Omega \cup \partial\Omega_1)$. From this, one obtains

$$\lim_{t \searrow \frac{1}{2}} (2t-1) \int_\Omega E^{2t-2} |\nabla E|^2 dx = \mu(\Omega_1) > 0,$$

which shows that $E^{\frac{1}{2}} \notin H_0^1(\Omega \cup \partial\Omega_1)$. To see the constant is optimal one uses the same minimizing sequence as in the previous theorem. To see the constant is not attained one uses the fact that $E^{\frac{1}{2}} \notin H_0^1(\Omega \cup \partial\Omega_1)$. □

REMARK 10.4.4. These inequalities have analogous weighted versions and using the methods developed earlier one easily obtains results concerning improvements.

EXAMPLE 10.4.5. Assume that $0 \in \Omega_1 \subset\subset B_R \subset \mathbb{R}^2$ where Ω_1 is connected and B_R is the open ball centered at 0 with radius R. In addition we assume that $x \cdot \nu(x) \geq 0$ on $\partial\Omega_1$ where ν is the outward pointing normal. Define $\Omega := B_R \setminus \overline{\Omega_1}$, which we assume is connected, and set $E(x) := -\log(R^{-1}|x|)$. Then by the above mentioned results on annular domains one has

$$\int_\Omega |\nabla u|^2 dx \geq \frac{1}{4} \int_\Omega \frac{u^2}{|x|^2 \log^2(R^{-1}|x|)} dx,$$

for all $u \in H_0^1(\Omega \cup \Omega_1)$. Moreover the constant is optimal and not attained.

10.5. Further comments

Results presented in this chapter are all due to C. Cowan [**95**]. His characterization of improving potentials follow the approach of Ghoussoub-Moradifam [**162**], but since he is not restricted to radial potentials, the formulation is in terms of partial differential inequalities. Cowan also considers improvements in the case of boundary weights, which were already started by H. Brezis and M. Marcus [**61**] who had shown that if Ω is a convex subset of $\mathbb{R}^n$, then

$$\int_\Omega |\nabla u|^2 dx \geq \frac{1}{4} \int_\Omega \frac{u^2}{\delta^2} dx + \frac{1}{4\mathrm{diam}^2(\Omega)} \int_\Omega u^2 dx \qquad u \in H_0^1(\Omega).$$

The inequality of Avkhadiev and Wirths [**32**], which is extended in Theorem 10.2.2 by Cowan [**95**] is actually a refinement of this result.

Open problems (10): 1) Develop suitable characterizations for a pair of functions $(W, \tilde{W})$ (radial or not) in order for the following to hold for any $u \in H_0^1(\Omega)$:

$$\int_\Omega W(x) |\nabla u|_A^2 dx \geq \int_\Omega \tilde{W}(x) u^2 dx.$$

Note that the above chapter dealt with the cases where $W \equiv 1$ and $\tilde{W}(x) = \frac{1}{4} \frac{|\nabla E|_A^2}{E^2} + V(x)$, where $-\mathrm{div}(A\nabla E) dx$ is a nonnegative nonzero finite measure on Ω.

2) Find explicit (inverse power type and logarithmic) improvements for the multipolar Hardy inequality.

CHAPTER 11

Regularity and Stability of Solutions in Non-Self-Adjoint Problems

Hardy-type inequalities for general elliptic operators are used here to study various nonlinear equations involving advective terms. We consider first the regularity of the extremal solution of the nonlinear eigenvalue problem

$$\begin{cases} -\Delta u + c(x) \cdot \nabla u = \frac{\lambda}{(1-u)^2} & \text{in } \Omega, \\ u = 0 & \text{on } \partial\Omega, \end{cases}$$

where Ω is a bounded domain in $\mathbb{R}^N$ and $c(x)$ is a smooth bounded vector field on $\bar{\Omega}$.

We then study the existence vs. non-existence of non-trivial entire semi-stable solutions of equations of the form

$$-\text{div}(\omega_1 \nabla u) = \omega_2 f(u) \quad \text{in } \mathbb{R}^n,$$

where ω_1, ω_2 are two positive smooth weights. We consider the cases $f(u) = e^u, u^p$ where $p > 1$ and $-u^{-p}$ where $p > 0$. We present various non-existence results which depend on the dimension N and also on p and on the behaviour of ω_1, ω_2 near infinity. The class of weights $\omega_1(x) = (|x|^2 + 1)^{\frac{\alpha}{2}}$ and $\omega_2(x) = (|x|^2 + 1)^{\frac{\beta}{2}} g(x)$ where $g(x)$ is a positive function with a finite limit at ∞, are considered. For this class of weights non-existence results are shown to be optimal by using various generalized Hardy inequalities established in the last two chapters.

11.1. Variational formulation of stability for non-self-adjoint eigenvalue problems

Consider again the following second order nonlinear Dirichlet boundary value problem,

$$(P) \quad \begin{cases} -\Delta u = f(u) & \text{in } \Omega \\ u = 0 & \text{on } \partial\Omega, \end{cases}$$

where Ω is a bounded domain in $\mathbb{R}^n$, $n \geq 2$, and where f is a C^1-function on the real line. Recall that a solution of (P) is said to be *semi-stable*, if the linearized operator at u, i.e., the Schrödinger operator $-\Delta - f'(u)$ is non-negative. In other words, the first eigenvalue of $-\Delta - f'(u)$ is non-negative, which translates into the following variational formulation of semi-stability:

(11.1) $\quad \int_\Omega \lambda f'(u)\psi^2 dx \leq \int_\Omega |\nabla \psi|^2 dx \quad$ for all $\psi \in H_0^1(\Omega)$.

Suppose now that (P) involves an additional smooth bounded advection term c on Ω, that is we are dealing with the following boundary value problem:

$$(Q) \quad \begin{cases} -\Delta u + c(x) \cdot \nabla u = f(u) & \text{in } \Omega \\ u = 0 & \text{on } \partial\Omega. \end{cases}$$

We can always define the notion of a *semi-stable solution u*, by stating that the first eigenvalue of the operator $-\Delta + c(x) \cdot \nabla - f'(u)$ is non-negative. In other words, there exists a principal eigenpair (φ, K) for the operator $-\Delta + c(x) \cdot \nabla - f'(u)$ such that $\varphi > 0$ in Ω, $K \geq 0$, and

(11.2) $\quad\begin{cases} -\Delta\varphi + c(x) \cdot \nabla\varphi - f'(u)\varphi = K\varphi & \Omega, \\ \varphi = 0 & \partial\Omega. \end{cases}$

Note however that there is no obvious variational formulation for the stability of u in this problem. Indeed, it suffices to take a divergence-free advection c to see that there is no immediate analogue to (11.1). On the other hand, if c is a pure potential, i.e., $c(x) = \nabla\gamma(x)$ for some smooth function γ on $\bar{\Omega}$, then (Q) can be rewritten as

(11.3) $\quad\quad\quad\quad -\operatorname{div}(e^\gamma \nabla u) = e^\gamma f(u) \quad\quad \text{in } \Omega,$

and the semi-stability condition on the minimal solution u of (Q) translates into the following variational formulation:

(11.4) $\quad\quad \int_\Omega e^{-\gamma} f'(u)\psi^2 \, dx \leq \int_\Omega e^{-\gamma} |\nabla\psi|^2 \, dx \quad\text{ for all } \psi \in H_0^1(\Omega).$

Then, with slight modifications, one can use the standard variational approach for studying semi-stable solutions for (P) to obtain analogous results for semi-stable solutions for (Q).

The question is therefore whether there is a variational formulation whenever c is a vector field with a divergence free component. To address this problem, we shall need the following version of the Hodge decomposition for general vector fields c.

LEMMA 11.1.1. *Any vector field $c \in C^\infty(\bar{\Omega}, \mathbb{R}^n)$ can be decomposed as $c(x) = -\nabla\gamma + a(x)$ where γ is a smooth scalar function and $a(x)$ is a smooth bounded vector field such that $\operatorname{div}(e^\gamma a) = 0$.*

Proof: By the Krein-Rutman theory, the linear eigenvalue problem

(11.5) $\quad\begin{cases} \Delta\alpha + \operatorname{div}(\alpha c) = \mu\alpha & \Omega, \\ (\nabla\alpha + \alpha c) \cdot \nu = 0 & \partial\Omega, \end{cases}$

where ν is the unit outer normal on $\partial\Omega$, has a positive solution α in Ω when μ is the principal eigenvalue. Integrating the equation over Ω, one sees that $\mu = 0$. The positivity of α on the boundary follows from the boundary condition and the maximum principle. In other words, we have that $\Delta\alpha + \operatorname{div}(\alpha c) = 0$ on Ω, and $\alpha > 0$ on $\bar{\Omega}$. Now define $\gamma := \log(\alpha)$ and $a := c + \nabla\gamma$. An easy computation shows that $\operatorname{div}(e^\gamma a) = 0$. $\quad\square$

We shall now use the following general Hardy inequality established in Chapter 9 (Theorem 9.1.1), to make up for the lack of a variational characterization for semi-stable solutions.

LEMMA 11.1.2. *Let $A(x)$ denote a uniformly positive definite $n \times n$ matrix with smooth coefficients defined on Ω. Suppose E is a smooth positive function on Ω and fix a constant β with $1 \leq \beta \leq 2$. Then, for all $\psi \in H_0^1(\Omega)$ we have*

(11.6) $\quad \int_\Omega |\nabla\psi|_A^2 \geq \dfrac{\beta(2-\beta)}{4} \int_\Omega \dfrac{|\nabla E|_A^2}{E^2}\psi^2 + \dfrac{\beta}{2} \int_\Omega \dfrac{-\operatorname{div}(A\nabla E)}{E}\psi^2,$

where $\int_\Omega |\nabla\psi|_A^2 = \int_\Omega A(x)\nabla\psi \cdot \nabla\psi \, dx$.

We can now deduce the following energy inequality associated to a semi-stable solution.

THEOREM 11.1.3. *Suppose that u is a semi-stable solution for (Q), with a principal eigenpair (φ_0, K_0) of (11.2) such that $\varphi_0 > 0$ on Ω and $K_0 \geq 0$. Let a and γ be the components of the vector field c in the Hodge decomposition. Then, for $1 \leq \beta \leq 2$ we have for all $\psi \in H_0^1(\Omega)$,*
(11.7)
$$\int_\Omega e^\gamma |\nabla \psi|^2 \geq \frac{\beta(2-\beta)}{4} \int_\Omega \frac{e^\gamma |\nabla \varphi_0|^2}{\varphi_0^2} \psi^2 + \frac{\beta}{2} \int_\Omega e^\gamma f'(u) \psi^2 - \frac{\beta}{2} \int_\Omega \frac{e^\gamma a \cdot \nabla \varphi_0}{\varphi_0} \psi^2.$$

Proof: Note that (11.2) can be rewritten as
$$(11.8) \qquad -\operatorname{div}(e^\gamma \nabla \varphi_0) + e^\gamma a \cdot \nabla \varphi_0 = e^\gamma \left(f'(u) + K_0 \right) \varphi_0 \qquad \text{in } \Omega,$$

where as mentioned above we are using the decomposition $c = -\nabla \gamma + a$. Apply now Lemma 11.1.2 with $E := \varphi_0$ and $A(x) = e^\gamma I$ (where I is the identity matrix) and use (11.6) along with (11.8) to obtain the desired result. Note that we have dropped the nonnegative term involving K. □

11.2. Regularity of semi-stable solutions in non-self-adjoint boundary value problems

We consider in this section, the regularity of the extremal solution of the non-linear eigenvalue problem

$$(S_\lambda) \qquad \begin{cases} -\Delta u + c(x) \cdot \nabla u = \frac{\lambda}{(1-u)^2} & \text{in } \Omega, \\ u = 0 & \text{on } \partial\Omega, \end{cases}$$

where Ω is a smooth bounded domain in $\mathbb{R}^n$ and $c(x)$ is a smooth bounded vector field on $\bar{\Omega}$.

One can again show the existence of a positive finite critical parameter λ^* such that for $0 < \lambda < \lambda^*$ there exists a smooth minimal solution u_λ of (S_λ), while there are no smooth solutions of (S_λ) for $\lambda > \lambda^*$. Moreover, the minimal solutions are also *semi-stable* in the sense that the principal eigenvalue of the corresponding linearized operator

$$L_{u,\lambda,c} := -\Delta + c(x) \cdot \nabla - \frac{2\lambda}{(1-u_\lambda)^3}$$

in $H_0^1(\Omega)$ is non-negative. Our main result concerns the regularity of the extremal solution in low dimensions.

THEOREM 11.2.1. *If $1 \leq n \leq 7$, then the extremal solution u^* of (S_{λ^*}) is smooth.*

Consider $c = -\nabla \gamma + a$ to be the decomposition of c described in Lemma 11.1.1. We shall need the following estimate.

LEMMA 11.2.1. *For $0 < \lambda < \lambda^*$, $1 < \beta < 2$ and $0 < t < \beta + \sqrt{\beta^2 + \beta}$, we have the following estimate:*
(11.9)
$$\lambda \left(\beta - \frac{t^2}{2t+1} \right) \int_\Omega \frac{e^\gamma}{(1-u_\lambda)^{2t+3}} \leq 2\beta \lambda \int_\Omega \frac{e^\gamma}{(1-u_\lambda)^{t+3}} + \frac{\beta \|a\|_{L^\infty}^2}{4(2-\beta)} \int_\Omega \frac{e^\gamma}{(1-u_\lambda)^{2t}}.$$

PROOF. Fix $0 < \beta < 2$, let $0 < t$ and let u_λ denote the minimal solution associated with (S_λ). Let $(\varphi_\lambda, K_\lambda)$ denote the principal eigenpair associated with the linearization of (S_λ) at u_λ. Note that $0 < \varphi_\lambda$ in Ω, and $0 \leq K_\lambda$. Use now Theorem 11.1.3 with $f'(u_\lambda) = \frac{2\lambda}{(1-u_\lambda)^3}$ and φ_λ, to obtain for all $\psi \in H_0^1(\Omega)$,

(11.10)
$$\int_\Omega e^\gamma |\nabla \psi|^2 \geq \frac{\beta(2-\beta)}{4} \int_\Omega \frac{e^\gamma |\nabla \varphi_\lambda|^2}{\varphi_\lambda^2} \psi^2 + \beta \lambda \int_\Omega \frac{e^\gamma}{(1-u_\lambda)^3} \psi^2 - \frac{\beta}{2} \int_\Omega \frac{e^\gamma a \cdot \nabla \varphi_\lambda}{\varphi_\lambda} \psi^2.$$

Put $\psi := \frac{1}{(1-u)^t} - 1$ into (11.10) to obtain

$$t^2 \int_\Omega \frac{e^\gamma |\nabla u|^2}{(1-u)^{2t+2}} \geq \beta \lambda \int_\Omega \frac{e^\gamma}{(1-u)^3} \left(\frac{1}{(1-u)^t} - 1 \right)^2$$
$$+ \frac{\beta}{2} \int_\Omega e^\gamma \left(\frac{(2-\beta)}{2} \frac{|\nabla \varphi|^2}{\varphi^2} - \frac{a \cdot \nabla \varphi}{\varphi} \right) \psi^2.$$

Now note that (S_λ) can be rewritten as

$$-\operatorname{div}(e^\gamma \nabla u) + e^\gamma a \cdot \nabla u = \frac{\lambda e^\gamma}{(1-u)^2} \quad \text{in } \Omega,$$

and test this on $\bar{\varphi} := \frac{1}{(1-u)^{2t+1}} - 1$ to obtain

$$(2t+1) \int_\Omega \frac{e^\gamma |\nabla u|^2}{(1-u)^{2t+2}} + H = \lambda \int_\Omega \frac{e^\gamma}{(1-u)^2} \left(\frac{1}{(1-u)^{2t+1}} - 1 \right),$$

where

$$H := \int_\Omega e^\gamma a \cdot \nabla u \left(\frac{1}{(1-u)^{2t+1}} - 1 \right).$$

One easily sees that $H = 0$ after considering the fact H can be rewritten in the form $\int_\Omega (e^\gamma a) \cdot \nabla G(u)$ for an appropriately chosen function G with $G(0) = 0$. Combining the above two inequalities and dropping some positive terms gives

$$\lambda \left(\beta - \frac{t^2}{2t+1} \right) \int_\Omega \frac{e^\gamma}{(1-u)^{2t+3}} \leq 2\beta\lambda \int_\Omega \frac{e^\gamma}{(1-u)^{3+t}} + \frac{\beta}{2} \int_\Omega e^\gamma \Lambda(x) \left(\frac{1}{(1-u)^t} - 1 \right)^2,$$

where

$$\Lambda(x) := \frac{a \cdot \nabla \varphi}{\varphi} - \frac{(2-\beta)}{2} \frac{|\nabla \varphi|^2}{\varphi^2}.$$

Simple calculus shows that

$$\sup_\Omega \Lambda(x) \leq \frac{\|a\|_{L^\infty}^2}{2(2-\beta)},$$

which, after substituting into the above inequality, completes the proof of Lemma 11.2.1. $\square$

LEMMA 11.2.2. *With the above notation, there exists a constant C independent of λ such that for all $p < p_0 := \frac{7}{2} + \sqrt{6}$, we have*

(11.11) $\qquad \|\frac{1}{(1-u_\lambda)^2}\|_p \leq C$ *for all* $0 \leq \lambda \leq \lambda^*$.

PROOF. Note now that the restriction $t < \beta + \sqrt{\beta^2 + \beta}$ is needed to ensure that the coefficient $\beta - \frac{t^2}{2t+1}$ is positive. It follows then that $\frac{1}{(1-u_\lambda)^2}$ is uniformly bounded (in λ) in $L^p(\Omega)$ for all $p < p_0 := \frac{7}{2} + \sqrt{6}$ and after passing to limits we have the same result for the extremal solution u^*. $\square$

To conclude the proof of Theorem 11.2.1, it suffices to note the following result.

LEMMA 11.2.3. *Suppose the extremal solution u^* is such that $\frac{1}{(1-u^*)^2} \in L^{\frac{3n}{4}}(\Omega)$, then u^* is smooth.*

Proof: First note that by elliptic regularity one has $u^* \in W^{2,\frac{3n}{4}}(\Omega)$ and after applying the Sobolev embedding theorem one has $u^* \in C^{0,\frac{2}{3}}(\overline{\Omega})$. Now suppose $\|u\|_{L^\infty} = 1$ so that there is some $x_0 \in \Omega$ such that $u(x_0) = 1$. Then $\frac{1}{1-u(x)} \geq \frac{C}{|x-x_0|^{\frac{2}{3}}}$ and hence

$$+\infty > \int_\Omega \frac{1}{((1-u^*)^2)^{\frac{3n}{4}}} \geq C \int_\Omega \frac{1}{|x|^n} = +\infty,$$

which is a contradiction. It follows that $\frac{1}{(1-u^*)^2} \in L^\infty(\Omega)$, and u^* is therefore smooth. $\square$

Proof of Theorem 11.2.1: Using Lemma 11.2.3 and the previous L^p-bound on $\frac{1}{(1-u^*)^2}$, one sees that u^* is smooth for $3 \leq n \leq 7$. $\square$

The same approach works on the following Gelfand problem with advection:

$$(T_\lambda) \quad \begin{cases} -\Delta u + a(x) \cdot \nabla u = \lambda e^u & \text{in } \Omega, \\ u = 0 & \text{on } \partial \Omega, \end{cases}$$

where Ω is a bounded domain in $\mathbb{R}^n$ with smooth boundary and where a is a smooth bounded vector field. For simplicity we assume that a is divergence free.

THEOREM 11.2.2. *Suppose $n \leq 9$, then the extremal solution associated with (T_λ) is smooth.*

Proof: Let $1 < t < \beta < 2$ and $0 < \lambda < \lambda^*$. Setting $\psi := e^{tu_\lambda} - 1$ and letting $\bar{\varphi}$ be a suitable multiple of $e^{2tu_\lambda} - 1$, follow the proof of Theorem 11.2.1 to obtain the inequality

$$\lambda(\beta - t) \int_\Omega e^{(2t+1)u_\lambda} dx \leq \beta \int_\Omega \left(\frac{a \cdot \nabla \varphi_\lambda}{\varphi_\lambda} - \frac{(\beta-2)}{2} \frac{|\nabla \varphi_\lambda|^2}{\varphi_\lambda^2} \right)(e^{tu_\lambda} - 1)^2$$
$$+ 2\beta\lambda \int_\Omega e^{(t+1)u_\lambda},$$

where $0 < \varphi_\lambda \in H_0^1(\Omega) \cap C^\infty(\Omega)$ is the positive principal eigenfunction associated with the linearized problem at u_λ. As before one can easily obtain the estimate that for all $1 < t < 2$ there is some $C_t < \infty$ such that

$\int_\Omega e^{(2t+1)u_\lambda} dx \leq C_t$ uniformly in λ.

Using elliptic regularity theory along with the Sobolev imbedding theorem one sees that u_λ is uniformly bounded in $L^\infty(\Omega)$ provided $n \leq 9$ which gives the desired result after passing to limits. $\square$

11.3. Liouville type theorems for general equations in divergence form

In this section, we are interested in the existence versus non-existence of semi-stable sub- and super-solutions of equations of the form

(11.12) $\qquad -\text{div}(\omega_1(x)\nabla u) = \omega_2(x) f(u) \qquad \text{in } \mathbb{R}^n,$

where $f(u)$ is one of the following non-linearities: e^u, u^p where $p > 1$ and $-u^{-p}$ where $p > 0$. We shall consider "weights" ω_1, ω_2 that are smooth positive functions

(ω_2 could be zero at -say- a point) and which satisfy various growth conditions at ∞.

Say that a C^2 sub/super-solution u of (11.12) is *semi-stable* provided

$$\tag{11.13} \int_{\mathbb{R}^n} \omega_2(x) f'(u) \psi^2 \, dx \leq \int_{\mathbb{R}^n} \omega_1(x) |\nabla \psi|^2 \, dx \qquad \forall \psi \in C_c^2(\mathbb{R}^n).$$

Note that (11.12) can be re-written as

$$\tag{11.14} -\Delta u + \nabla \gamma(x) \cdot \nabla u = \frac{\omega_2(x)}{\omega_1(x)} f(u) \qquad \text{in } \mathbb{R}^n,$$

where $\gamma = -\log(\omega_1)$.

THEOREM 11.3.1. *Let $\omega_1(x) = (|x|^2 + 1)^{\frac{\alpha}{2}}$ and $\omega_2(x) = g(x)(|x|^2 + 1)^{\frac{\beta}{2}}$, where $g(x)$ is smooth, positive except possibly at a point, and such that $\lim_{|x| \to \infty} g(x) = C \in (0, \infty)$.*

(1) *If $n + \alpha - 2 < 4(\beta - \alpha + 2)$, then there is no semi-stable sub-solution for*

$$\tag{11.15} -\operatorname{div}(\omega_1 \nabla u) = \omega_2 e^u \qquad \text{in } \mathbb{R}^n.$$

(2) *If $n + \alpha - 2 < \frac{2(\beta - \alpha + 2)}{p - 1} \left(p + \sqrt{p(p-1)} \right)$ where $p > 1$, then there is no positive semi-stable sub-solution for*

$$\tag{11.16} -\operatorname{div}(\omega_1 \nabla u) = \omega_2 u^p \qquad \text{in } \mathbb{R}^n.$$

(3) *If $n + \alpha - 2 < \frac{2(\beta - \alpha + 2)}{p + 1} \left(p + \sqrt{p(p+1)} \right)$ where $p > 0$, then there is no positive semi-stable super-solution for*

$$\tag{11.17} -\operatorname{div}(\omega_1 \nabla u) = \omega_2 u^{-p} \qquad \text{in } \mathbb{R}^n.$$

(4) *Furthermore 1), 2) and 3) are optimal in the sense that if $n + \alpha - 2 > 0$ and the remaining inequality is reversed (e.g., $n + \alpha - 2 > 4(\beta - \alpha + 2)$) in (1), then we can find a suitable function $g(x)$ which satisfies the above properties and a semi-stable sub/super-solution u for the appropriate equation.*

Denote by C_R the ring $\{x; R < |x| < 2R\}$, and define the following quantities.

$$I_1(t) := R^{-4t-2} \int_{C_R} \frac{\omega_1^{2t+1}}{\omega_2^{2t}} \, dx,$$

$$J_1(t) := R^{-2t-1} \int_{C_R} \frac{|\nabla \omega_1|^{2t+1}}{\omega_2^{2t}} \, dx,$$

$$I_2(t) := R^{-\frac{2(2t+p-1)}{p-1}} \int_{C_R} \left(\frac{w_1^{p+2t-1}}{w_2^{2t}} \right)^{\frac{1}{p-1}} dx,$$

$$J_2(t) := R^{-\frac{p+2t-1}{p-1}} \int_{C_R} \left(\frac{|\nabla w_1|^{p+2t-1}}{w_2^{2t}} \right)^{\frac{1}{p-1}} dx,$$

$$I_3(t) := R^{-2\frac{p+2t+1}{p+1}} \int_{C_R} \left(\frac{w_1^{p+2t+1}}{w_2^{2t}} \right)^{\frac{1}{p+1}} dx,$$

$$J_3(t) := R^{-\frac{p+2t+1}{p+1}} \int_{C_R} \left(\frac{|\nabla w_1|^{p+2t+1}}{w_2^{2t}} \right)^{\frac{1}{p+1}} dx.$$

11.3. LIOUVILLE TYPE THEOREMS

LEMMA 11.3.1. *Let ω_1 and ω_2 be smooth positive functions on $\mathbb{R}^n$.*
(1) *If $I_1(t), J_1(t) \to 0$ as $R \to +\infty$ for some $0 < t < 2$, then there is no semi-stable sub-solution for equation (11.15).*
(2) *If $I_2(t), J_2(t) \to 0$ as $R \to +\infty$ for some $p - \sqrt{p(p-1)} < t < p + \sqrt{p(p-1)}$ where $p > 1$ (resp., $0 < t < \frac{1}{2}$), then there is no positive semi-stable sub-solution (resp., super-solution) for equation (11.16).*
(3) *If $I_3(t), J_3(t) \to 0$ as $R \to +\infty$ for some $0 < t < p + \sqrt{p(p+1)}$, where $p > 0$, then there is no positive semi-stable super-solution for equation (11.17).*

Proof: (1) Suppose u is a semi-stable sub-solution of (11.15) with $I_1(t), J_1(t) \to 0$ as $R \to +\infty$ and let $0 \le \varphi \le 1$ denote a smooth compactly supported function. Put $\psi := e^{tu}\varphi$ into (11.13), where $0 < t < 2$, to arrive at

$$\int_{\mathbb{R}^n} \omega_2 e^{(2t+1)u} \varphi^2 \le t^2 \int_{\mathbb{R}^n} \omega_1 e^{2tu} |\nabla u|^2 \varphi^2$$
$$+ \int_{\mathbb{R}^n} \omega_1 e^{2tu} |\nabla \varphi|^2 + 2t \int_{\mathbb{R}^n} \omega_1 e^{2tu} \varphi \nabla u \cdot \nabla \varphi.$$

Multiply (11.15) by $e^{2tu}\varphi^2$ and integrate by parts to arrive at

$$2t \int_{\mathbb{R}^n} \omega_1 e^{2tu} |\nabla u|^2 \varphi^2 \le \int_{\mathbb{R}^n} \omega_2 e^{(2t+1)u} \varphi^2 - 2 \int_{\mathbb{R}^n} \omega_1 e^{2tu} \varphi \nabla u \cdot \nabla \varphi.$$

If we equate like-terms we arrive at

$$\frac{(2-t)}{2} \int_{\mathbb{R}^n} \omega_2 e^{(2t+1)u} \varphi^2 \le \int_{\mathbb{R}^n} \omega_1 e^{2tu} \left(|\nabla \varphi|^2 - \frac{\Delta \varphi}{2} \right) dx$$
(11.18)
$$- \frac{1}{2} \int_{\mathbb{R}^n} e^{2tu} \varphi \nabla \omega_1 \cdot \nabla \varphi.$$

Now substitute φ^m into this inequality for φ where m is a large enough integer to obtain

$$\frac{(2-t)}{2} \int_{\mathbb{R}^n} \omega_2 e^{(2t+1)u} \varphi^{2m} \le C_m \int_{\mathbb{R}^n} \omega_1 e^{2tu} \varphi^{2m-2} \left(|\nabla \varphi|^2 + \varphi |\Delta \varphi| \right)$$
(11.19)
$$- D_m \int_{\mathbb{R}^n} e^{2tu} \varphi^{2m-1} \nabla \omega_1 \cdot \nabla \varphi,$$

where C_m and D_m are positive constants just depending on m. We now estimate the terms on the right.

$$\int_{\mathbb{R}^n} \omega_1 e^{2tu} \varphi^{2m-2} |\nabla \varphi|^2 = \int_{\mathbb{R}^n} \omega_2^{\frac{2t}{2t+1}} e^{2tu} \varphi^{2m-2} \frac{\omega_1}{\omega_2^{\frac{2t}{2t+1}}} |\nabla \varphi|^2$$
$$\le \left(\int_{\mathbb{R}^n} \omega_2 e^{(2t+1)u} \varphi^{(2m-2)\frac{(2t+1)}{2t}} dx \right)^{\frac{2t}{2t+1}}$$
$$\left(\int_{\mathbb{R}^n} \frac{\omega_1^{2t+1}}{\omega_2^{2t}} |\nabla \varphi|^{2(2t+1)} \right)^{\frac{1}{2t+1}}.$$

For fixed $0 < t < 2$, take m large enough so that $(2m-2)\frac{(2t+1)}{2t} \ge 2m$. Since $0 \le \varphi \le 1$, this allows to replace φ^{2m-2} in the first term on the right by φ^{2m} and

hence we obtain
(11.20)
$$\int_{\mathbb{R}^n} \omega_1 e^{2tu} \varphi^{2m-2}|\nabla\varphi|^2 \leq \left(\int_{\mathbb{R}^n} \omega_2 e^{(2t+1)u} \varphi^{2m} dx\right)^{\frac{2t}{2t+1}} \left(\int_{\mathbb{R}^n} \frac{\omega_1^{2t+1}}{\omega_2^{2t}}|\nabla\varphi|^{2(2t+1)}\right)^{\frac{1}{2t+1}}$$

We now take the test functions φ to be such that $0 \leq \varphi \leq 1$ with φ supported in the ball B_{2R} with $\varphi = 1$ on B_R and $|\nabla\varphi| \leq \frac{C}{R}$ where $C > 0$ is independent of R. Using this choice of φ we obtain

(11.21)
$$\int_{\mathbb{R}^n} \omega_1 e^{2tu} \varphi^{2m-2}|\nabla\varphi|^2 \leq \left(\int_{\mathbb{R}^n} \omega_2 e^{(2t+1)u} \varphi^{2m}\right)^{\frac{2t}{2t+1}} I_1(t)^{\frac{1}{2t+1}}.$$

One similarly shows that
$$\int_{\mathbb{R}^n} \omega_1 e^{2tu} \varphi^{2m-1}|\Delta\varphi| \leq \left(\int_{\mathbb{R}^n} \omega_2 e^{(2t+1)u} \varphi^{2m}\right)^{\frac{2t}{2t+1}} I_1(t)^{\frac{1}{2t+1}}.$$

By combining the above, we obtain

$$\frac{(2-t)}{2}\int_{\mathbb{R}^n} \omega_2 e^{(2t+1)u} \varphi^{2m} \leq C_m \left(\int_{\mathbb{R}^n} \omega_2 e^{(2t+1)u} \varphi^{2m} dx\right)^{\frac{2t}{2t+1}} I_1(t)^{\frac{1}{2t+1}}$$
(11.22)
$$-D_m \int_{\mathbb{R}^n} e^{2tu} \varphi^{2m-1} \nabla\omega_1 \cdot \nabla\varphi.$$

We now estimate this last term. A similar argument using Hölder's inequality shows that
$$\int_{\mathbb{R}^n} e^{2tu}\varphi^{2m-1}|\nabla\omega_1||\nabla\varphi| \leq \left(\int_{\mathbb{R}^n} \omega_2 \varphi^{2m} e^{(2t+1)u} dx\right)^{\frac{2t}{2t+1}} J_1(t)^{\frac{1}{2t+1}}.$$

Combining this with the above estimates gives that

(11.23) $$(2-t)\left(\int_{\mathbb{R}^n} \omega_2 e^{(2t+1)u} \varphi^{2m} dx\right)^{\frac{1}{2t+1}} \leq I_1(t)^{\frac{1}{2t+1}} + J_1(t)^{\frac{1}{2t+1}}.$$

Send $R \to +\infty$ and use the fact that $I_1(t), J_1(t)$ go to 0 to see that $\int_{\mathbb{R}^n} \omega_2 e^{(2t+1)u} = 0$, which is clearly a contradiction. Hence there is no semi-stable sub-solution of (11.15).

(2) Suppose that $u > 0$ is a semi-stable sub-solution (resp., super-solution) of (11.16). A similar calculation as in (1) shows that for $p - \sqrt{p(p-1)} < t < p + \sqrt{p(p-1)}$, (resp., $0 < t < \frac{1}{2}$) one has

$$(p - \frac{t^2}{2t-1})\int_{\mathbb{R}^n} \omega_2 u^{2t+p-1} \varphi^{2m} \leq D_m \int_{\mathbb{R}^n} \omega_1 u^{2t} \varphi^{2(m-1)}(|\nabla\varphi|^2 + \varphi|\Delta\varphi|)$$
(11.24)
$$+ C_m \frac{(1-t)}{2(2t-1)}\int_{\mathbb{R}^n} u^{2t} \varphi^{2m-1} \nabla\omega_1 \cdot \nabla\varphi.$$

Apply Hölder's argument as in (1) but this time, the terms I_2 and J_2 will appear on the right hand side of the resulting equation. This shift from a sub-solution to a super-solution depending on whether $t > \frac{1}{2}$ or $t < \frac{1}{2}$ is a result of the change in the sign of $2t - 1$ at $t = \frac{1}{2}$. We leave the details for the reader.

(3) This case is also similar to (1) and (2). □

To prove Theorem 11.3.1, we shall also need the following weighted Hardy inequality (9.25) given in Chapter 9.

LEMMA 11.3.2. *(1) Suppose $E > 0$ is a smooth function and $\tau < \frac{1}{2}$. Then for all $\varphi \in C_c^\infty(\mathbb{R}^n)$, we have*

(11.25) $$\int_{\mathbb{R}^n} E^{2\tau} |\nabla \varphi|^2 \, dx \geq (\tau - \frac{1}{2})^2 \int_{\mathbb{R}^n} E^{2\tau - 2} |\nabla E|^2 \varphi^2 \, dx$$
$$+ (\frac{1}{2} - \tau) \int_{\mathbb{R}^n} (-\Delta E) E^{2\tau - 1} \varphi^2 \, dx.$$

(2) In particular, by applying (11.25) to $E := (1 + |x|^2)^{\frac{\alpha}{2(1-\alpha-2t)}}$ and $\tau = \frac{1-\alpha-2t}{2}$, we get for all $\varphi \in C_c^\infty$,

(11.26)
$$\int_{\mathbb{R}^n} (1 + |x|^2)^{\frac{\alpha}{2}} |\nabla \varphi|^2 \, dx \geq (t + \frac{\alpha}{2})^2 \int_{\mathbb{R}^n} |x|^2 (1 + |x|^2)^{-2 + \frac{\alpha}{2}} \varphi^2 \, dx$$
$$+ (t + \frac{\alpha}{2}) \int_{\mathbb{R}^n} (n - 2(t+1) \frac{|x|^2}{1 + |x|^2}) (1 + |x|^2)^{-1 + \frac{\alpha}{2}} \varphi^2 \, dx.$$

Proof of Theorem 11.3.1: We note first that if $n + \alpha - 2 < 0$ then one can easily see that there is no semi-stable sub-solution of (11.15), no positive sub-solution for (11.16), nor a positive semi-stable super-solution (11.17). Indeed, more generally, if ω_1 has enough integrability then it is immediate that if u is a semi-stable solution of (11.12) and if f is increasing, then

(11.27) $$\int_{\mathbb{R}^n} \omega_2 f'(u) \, dx = 0.$$

To see this let $0 \leq \psi \leq 1$ be supported in a ball of radius $2R$ centered at the origin (B_{2R}) with $\psi = 1$ on B_R and such that $|\nabla \psi| \leq \frac{C}{R}$ where $C > 0$ is independent of R. Putting this ψ into (11.13) one obtains

$$\int_{B_R} \omega_2 f'(u) \, dx \leq \frac{C}{R^2} \int_{R < |x| < 2R} \omega_1 \, dx,$$

and so if the right hand side goes to zero as $R \to \infty$ we have the desired result.

So we now assume that $n + \alpha - 2 > 0$. A direct computation shows that for such ω_1 and ω_2, the quantities I and J are just multiples of each other in all three cases so it suffices to show that $\lim_{R \to \infty} I_1(t) = 0$.

Starting with case (1), we have for $R > 1$ that

$$I_1(t) \leq \frac{C}{R^{4t+2}} \int_{R < |x| < 2R} |x|^{\alpha(2t+1) - 2t\beta} \leq \frac{C}{R^{4t+2}} R^{n + \alpha(2t+1) - 2t\beta},$$

and so in order to show the non-existence, we want to find some $0 < t < 2$ such that $4t + 2 > n + \alpha(2t+1) - 2t\beta$, which is equivalent to $2t(\beta - \alpha + 2) > (n + \alpha - 2)$. Now recall that we are assuming that $0 < n + \alpha - 2 < 4(\beta - \alpha + 2)$ and hence we have the desired result by taking $t < 2$ but sufficiently close. The proof of the non-existence results for (2) and (3) are similar and we omit the details.

(4) We now show that the above results are optimal meaning that when $n + \alpha - 2 > 0$, and if the above conditions are not satisfied then we have existence of semi-stable sub/super-solutions. We shall consider the cases $\beta - \alpha + 2 < 0$ and $\beta - \alpha + 2 > 0$ separately.

• Case where $\beta - \alpha + 2 < 0$: Here we take $u \equiv 0$ in the case of (11.15) and $u \equiv 1$ in the case of (11.16) and (11.17). In addition we take $g(x) = \varepsilon$. It is clear that in all cases u is the appropriate sub or super-solution. The only thing one needs to check is the stability. In all cases this reduces to trying to show that we have

(11.28) $\quad \sigma \int_{\mathbb{R}^n} (1+|x|^2)^{\frac{\alpha}{2}-1} \varphi^2 \, dx \leq \int_{\mathbb{R}^n} (1+|x|^2)^{\frac{\alpha}{2}} |\nabla \varphi|^2 \, dx \quad$ for all $\varphi \in C_c^\infty$,

where σ is some small positive constant. Actually σ will be either ε or $p\varepsilon$ depending on which equation were are examining. To show (11.28), we use Lemma 11.3.2 and we drop a few positive terms to arrive at

$$\int_{\mathbb{R}^n} (1+|x|^2)^{\frac{\alpha}{2}} |\nabla \varphi|^2 \geq (t+\frac{\alpha}{2}) \int_{\mathbb{R}^n} \left(n - 2(t+1)\frac{|x|^2}{1+|x|^2} \right) (1+|x|^2)^{-1+\frac{\alpha}{2}},$$

which holds for all $\varphi \in C_c^\infty$ and $t, \alpha \in \mathbb{R}$. Now, since $n+\alpha-2 > 0$, we can choose t such that $-\frac{\alpha}{2} < t < \frac{n-2}{2}$. So, the integrand function in the right hand side is positive and since for small enough σ we have

$$\sigma \leq (t+\frac{\alpha}{2})(n - 2(t+1)\frac{|x|^2}{1+|x|^2}) \quad \text{for all } x \in \mathbb{R}^n,$$

we get the required stability.

• Case where $\beta - \alpha + 2 > 0$): For equation (11.15) we take $u(x) = -\frac{\beta-\alpha+2}{2}\log(1+|x|^2)$ and $g(x) := (\beta - \alpha + 2)(n + (\alpha - 2)\frac{|x|^2}{1+|x|^2})$. By a computation one sees that u is a sub-solution of (11.15) and hence we need to only show the stability, which amounts to verifying that

(11.29) $\quad \int_{\mathbb{R}^n} \frac{g(x)\psi^2}{(1+|x|^2)^{-\frac{\alpha}{2}+1}} \leq \int_{\mathbb{R}^n} \frac{|\nabla \psi|^2}{(1+|x|^2)^{-\frac{\alpha}{2}}} \quad$ for all $\psi \in C_c^\infty$.

To show this we use again Lemma 11.3.2, but with an appropriate t so that $-\frac{\alpha}{2} \leq t \leq \frac{n-2}{2}$ such that for all $x \in \mathbb{R}^n$ we have

$$(\beta - \alpha + 2)\left(n + (\alpha - 2)\frac{|x|^2}{1+|x|^2} \right) \leq (t+\frac{\alpha}{2})^2 \frac{|x|^2}{(1+|x|^2} $$
$$+ (t+\frac{\alpha}{2})\left(n - 2(t+1)\frac{|x|^2}{1+|x|^2} \right).$$

With a simple calculation one sees that we just need to have

$$(\beta - \alpha + 2) \leq (t+\frac{\alpha}{2})$$
$$(\beta - \alpha + 2)(n + \alpha - 2) \leq (t+\frac{\alpha}{2})\left(n - t - 2 + \frac{\alpha}{2} \right).$$

If one takes $t = \frac{n-2}{2}$ in the case where $n \neq 2$ and t close to zero in the case for $n = 2$ one easily sees that both inequalities above hold, after considering all constraints on α, β and n.

We now consider the case of equation (11.16). Here one takes $g(x) := \frac{\beta-\alpha+2}{p-1}(n + (\alpha - 2 - \frac{\beta-\alpha+2}{p-1})\frac{|x|^2}{1+|x|^2})$ and $u(x) = (1+|x|^2)^{-\frac{\beta-\alpha+2}{2(p-1)}}$. Using essentially the same approach as in (11.15) one shows that u is a semi-stable sub-solution of (11.16) with this choice of g.

For the case of equation (11.17) we take $u(x) = (1+|x|^2)^{\frac{\beta-\alpha+2}{2(p+1)}}$ and

$$g(x) := \frac{\beta-\alpha+2}{p+1}(n + (\alpha - 2 + \frac{\beta-\alpha+2}{p+1})\frac{|x|^2}{1+|x|^2})$$

□

11.4. Further remarks

The results of section 11.1 and 11.2 appeared in the paper of Cowan-Ghoussoub [**100**]. The results of section 11.3 are due to Cowan and Fazly [**96**]. The work on non-linear non-selfadjoint eigenvalue problems was motivated by the paper of Berestycki et al [**48**] regarding an explosion problem with a flow. The method of Cowan-Ghoussoub also applies whenever one deals with an explicit convex non-linearity such as $f(u) = (1+u)^p$. However, many of the qualitative properties of the explosion threshold $\lambda^*(\mathbf{c})$ in terms of the geometry and the amplitude of the flow $\mathbf{c}$ for general non-linearities remain elusive. In particular, the following questions are of interest.

Open problem (11): Consider a non-selfadjoint eigenvalue problems of the form

$$(11.30) \quad \begin{cases} -\Delta u + \mathbf{c}(\mathbf{x}) \cdot \nabla u = \lambda f(u) & \text{in } \Omega, \\ u = 0 & \text{on } \partial\Omega, \end{cases}$$

where $f(u)$ is an appropriate convex nonlinearity of type (R) or (S).

(1) Does the presence of an advection change the critical dimension of the problem?
(2) Do the general regularity results obtained in the absence of an advection term by Nedev [**227**] (for general convex f in dimensions 2 and 3) and those of Cabré and Capella [**71**] (for general f on the unit ball in $\mathbb{R}^n$ for $n \leq 9$) extend to the case of a non-selfadjoint eigenvalue problem?

Open problem (12): Does there exist (for a general domain Ω) a flow $\mathbf{c}$ that minimizes the explosion threshold $\lambda^*(\mathbf{c})$ of (11.30) among all incompressible flows?

Open problem (13): Find the best constant $C(\Omega, n, p) > 0$ which does not depend on the flow $\mathbf{c}$ such that $\|u\|_\infty \leq C\|f\|_p$ for any solution u of

$$(11.31) \quad \begin{cases} -\Delta u + \mathbf{c}(\mathbf{x}) \cdot \nabla u = \lambda f(x) & \text{in } \Omega, \\ u = 0 & \text{on } \partial\Omega. \end{cases}$$

Part 4

Mass Transport and Optimal Geometric Inequalities

CHAPTER 12

A General Comparison Principle for Interacting Gases

Brenier's solution of the Monge problem with quadratic cost allows one to show that certain natural free energy functionals are convex on the geodesics of optimal mass transport joining two probability densities. This convexity property translates into an inequality relating the relative total energy – internal, potential and interactive – of the initial and final configurations (probability densities), to their entropy production, their Wasserstein distance, and their barycenters. Once this general comparison principle is established, various – new and old – inequalities follow by simply considering different examples of internal, potential and interactive energies. The framework is remarkably encompassing as it contains most known geometrical – Gaussian and Euclidean – inequalities, while allowing a direct and unified way for computing best constants and extremals.

12.1. Mass transport with quadratic cost

Given two compactly supported probability densities $\rho_0, \rho_1 \geq 0$ on $\mathbb{R}^n$, with $X := \text{support}(\rho_0)$ and $Y := \text{support}(\rho_1)$, we define the Wasserstein distance $W(\rho_0, \rho_1)$ between them by the formula

$$(12.1) \qquad W(\rho_0, \rho_1)^2 = \inf \left\{ \int_X |x - s(x)|^2 dx; \, s \in S(\rho_0, \rho_1) \right\}$$

where $S(\rho_0, \rho_1)$ is the class of all Borel measurable maps $s : X \to Y$ that "push" ρ_0 into ρ_1, i.e., those which satisfy the change of variables formula,

$$(12.2) \qquad \int_Y h(y)\rho_1(y)dy = \int_X h(s(x))\rho_0(x)dx, \quad \text{for every } h \in C(Y).$$

Whether the infimum describing the Wasserstein distance $W(\rho_0, \rho_1)$ is achieved by an optimal map $\bar{s}$ is a variation on the original mass transport problem of G. Monge, who inquired about finding the optimal way for rearranging ρ_0 into ρ_1 against the cost function $c(x) = |x|$. Our cost function here $c(x) = \frac{1}{2}|x|^2$ is quadratic, and the existence, uniqueness and characterization of an optimal map that we give below, was established by Y. Brenier.

THEOREM 12.1.1. *There exists a unique optimal map $\bar{s}$ in $S(\rho_0, \rho_1)$, where the infimum in (12.1) is achieved. Moreover, the map $\bar{s} : X \to Y$ is one-to-one and onto a.e., and is equal to $\nabla \varphi$ a.e on X, for some convex function $\varphi : \mathbb{R}^n \to \mathbb{R}$.*

For the proof, we first consider the following "dual problem":

$$(12.3) \qquad B(\rho_0, \rho_1) := \inf \left\{ L[\varphi, \psi]; \, \varphi(x) + \psi(y) \geq x.y \text{ for } x \in X, y \in Y \right\},$$

where L is defined on any pair of continuous functions (φ, ψ) by

$$(12.4) \qquad L[\varphi, \psi] := \int_X \varphi(x) \rho_0(x) \, dx + \int_Y \psi(y) \rho_1(y) \, dy.$$

We first prove that the infimum in (12.3) is attained.

LEMMA 12.1.2. *There exist Legendre dual convex functions $(\bar{\varphi}, \bar{\psi})$ on $\mathbb{R}^n$ where the minimization problem (12.3) is attained.*

Proof: Note that if the pair (φ, ψ) satisfies the constraint $\varphi(x) + \psi(y) \geq x.y$ on $X \times Y$, then the functions

$$(12.5) \qquad \hat{\varphi}(x) := \max_{y \in Y}(x.y - \psi(y)) \quad \text{and} \quad \hat{\psi}(y) := \max_{x \in X}(x.y - \hat{\varphi}(x)),$$

which satisfy $\varphi \geq \hat{\varphi}, \psi \geq \hat{\psi}$, also satisfy the constraint $\hat{\varphi}(x) + \hat{\psi}(y) \geq x.y$ on $X \times Y$. Moreover, since $\rho_0, \rho_1 \geq 0$, we have that $L[\hat{\varphi}, \hat{\psi}] \leq L[\varphi, \psi]$. Consequently in seeking minimizers of L we may restrict our attention to convex dual pairs (φ, ψ) that are uniformly Lipschitz continuous. We can also assume that such a minimizing sequence (φ_n, ψ_n) satisfies $\inf_{B_R} \psi_n = 0$ on a ball B_R containing Y, since adding or substracting a constant to one of the functions does not affect the constraint nor the value of the functional. We can therefore assume the sequence to be uniformly bounded and use the Arzela-Ascoli theorem to show that a subsequence of (φ_n, ψ_n) converges to an optimal convex dual pair. $\square$

Proof of Theorem 12.1.1: Consider the pair of functions $(\bar{\varphi}, \bar{\psi})$ obtained in the above lemma, that we can clearly suppose to be convex on all of $\mathbb{R}^n$ and therefore differentiable a.e. We now show that $\bar{s}(x) := \nabla \bar{\varphi}(x)$ for a.e. $x \in X$, satisfies the claims in Theorem 12.1.1.

To prove that $\bar{s} : X \to Y$ is a.e. one-to-one and onto, it suffices to note that $\bar{\varphi}$ and $\bar{\psi}$ are dual convex functions, and therefore $y \in \partial \bar{\varphi}(x)$ if and only if $x \in \partial \bar{\psi}(y)$. So, $y = \nabla \bar{\varphi}(\nabla \bar{\psi}(y))$ and $x = \nabla \bar{\psi}(\nabla \bar{\varphi}(x))$ a.e.

Now we show that $\bar{s}$ pushes ρ_0 into ρ_1. For that, fix $\tau > 0$ and define the following perturbations of $\bar{\psi}$ (resp., $\bar{\varphi}$) on Y (resp., on X).

$$(12.6) \qquad \psi_\tau(y) := \bar{\psi}(y) + \tau h(y) \quad \text{and} \quad \varphi_\tau(x) := \max_{y \in Y}(x.y - \psi_\tau(y)).$$

Then $(\psi_\tau, \varphi_\tau)$ satisfies the constraint in (12.3). Since $(\bar{\varphi}, \bar{\psi})$ is a minimizer for (12.3), we have $L[\bar{\varphi}, \bar{\psi}] \leq L[\varphi_\tau, \psi_\tau]$ and so for every $\tau > 0$,

$$(12.7) \qquad 0 \leq \frac{1}{\tau}\left(L[\varphi_\tau, \psi_\tau] - L[\bar{\varphi}, \bar{\psi}]\right)$$

$$= \int_X \left[\frac{\varphi_\tau(x) - \varphi(x)}{\tau}\right] \rho_0(x) dx + \int_Y h(y) \rho_1(y) dy.$$

In order to use Lebesgue's dominated convergence theorem, we need to find an appropriate bound for $(\varphi_\tau - \varphi)/\tau$. For that take any $x \in X$ where $\nabla \bar{\varphi}(x)$ exists, and pick $y_\tau \in Y$ so that

$$(12.8) \qquad \varphi_\tau(x) - \bar{\varphi}(x) = x.y_\tau - \psi(y_\tau) - \tau h(y_\tau) - \bar{\varphi}(x) \leq -\tau h(y_\tau).$$

Since $\bar{s}(x) \in Y$ we have $\bar{\varphi}(x) = x.\bar{s}(x) - \bar{\psi}(\bar{s}(x))$ and

$$(12.9) \qquad \varphi_\tau(x) - \bar{\varphi}(x) \geq x.\bar{s}(x) - \bar{\psi}(\bar{s}(x)) - \tau h(\bar{s}(x)) - \bar{\varphi}(x) = -\tau h(\bar{s}(x)).$$

Hence,

(12.10) $$-h(\bar{s}(x)) \leq \frac{\varphi_\tau(x) - \bar{\varphi}(x)}{\tau} \leq -h(y_\tau).$$

By noting the inequality $|\varphi_\tau - \bar{\varphi}| \leq \tau ||h||_{L^\infty}$, we deduce that if $\tau \to 0$, then $\varphi_\tau \to \bar{\varphi}$ uniformly. Therefore, from (10) we can see that $y_\tau \to \bar{s}(x)$. In light of Lebesgue's dominated convergence theorem, we have that

$$\int_X h(\bar{s}(x))\rho_0(x)dx \leq \int_Y h(y)\rho_1(y)dy.$$

Replacing h by $-h$, we conclude that equality holds.

In order to show that $\bar{s}$ is optimal for (12.1), take any s in $S(\rho_0, \rho_1)$ and note that $\int_X \bar{\psi}(s(x))f(x)dx = \int_Y \bar{\psi}(y)g(y)dy$. Now for every $x \in X$, we have $\bar{\varphi}(x) + \bar{\psi}(s(x)) \geq x.s(x)$ and by the definition of $\bar{s}$, we have $\bar{\varphi}(x) + \bar{\psi}(\bar{s}(x)) = x.\bar{s}(x)$. It follows that

$$\bar{\psi}(s(x)) - \bar{\psi}(\bar{s}(x)) \leq x.(s(x) - \bar{s}(x)),$$

and consequently $0 \leq \int_X x.(s(x) - \bar{s}(x))\rho_0(x)dx$, which means that $\bar{s}$ is optimal.

12.2. A comparison principle between configurations of interacting gases

Let $F : [0, \infty) \to \mathbb{R}$ be a differentiable function on $(0, \infty)$ and let V and W be C^2-real valued functions on $\mathbb{R}^n$. For any open convex subset $\Omega \subset \mathbb{R}^n$, we consider the set of probability densities over Ω that we denote by

$$\mathcal{P}_a(\Omega) = \{\rho : \Omega \to \mathbb{R};\ \rho \geq 0 \text{ and } \int_\Omega \rho(x)dx = 1\}.$$

The *Free Energy Functional* associated to F, V, W is defined on $\mathcal{P}_a(\Omega)$ by

$$H_V^{F,W}(\rho) := \int_\Omega \left[F(\rho) + \rho V + \frac{1}{2}(W \star \rho)\rho\right] dx,$$

and is therefore the sum of the internal energy

$$H^F(\rho) := \int_\Omega F(\rho)dx,$$

the potential energy

$$H_V(\rho) := \int_\Omega \rho V dx,$$

and the interaction energy

$$H^W(\rho) := \frac{1}{2}\int_\Omega \rho(W \star \rho)\, dx.$$

Of importance is also the concept of *relative energy* of ρ_0 with respect to ρ_1 defined as

$$H_V^{F,W}(\rho_0|\rho_1) := H_V^{F,W}(\rho_0) - H_V^{F,W}(\rho_1),$$

where ρ_0 and ρ_1 are two probability densities.

The *relative entropy production* of ρ with respect to ρ_V is normally defined as

$$I_2(\rho|\rho_V) = \int_\Omega \rho \left|\nabla\left(F'(\rho) + V + W \star \rho\right)\right|^2 dx$$

in such a way that if ρ_V is a probability density that satisfies

$$\nabla\left(F'(\rho_V) + V + W \star \rho_V\right) = 0 \quad \text{a.e.}$$

then
$$I_2(\rho|\rho_V) = \int_\Omega \rho |\nabla (F'(\rho) - F'(\rho_V) + W \star (\rho - \rho_V)|^2 \, dx.$$
Our notation for the density ρ_V reflects our emphasis here on its dependence on the confinement potential, though it obviously also depends on F and W.

The *barycentre* (or centre of mass) of a probability density ρ, denoted by
$$b(\rho) := \int_{\mathbb{R}^n} x\rho(x) dx$$
will also play a role in the presence of an interactive potential.

We shall also deal with non-quadratic versions of the entropy. For that we call *Young function*, any strictly convex C^1-function $c : \mathbb{R}^n \to \mathbb{R}$ such that $c(0) = 0$ and $\lim_{|x| \to \infty} \frac{c(x)}{|x|} = \infty$. We denote by c^* its Legendre conjugate defined by
$$c^*(y) = \sup_{z \in \mathbb{R}^n} \{y \cdot z - c(z)\}.$$
For any probability density ρ on Ω, we define the *generalized relative entropy production-type function of ρ with respect to ρ_V measured against c^** by
$$\mathcal{I}_{c^*}(\rho|\rho_V) := \int_\Omega \rho c^* \left(-\nabla (F'(\rho) + V + W \star \rho) \right) \, dx,$$
which is closely related to the *generalized relative entropy production function of ρ with respect to ρ_V measured against c^** defined as:
$$I_{c^*}(\rho|\rho_V) := \int_\Omega \rho \nabla (F'(\rho) + V + W \star \rho) \cdot \nabla c^* (\nabla (F'(\rho) + V + W \star \rho)) \, dx.$$
Indeed, the convexity inequality $c^*(z) \leq z \cdot \nabla c^*(z)$ satisfied by any Young function c, readily implies that $\mathcal{I}_{c^*}(\rho|\rho_V) \leq I_{c^*}(\rho|\rho_V)$. Note that when $c(x) = \frac{|x|^2}{2}$, we have
$$I_{c^*}(\rho|\rho_V) =: I_2(\rho|\rho_V) = \int_\Omega \rho \left| \nabla (F'(\rho) + V + W \star \rho) \right|^2 dx = 2\mathcal{I}_{c^*}(\rho|\rho_V).$$
In this case, we shall denote $\mathcal{I}_{c^*}(\rho|\rho_V)$ by $\mathcal{I}_2(\rho|\rho_V)$.

The following general inequality is behind all the geometric inequalities that appear in this part of the book. It relates the free energies of two arbitrary probability densities, their Wasserstein distance, their barycenters and their relative entropy production functional. The fact that it yields many admittedly powerful geometric inequalities is remarkable.

THEOREM 12.2.1. **(Basic comparison principle for interactive gases)** *Let $F : [0, \infty) \to \mathbb{R}$ be differentiable function on $(0, \infty)$ with $F(0) = 0$ and $x \mapsto x^n F(x^{-n})$ convex and non-increasing, and let $P_F(x) := xF'(x) - F(x)$ be its associated pressure function. Let $V : \mathbb{R}^n \to \mathbb{R}$ be a C^2-confinement potential with $D^2V \geq \lambda I$, and let W be an even C^2-interaction potential with $D^2W \geq \nu I$ where $\lambda, \nu \in \mathbb{R}$, and I denotes the identity map.*

Then, for any Young function $c : \mathbb{R}^n \to \mathbb{R}$, and any pair of probability densities ρ_0 and ρ_1 on an open, bounded and convex subset Ω of $\mathbb{R}^n$ satisfying $\operatorname{supp} \rho_0 \subset \Omega$ and $P_F(\rho_0) \in W^{1,\infty}(\Omega)$, the following inequality holds:
(12.11)
$$H^{F,W}_{V+c}(\rho_0|\rho_1) + \frac{\lambda+\nu}{2} W_2^2(\rho_0, \rho_1) - \frac{\nu}{2}|b(\rho_0) - b(\rho_1)|^2 \leq H^{-nP_F, 2x \cdot \nabla W}_{c + \nabla V \cdot x}(\rho_0) + \mathcal{I}_{c^*}(\rho_0|\rho_V).$$

Furthermore, equality holds in (12.11) whenever $\rho_0 = \rho_1 = \rho_{V+c}$, where ρ_{V+c} is the probability density that satisfies

(12.12) $$\nabla\left(F'(\rho_{V+c}) + V + c + W \star \rho_{V+c}\right) = 0 \quad \text{a.e.}$$

Here is the lemma behind the main inequality (12.11).

LEMMA 12.2.2. *Let $\Omega \subset \mathbb{R}^n$ be open, bounded and convex, and let ρ_0 and ρ_1 be probability densities on Ω, with $\operatorname{supp}\rho_0 \subset \Omega$, and $P_F(\rho_0) \in W^{1,\infty}(\Omega)$. Let T be the optimal map that pushes $\rho_0 \in \mathcal{P}_a(\Omega)$ forward to $\rho_1 \in \mathcal{P}_a(\Omega)$ given by Theorem 12.1.1. Then, the following inequalities hold:*

(1) *If $F : [0, \infty) \to \mathbb{R}$ is differentiable on $(0, \infty)$, $F(0) = 0$ and $x \mapsto x^n F(x^{-n})$ is convex and non-increasing, then the following inequality holds for the internal energy:*

(12.13) $$\mathrm{H}^F(\rho_1) - \mathrm{H}^F(\rho_0) \geq \int_\Omega \rho_0 (T - I) \cdot \nabla (F'(\rho_0))\, dx.$$

(2) *If $V : \mathbb{R}^n \to \mathbb{R}$ is such that $D^2 V \geq \lambda I$ for some $\lambda \in \mathbb{R}$, then the potential energy satisfies*

(12.14) $$\mathrm{H}_V(\rho_1) - \mathrm{H}_V(\rho_0) \geq \int_\Omega \rho_0 (T - I) \cdot \nabla V\, dx + \frac{\lambda}{2} W_2^2(\rho_0, \rho_1).$$

(3) *If $W : \mathbb{R}^n \to \mathbb{R}$ is even and if $D^2 W \geq \nu I$ for some $\nu \in \mathbb{R}$, then the interaction energy satisfies*

(12.15) $$\mathrm{H}^W(\rho_1) - \mathrm{H}^W(\rho_0) \geq \int_\Omega \rho_0 (T - I) \cdot \nabla(W \star \rho_0)\, dx$$
$$+ \frac{\nu}{2}\left(W_2^2(\rho_0, \rho_1) - |\mathrm{b}(\rho_0) - \mathrm{b}(\rho_1)|^2\right).$$

Proof: If $T = \nabla \psi$ with ψ convex is the optimal map that pushes $\rho_0 \in \mathcal{P}_a(\Omega)$ forward to $\rho_1 \in \mathcal{P}_a(\Omega)$ for the quadratic cost $d(x) = |x|^2$, then $\nabla T(x)$ is diagonalizable with positive eigenvalues for ρ_0 a.e., and the Monge-Ampère equation

(12.16) $$0 \neq \rho_0(x) = \rho_1(T(x)) \det \nabla T(x)$$

holds for ρ_0 a.e. So, $\rho_1(T(x)) \neq 0$ for ρ_0 a.e. Here, $\nabla T(x) = \nabla^2 \psi(x)$ denotes the derivative in the sense of Aleksandrov of ψ.

To prove (1), set $A(x) = x^n F(x^{-n})$. Since A is non-increasing by assumption, then P_F is non-negative and $x \mapsto \frac{F(x)}{x}$ is also non-increasing. We use that $F(0) = 0$, $T_\# \rho_0 = \rho_1$ and (12.16), to write

$$\mathrm{H}^F(\rho_1) = \int_{[\rho_1 \neq 0]} \frac{F(\rho_1(y))}{\rho_1(y)} \rho_1(y)\, dy = \int_\Omega \frac{F(\rho_1(Tx))}{\rho_1(Tx)} \rho_0(x)\, dx$$

(12.17) $$= \int_\Omega F\left(\frac{\rho_0(x)}{\det \nabla T(x)}\right) \det \nabla T(x)\, dx.$$

Comparing the geometric mean $(\det \nabla T(x))^{1/n}$ to the arithmetic mean $\frac{\operatorname{tr} \nabla T(x)}{d}$, we have that

$$\frac{1}{\det \nabla T(x)} \geq \left(\frac{n}{\operatorname{tr} \nabla T(x)}\right)^n,$$

then we use that $x \mapsto \frac{F(x)}{x}$ is non-decreasing, to get that

(12.18) $$F\left(\frac{\rho_0(x)}{\det \nabla T(x)}\right) \det \nabla T(x) \geq \Lambda^n F\left(\frac{\rho_0(x)}{\Lambda^n}\right) = \rho_0(x) A\left(\frac{\Lambda}{\rho_0(x)^{1/n}}\right),$$

where
$$\Lambda := \frac{\operatorname{tr} \nabla T(x)}{n}.$$

Next, we use that $A'(x) = -nx^{n-1} P_F(x^{-n})$ and the fact that A is convex, to obtain that

$$\rho_0(x) A\left(\frac{\Lambda}{\rho_0(x)^{1/n}}\right) \geq \rho_0(x) \left[A\left(\frac{1}{\rho_0(x)^{1/n}}\right) + A'\left(\frac{1}{\rho_0(x)^{1/n}}\right) \left(\frac{\Lambda - 1}{\rho_0(x)^{1/n}}\right) \right]$$
$$= \rho_0(x) \left[\frac{F(\rho_0(x))}{\rho_0(x)} - n(\Lambda - 1) \frac{P_F(\rho_0(x))}{\rho_0(x)} \right]$$
(12.19)
$$= F(\rho_0(x)) - P_F(\rho_0(x)) \operatorname{tr}(\nabla T(x) - I).$$

We combine (12.17) - (12.19), to conclude that

$$H^F(\rho_1) - H^F(\rho_0) \geq -\int_\Omega P_F(\rho_0(x)) \operatorname{tr}(\nabla T(x) - I) \, dx$$
$$= -\int_\Omega P_F(\rho_0(x)) \operatorname{div}(T(x) - I) \, dx$$
(12.20)
$$\geq \int_\Omega \rho_0 (T - I) \cdot \nabla (F'(\rho_0)) \, dx.$$

(2) To prove (12.14), use the fact that $D^2 V \geq \lambda I$, that is,

(12.21)
$$V(b) - V(a) \geq \nabla V(a) \cdot (b - a) + \frac{\lambda}{2} |a - b|^2$$

for all $a, b \in \mathbb{R}^n$, and set $a = x$ and $b = T(x)$ in (12.21), where $T_\# \rho_0 = \rho_1$ is the optimal mass transport map.

To prove (3) we write the interaction energy as follows:

(12.22)
$$H^W(\rho_1) = \frac{1}{2} \int_{\Omega \times \Omega} W(x - y) \rho_1(x) \rho_1(y) \, dxdy$$
$$= \frac{1}{2} \int_{\Omega \times \Omega} W(T(x) - T(y)) \rho_0(x) \rho_0(y) \, dxdy$$
$$= \frac{1}{2} \int_{\Omega \times \Omega} W(x - y + (T - I)(x) - (T - I)(y)) \rho_0(x) \rho_0(y) \, dxdy$$
$$\geq \frac{1}{2} \int_{\Omega \times \Omega} [W(x - y) + \nabla W(x - y) \cdot ((T - I)(x) - (T - I)(y))] \rho_0(x) \rho_0(y) \, dxdy$$
$$+ \frac{\nu}{4} \int_{\Omega \times \Omega} |(T - I)(x) - (T - I)(y)|^2 \rho_0(x) \rho_0(y) \, dxdy$$
$$= H^W(\rho_0) + \frac{1}{2} \int_{\Omega \times \Omega} \nabla W(x - y) \cdot ((T - I)(x) - (T - I)(y)) \rho_0(x) \rho_0(y) \, dxdy$$
$$+ \frac{\nu}{4} \int_{\Omega \times \Omega} |(T - I)(x) - (T - I)(y)|^2 \rho_0(x) \rho_0(y) \, dxdy,$$

where we used above that $D^2 W \geq \nu I$. The last term of the subsequent inequality can be written as:

$$\int_{\Omega \times \Omega} |(T-I)(x) - (T-I)(y)|^2 \rho_0(x)\rho_0(y)\,dxdy$$

$$= 2\int_\Omega |(T-I)(x)|^2 \rho_0(x)\,dx - 2\left|\int_{\mathbb{R}^n}(T-I)(x)\rho_0(x)\,dx\right|^2$$

(12.23) $$= 2\int_\Omega |(T-I)(x)|^2 \rho_0(x)\,dx - 2|\mathrm{b}(\rho_1) - \mathrm{b}(\rho_0)|^2.$$

And since ∇W is odd (because W is even), we get for the second term of (12.22)

$$\int_{\Omega \times \Omega}[\nabla W(x-y)\cdot((T-I)(x)-(T-I)(y))]\rho_0(x)\rho_0(y)\,dxdy$$

$$= 2\int_{\Omega \times \Omega} \nabla W(x-y)\cdot(T-I)(x)\rho_0(x)\rho_0(y)\,dxdy$$

(12.24) $$= 2\int_{\Omega \times \Omega} \rho_0(T-I)\cdot\nabla(W\star\rho_0)\,dx.$$

Combining (12.22) - (12.24), we obtain that

$$\mathrm{H}^W(\rho_1) - \mathrm{H}^W(\rho_0)$$
$$\geq \int_{\Omega \times \Omega} \rho_0(T-I)\cdot\nabla(W\star\rho_0)\,dx + \frac{\nu}{2}\left(\int_\Omega |(T-I)(x)|^2\rho_0\,dx - |\mathrm{b}(\rho_0) - \mathrm{b}(\rho_1)|^2\right).$$

This proves (12.15). $\square$

Proof of Theorem 12.2.1: Adding (12.13), (12.14) and (12.15), one gets

(12.25) $$\mathrm{H}_V^{F,W}(\rho_0) - \mathrm{H}_V^{F,W}(\rho_1) + \frac{\lambda+\nu}{2}W_2^2(\rho_0,\rho_1) - \frac{\nu}{2}|\mathrm{b}(\rho_0) - \mathrm{b}(\rho_1)|^2$$
$$\leq \int_\Omega (x - Tx)\cdot \rho_0 \nabla\left(F'(\rho_0) + V + W\star\rho_0\right)\,dx.$$

Since $\rho_0 \nabla(F'(\rho_0)) = \nabla(P_F(\rho_0))$, we integrate $\int_\Omega \rho_0 \nabla(F'(\rho_0))\cdot x\,dx$ by parts and obtain that

$$\int_\Omega x\cdot\nabla(F'(\rho_0) + V + W\star\rho_0)\rho_0 = \mathrm{H}_{x\cdot\nabla V}^{-nP_F,\,2x\cdot\nabla W}(\rho_0).$$

This leads to

(12.26) $$\mathrm{H}_V^{F,W}(\rho_0) - \mathrm{H}_V^{F,W}(\rho_1) + \frac{\lambda+\nu}{2}W_2^2(\rho_0,\rho_1) - \frac{\nu}{2}|\mathrm{b}(\rho_0) - \mathrm{b}(\rho_1)|^2$$
$$\leq \mathrm{H}_{x\cdot\nabla V}^{-nP_F,\,2x\cdot\nabla W}(\rho_0) - \int_\Omega \rho_0 \nabla(F'(\rho_0) + V + W\star\rho_0)\cdot T(x)\,dx.$$

Now, use Young's inequality to get

(12.27) $$-\nabla\left(F'(\rho_0(x)) + V(x) + (W\star\rho_0)(x)\right)\cdot T(x)$$
$$\leq c(T(x)) + c^\star\left(-\nabla(F'(\rho_0(x)) + V(x) + (W\star\rho_0)(x))\right)$$

and deduce that

(12.28) $$\mathrm{H}_V^{F,W}(\rho_0) - \mathrm{H}_V^{F,W}(\rho_1) + \frac{\lambda+\mu}{2}W_2^2(\rho_0,\rho_1) - \frac{\nu}{2}|\mathrm{b}(\rho_0) - \mathrm{b}(\rho_1)|^2$$
$$\leq \mathrm{H}_{x\cdot\nabla V}^{-nP_F,\,2x\cdot\nabla W}(\rho_0) + \int_\Omega \rho_0 c^\star\left(-\nabla(F'(\rho_0) + V + W\star\rho_0)\right) + \int_\Omega c(Tx)\rho_0\,dx.$$

Use again that T pushes ρ_0 forward to ρ_1, to rewrite the last integral on the right hand side of (12.28) as $\int_\Omega c(y)\rho_1(y)dy$ to obtain (12.11).

Now, set $\rho_0 = \rho_1 := \rho_{V+c}$ in (12.26). T is then the identity, and equality then holds in (12.26). Therefore, equality holds in (12.11) whenever equality holds in (12.27) with $T(x) = x$. This occurs since (12.12) is satisfied. □

By choosing $\rho_0 := \rho$ and $\rho_1 := \rho_{V+c}$ in (12.11), we obtain the following result.

COROLLARY 12.2.1. *Under the hypothesis of Theorem 12.2.1, we have for any probability density ρ on Ω with $\operatorname{supp}\rho \subset \Omega$ and $P_F(\rho) \in W^{1,\infty}(\Omega)$,*

$$(12.29) \quad \mathrm{H}_{V - x \cdot \nabla V}^{F + nP_F,\, W - 2x \cdot \nabla W}(\rho) + \frac{\lambda + \nu}{2} W_2^2(\rho, \rho_{V+c}) - \frac{\nu}{2}|\mathrm{b}(\rho) - \mathrm{b}(\rho_{V+c})|^2$$
$$\leq \mathcal{I}_{c^*}(\rho|\rho_V) - \mathrm{H}^{P_F, W}(\rho_{V+c}) + K_{V+c},$$

where ρ_{V+c} is the solution of the equation

$$(12.30) \qquad \nabla\left(F'(\rho_{V+c}) + V + c + W \star \rho_{V+c}\right) = 0 \quad a.e.$$

such that $\int_\Omega \rho_{V+c} = 1$, and K_{V+c} is the constant

$$(12.31) \qquad K_{V+c} := F'(\rho_{V+c}) + V + c + W \star \rho_{V+c}.$$

COROLLARY 12.2.2. *Under the hypothesis of Theorem 12.2.1, assume that W is convex, and that V is strictly convex, then we have for any probability density ρ such that $\operatorname{supp}\rho \subset \Omega$ and $P_F(\rho) \in W^{1,\infty}(\Omega)$,*

$$(12.32) \qquad \mathrm{H}_{-V^*(\nabla V)}^{F + nP_F,\, W - 2x \cdot \nabla W}(\rho) \leq \mathcal{I}_{c^*}(\rho|\rho_V) + K_{V+c},$$

where V^* is the Legendre transform of V and K_{V+c} is as in (12.31).

Proof: If W is convex, then $\nu \geq 0$ and the barycentric term can also be omitted. If V is also convex, hence $\lambda + \nu \geq 0$, then the term involving the Wasserstein distance can be omitted from Equation (12.29). Finally, if V is strictly convex, then we have the identity $V(x) - x \cdot \nabla V(x) = -V^*(\nabla V(x))$ in such a way that a correcting "moment" appears in the inequality:

$$(12.33) \qquad \mathrm{H}_{-V^*(\nabla V)}^{F + nP_F,\, W - 2x \cdot \nabla W}(\rho) \leq \mathcal{I}_{c^*}(\rho|\rho_V) - \mathrm{H}^{P_F}(\rho_{V+c}) + K_{V+c}.$$

On the other hand, under the condition on F, the pressure P_F is always positive and can therefore be removed to obtain the claimed inequality.

The following is now immediate, but we single it out for its applications in the next chapter.

COROLLARY 12.2.3. *Under the hypothesis of Theorem 12.2.1, assume that $W \equiv 0$, and that V is strictly convex, then we have for any probability density ρ such that $\operatorname{supp}\rho \subset \Omega$ and $P_F(\rho) \in W^{1,\infty}(\Omega)$,*

$$(12.34) \qquad \mathrm{H}_{-V^*(\nabla V)}^{F + nP_F}(\rho) \leq \mathcal{I}_{c^*}(\rho|\rho_V) + K_{V+c},$$

where ρ_{V+c} is the solution of the equation

$$(12.35) \qquad \nabla\left(F'(\rho_{V+c}) + V + c\right) = 0 \quad a.e.$$

such that $\int_\Omega \rho_{V+c} = 1$, and K_{V+c} is the constant

$$(12.36) \qquad K_{V+c} := F'(\rho_{V+c}) + V + c.$$

COROLLARY 12.2.4. *Under the hypothesis of Theorem 12.2.1, assume that $W \equiv 0$. Then the following inequalities hold:*

(1) *for all probability densities ρ_0 and ρ_1 on Ω, satisfying $\operatorname{supp} \rho_0 \subset \Omega$, $\rho_0 > 0$ a.e. on Ω and $P_F(\rho_0) \in W^{1,\infty}(\Omega)$, we have*

$$-\mathrm{H}^{F}_{V+c}(\rho_1) + \frac{\lambda}{2} W_2^2(\rho_0, \rho_1) \leq -\mathrm{H}^{F+nP_F}_{V - x \cdot \nabla V}(\rho_0) + \mathcal{I}_{c^*}(\rho_0 | \rho_V) \tag{12.37}$$

where ρ_V is defined by $\nabla(F'(\rho_V) + V) = 0$ a.e. Furthermore, equality holds in (12.37) whenever $\rho_0 = \rho_1 = \rho_{V+c}$ where the latter is defined in (12.35)

(2) *If V is strictly convex, then we have for all probability densities ρ_0 and ρ_1 on Ω, satisfying $\operatorname{supp} \rho_0 \subset \Omega$, $\rho_0 > 0$ a.e. on Ω and $P_F(\rho_0) \in W^{1,\infty}(\Omega)$,*

$$-\mathrm{H}^{F}_{V+c}(\rho_1) \leq -\mathrm{H}^{F+nP_F}_{-V^*(\nabla V)}(\rho_0) + \mathcal{I}_{c^*}(\rho_0 | \rho_V) \tag{12.38}$$

12.3. Further comments

Recent advances in the theory of mass transport are due to Brenier [**59**], Gangbo-McCann [**149**], and many others. For a survey, see Villani [**268**] and [**269**]. Missing from this chapter is a fundamental result of Caffarelli [**74**], which implies that if the initial measure ρ_0 is the standard Gaussian probability measure on $\mathbb{R}^n$, that is $\frac{d\rho}{dx} = \frac{1}{(2\pi)^{n/2}} e^{-\frac{|x|^2}{2}}$ and if the target measure ρ_1 is such that $\frac{d\rho_1}{d\rho_0}$ exists and is a log-concave function, then the Brenier map $T := \nabla \varphi$ which transport ρ_0 into ρ_1 is a contaction, i.e. $|Tx - Ty| \leq |x - y|$ for all $x, y \in \mathbb{R}^n$.

Starting with McCann's proof and generalization of the Brunn-Minkowski's inequality [**217**], these advances in the Monge-Kantorovich theory have – among other things – led to new and quite natural proofs for a wide range of geometric inequalities. The key idea behind this approach is the concept of *displacement convexity* introduced by McCann [**217**], which describes the evolution of an energy functional along optimal transport.

While this chapter reflects this spirit, we describe here a basic framework proposed by Agueh-Ghoussoub-Kang [**17**] to which most geometric inequalities belong, and a general inequality from which most of them follow. Besides the obvious pedagogical relevance of a streamlined approach, we find it interesting and intriguing that most of these inequalities appear as different manifestations of one basic principle in the theory of interacting gases that compares the different types of – internal, potential and interactive – energies of two states of a system after one is transported "at minimal cost" into another. For generalized cost functions, and when the confining potential $V = 0$, the convexity principle was first obtained by Otto [**233**] for the Tsallis entropy functionals and by Agueh [**14**] in general. The case of a nonzero confinement potential V and an interaction potential W were conidered by Cordero-Erausquin, Gangbo and Houdré [**105**], and Cordero-Erausquin, Nazaret and C. Villani [**103**]. We also refer to the book of Ambrosio-Gigli-Savaré [**21**] for a penetrating analysis of displacement convexity and its ramifications.

Open problem (14): Characterize and identify all functions $F : [0, +\infty) \to \mathbb{R}$ such that the corresponding internal energy H^F is convex along Wasserstein geodesics $(\rho_t)_t$ connecting any two probability densities ρ_0 and ρ_1 on Ω. In other words, what are the functions F such that $t \to H^F(\rho_t) = \int_\Omega F(\rho_t(x)) \, dx$ is convex on $[0, 1]$, where the path $\rho_t := \nabla \varphi_t \# \rho_0$ with $\varphi_t = (1-t)\frac{|x|^2}{2} + t\varphi$ and φ being the convex function associated to the transport of ρ_0 to ρ_1 given in Theorem 12.1.1.

CHAPTER 13

Optimal Euclidean Sobolev Inequalities

It is shown that most Euclidean Sobolev inequalities follow from the following general formula which relates the internal energy of a probability density ρ on a domain $\Omega \subset \mathbb{R}^n$ to the corresponding entropy production,

$$(13.1) \qquad \int_\Omega [(1-n)F(\rho) + n\rho F'(\rho)]\, dx \leq \int_\Omega \rho c^\star \left(-\nabla(F' \circ \rho)\right)\, dx + K_c.$$

Here $F : [0, \infty) \to \mathbb{R}$ is any differentiable function on $(0, \infty)$ with $F(0) = 0$ and $x \mapsto x^n F(x^{-n})$ convex and non-increasing, c is any Young function on $\mathbb{R}^n$, while the constant K_c can be evaluated from F and c.

We also establish and apply the following duality formula between any pair of probability densities ρ_0 and ρ_1 on Ω such that $F(\rho_0)$ and $F'(\rho_0)$ are in $W^{1,\infty}(\Omega)$:

$$(13.2) \qquad -\int_\Omega [F(\rho_1) + c(x)\rho_1]\, dx \leq \int_\Omega [(n-1)F(\rho_0) - n\rho F'(\rho_0)]\, dx + \int_\Omega \rho c^\star \left(-\nabla(F' \circ \rho_0)\right)\, dx.$$

While the right-hand term expresses the gap in the above Sobolev-type estimate, the left-hand side points to the presence of a dual "moment inequality". This duality leads to a remarkable correspondence between ground state solutions of certain quasilinear (or semi-linear) equations and stationary solutions of (non-linear) Fokker-Planck type equations.

13.1. A general Sobolev inequality

We consider a system, where there is no potential nor interaction energies, i.e., where $V = W = 0$. Let $\Omega \subset \mathbb{R}^n$ be open and convex, and consider again the set $\mathcal{P}_a(\Omega)$ of probability densities over Ω. We associate again to a function $F : [0, \infty) \to \mathbb{R}$, the *internal energy* functional $\mathrm{H}^F(\rho) := \int_\Omega F(\rho)\,dx$ on $\mathcal{P}_a(\Omega)$. The following result is an immediate corollary of Theorem 12.2.1 applied in the case where $V = W \equiv 0$.

THEOREM 13.1.1. *Let $F : [0, \infty) \to \mathbb{R}$ be a differentiable function on $(0, \infty)$ with $F(0) = 0$ and $x \mapsto x^n F(x^{-n})$ convex and non-increasing, and let $P_F(x) := xF'(x) - F(x)$ be its associated pressure function. For a given Young function c, we denote by ρ_c the probability density satisfying the equation*

$$(13.3) \qquad \nabla\left(F'(\rho_c) + c\right) = 0 \quad a.e.,$$

and by K_c be the unique constant such that

$$(13.4) \qquad K_c := F'(\rho_c) + c.$$

If Ω is an open, bounded and convex subset of $\mathbb{R}^n$, then for any probability density ρ such that $\operatorname{supp}\rho \subset \Omega$ and $P_F(\rho) \in W^{1,\infty}(\Omega)$, we have

$$(13.5) \qquad \mathrm{H}^{F+nP_F}(\rho) \leq \int_\Omega \rho c^\star \left(-\nabla(F' \circ \rho)\right)\, dx - \mathrm{H}^{P_F}(\rho_c) + K_c.$$

REMARK 13.1.2. Note that it is customary to view the quantity $\mathcal{I}_{c^*}(\rho) := \int_\Omega \rho c^*\left(-\nabla\left(F'(\rho)\right)\right) dx$ as a relative entropy with respect to ρ_∞, where ρ_∞ is the probability density on Ω such that

(13.6) $$\nabla(F'(\rho_\infty)) = 0 \text{ a.e. on } \Omega.$$

We therefore define the *generalized relative entropy production-type function of ρ with respect to ρ_∞, measured against c^** as

(13.7) $$\mathcal{I}_{c^*}(\rho|\rho_\infty) := \int_\Omega \rho c^*\left(-\nabla\left(F'(\rho)\right)\right) dx.$$

Since the internal energy associated to the pressure $H^{P_F}(\rho)$ is non-negative, the inequality then says that the internal energy is –up to a constant– dominated by the entropy production, that is

(13.8) $$H^{F+nP_F}(\rho) \leq \mathcal{I}_{c^*}(\rho|\rho_\infty) + K_c.$$

Note that when $c(x) = \frac{|x|^2}{2}$, the inequality becomes

(13.9) $$H^{F+nP_F}(\rho) \leq \int_\Omega \rho \left|\nabla\left(F'(\rho)\right)\right|^2 dx + K_2,$$

where K_2 is the unique constant such that $K_2 = F'(\rho_2(x)) + \frac{1}{2}|x|^2$ and where ρ_2 is the probability density solution of $\nabla(F'(\rho_2)) + x = 0$ a.e. on Ω.

We shall show in the next sections how this inequality, once applied to various admissible functionals F, yields several classical Euclidean Sobolev inequalities.

13.2. Sobolev and Gagliardo-Nirenberg inequalities

COROLLARY 13.2.1. *(Gagliardo-Nirenberg inequalities) Let $1 < p < n$ and $r \in \left(0, \frac{np}{n-p}\right)$ such that $r \neq p$. Set $\gamma := \frac{1}{r} + \frac{1}{q}$, where $\frac{1}{p} + \frac{1}{q} = 1$. Then, for any $f \in W^{1,p}(\mathbb{R}^n)$ we have*

(13.10) $$\|f\|_r \leq C(p,r) \|\nabla f\|_p^\theta \|f\|_{r\gamma}^{1-\theta},$$

where θ is given by

(13.11) $$\frac{1}{r} = \frac{\theta}{p^*} + \frac{1-\theta}{r\gamma},$$

$p^* = \frac{np}{n-p}$ *and where the best constant $C(p,r) > 0$ can be obtained by scaling.*

Proof: Let $F(x) = \frac{x^\gamma}{\gamma - 1}$, where $1 \neq \gamma > 1 - \frac{1}{n}$, which follows from the fact that $p \neq r \in \left(0, \frac{np}{n-p}\right)$. For this value of γ, the function F satisfies the conditions of Theorem 13.1.1.

Let $c(x) = \frac{r\gamma}{q}|x|^q$ so that $c^*(x) = \frac{1}{p(r\gamma)^{p-1}}|x|^p$. Inequality (13.5) then gives for all $f \in C_c^\infty(\mathbb{R}^n)$ such that $\|f\|_r = 1$,

(13.12) $$\left(\frac{1}{\gamma - 1} + n\right) \int_{\mathbb{R}^n} |f|^{r\gamma} \leq \frac{r\gamma}{p} \int_{\mathbb{R}^n} |\nabla f|^p - H^{P_F}(\rho_\infty) + K_q,$$

where $\rho_\infty = h_\infty^r$ satisfies

(13.13) $$-\nabla h_\infty(x) = x|x|^{q-2} h_\infty^{\frac{r}{p}}(x) \text{ a.e.}$$

and where K_q insures that $\int h_\infty^r = 1$. The constants on the right hand side of (13.12) are not easy to calculate, so one can obtain θ and the best constant by a standard scaling procedure. Namely, write (13.12) as

$$(13.14) \qquad \frac{r\gamma}{p} \frac{\|\nabla f\|_p^p}{\|f\|_r^p} - \left(\frac{1}{\gamma-1} + n\right) \frac{\|f\|_{r\gamma}^{r\gamma}}{\|f\|_r^{r\gamma}} \geq H^{P_F}(\rho_\infty) - K_q =: C,$$

for some constant C. Then apply (13.14) to $f_\lambda(x) = f(\lambda x)$ for $\lambda > 0$. A minimization over λ gives the required constant.

The limiting case where r is the critical Sobolev exponent $r = p^* = \frac{np}{n-p}$ (and then $\gamma = 1 - \frac{1}{n}$) leads to the Sobolev inequalities:

COROLLARY 13.2.2. *(Sobolev inequalities) If $1 < p < n$, then for any $f \in W^{1,p}(\mathbb{R}^n)$,*

$$(13.15) \qquad \|f\|_{p^*} \leq C(p,n) \|\nabla f\|_p$$

for some constant $C(p,n) > 0$.

Proof: It follows directly from (13.12) by using $\gamma = 1 - \frac{1}{n}$ and $r = p^*$. Note however that the scaling argument cannot be used here to compute the best constant $C(p,n)$ in (13.15), since $\|\nabla f_\lambda\|_p^p = \lambda^{p-n} \|\nabla f\|_p^p$ and $\|f_\lambda\|_r^p = \lambda^{p-n} \|f\|_r^p$ scale the same way in (13.14). Instead, one can proceed directly from (13.12) to have that

$$\|f\|_{p^*} = 1 \leq \left(\frac{r\gamma}{p[H^{P_F}(\rho_\infty) - K_p]}\right)^{1/p} \|\nabla f\|_p = \left(\frac{p^*(n-1)}{np[H^{P_F}(\rho_\infty) - K_p]}\right)^{1/p} \|\nabla f\|_p,$$

which shows that

$$(13.16) \qquad C(p,n) = \left(\frac{p^*(n-1)}{np[H^{P_F}(\rho_\infty) - K_p]}\right)^{1/p},$$

where $\rho_\infty = h_\infty^{p^*} = \left(\frac{p^*}{nq}|x|^q - \frac{K_p}{n-1}\right)^{-n}$ is obtained from (13.13), and K_p can be found using that ρ_∞ is a probability density,

$$(13.17) \qquad K_p = (1-n) \left[\int_{\mathbb{R}^n} \left(\frac{p^*}{nq}|x|^q + 1\right)^{-n} dx\right]^{p/n}.$$

13.3. Euclidean Log-Sobolev inequalities

COROLLARY 13.3.1. *(General Euclidean Log-Sobolev inequality) Let $\Omega \subset \mathbb{R}^n$ be open bounded and convex, and let $c : \mathbb{R}^n \to \mathbb{R}$ be a Young functional such that its conjugate c^* is p-homogeneous for some $p > 1$, and let $\sigma_c := \int_{\mathbb{R}^n} e^{-c} dx$. Then,*

$$(13.18) \qquad \int_{\mathbb{R}^n} \rho \log \rho \, dx \leq \frac{n}{p} \log \left(\frac{p}{ne^{p-1}\sigma_c^{p/n}} \int_{\mathbb{R}^n} \rho c^* \left(-\frac{\nabla \rho}{\rho}\right) dx\right),$$

for all probability densities ρ in $\mathcal{P}_a(\Omega)$ such that $\rho \in W^{1,\infty}(\mathbb{R}^n)$.

Moreover, equality holds in (13.18) if $\rho(x) = K_\lambda e^{-\lambda^q c(x)}$ for some $\lambda > 0$, where $K_\lambda = \left(\int_{\mathbb{R}^n} e^{-\lambda^q c(x)} dx\right)^{-1}$ and q is the conjugate of p $(\frac{1}{p} + \frac{1}{q} = 1)$.

Proof: Use $F(x) = x\log(x)$ in (13.1.1). Note that $P_F(x) = x$, hence $H^{P_F}(\rho) = 1$ for any $\rho \in \mathcal{P}_a(\mathbb{R}^n)$ and $\rho_c(x) = \frac{e^{-c(x)}}{\sigma_c}$. We then have for any $\rho \in \mathcal{P}_a(\mathbb{R}^n) \cap W^{1,\infty}(\mathbb{R}^n)$ such that $\operatorname{supp}\rho \subset \Omega$,

$$(13.19) \qquad \int_\Omega \rho \log\rho \, dx \le \int_{\mathbb{R}^n} \rho c^\star\left(-\frac{\nabla\rho}{\rho}\right) dx - n - \log\left(\int_{\mathbb{R}^n} e^{-c(x)} dx\right),$$

with equality when $\rho = \rho_c$.

Now assume that $c^\star$ is p-homogeneous and set $\Gamma_\rho^c = \int_{\mathbb{R}^n} \rho c^\star\left(-\frac{\nabla\rho}{\rho}\right) dx$. Using that $c_\lambda(x) := c(\lambda x)$ in (13.19), we get for $\lambda > 0$ that

$$(13.20) \qquad \int_{\mathbb{R}^n} \rho \log\rho \, dx \le \int_{\mathbb{R}^n} \rho c^\star\left(-\frac{\nabla\rho}{\lambda\rho}\right) dx + n \log\lambda - n - \log\sigma_c,$$

for all $\rho \in \mathcal{P}_a(\Omega)$ satisfying $\operatorname{supp}\rho \subset \Omega$ and $\rho \in W^{1,\infty}(\Omega)$. Equality holds in (13.20) if $\rho_\lambda(x) = \left(\int_{\mathbb{R}^n} e^{-\lambda^q c(x)} dx\right)^{-1} e^{-\lambda^q c(x)}$. Hence

$$\int_{\mathbb{R}^n} \rho \log\rho \, dx \le -n - \log\sigma_c + \inf_{\lambda > 0}(G_\rho(\lambda)),$$

where

$$G_\rho(\lambda) = n\log(\lambda) + \frac{1}{\lambda^p}\int_{\mathbb{R}^n} \rho c^\star\left(-\frac{\nabla\rho}{\rho}\right) = n\log(\lambda) + \frac{\Gamma_\rho^c}{\lambda^p}.$$

The infimum of $G_\rho(\lambda)$ over $\lambda > 0$ is attained at $\bar\lambda_\rho = \left(\frac{p}{n}\Gamma_\rho^c\right)^{1/p}$. Hence

$$\begin{aligned}
\int_{\mathbb{R}^n} \rho \log\rho \, dx &\le G_\rho(\bar\lambda_\rho) - n - \log(\sigma_c) \\
&= \frac{n}{p}\log\left(\frac{p}{n}\Gamma_\rho^c\right) + \frac{n}{p} - n - \log(\sigma_c) \\
&= \frac{n}{p}\log\left(\frac{p}{n e^{p-1}\sigma_c^{p/n}}\Gamma_\rho^c\right),
\end{aligned}$$

for all probability densities $\rho \in \mathcal{P}_a(\Omega) \cap W^{1,\infty}(\Omega)$.

COROLLARY 13.3.2. (Optimal Euclidean p-Log Sobolev inequality) *The inequality*

$$(13.21) \qquad \int_{\mathbb{R}^n} |f|^p \log(|f|^p) \, dx \le \frac{n}{p}\log\left(C_p \int_{\mathbb{R}^n} |\nabla f|^p \, dx\right),$$

holds for all $p \ge 1$, and for all $f \in W^{1,p}(\mathbb{R}^n)$ such that $\|f\|_p = 1$, where

$$(13.22) \qquad C_p := \begin{cases} \left(\frac{p}{n}\right)\left(\frac{p-1}{e}\right)^{p-1} \pi^{-\frac{p}{2}} \left[\frac{\Gamma(\frac{n}{2}+1)}{\Gamma(\frac{n}{q}+1)}\right]^{\frac{p}{n}} & \text{if } p > 1, \\ \frac{1}{n\sqrt\pi}\left[\Gamma(\frac{n}{2}+1)\right]^{\frac{1}{n}} & \text{if } p = 1, \end{cases}$$

and q is the conjugate of p ($\frac{1}{p} + \frac{1}{q} = 1$).

For $p > 1$, there exists $\lambda > 0$ and $\bar x \in \mathbb{R}^n$ such that equality holds in (13.21) for $f(x) = Ke^{-\lambda^q \frac{|x - \bar x|^q}{q}}$, where $K = \left(\int_{\mathbb{R}^n} e^{-(p-1)|\lambda x|^q} dx\right)^{-1/p}$.

Proof: First assume that $p > 1$, and set $c(x) = (p-1)|x|^q$ and $\rho = |f|^p$ in (13.18), where $f \in C_c^\infty(\mathbb{R}^n)$ and $\|f\|_p = 1$. We then have that $c^\star(x) = \frac{|x|^p}{p^p}$, and

$$\int_{\mathbb{R}^n} \rho c^\star\left(-\frac{\nabla\rho}{\rho}\right) dx = \int_{\mathbb{R}^n} |\nabla f|^p dx.$$

Therefore, (13.18) reads as

(13.23) $$\int_{\mathbb{R}^n} |f|^p \log(|f|^p) dx \leq \frac{n}{p} \log\left(\frac{p}{ne^{p-1}\sigma_c^{p/n}} \int_{\mathbb{R}^n} |\nabla f|^p dx\right).$$

Now, it suffices to note that

(13.24) $$\sigma_c := \int_{\mathbb{R}^n} e^{-(p-1)|x|^q} dx = \frac{\pi^{\frac{n}{2}} \Gamma\left(\frac{n}{q}+1\right)}{(p-1)^{\frac{n}{q}} \Gamma\left(\frac{n}{2}+1\right)}.$$

To prove the case where $p = 1$, it is sufficient to apply the above to $p_\epsilon = 1 + \epsilon$ for some arbitrary $\epsilon > 0$. Note that

$$C_{p_\epsilon} = \left(\frac{1+\epsilon}{n}\right)\left(\frac{\epsilon}{e}\right)^\epsilon \pi^{-\frac{1+\epsilon}{2}} \left[\frac{\Gamma(\frac{n}{2}+1)}{\Gamma(\frac{n\epsilon}{1+\epsilon}+1)}\right]^{\frac{1+\epsilon}{n}},$$

so that when ϵ go to 0, we have

$$\lim_{\epsilon \to 0} C_{p_\epsilon} = \frac{1}{n\sqrt{\pi}} \left[\Gamma\left(\frac{n}{2}+1\right)\right]^{\frac{1}{n}} = C_1.$$

13.4. A remarkable duality

In this section, we apply Corollary 12.2.4 when $V := 0$, to obtain an intriguing duality between two types of functional inequalities.

THEOREM 13.4.1. *Let $F : [0, \infty) \to \mathbb{R}$ be differentiable function on $(0, \infty)$ with $F(0) = 0$ and $x \mapsto x^n F(x^{-n})$ convex and non-increasing, let $P_F(x) := xF'(x) - F(x)$ be its associated pressure function, and let c be a Young function. If Ω is an open bounded and convex subset of $\mathbb{R}^n$, then for any probability densities $\rho_0, \rho_1 \in \mathcal{P}_a(\Omega)$ such that $\operatorname{supp} \rho_0 \subset \Omega$ and $P_F(\rho_0) \in W^{1,\infty}(\Omega)$, we have*

(13.25) $$-H_c^F(\rho_1) \leq -H^{F+nP_F}(\rho_0) + \int_\Omega \rho_0 c^\star(-\nabla(F' \circ \rho_0)) dx.$$

Moreover, equality holds whenever $\rho_0 = \rho_1 = \rho_c$ where ρ_c is a probability density on Ω satisfying the equation

(13.26) $$\nabla(F'(\rho_c) + c) = 0 \quad a.e.$$

This yields the following duality.

COROLLARY 13.4.1. *Let $F : [0, \infty) \to \mathbb{R}$ be differentiable on $(0, \infty)$ such that $F(0) = 0$ and $x \mapsto x^n F(x^{-n})$ is convex and non-increasing, and set $G_F(x) := (1-n)F(x) + nxF'(x)$. Let c be any Young function such that its Legendre transform $c^\star$ is p-homogeneous for some $p > 1$. Let $\psi : \mathbb{R} \to [0, \infty)$ be a differentiable function chosen in such a way that*

(13.27) $$\psi(0) = 0 \text{ and } |\psi^{\frac{1}{p}}(F' \circ \psi)'| = 1.$$

For $\Omega \subset \mathbb{R}^n$ open bounded and convex, consider the following two extremal problems:

(13.28) $$D_\infty := \sup\{-\int_\Omega [F(\rho) + c\rho]\,dx; \rho \in \mathcal{P}_a(\Omega)\}$$

and

(13.29) $$P_\infty := \inf\{\int_\Omega [c^*(-\nabla f) - G_F \circ \psi(f)]\,dx; f \in C_0^\infty(\Omega), \int_\Omega \psi(f) = 1\}.$$

(1) The following inequality then holds:

(13.30) $$D_\infty \leq P_\infty.$$

(2) If there exists $\bar{f}$ that satisfies

(13.31) $$-(F' \circ \psi)'(\bar{f})\nabla \bar{f}(x) = \nabla c(x) \text{ a.e.}$$

then D_∞ is attained at $\bar{f}$, P_∞ is attained at $\bar{\rho} = \psi(\bar{f})$, and $D_\infty = P_\infty$.

(3) Moreover, $\bar{f}$ solves

(13.32) $$\begin{aligned} \operatorname{div}\{\nabla c^*(-\nabla f)\} - (G_F \circ \psi)'(f) = \lambda \psi'(f) & \quad \text{in } \Omega \\ \nabla c^*(-\nabla f) \cdot \nu = 0 & \quad \text{on } \partial\Omega, \end{aligned}$$

for some $\lambda \in \mathbb{R}$, while $\bar{\rho}$ is a stationary solution of

(13.33) $$\begin{aligned} \tfrac{\partial \rho}{\partial t} = \operatorname{div}\{\rho \nabla (F'(\rho) + c)\} & \quad \text{in } (0,\infty) \times \Omega \\ \rho \nabla (F'(\rho) + c) \cdot \nu = 0 & \quad \text{on } (0,\infty) \times \partial\Omega. \end{aligned}$$

Proof: Assume that c^* is p-homogeneous and let $Q''(x) = x^{\frac{1}{q}} F''(x)$. Let

$$J(\rho) := -\int_\Omega [F(\rho(y)) + c(y)\rho(y)]\,dy$$

and

$$\tilde{J}(\rho) := -\int_\Omega (F + nP_F)(\rho(x))\,dx + \int_\Omega c^*(-\nabla(Q'(\rho(x))))\,dx.$$

Equation (13.25) then becomes

(13.34) $$J(\rho_1) \leq \tilde{J}(\rho_0)$$

for all probability densities ρ_0, ρ_1 on Ω such that $\operatorname{supp}\rho_0 \subset \Omega$ and $P_F(\rho_0) \in W^{1,\infty}(\Omega)$. If $\bar{\rho}$ satisfies

$$-\nabla(F'(\bar{\rho}(x))) = \nabla c(x) \text{ a.e.},$$

then equality holds in (13.34), and $\bar{\rho}$ is an extremal of the variational problems

$$\sup\{J(\rho); \rho \in \mathcal{P}_a(\Omega)\} = \inf\{\tilde{J}(\rho); \rho \in \mathcal{P}_a(\Omega), P_F(\rho) \in W^{1,\infty}(\Omega)\}.$$

In particular, $\bar{\rho}$ is a solution of

(13.35) $$\begin{aligned} \operatorname{div}\{\rho \nabla(F'(\rho) + c)\} = 0 & \quad \text{in } \Omega \\ \rho \nabla(F'(\rho) + c) \cdot \nu = 0 & \quad \text{on } \partial\Omega. \end{aligned}$$

Suppose now $\psi : \mathbb{R} \to [0,\infty)$ differentiable, $\psi(0) = 0$ and that $\bar{f} \in C_0^\infty(\Omega)$ satisfies $-(F' \circ \psi)'(\bar{f})\nabla \bar{f}(x) = \nabla c(x)$ a.e. Then equality holds in (13.34), and $\bar{f}$ and $\bar{\rho} = \psi(\bar{f})$ are extremals of the following variational problems

$$\inf\{I(f); f \in C_0^\infty(\Omega), \int_\Omega \psi(f) = 1\} = \sup\{J(\rho); \rho \in \mathcal{P}_a(\Omega)\}$$

where
$$I(f) = \tilde{J}(\psi(f)) = -\int_\Omega (F + nP_F) \circ \psi(f) dx + \int_\Omega c^*(-\nabla(Q' \circ \psi(f))) dx.$$

If now ψ is such that $|\psi^{\frac{1}{p}}(F' \circ \psi)'| = 1$, then $|(Q' \circ \psi)'| = 1$ and
$$I(f) = -\int_\Omega (F + nP_F) \circ \psi(f) dx + \int_\Omega c^*(-\nabla f)) dx,$$

because c^* is p-homogeneous. This proves (13.30). The Euler-Lagrange equation of the variational problem
$$\inf\left\{\int_\Omega c^*(-\nabla(f)) - (F + nP_F) \circ \psi(f) dx; \int_\Omega \psi(f) dx = 1\right\}$$

reads as

(13.36)
$$\begin{aligned} \text{div}\{\nabla c^*(-\nabla f)\} - (G_F \circ \psi)'(f) = \lambda \psi'(f) & \quad \text{in } \Omega \\ \nabla c^*(-\nabla f) \cdot \nu = 0 & \quad \text{on } \partial\Omega, \end{aligned}$$

where $\lambda \in \mathbb{R}$ is a Lagrange multiplier and $G(x) = (1-n)F(x) + nxF'(x)$. This proves (13.32). The proof that the maximizer $\bar\rho$ of
$$\sup\{-\int_\Omega (F(\rho) + c\rho) \, dx; \, \rho \in \mathcal{P}_a(\Omega)\}$$

is a stationary solution of (13.33) is straightforward. □

We now apply Corollary 13.4.1 to the functions $F(x) = x \log x$, $\psi(x) = |x|^p$ and $c(x) = (p-1)|\mu x|^q$, with $\mu > 0$ and $c^*(x) = \frac{1}{p}\left|\frac{x}{\mu}\right|^p$ and $\frac{1}{p} + \frac{1}{q} = 1$, to derive a duality between stationary solutions of Fokker-Planck equations, and ground state solutions of some semi-linear equations. We note here that $|\psi^{\frac{1}{p}}(F' \circ \psi)| = p$, as opposed to 1. We then obtain the following duality.

COROLLARY 13.4.2. *Let $p > 1$ and let q be its conjugate ($\frac{1}{p} + \frac{1}{q} = 1$). For any $f \in W^{1,p}(\mathbb{R}^n)$ such that $\|f\|_p = 1$, any probability density ρ such that $\int_{\mathbb{R}^n} \rho(x)|x|^q dx < \infty$, and any $\mu > 0$, we have*

(13.37) $$J_\mu(\rho) \leq I_\mu(f),$$

where
$$J_\mu(\rho) := -\int_{\mathbb{R}^n} \rho \log(\rho) \, dy - (p-1)\int_{\mathbb{R}^n} |\mu y|^q \rho(y) \, dy,$$

and
$$I_\mu(f) := -\int_{\mathbb{R}^n} |f|^p \log(|f|^p) + \int_{\mathbb{R}^n} \left|\frac{\nabla f}{\mu}\right|^p - n.$$

Furthermore, if $h \in W^{1,p}(\mathbb{R}^n)$ is such that $h \geq 0$, $\|h\|_p = 1$, and
$$\nabla h(x) = -\mu^q x |x|^{q-2} h(x) \quad a.e.,$$

then
$$J_\mu(h^p) = I_\mu(h).$$

Therefore, h (resp., $\rho = h^p$) is an extremum of the variational problem:
$$\sup\{J_\mu(\rho) : \rho \in W^{1,1}(\mathbb{R}^n), \|\rho\|_1 = 1\} = \inf\{I_\mu(f) : f \in W^{1,p}(\mathbb{R}^n), \|f\|_p = 1\}.$$

It follows that h satisfies the Euler-Lagrange equation corresponding to the constraint minimization problem, i.e., h is a solution of

(13.38) $$\mu^{-p}\Delta_p f + pf|f|^{p-2}\log(|f|) = \lambda f|f|^{p-2},$$

where λ is a Lagrange multiplier. On the other hand, $\rho = h^p$ is a stationary solution of the Fokker-Planck equation:

(13.39) $$\frac{\partial u}{\partial t} = \Delta u + \text{div}(p\mu^q|x|^{q-2}xu).$$

In particular, we have the duality

$$\sup_{\rho \in W^{1,1}, \|\rho\|_1 = 1} \left(n - \int_{\mathbb{R}^n} \rho \log \rho - \int_{\mathbb{R}^n} |x|^2 \rho \right)$$
$$= \inf_{f \in W^{1,2}, \|f\|_2 = 1} \left(\int_{\mathbb{R}^n} |\nabla f|^2 dx - \int_{\mathbb{R}^n} f^2 \log(f^2) dx \right).$$

We can also apply Corollary 13.4.1 to recover the duality associated to the Gagliardo-Nirenberg inequalities.

COROLLARY 13.4.3. *Let* $1 < p < n$, *and* $r \in \left(0, \frac{np}{n-p}\right]$ *such that* $r \neq p$. *Set* $\gamma := \frac{1}{r} + \frac{1}{q}$, *where* $\frac{1}{p} + \frac{1}{q} = 1$. *Then, for any* $f \in W^{1,p}(\mathbb{R}^n)$ *such that* $\|f\|_r = 1$, *for any probability density* ρ *and for any* $\mu > 0$, *we have*

(13.40) $$J_\mu(\rho) \leq I_\mu(f),$$

where

$$J_\mu(\rho) := -\frac{1}{\gamma - 1} \int_{\mathbb{R}^n} \rho^\gamma - \frac{r\gamma\mu^q}{q} \int_{\mathbb{R}^n} |y|^q \rho(y)(y)\, dy,$$

and

$$I_\mu(f) := -\left(\frac{1}{\gamma - 1} + n\right) \int_{\mathbb{R}^n} |f|^{r\gamma} + \frac{r\gamma}{p\mu^p} \int_{\mathbb{R}^n} |\nabla f|^p.$$

Furthermore, if $h \in W^{1,p}(\mathbb{R}^n)$ *is such that* $h \geq 0$, $\|h\|_r = 1$, *and*

$$\nabla h(x) = -\mu^q x|x|^{q-2} h^{\frac{r}{p}}(x) \quad a.e.,$$

then

$$J_\mu(h^r) = I_\mu(h).$$

Therefore, h (resp., $\rho = h^r$) is an extremum of the variational problems

$$\sup\{J_\mu(\rho) : \rho \in W^{1,1}(\mathbb{R}^n), \|\rho\|_1 = 1\} = \inf\{I_\mu(f) : f \in W^{1,p}(\mathbb{R}^n), \|f\|_r = 1\}.$$

Proof: Again, the proof follows from Corollary 13.4.1, by using now $\psi(x) = |x|^r$ and $F(x) = \frac{x^\gamma}{\gamma - 1}$, where $1 \neq \gamma \geq 1 - \frac{1}{n}$, which follows from the fact that $p \neq r \in \left(0, \frac{np}{n-p}\right]$. Indeed, for this value of γ, the function F satisfies the conditions of Corollary 13.4.1. The Young function is now $c(x) = \frac{r\gamma}{q}|\mu x|^q$, that is $c^*(x) = \frac{1}{p(r\gamma)^{p-1}}\left|\frac{x}{\mu}\right|^p$, and the condition $|\psi^{\frac{1}{p}}(F' \circ \psi)'| = K$ holds with $K = r\gamma$. Moreover, if $h \geq 0$ satisfies (13.31), which is here,

$$-\nabla h(x) = \mu^q x|x|^{q-2} h^{\frac{r}{p}}(x) \quad a.e.,$$

then h is extremal in the minimization problem defined in Corollary 13.4.3.

As above, we also note that h satisfies the Euler-Lagrange equation corresponding to the constraint minimization problem, that is, h is a solution of

$$\mu^{-p}\Delta_p f + \left(\frac{1}{\gamma-1}+n\right)f|f|^{r\gamma-2} = \lambda f|f|^{r-2}, \tag{13.41}$$

where λ is a Lagrange multiplier. On the other hand, $\rho = h^r$ is a stationary solution of the evolution equation:

$$\frac{\partial u}{\partial t} = \Delta u^\gamma + \mathrm{div}(r\gamma\mu^q|x|^{q-2}xu). \tag{13.42}$$

As a consequence one gets the following inequality that is dual to the Sobolev inequality.

COROLLARY 13.4.4. *The following duality holds for $n \geq 3$.*

$$\sup\left\{\frac{n(n-2)}{n-1}\int_{\mathbb{R}^n}\rho^{\frac{n-1}{n}}dx - \int_{\mathbb{R}^n}|x|^2\rho(x)dx;\ \int_{\mathbb{R}^n}\rho(x)\ dx = 1\right\} \tag{13.43}$$
$$= \inf\left\{\int_{\mathbb{R}^n}|\nabla f|^2 dx;\ f \in C_0^\infty(\mathbb{R}^n),\ \int_{\mathbb{R}^n}|f|^{2^*}dx = 1\right\}.$$

Proof: Take $\mu = 1, p = 2, \gamma = 1 - \frac{1}{n}$ and then $r = 2^* = \frac{2n}{n-2}$ to be the critical Sobolev exponent, then Corollary 13.4.3 yields a duality between solutions of (13.41), which is here the Yamabe equation:

$$-\Delta f = \lambda f|f|^{2^*-2},$$

(where λ is the Lagrange multiplier due to the constraint $\|f\|_{2^*} = 1$), and stationary solutions of (13.42), which is here the rescaled fast diffusion equation:

$$\frac{\partial u}{\partial t} = \Delta u^{1-\frac{1}{n}} + \mathrm{div}\left(\frac{2n-2}{n-2}xu\right).$$

13.5. Further remarks and comments

M. Gromov was first to use mass transport –though a different one from Brenier's– to establish the isoperimetric inequality (See the appendix of the book of Milman-Schechtman [**220**]). Later, D. Cordero-Erausquin, B. Nazaret and C. Villani [**103**] used Brenier's mass transport to establish Sobolev-type inequalities and corresponding duality formulae. They also dealt with the Gagliardo-Nirenberg inequalities and obtained best constant results that Del Pino-Dolbeault [**113**] had obtained earlier by analyzing certain evolution equations describing porous media. Theorem 12.2.1 of Agueh-Ghoussoub-Kang is essentially a generalization of their approach. It is therefore not surprising that all the results of this chapter follow from Theorem 12.2.1 in the case where the confinement potential V and the convolution potential W are trivial.

The optimal Euclidean p-Log Sobolev inequalities were obtained by Agueh-Ghoussoub-Kang [**17**] for all $p \geq 1$. They were first established by Beckner in [**42**] for $p = 1$, and by Del-Pino and Dolbeault [**113**] for $1 < p < n$. The general case (i.e., for $p \geq 1$) was also established around the same time and independently by I. Gentil [**153**] who used the Prékopa-Leindler inequality and the Hopf-Lax semigroup associated to the Hamilton-Jacobi equation.

The remarkable duality exhibited in section 13.4 between Euclidean Sobolev-type inequalities and certain functional inequalities on probability densities first appeared in the work of Cordero-Erausquin-Nazaret-Villani [**103**] in their proof of Corollary 13.4.3. This is what motivated Theorem 12.2.1 and its particular case described in Corollary 13.4.1. Corollary 13.4.2 originated in [**17**].

This duality points to a remarkable correspondence between Fokker-Planck evolution equations and certain quasilinear or semi-linear equations which appear as Euler-Lagrange equations of the entropy production functionals. Behind this correspondence lies a non-trivial "change of variable" that is given by the solution of the Monge transport problem. It essentially maps the solutions of the evolution equation associated to (13.32) to those of the Fokker-Planck equations (13.33).

One can also use mass transportation to study optimal Sobolev trace inequalities on the half-space as was done by Nazaret [**226**] when he proved the following conjecture of J. F. Escobar: If $n \geq 3$ and $1 < p < n$, then

$$(13.44) \qquad \|u\|_{L^{p^*(n)}(\partial \mathbb{R}_-^n)} \leq C_n(p) \|\nabla u\|_{L^p(\mathbb{R}_-^n)} \quad \text{for all } u \in W^{1,p}(\mathbb{R}_-^n),$$

It should be interesting to establish a general duality formula such as (13.25) which also involves the boundary terms. The following open question is well known.

Open problem (15): Determine the best constants $C(p, s, r)$ and –if possible– the extremals for the Gagliardo-Nirenberg inequalities in the remaining cases, that is for the following inequalities,

$$(13.45) \qquad \|f\|_{L^r} \leq C(p, s, r) \cdot \|\nabla f\|_{L^p}^\theta \|f\|_{L^s}^{1-\theta}, \quad \forall f \in W^{1,p}(\mathbb{R}^n),$$

where $1 < p < n$, $s < r < p^* := \frac{np}{n-p}$ and where θ is given by

$$\frac{1}{r} = \frac{\theta}{p^*} + \frac{1-\theta}{s}.$$

Note that all the proofs known so far, including the one via mass transport, cover only the case when $s = 1 + \frac{r(p-1)}{p}$. See Agueh [**15**] for more information on this elusive problem.

CHAPTER 14

Geometric Inequalities

We establish the so-called HWBI inequalities relating the total energy of two arbitrary probability densities, their Wasserstein distance, their barycenters and their entropy production functional. This gives a unified approach for – extensions of– various powerful inequalities by Gross, Bakry-Emery, Talagrand, Otto-Villani, Cordero-Erausquin et al., and others. As expected, such inequalities also lead to exponential rates of convergence to equilibria for solutions of Fokker-Planck and McKean-Vlasov type equations.

14.1. Quadratic case of the comparison principle and the HWBI inequality

We continue to use last chapter's terminology for the relative internal energy $\mathrm{H}_U^{F,W}(\rho_0|\rho_1)$ between two probability densities ρ_0 and ρ_1, where F is a function on $[0,+\infty)$, $U:\mathbb{R}^n \to \mathbb{R}$ (resp., $W:\mathbb{R}^n \to \mathbb{R}$) is a confinement potential (resp., an interaction potential. The relative entropy production of ρ with respect to the potential V is denoted by $I_2(\rho|\rho_V)$, and $b(\rho)$ will again be the barycenter of ρ.

Inequality (12.11) simplifies considerably when we consider quadratic Young functions of the form $c(x) := c_\sigma(x) = \frac{1}{2\sigma}|x|^2$ for $\sigma > 0$. After scaling σ, we shall obtain the following fundamental inequality.

THEOREM 14.1.1. *(HWBI inequality)* Let $F : [0,\infty) \to \mathbb{R}$ be a differentiable function on $(0,\infty)$ with $F(0) = 0$ and $x \mapsto x^n F(x^{-n})$ convex and non-increasing, and let $P_F(x) := xF'(x) - F(x)$ be its associated pressure function. Let $U : \mathbb{R}^n \to \mathbb{R}$ be a C^2-confinement potential with $D^2 U \geq \mu I$, and let W be an even C^2-interaction potential with $D^2 W \geq \nu I$ where $\mu, \nu \in \mathbb{R}$.

If Ω is an open, bounded and convex subset of $\mathbb{R}^n$, then for all probability densities ρ_0 and ρ_1 on Ω such that $\operatorname{supp}\rho_0 \subset \Omega$ and $P_F(\rho_0) \in W^{1,\infty}(\Omega)$, we have
(14.1)
$$\mathrm{H}_U^{F,W}(\rho_0|\rho_1) \leq W_2(\rho_0,\rho_1)\sqrt{I_2(\rho_0|\rho_U)} - \frac{\mu+\nu}{2}W_2^2(\rho_0,\rho_1) + \frac{\nu}{2}|b(\rho_0) - b(\rho_1)|^2.$$

The proof of Theorem 14.1.1 relies on the following proposition.

PROPOSITION 14.1.1. Let Ω, F, U and W satisfy the hypothesis of Theorem 14.1.1 and fix $\sigma > 0$. Then, we have the following inequality between any pair of probability densities ρ_0 and ρ_1 on Ω, such that $\operatorname{supp}\rho_0 \subset \Omega$, and $P_F(\rho_0) \in W^{1,\infty}(\Omega)$,

(14.2) $\mathrm{H}_U^{F,W}(\rho_0|\rho_1) + \frac{1}{2}(\mu+\nu-\frac{1}{\sigma})W_2^2(\rho_0,\rho_1) - \frac{\nu}{2}|b(\rho_0) - b(\rho_1)|^2 \leq \frac{\sigma}{2}I_2(\rho_0|\rho_U).$

Proof: Use (12.11) with $c(x) = \frac{1}{2\sigma}|x|^2$, $V = U - c$ and $\lambda = \mu - \frac{1}{\sigma}$ to obtain

(14.3)
$$\mathrm{H}_U^{F,W}(\rho_0|\rho_1) + \frac{1}{2}(\mu+\nu-\frac{1}{\sigma})W_2^2(\rho_0,\rho_1) - \frac{\nu}{2}|b(\rho_0) - b(\rho_1)|^2$$
$$\leq \mathrm{H}_{c+\nabla(U-c)\cdot x}^{-nP_F,2x\cdot\nabla W}(\rho_0) + \int_\Omega \rho_0 c^*\left(-\nabla\left(F'(\rho_0) + U - c + W \star \rho_0\right)\right)\,\mathrm{d}x.$$

By elementary computations, we have

$$\int_\Omega \rho_0 c^* \left(-\nabla \left(F' \circ \rho_0 + U - c + W \star \rho_0\right)\right) dx$$
$$= \frac{\sigma}{2}\int_\Omega \rho_0 \left|\nabla \left(F'(\rho_0) + U + W \star \rho_0\right)\right|^2 dx + \frac{1}{2\sigma}\int_\Omega \rho_0 |x|^2 dx - \int_\Omega \rho_0 x \cdot \nabla \left(F'(\rho_0)\right) dx$$
$$- \int_\Omega \rho_0 x \cdot \nabla U \, dx - \int_\Omega \rho_0 x \cdot \nabla(W \star \rho_0) \, dx,$$

and

$$\mathrm{H}^{-nP_F, 2x \cdot \nabla W}_{c+\nabla(U-c)\cdot x}(\rho_0) = -\mathrm{H}^{nP_F}(\rho_0) + \int_\Omega \rho_0 x \cdot \nabla(W \star \rho_0) \, dx + \int_\Omega \rho_0 x \cdot \nabla U \, dx - \frac{1}{2\sigma}\int_\Omega |x|^2 \rho_0 \, dx.$$

By combining the last 2 identities, we can rewrite the right hand side of (14.3) as

(14.4)
$$\mathrm{H}^{-nP_F, 2x \cdot \nabla W}_{c+\nabla(U-c)\cdot x}(\rho_0) + \int_\Omega \rho_0 c^* \left(-\nabla(F' \circ \rho_0 + U - c + W \star \rho_0)\right) dx$$
$$= \frac{\sigma}{2}\int_\Omega \rho_0 \left|\nabla \left(F'(\rho_0) + U + W \star \rho_0\right)\right|^2 dx - \int_\Omega \rho_0 x \cdot \nabla \left(F' \circ \rho_0\right) dx - \int_\Omega nP_F(\rho_0) \, dx$$
$$= \frac{\sigma}{2}\int_\Omega \rho_0 \left|\nabla \left(F'(\rho_0) + U + W \star \rho_0\right)\right|^2 dx + \int_\Omega \mathrm{div}\,(\rho_0 x) F'(\rho_0) \, dx - \int_\Omega nP_F(\rho_0) \, dx$$
$$= \frac{\sigma}{2}\int_\Omega \rho_0 \left|\nabla \left(F'(\rho_0) + U + W \star \rho_0\right)\right|^2 dx + n\int_\Omega \rho_0 F'(\rho_0) \, dx + \int_\Omega x \cdot \nabla F(\rho_0) \, dx$$
$$\quad - \int_\Omega nP_F(\rho_0) \, dx$$
$$= \frac{\sigma}{2}\int_\Omega \rho_0 \left|\nabla \left(F'(\rho_0) + U + W \star \rho_0\right)\right|^2 dx + \int_\Omega x \cdot \nabla F(\rho_0) \, dx + n\int_\Omega F \circ \rho_0 \, dx$$
$$= \frac{\sigma}{2}\int_\Omega \rho_0 \left|\nabla \left(F'(\rho_0) + U + W \star \rho_0\right)\right|^2 dx.$$

Inserting (14.4) into (14.3), we conclude (14.2).

Proof of Theorem 14.1.1: Rewrite (14.2) as

(14.5)
$$\mathrm{H}^{F,W}_U(\rho_0|\rho_1) + \frac{\mu+\nu}{2}W_2^2(\rho_0,\rho_1) - \frac{\nu}{2}|\mathrm{b}(\rho_0)-\mathrm{b}(\rho_1)|^2 \leq \frac{1}{2\sigma}W_2^2(\rho_0,\rho_1) + \frac{\sigma}{2}I_2(\rho_0|\rho_U),$$

then minimize the right hand side of (14.5) over $\sigma > 0$. The minimum is obviously achieved at $\bar\sigma = \frac{W_2(\rho_0,\rho_1)}{\sqrt{I_2(\rho_0|\rho_U)}}$. This yields (14.1).

Setting $W = 0$ (and then $\nu = 0$) in Theorem 14.1.1, we obtain in particular, the following so-called HWI inequality.

COROLLARY 14.1.1. *(HWI inequalities) Under the hypothesis on Ω and F in Theorem 14.1.1, let $U : \mathbb{R}^n \to \mathbb{R}$ be a C^2-function with $D^2 U \geq \mu I$, where $\mu \in \mathbb{R}$. Then we have for all probability densities ρ_0 and ρ_1 on Ω, satisfying $\mathrm{supp}\,\rho_0 \subset \Omega$, and $P_F(\rho_0) \in W^{1,\infty}(\Omega)$,*

(14.6) $$\mathrm{H}^F_U(\rho_0|\rho_1) \leq W_2(\rho_0,\rho_1)\sqrt{I(\rho_0|\rho_U)} - \frac{\mu}{2}W_2^2(\rho_0,\rho_1).$$

If $U + W$ is uniformly convex (i.e., $\mu + \nu > 0$) inequality (14.2) yields the following extensions of the Log-Sobolev inequality:

COROLLARY 14.1.2. (Log-Sobolev inequalities with interaction potentials) *In addition to the hypothesis on Ω, F, U and W in Theorem 14.1.1, assume $\mu + \nu > 0$. Then for all probability densities ρ_0 and ρ_1 on Ω satisfying $\operatorname{supp}\rho_0 \subset \Omega$, and $P_F(\rho_0) \in W^{1,\infty}(\Omega)$, we have*

$$\text{(14.7)} \qquad \mathrm{H}_U^{F,W}(\rho_0|\rho_1) - \frac{\nu}{2}|\mathrm{b}(\rho_0) - \mathrm{b}(\rho_1)|^2 \leq \frac{1}{2(\mu+\nu)} I_2(\rho_0|\rho_U).$$

In particular, if $\mathrm{b}(\rho_0) = \mathrm{b}(\rho_1)$, then

$$\text{(14.8)} \qquad \mathrm{H}_U^{F,W}(\rho_0|\rho_1) \leq \frac{1}{2(\mu+\nu)} I_2(\rho_0|\rho_U).$$

Furthermore, if W is convex (in particular if $W \equiv 0$), then we have the following energy-entropy inequality.

$$\text{(14.9)} \qquad \mathrm{H}_U^{F,W}(\rho_0|\rho_1) \leq \frac{1}{2\mu} I_2(\rho_0|\rho_U).$$

Proof: (14.7) follows easily from (14.2) by choosing $\sigma = \frac{1}{\mu+\nu}$, and (14.9) follows from (14.7), using $\nu = 0$ because W is convex.

One can also deduce the following generalization of Talagrand's inequality.

COROLLARY 14.1.3. (Generalized Talagrand Inequality with interaction potentials) *In addition to the hypothesis on Ω, F, U and W in Theorem 14.1.1, assume $\mu + \nu > 0$. Then for all probability densities ρ on Ω, we have*

$$\text{(14.10)} \qquad \frac{\nu+\mu}{2} W_2^2(\rho, \rho_U) - \frac{\nu}{2}|\mathrm{b}(\rho) - \mathrm{b}(\rho_U)|^2 \leq \mathrm{H}_U^{F,W}(\rho|\rho_U).$$

In particular, if $\mathrm{b}(\rho) = \mathrm{b}(\rho_U)$, we have that

$$\text{(14.11)} \qquad W_2(\rho, \rho_U) \leq \sqrt{\frac{2\mathrm{H}_U^{F,W}(\rho|\rho_U)}{\mu+\nu}}.$$

Furthermore, if W is convex, then the following inequality holds:

$$\text{(14.12)} \qquad W_2(\rho, \rho_U) \leq \sqrt{\frac{2\mathrm{H}_U^{F,W}(\rho|\rho_U)}{\mu}}.$$

Proof: (14.10) follows from (14.2) if we use $\rho_0 := \rho_U$, $\rho_1 := \rho$. Notice that $I_2(\rho_U|\rho_U) = 0$, and then let σ go to ∞. (14.12) follows from (14.10), where we could use that $\nu = 0$ because W is convex.

14.2. Gaussian inequalities

Proposition 14.1.1 applied to $F(x) = x \log x$ when $W = 0$, yields the following improvement of the Log-Sobolev inequality of L. Gross.

COROLLARY 14.2.1. *Let $U : \mathbb{R}^n \to \mathbb{R}$ be a C^2-function with $D^2 U \geq \mu I$ where $\mu \in \mathbb{R}$, and denote by ρ_U the normalized Gaussian $\frac{e^{-U}}{\sigma_U}$, where $\sigma_U = \int_{\mathbb{R}^n} e^{-U}\, dx$. Then for any $\sigma > 0$, the following holds for any nonnegative function f such that $f\rho_U \in W^{1,\infty}(\mathbb{R}^n)$ and $\int_{\mathbb{R}^n} f \rho_U\, dx = 1$,*

$$\text{(14.13)} \qquad \int_{\mathbb{R}^n} f \log \mathrm{f}\, \rho_U\, dx + \frac{1}{2}(\mu - \frac{1}{\sigma}) \mathrm{W}_2^2(\mathrm{f}\rho_\mathrm{U}, \rho_\mathrm{U}) \leq \frac{\sigma}{2} \int_{\mathbb{R}^n} \frac{|\nabla \mathrm{f}|^2}{\mathrm{f}} \rho_U\, dx.$$

Proof: First assume that f has compact support, and set $F(x) = x \log f$, $\rho_0 = f\rho_U$, $\rho_1 = \rho_U$ and $W = 0$ in (14.2). We then get

$$(14.14) \quad H_U^F(f\rho_U|\rho_U) + \frac{1}{2}(\mu - \frac{1}{\sigma})W_2^2(f\rho_U, \rho_U) \leq \frac{\sigma}{2}\int_{\mathbb{R}^n}\left|\frac{\nabla(f\rho_U)}{f\rho_U} + U\right|^2 f\rho_U \, dx.$$

By simple computations,

$$(14.15) \quad \frac{\nabla(f\rho_U)}{f\rho_U} = \frac{\nabla f}{f} - \nabla U,$$

and

$$(14.16) \quad \begin{aligned} H_U^{F,W}(f\rho_U|\rho_U) &\leq \int_{\mathbb{R}^n} [f\rho_U \log(f\rho_U) + Uf\rho_U - \rho_U \log\rho_U - U\rho_U] \, dx \\ &= \int_{\mathbb{R}^n} f\rho_U \log f \, dx + \log\sigma_U \int_{\mathbb{R}^n}(\rho_U - f\rho_U) \, dx \\ &= \int_{\mathbb{R}^n} f \log(f)\rho_U \, dx. \end{aligned}$$

Combining (14.14) - (14.16), we get (14.13). One can then finish the proof using a standard approximation argument.

COROLLARY 14.2.2. (Otto-Villani's HWI inequality) *Let* $U : \mathbb{R}^n \to \mathbb{R}$ *be a* C^2*-uniformly convex function with* $D^2 U \geq \mu I$, *where* $\mu > 0$, *and denote by* ρ_U *the normalized Gaussian* $\frac{e^{-U}}{\sigma_U}$, *where* $\sigma_U = \int_{\mathbb{R}^n} e^{-U} \, dx$. *Then, for any nonnegative function* f *such that* $f\rho_U \in W^{1,\infty}(\mathbb{R}^n)$ *and* $\int_{\mathbb{R}^n} f\rho_U \, dx = 1$,

$$(14.17) \quad \int_{\mathbb{R}^n} f \log f \rho_U \, dx \leq W_2(\rho_U, f\rho_U)\sqrt{I(f\rho_U|\rho_U)} - \frac{\mu}{2}W_2^2(f\rho_U, \rho_U),$$

where

$$I(f\rho_U|\rho_U) = \int_{\mathbb{R}^n} \frac{|\nabla f|^2}{f} \rho_U \, dx.$$

Proof: It is similar to the proof of Theorem 14.1.1. Rewrite (14.13) as

$$\int_{\mathbb{R}^n} f \log(f) \rho_U \, dx + \frac{\mu}{2} W_2^2(f\rho_U, \rho_U) \leq \frac{\mu}{2\sigma} W_2^2(f\rho_U, \rho_U) + \frac{\sigma}{2} I(f\rho_U|\rho_U),$$

and show that the minimum over $\sigma > 0$ of the right hand side is attained at $\bar{\sigma} = \frac{W_2(f\rho_U, \rho_U)}{\sqrt{I(f\rho_U|\rho_U)}}$.

Now, setting $f := g^2$ and $\sigma := \frac{1}{\mu}$ in (14.17), one obtains

COROLLARY 14.2.3. (Log Sobolev inequality with general convex potential) *Let* $U : \mathbb{R}^n \to \mathbb{R}$ *be a* C^2*-uniformly convex function with* $D^2 U \geq \mu I$ *where* $\mu > 0$, *and denote by* ρ_U *the normalized Gaussian* $\frac{e^{-U}}{\sigma_U}$, *where* $\sigma_U = \int_{\mathbb{R}^n} e^{-U} \, dx$. *Then, for any function* g *such that* $g^2 \rho_U \in W^{1,\infty}(\mathbb{R}^n)$ *and* $\int_{\mathbb{R}^n} g^2 \rho_U \, dx = 1$, *we have*

$$(14.18) \quad \int_{\mathbb{R}^n} g^2 \log(g^2) \rho_U \, dx \leq \frac{2}{\mu} \int_{\mathbb{R}^n} |\nabla g|^2 \rho_U \, dx.$$

The above Log-Sobolev inequality implies the following Poincaré's inequality.

COROLLARY 14.2.4. (Poincaré's inequality) *Let* $U : \mathbb{R}^n \to \mathbb{R}$ *be a* C^2*-uniformly convex function with* $D^2 U \geq \mu I$ *where* $\mu > 0$, *and denote by* ρ_U *the normalized*

Gaussian $\frac{e^{-U}}{\sigma_U}$, where $\sigma_U = \int_{\mathbb{R}^n} e^{-U}\, dx$. Then, for any function f such that $f\rho_U \in W^{1,\infty}(\mathbb{R}^n)$ and $\int_{\mathbb{R}^n} f\rho_U\, dx = 0$, we have

(14.19)
$$\int_{\mathbb{R}^n} f^2 \rho_U\, dx \leq \frac{1}{\mu} \int_{\mathbb{R}^n} |\nabla f|^2 \rho_U\, dx.$$

Proof: From (14.18), we have that

(14.20)
$$\int_{\mathbb{R}^n} f_\epsilon \log(f_\epsilon)\, \rho_U\, dx \leq \frac{1}{2\mu} \int_{\mathbb{R}^n} \frac{|\nabla f_\epsilon|^2}{f_\epsilon} \rho_U\, dx,$$

where $f_\epsilon = 1 + \epsilon f$ for some $\epsilon > 0$. Using that $\int_{\mathbb{R}^n} f\rho_U\, dx = 0$, we have for small ϵ,

(14.21)
$$\int_{R^n} f_\epsilon \log(f_\epsilon)\rho_U\, dx = \frac{\epsilon^2}{2} \int_{\mathbb{R}^n} f^2 \rho_U\, dx + o(\epsilon^3),$$

and

(14.22)
$$\int_{\mathbb{R}^n} \frac{|\nabla f_\epsilon|^2}{f_\epsilon} \rho_U\, dx = \epsilon^2 \int_{\mathbb{R}^n} |\nabla f|^2 \rho_U\, dx + o(\epsilon^3).$$

We combine (14.20) - (14.22) to have that

(14.23)
$$\int_{\mathbb{R}^n} f^2 \rho_U\, dx \leq \frac{1}{\mu} \int_{\mathbb{R}^n} |\nabla f|^2 \rho_U\, dx + o(\epsilon).$$

We let ϵ go to 0 in (14.23) to conclude (14.19).

If we apply Corollary 14.1.3 to $F(x) = x \log x$ when $W = 0$, we obtain the following extension of Talagrand's inequality established by Otto and Villani.

COROLLARY 14.2.5. *((Talagrand's inequality with general convex potential) Let $U : \mathbb{R}^n \to \mathbb{R}$ be a C^2-uniformly convex function with $D^2 U \geq \mu I$ where $\mu > 0$, and denote by ρ_U the normalized Gaussian $\frac{e^{-U}}{\sigma_U}$, where $\sigma_U = \int_{\mathbb{R}^n} e^{-U}\, dx$. Then, for any nonnegative function f such that $\int_{\mathbb{R}^n} f\rho_U\, dx = 1$, we have*

(14.24)
$$W_2(f\rho_U, \rho_U) \leq \sqrt{\frac{2}{\mu} \int_{\mathbb{R}^n} f \log(f)\rho_U\, dx}.$$

In particular, if $f = \frac{I_B}{\gamma(B)}$ for some measurable subset B of $\mathbb{R}^n$, where $d\gamma(x) = \rho_U(x)dx$ and I_B is the characteristic function of B, we obtain the following inequality in the concentration of measures in Gauss space.

COROLLARY 14.2.6. *(Concentration of measure inequality) Let $U : \mathbb{R}^n \to \mathbb{R}$ be a C^2-uniformly convex function with $D^2 U \geq \mu I$ where $\mu > 0$, and denote by γ the normalized Gaussian measure with density $\rho_U = \frac{e^{-U}}{\sigma_U}$, where $\sigma_U = \int_{\mathbb{R}^n} e^{-U}\, dx$. Then, for any ϵ-neighborhood B_ϵ of a measurable set B in $\mathbb{R}^n$, we have*

(14.25)
$$\gamma(B_\epsilon) \geq 1 - e^{-\frac{\mu}{2}\left(\epsilon - \sqrt{\frac{2}{\mu}\log\left(\frac{1}{\gamma(B)}\right)}\right)^2},$$

where $\epsilon \geq \sqrt{\frac{2}{\mu} \log\left(\frac{1}{\gamma(B)}\right)}$.

Proof: Using $f = f_B = \frac{I_B}{\gamma(B)}$ in (14.24), we have that

$$W_2(f_B \rho_U, \rho_U) \leq \sqrt{\frac{2}{\mu} \log\left(\frac{1}{\gamma(B)}\right)},$$

and then, from the triangle inequality,

$$(14.26) \quad W_2(f_B \rho_U, f_{\mathbb{R}^n \setminus B_\epsilon} \rho_U) \leq \sqrt{\frac{2}{\mu} \log\left(\frac{1}{\gamma(B)}\right)} + \sqrt{\frac{2}{\mu} \log\left(\frac{1}{1 - \gamma(B_\epsilon)}\right)}.$$

But since $|x - y| \geq \epsilon$ for all $(x, y) \in B \times (\mathbb{R}^n \setminus B_\epsilon)$, we have that

$$(14.27) \quad W_2(f_B \rho_U, \rho_U) \geq \epsilon.$$

We combine (14.26) and (14.27) to deduce that

$$\log\left(\frac{1}{1 - \gamma(\mathbb{R}^n \setminus B_\epsilon)}\right) \geq \frac{\mu}{2}\left(\epsilon - \sqrt{\frac{2}{\mu} \log\left(\frac{1}{\gamma(B)}\right)}\right)^2,$$

which leads to (14.25).

14.3. Trends to equilibrium in Fokker-Planck equations

We now use Corollary 14.1.3 to recover rates of convergence for solutions to equation

$$(14.28) \quad \begin{cases} \frac{\partial \rho}{\partial t} = \operatorname{div}\{\rho \nabla (F'(\rho) + V + W \star \rho)\} & \text{in } (0, \infty) \times \mathbb{R}^n \\ \rho(t = 0) = \rho_0 & \text{in } \{0\} \times \mathbb{R}^n. \end{cases}$$

Here we consider the case where $V + W$ is uniformly convex and W convex, and the case when only $V + W$ is uniformly convex but the barycenter $b(\rho(t))$ of any solution $\rho(t, x)$ of (14.28) is invariant in t.

COROLLARY 14.3.1. (Trend to equilibrium) *Let $F : [0, \infty) \to \mathbb{R}$ be strictly convex, differentiable on $(0, \infty)$ and satisfies $F(0) = 0$, $\lim_{x \to \infty} \frac{F(x)}{x} = \infty$, and $x \mapsto x^n F(x^{-n})$ is convex and non-increasing. Let V (resp., W): $\mathbb{R}^n \to [0, \infty)$ be a C^2-confinement (resp., interaction) potential with $D^2 V \geq \lambda I$ and $D^2 W \geq \nu I$, where $\lambda, \nu \in \mathbb{R}$. Assume that the initial probability density ρ_0 has finite total energy. Then*

(1) *If $V + W$ is uniformly convex (i.e. $\lambda + \nu > 0$) and W is convex (i.e. $\nu \geq 0$), then, for any solution ρ of (14.28), such that $\mathrm{H}_V^{F,W}(\rho(t)) < \infty$, we have:*

$$(14.29) \quad \mathrm{H}_V^{F,W}(\rho(t) | \rho_V) \leq e^{-2\lambda t} \mathrm{H}_V^{F,W}(\rho_0 | \rho_V),$$

and

$$(14.30) \quad W_2(\rho(t), \rho_V) \leq e^{-\lambda t} \sqrt{\frac{2 \mathrm{H}_V^{F,W}(\rho_0 | \rho_V)}{\lambda}}.$$

(2) *If $V + W$ is uniformly convex (i.e. $\lambda + \nu > 0$) and if we assume that the barycenter $b(\rho(t))$ of any solution $\rho(t, x)$ of (14.28) is invariant in t, then, for any solution ρ of (14.28) such that $\mathrm{H}_V^{F,W}(\rho(t)) < \infty$, we have:*

$$(14.31) \quad \mathrm{H}_V^{F,W}(\rho(t) | \rho_V) \leq e^{-2(\lambda + \nu)t} \mathrm{H}_V^{F,W}(\rho_0 | \rho_V),$$

and

$$(14.32) \quad W_2(\rho(t), \rho_V) \leq e^{-2(\lambda+\nu)t} \sqrt{\frac{2 \mathrm{H}_V^{F,W}(\rho_0 | \rho_V)}{\lambda + \nu}}.$$

Proof: Under the assumptions on F, V and W in Corollary 14.3.1, the total energy $\mathrm{H}_V^{F,W}$ – which is a Lyapunov functional for (14.28) – has a unique minimizer ρ_V defined by
$$\rho_V \nabla \left(F'(\rho_V) + V + W \star \rho_V\right) = 0 \quad \text{a.e.}$$
If ρ is a – smooth – solution of (14.28), we then have the following energy dissipation equation

(14.33) $$\frac{d}{dt} \mathrm{H}_V^{F,W}\left(\rho(t)|\rho_V\right) = -I_2\left(\rho(t)|\rho_V\right).$$

Combining (14.33) with (14.9), we have that

(14.34) $$\frac{d}{dt} \mathrm{H}_V^{F,W}\left(\rho(t)|\rho_V\right) \leq -2\lambda \mathrm{H}_V^{F,W}\left(\rho(t)|\rho_V\right).$$

Now integrate (14.34) over $[0,t]$ to conclude (14.29). Estimate (14.30) follows directly from (14.12) and (14.29).
To prove (14.31), we use (14.33) and (14.8) to have that

(14.35) $$\frac{d}{dt} \mathrm{H}_V^{F,W}\left(\rho(t)|\rho_V\right) \leq -2(\lambda+\nu)\mathrm{H}_V^{F,W}\left(\rho(t)|\rho_V\right).$$

We integrate (14.35) over $[0,t]$ to obtain (14.31). As before, (14.32) is a consequence of (14.31) and (14.11).

Examples:
(1) If $W = 0$ and $F(x) = x \log x$ in which case (14.28) is the linear Fokker-Planck equation $\frac{\partial \rho}{\partial t} = \Delta \rho + \mathrm{div}(\rho \nabla V)$, Corollary 14.3.1 gives an exponential decay in relative entropy of solutions of this equation to the Gaussian density $\rho_V = \frac{e^{-V}}{\sigma_V}$, $\sigma_V = \int_{\mathbb{R}^n} e^{-V}\, dx$, at the rate 2λ when $D^2 V \geq \lambda I$ for some $\lambda > 0$, and an exponential decay in the Wasserstein distance, at the rate λ.
(2) If $W = 0$, $F(x) = \frac{x^m}{m-1}$ where $1 \neq m \geq 1 - \frac{1}{n}$, and $V(x) = \lambda \frac{|x|^2}{2}$ for some $\lambda > 0$, in which case (14.28) is the rescaled porous medium equation ($m > 1$), or fast diffusion equation ($1 - \frac{1}{n} \leq m < 1$), that is $\frac{\partial \rho}{\partial t} = \Delta \rho^m + \mathrm{div}(\lambda x \rho)$, Corollary 14.3.1 gives an exponential decay in relative entropy of solutions of this equation to the Barenblatt-Prattle profile
$$\rho_V(x) = \left[\left(C + \frac{\lambda(1-m)}{2m}|x|^2\right)^{\frac{1}{m-1}}\right]^+$$
(where $C > 0$ is such that $\int_{\mathbb{R}^n} \rho(x)\, dx = 1$) at the rate 2λ, and an exponential decay in the Wasserstein distance at the rate λ.

14.4. Further comments

The HWBI inequality, which relates the relative total energy H of two probability densities, to their Wasserstein distance W, the Fisher information I, as well as to the distance between their barycenters B, was established by Agueh-Ghoussoub-Kang [**17**]. It is an extension of the so-called HWI inequality first established by Otto and Villani [**234**] in the case of the classical Tsallis entropy $F(x) = x \log x$ and in the absence of the convolution term $W \equiv 0$. The latter was already known to comprise various powerful inequalities by Gross [**175**], Bakry-Emery [**36**], Talagrand, [**263**], Cordero-Erausquin et al. [**104**] and others. Related results were also obtained by Blower [**52**], and Bobkov-Ledoux [**53**]. A comprehensive survey is given in the book of Ledoux [**200**]. Extensions to generalized entropy functions F were also given by Carillo, McCann and Villani in [**79**]. The case of a nonzero confinement potential V and an interaction potential W was also considered in Cordero-Erausquin, Gangbo and Houdré [**105**]. Generalization of Log-Sobolev and Talagrand's inequalities to general entropy functions F then followed.

Functional inequalities such as the ones we exhibited above are closely related to geometric inequalities, which normally deal with shapes and bodies, such as the Brunn-Minkowski inequality, Santalo's inequality, mixed volumes and Alexandrov-Fenchel inequalities. For these geometric aspects and their applications, we refer to the books of Ledoux [**200**], Pisier [**240**], and Milman-Schechtman [**220**].

As expected, such inequalities also lead to exponential rates of convergence to equilibria for solutions of Fokker-Planck and McKean-Vlasov type equations. For a background and other cases of convergence to equilibrium for this equation, we refer to [**79**] and the references therein. The books of Villani [**268**] and [**269**] contain much of the history of this fascinating development.

We have seen in Chapter 2 that the Hardy inequality is formally stronger than the basic Sobolev inequality. This leads to the following question.

Open problem (16): Can the Hardy inequality be described as another manifestation of a mass transport phenomenon? Can it be associated to a convexity property of non-homogenous internal energies of the form $H^F(\rho) = \int_\Omega F(x, \rho(x))\, dx$ along the Wasserstein geodesics?

Part 5

Hardy-Rellich-Sobolev Inequalities

CHAPTER 15

The Hardy-Sobolev Inequalities

This chapter deals with Hardy-Sobolev, Caffarelli-Kohn-Nirenberg and Hardy-Rellich-Sobolev type inequalities. All these can be obtained by simply interpolating via Hölder's inequalities many of the previously obtained inequalities. We also address the problem of estimating the best constants and whether they are attained. The best constant in the Hardy-Sobolev inequality, that is

$$\mu_s(\Omega) := \inf\left\{\int_\Omega |\nabla u|^2 dx;\ u \in H_0^1(\Omega) \text{ and } \int_\Omega \frac{|u|^{2^*(s)}}{|x|^s}\,dx = 1\right\},$$

where $0 < s < 2$ and $2^*(s) = \frac{2(n-s)}{n-2}$, is never attained when 0 is in the interior of the domain Ω, unless the latter is the whole space $\mathbb{R}^n$ in which case explicit extremals are given. This is not the case when Ω is half-space $\mathbb{R}^n_-$, where only the symmetry of the extremals is shown. Unless $s = 0$, much less is known about the extremals in the Hardy-Rellich-Sobolev inequality,

$$\nu_s(\Omega) := \inf\left\{\int_\Omega |\Delta u|^2 dx;\ u \in H_0^2(\Omega) \text{ and } \int_\Omega \frac{|u|^{2^{**}(s)}}{|x|^s}\,dx = 1\right\},$$

where $0 < s < 4$ and $2^{**}(s) = \frac{2(n-s)}{n-4}$, even when $\Omega = \mathbb{R}^n$.

15.1. Interpolating between Hardy's and Sobolev inequalities

The starting point of the next group of inequalities is the following result.

THEOREM 15.1.1. *(Sobolev-Hardy Inequality) Assume that $1 < p < n$ and $0 \leq s \leq p$. Setting $p^*(s) := \frac{n-s}{n-p}p$, then for any domain Ω in $\mathbb{R}^n$, there exists a constant $C := C(p, s, \Omega) > 0$ such that*

(15.1) $\quad (\int_\Omega \frac{|u|^{p^*(s)}}{|x|^s})^{\frac{1}{p^*(s)}}\,dx \leq C(\int_\Omega |\nabla u|^p)^{\frac{1}{p}}\,dx \quad$ *for all $u \in W_0^{1,p}(\Omega)$.*

If Ω is bounded, then the inequality holds with $p^*(s)$ replaced by any q with $p \leq q \leq p^*(s)$.

Proof: Note that for $s = 0$ (resp., $s = p$) this is just the *Sobolev* (resp., the *Hardy*) inequality. We therefore have to only consider the case where $0 < s < p$. Note also that since $0 \leq s \leq p$, we have that $p^*(s) \geq p$. By applying Hölder's inequality, then Hardy's and Sobolev's, we have

$$\int_\Omega \frac{|u|^{p^*(s)}}{|x|^s}\,dx = \int_\Omega \frac{|u|^s}{|x|^s}\cdot |u|^{p^*(s)-s}\,dx$$

$$\leq (\int_\Omega |\frac{|u|^p}{|x|^p})^{\frac{s}{p}}\,dx)(\int_\Omega |u|^{(p^*(s)-s)\frac{p}{p-s}})^{\frac{p-s}{p}}\,dx$$

$$= (\int_\Omega |\frac{|u|^p}{|x|^p})^{\frac{s}{p}}\,dx)(\int_\Omega |u|^{p^*})^{\frac{p-s}{p}}\,dx$$

$$\leq (C_1\int_\Omega |\nabla u|^p)^{\frac{s}{p}}\,dx)(C_2 \int_\Omega |\nabla u|^p)^{\frac{p^*}{p}\cdot\frac{p-s}{p}}\,dx$$

$$= C(\int_\Omega |\nabla u|^p)^{\frac{n-s}{n-p}}\,dx.$$

REMARK 15.1.2. If Ω is the whole space, one can show that the conditions $p \leq q = p^*(s) := \frac{n-s}{n-p}p$ are also necessary for the above inequality to hold. Indeed, a standard scaling argument shows that q must be equal to $p^*(s)$. On the other hand, if we insert into the inequality the following function (ρ and $\theta \in S^{n-1}$ being the polar coordinates),

$$u(x) = \begin{cases} 0 & \text{for } |x| \geq 1 \\ |x|^{\frac{p-n}{p}}\log\frac{1}{|x|} & \text{for } \varepsilon \leq |x| < 1 \\ \varepsilon^{\frac{p-n}{p}}\log\frac{1}{\varepsilon} & \text{for } |x| \leq \varepsilon, \end{cases}$$

and since

$$\frac{du(x)}{d\rho} = \begin{cases} 0 & |x| \geq 1 \\ 0 & |x| \leq \varepsilon \\ (1-\frac{n}{p})\rho^{-\frac{n}{p}}\log\frac{1}{\rho} - \rho^{-\frac{n}{p}} & \varepsilon \leq |x| < 1, \end{cases}$$

we get

$$\int_{\mathbb{R}^n}|\nabla u|^p \sim \int_\varepsilon^1 \rho^{-1}(1+(\frac{n}{p}-1)\log\frac{1}{\rho})^p d\rho.$$

By L'Hospital's rule, we have

$$\lim_{\varepsilon \to 0} \frac{\int_\varepsilon^1 \rho^{-1}(1+(\frac{n}{p}-1)\log\frac{1}{\rho})^p d\rho}{\log^{1+p}\frac{1}{\varepsilon}} = \frac{\frac{n}{p}-1}{1+p},$$

and also

$$\int_{\mathbb{R}^n}\frac{|u|^q}{|x|^s} \sim \int_\varepsilon^1 \rho^{-s}\log^q\frac{1}{\rho}\rho^{\frac{p-n}{p}q}\rho^{n-1} = \int_\varepsilon^1 \frac{1}{\rho}\log^q\frac{1}{\rho} \sim \log^{1+q}\frac{1}{\varepsilon}.$$

Thus from the inequality

$$\log^{1+\frac{1}{q}}\frac{1}{\varepsilon} \leq \log^{1+\frac{1}{p}}\frac{1}{\varepsilon},$$

we get that $q \geq p$. □

Let again Ω be a domain in $\mathbb{R}^n$, $a \in \mathbb{R}$ and define $\mathcal{D}_a^{1,2}(\Omega)$ to be the completion of $C_c^\infty(\Omega)$ with respect to the norm

(15.2) $$\|u\|_a^2 = \int_\Omega |x|^{-2a}|\nabla u|^2 dx.$$

The following Caffarelli-Kohn-Nirenberg inequalities can be obtained directly from the Hardy-Sobolev inequality.

COROLLARY 15.1.1. *Assume*

(15.3) $$-\infty < a < \frac{n-2}{2} \quad \text{and} \quad 0 \leq b - a \leq 1.$$

Then for any domain Ω in $\mathbb{R}^n$, there exists a constant $C > 0$ such that for all $u \in \mathcal{D}_a^{1,2}(\Omega)$,

(15.4) $$\left(\int_\Omega |x|^{-bq} |u|^q \right)^{\frac{2}{q}} dx \leq C \int_\Omega |x|^{-2a} |\nabla u|^2 dx,$$

where $q = \frac{2n}{n-2+2(b-a)}$.

Proof: Set

(15.5) $$w(x) = |x|^{-a} u(x) \quad \text{for} \quad x \in \Omega.$$

By a straightforward calculation, we have for any $u \in C_c^\infty(\Omega)$,

$$\int_\Omega |x|^{-2a} |\nabla u|^2 dx = \int_\Omega |x|^{-2a} (a^2 |x|^{2a-2} w^2(x) + 2a |x|^{2a-2} w(x) x \cdot \nabla w(x) + |x|^{2a} |\nabla w(x)|^2) dx$$

$$= \int_\Omega |\nabla w(x)|^2 dx + a^2 \int_\Omega \frac{w^2(x)}{|x|^2} dx + \int_\Omega 2a |x|^{-2} w(x) x \cdot \nabla w(x)$$

$$= \int_\Omega |\nabla w(x)|^2 dx + a^2 \int_\Omega \frac{w^2(x)}{|x|^2} dx + a \int_\Omega |x|^{-2} x \cdot \nabla(w^2) dx$$

$$= \int_\Omega |\nabla w(x)|^2 dx - \gamma \int_\Omega \frac{w^2(x)}{|x|^2} dx,$$

where

(15.6) $$\gamma = a(n - 2 - a),$$

and where the last equality is obtained by integration by parts. Now note that if $a < \frac{n-2}{2}$, then both $\int_\Omega |\nabla w|^2 dx$ and $\int_\Omega \frac{w^2}{|x|^2}$ are finite since by Hardy's inequality we have

$$(\frac{n-2}{2})^2 \int_\Omega \frac{w^2}{|x|^2} dx \leq |\nabla w|^2 dx.$$

In other words, if $a < \frac{n-2}{2}$, then

$$u \in \mathcal{D}_a^{1,2}(\Omega) \quad \text{if and only if} \quad w \in H_0^1(\Omega),$$

and furthermore,

(15.7) $$E_{a,b}(u) := \frac{\int_\Omega |x|^{-2a} |\nabla u|^2 dx}{\left(\int_\Omega |x|^{-bq} |u|^q \right)^{\frac{2}{q}}} = \frac{\int_\Omega |\nabla w|^2 dx - \gamma \int_\Omega \frac{w^2}{|x|^2} dx}{\left(\int_\Omega \frac{w^{2^*}}{|x|^s} dx \right)^{\frac{2}{2^*}}},$$

which readily implies that (15.4) follows from (15.1).

15.2. Best constants and extremals when 0 is in the interior of the domain

Denote by $\mu_{s,q}(\Omega)$ the best *Hardy-Sobolev* constant, i.e.,

(15.8) $$\mu_{s,q}(\Omega) = \inf \left\{ \frac{\int_\Omega |\nabla u|^p \, dx}{\left(\int_\Omega \frac{|u|^q}{|x|^s} \right)^{\frac{p}{q}} dx} ; \, u \in W_0^{1,p}(\Omega) \setminus \{0\} \right\}.$$

In the important case where $q = p^*(s)$, we shall simply denote $\mu_{s,p^*(s)}(\Omega)$ as $\mu_s(\Omega)$. Note that μ_0 is nothing but the best constant in the *Sobolev inequality* while μ_p is the best constant in the *Hardy inequality*, i.e.,

$$(15.9) \qquad \mu_p(\Omega) = \inf\left\{ \frac{\int_\Omega |\nabla u|^p \, dx}{(\int_\Omega \frac{|u|^p}{|x|^p})dx}; \; u \in W_0^{1,p}(\Omega) \setminus \{0\} \right\},$$

which is equal to $(\frac{p}{n-p})^p$, whenever Ω contains 0.

The Euler-Lagrange equation for the extremal solutions of (15.8) when $p = 2$ is

$$(15.10) \qquad \begin{cases} \Delta v + \frac{v^{q-1}}{|x|^s} = 0 & \text{in } \Omega \\ v > 0 & \text{on } \Omega, \\ v = 0 & \text{on } \partial\Omega. \end{cases}$$

Define now

$$(15.11) \qquad S(a,b,\Omega) = \inf_{u \in \mathcal{D}_a^{1,2}(\Omega) \setminus \{0\}} E_{a,b}(u).$$

The extremal functions for $S(a,b,\Omega)$ are the least-energy solutions of the Euler-Lagrange equations:

$$(15.12) \qquad \begin{cases} \operatorname{div}(|x|^{-2a}\nabla u) + |x|^{-bq} u^{q-1} = 0 & \text{in } \Omega \\ u > 0 & \text{in } \Omega \\ u = 0 & \text{on } \partial\Omega. \end{cases}$$

Also u is a solution of (15.12) if and only if $w(x) = |x|^{-a} u(x)$ is a positive solution of

$$(15.13) \qquad \begin{cases} \Delta w + \gamma \frac{w}{|x|^2} + \frac{w^{2^*(s)-1}}{|x|^s} = 0 & \text{in } \Omega \\ w = 0 & \text{on } \partial\Omega. \end{cases}$$

Therefore instead of studying solutions of (15.12) one can study the solutions of (15.13).

The following is an immediate application of the Pohozaev identity.

THEOREM 15.2.1. *If Ω is a star-shaped domain in $\mathbb{R}^n$ containing 0 and if $1 < p < n$, then for any scalars μ, ν, γ, the equation*

$$(15.14) \qquad \begin{cases} \operatorname{div}(|\nabla u|^{p-2}\nabla u) + \gamma \frac{u^{p-1}}{|x|^p} + \mu \frac{u^{q-1}}{|x|^s} + \lambda |u|^{r-1} u = 0 & \text{in } \Omega \\ u = 0 & \text{on } \partial\Omega, \end{cases}$$

has no non-trivial solution in $W_0^{1,p}(\Omega)$ whenever $r = p^* := \frac{np}{n-p}$ and $q = p^*(s) = \frac{n-s}{n-p}p$.

Proof: Since Ω is a star-shaped domain, then, if ν denotes the outwards normal to $\partial\Omega$, we must have that $\langle x, \nu \rangle > 0$ on $\partial\Omega$. We assume we have the necessary regularity in the following operations since otherwise, one can use an approximation argument as in *Guedda-Veron* [**171**].

Multiply the equation (15.14) by $\langle x, \nabla u\rangle$ on both sides and integrate by parts to get

$$\frac{p-1}{p}\int_{\partial\Omega}|\nabla u|^p\langle x,\ v\rangle+\frac{n-p}{p}\int_{\Omega}|\nabla u|^p$$
$$=\gamma\frac{n-p}{p}\int_{\Omega}\frac{|u|^p}{|x|^p}+\mu\frac{n-s}{q}\int_{\Omega}\frac{|u|^q}{|x|^s}+\lambda\frac{n}{r}\int_{\Omega}|u|^r.$$

On the other hand, multiply the equation by u and integrate to get

$$\int_{\Omega}|\nabla u|^p dx=\gamma\int_{\Omega}\frac{|u|^p}{|x|^p}dx+\mu\int_{\Omega}\frac{|u|^q}{|x|^s}dx+\lambda\int_{\Omega}|u|^r dx.$$

Putting the two identities together, we obtain

$$\frac{p-1}{p}\int_{\partial\Omega}|\nabla u|^p\langle x,\ v\rangle d\sigma=\mu(\frac{n-s}{q}-\frac{n-p}{p})\int_{\Omega}\frac{|u|^q}{|x|^s}dx+\lambda(\frac{n}{r}-\frac{n-p}{p})\int_{\Omega}|u|^r\ dx.$$

So if $r=\frac{np}{n-p}=p^*$ and $q=\frac{n-s}{n-p}p$, the problem has no non-trivial solution. □

An immediate corollary of the above result is that none of the above best constants $\mu_s(\Omega)$ or $S(\Omega,a,b)$ is attained on a star-shaped bounded domain Ω containing 0. Actually, one has the following result.

THEOREM 15.2.2. *Suppose $1<p<n$, $0\leq s<p$ and $q=p^*(s)$, then the following hold:*

(1) *$\mu_s(\Omega)$ is independent of the domain Ω whenever the latter contains 0 in its interior.*
(2) *$\mu_s(\Omega)$ is only attained when $\Omega=\mathbb{R}^n$ with the extremals being the functions*

(15.15) $$y_a(x)=(a\cdot(n-s)(\frac{n-p}{p-1})^{p-1})^{\frac{n-p}{p(p-s)}}(a+|x|^{\frac{p-s}{p-1}})^{\frac{p-n}{p-s}},$$

for some $a>0$. Moreover the functions y_a are the only positive radial solutions of

(15.16) $$-\mathrm{div}(|\nabla u|^{p-2}\nabla u)=\frac{u^{p^*(s)-1}}{|x|^s} \quad on\ \mathbb{R}^n.$$

Consequently,

(15.17) $$\mu_s(\int_{\mathbb{R}^n}\frac{|y_a|^q}{|x|^s})^{\frac{p}{q}}=\|\nabla y_a\|_p^p=\int_{\mathbb{R}^n}\frac{|y_a|^q}{|x|^s}=\mu_s^{\frac{n-s}{p-s}}.$$

Proof: We first show that the best constant $\mu_s(\mathbb{R}^n)$ is only attained at functions of the form

(15.18) $$u_s(x)=c(\lambda+|x|^{\frac{p-s}{p-1}})^{-\frac{n-p}{p-s}}\ (0\leq s<p),$$

where $\lambda>0$ is a constant. Indeed, for any f, we consider f^* to be its *Schwarz symmetrization* –or *rearrangement*– which was defined in Chapter 2. Recall that

$$\int_{\mathbb{R}^n}|\nabla f^*|^p\leq\int_{\mathbb{R}^n}|\nabla f|^p \quad\text{and}\quad \int_{\mathbb{R}^n}\frac{|f^*|^q}{|x|^t}\geq\int_{\mathbb{R}^n}\frac{|f|^q}{|x|^t},$$

whenever the above integrals are well defined. We may therefore restrict our discussion to radial symmetric functions and consider the following variational problem:

(15.19) $$\sup\{\int_0^\infty|g(r)|^q r^{n-s-1}dr;\ g\in C^1(0,\infty)\ \text{with}\ \int_0^\infty|g'(r)|^p r^{n-1}dr=1\}$$

The corresponding *Euler-Lagrange* equation is then

(15.20) $$(r^{n-1}|u'(r)|^{p-2}u'(r))' + kr^{n-s-1}|u|^{q-1} = 0,$$

and it can be easily verified that the functions u_s are solutions for any $\lambda > 0$. The rest follows from the following lemma of *Bliss*, which can be found in [28] or [51]).

LEMMA 15.2.3. *Let p_0 and q_0 be two constants such that $q_0 > p_0 > 1$. For any positive scalar J_0, consider*

$$\mathcal{C} = \{h : [0, \infty) \to \mathbb{R}; h \geq 0 \text{ and } \int_0^\infty h^{p_0}(x)dx = J_0\}.$$

The functional $I(h) := \int_0^\infty H^{q_0}(x)x^{\alpha-q_0}dx$, where $H(x) = \int_0^x h(t)dt$, then attains its maximum on $\mathcal{C}$ at functions of the form $h(x) = (\lambda x^\alpha + 1)^{-\frac{\alpha+1}{\alpha}}$, where $\alpha = \frac{q_0}{p_0} - 1$ and $\lambda > 0$.

This lemma coupled with the change of variables $x = r^{\frac{p-n}{p-1}}$ yields that the functional I attains its maximum at the functions u_s. Note that when $h(x) = (\lambda x^\alpha + 1)^{-\frac{\alpha+1}{\alpha}}$, then $H(x) = \int_0^x h(t)dt = (\lambda + x^{-\alpha})^{-\frac{1}{\alpha}}$, and if $q = \frac{n-s}{n-p}p$, then $\alpha = \frac{q}{p} - 1 = \frac{p-s}{n-p}$.

That $\mu_s(\Omega) = \mu_s(\mathbb{R}^n)$ whenever $0 \in \Omega$ follows from the fact that truncation and scaling permit to localize the extremals u_s and make them supported in Ω without changing their energy nor their integral against $|x|^{-s}dx$.

15.3. Symmetry of the extremals on half-space

We have seen that $\mu_s(\Omega) = \mu_s(\mathbb{R}^n)$ for any domain Ω containing 0 and that $\mu_s(\Omega)$ is never attained unless $\Omega = \mathbb{R}^n$. We shall see however in the next chapter, that if $0 \in \partial\Omega$, then $\mu_s(\Omega)$ may sometimes be attained and therefore the equation

(15.21) $$\begin{cases} -\Delta u = \frac{u^{2^*-1}}{|x|^s} & \text{in } \Omega \\ u > 0 & \text{in } \Omega \\ u = 0 & \text{on } \partial\Omega, \end{cases}$$

may have a solution. However, and in contrast to the case where $\Omega = \mathbb{R}^n$ where we have an explicit formula for the extremals, the corresponding solutions for half-space ($\Omega = \mathbb{R}^n_-$) are not known. One can however establish symmetry properties for such solutions, a property that will be needed in the next chapter.

THEOREM 15.3.1. *Let $n \geq 3$, $s \in (0, 2)$ and consider $u \in C^2(\mathbb{R}^n_-) \cap C^1(\overline{\mathbb{R}^n_-})$ such that*

(15.22) $$\begin{cases} -\Delta u = \frac{u^{2^*-1}}{|x|^s} & \text{in } \mathbb{R}^n_- \\ u > 0 & \text{in } \mathbb{R}^n_- \\ u = 0 & \text{on } \partial\mathbb{R}^n_-. \end{cases}$$

If for some $C > 0$ we have the bound

(15.23) $$u(x) \leq \frac{C}{(1+|x|)^{n-1}} \text{ for all } x \in \mathbb{R}^n_-,$$

then $u \circ \sigma = u$ for all isometries of $\mathbb{R}^n$ such that $\sigma(\mathbb{R}^n_-) = \mathbb{R}^n_-$.

In particular, there exists $v \in C^2(\mathbb{R}^\star_- \times \mathbb{R}) \cap C^1(\mathbb{R}_- \times \mathbb{R})$ such that for all $x_1 < 0$ and all $x' \in \mathbb{R}^{n-1}$, we have that $u(x_1, x') = v(x_1, |x'|)$.

15.3. SYMMETRY OF THE EXTREMALS ON HALF-SPACE

Proof: Denoting by $\vec{e}_1$ the first vector of the canonical basis of $\mathbb{R}^n$, we consider the open ball $D := B_{1/2}\left(-\frac{1}{2}\vec{e}_1\right)$ and define

$$v(x) := |x|^{2-n} u\left(\vec{e}_1 + \frac{x}{|x|^2}\right) \tag{15.24}$$

for all $x \in \overline{D} \setminus \{0\}$ and $v(0) = 0$.

We first claim that

$$v \in C^2(D) \cap C^1(\overline{D}) \text{ and } \frac{\partial v}{\partial \nu} < 0 \text{ on } \partial D, \tag{15.25}$$

where $\partial/\partial\nu$ denotes the outward normal derivative.

Indeed, the assumptions on u yield that $v \in C^2(D) \cap C^1(\overline{D} \setminus \{0\})$. Moreover, $v(x) > 0$ for all $x \in D$ and $v(x) = 0$ for all $x \in \partial D \setminus \{0\}$. It follows from (15.23) that there exists $C > 0$ such that

$$v(x) \leq C|x| \text{ for all } x \in \overline{D} \setminus \{0\}. \tag{15.26}$$

Since $v(0) = 0$, we have that $v \in C^0(\overline{D})$. The function v verifies the equation

$$-\Delta v = \frac{v^{2^\star - 1}}{|x + |x|^2 \vec{e}_1|^s} = \frac{v^{2^\star - 1}}{|x|^s |x + \vec{e}_1|^s} \text{ in } D. \tag{15.27}$$

Since $-\vec{e}_1 \in \partial D \setminus \{0\}$ and $v \in C^1(\overline{D} \setminus \{0\}) \cap C^0(\overline{D})$, there exists $C > 0$ such that

$$v(x) \leq C|x + \vec{e}_1| \text{ for all } x \in \overline{D}. \tag{15.28}$$

It then follows from (15.26), (15.27), (15.28) and standard elliptic theory that $v \in C^1(\overline{D})$. Since $v > 0$ in D, it follows from Hopf's Lemma that $\frac{\partial v}{\partial \nu} < 0$ on ∂D.

We now use the moving plane method to prove the symmetry of u by proving a symmetry property of v, which is defined on a ball. For any $\mu \geq 0$ and any $x = (x', x_n) \in \mathbb{R}^n$ ($x' \in \mathbb{R}^{n-1}$ and $x_n \in \mathbb{R}$), we let

$$x_\mu = (x', 2\mu - x_n) \text{ and } D_\mu = \{x \in D / x_\mu \in D\}.$$

It follows from Hopf's Lemma (See (15.25)) that there exists $\epsilon_0 > 0$ such that for any $\mu \in (\frac{1}{2} - \epsilon_0, \frac{1}{2})$, we have that $D_\mu \neq \emptyset$ and $v(x) \geq v(x_\mu)$ for all $x \in D_\mu$ such that $x_n \leq \mu$. We let $\mu \geq 0$. We say that (P_μ) holds if:

$$D_\mu \neq \emptyset \text{ and } v(x) \geq v(x_\mu) \text{ for all } x \in D_\mu \text{ such that } x_n \leq \mu.$$

We let

$$\lambda := \min\left\{\mu \geq 0; (P_\nu) \text{ holds for all } \nu \in \left(\mu, \frac{1}{2}\right)\right\}. \tag{15.29}$$

We claim that $\lambda = 0$. Indeed, otherwise we have $\lambda > 0$, $D_\lambda \neq \emptyset$ and that (P_λ) holds. We let

$$w(x) := v(x) - v(x_\lambda)$$

for all $x \in D_\lambda \cap \{x_n < \lambda\}$. Since (P_λ) holds, we have that $w(x) \geq 0$ for all $x \in D_\lambda \cap \{x_n < \lambda\}$. With the equation (15.27) of v and (P_λ), we get that

$$\begin{aligned}-\Delta w &= \frac{v(x)^{2^\star - 1}}{|x + |x|^2 \vec{e}_1|^s} - \frac{v(x_\lambda)^{2^\star - 1}}{|x_\lambda + |x_\lambda|^2 \vec{e}_1|^s} \\ &\geq v(x_\lambda)^{2^\star - 1}\left(\frac{1}{|x + |x|^2 \vec{e}_1|^s} - \frac{1}{|x_\lambda + |x_\lambda|^2 \vec{e}_1|^s}\right)\end{aligned}$$

for all $x \in D_\lambda \cap \{x_n < \lambda\}$. With straightforward computations, we have that
$$|x_\lambda|^2 - |x|^2 = 4\lambda(\lambda - x_n)$$
$$|x_\lambda + |x_\lambda|^2 \vec{e}_1|^2 - |x + |x|^2 \vec{e}_1|^2 = (|x_\lambda|^2 - |x|^2)\left(1 + |x_\lambda|^2 + |x|^2 + 2x_1\right)$$
for all $x \in \mathbb{R}^n$. It follows that $-\Delta w(x) > 0$ for all $x \in D_\lambda \cap \{x_n < \lambda\}$. Note that we have used that $\lambda > 0$. It then follows from Hopf's Lemma and the strong comparison principle that

(15.30) $\quad w > 0$ in $D_\lambda \cap \{x_n < \lambda\}$ and $\dfrac{\partial w}{\partial \nu} < 0$ on $D_\lambda \cap \{x_n = \lambda\}$.

By definition, there exists a sequence $(\lambda_i)_{i \in \mathbb{N}} \in \mathbb{R}$ and a sequence $(x^i)_{i \in \mathbb{N}} \in D$ such that $\lambda_i < \lambda$, $x^i \in D_{\lambda_i}$, $(x^i)_n < \lambda_i$, $\lim_{i \to +\infty} \lambda_i = \lambda$ and

(15.31) $\quad v(x^i) < v((x^i)_{\lambda_i})$

for all $i \in \mathbb{N}$. Up to extraction a subsequence, we assume that there exists $x \in \overline{D_\lambda} \cap \{x_n \leq \lambda\}$ such that $\lim_{i \to +\infty} x^i = x$ with $x_n \leq \lambda$. Passing to the limit as $i \to +\infty$ in (15.31), we get that $v(x) \leq v(x_\lambda)$. It follows from this last inequality and (15.30) that $v(x) - v(x_\lambda) = w(x) = 0$, and then $x \in \partial(D_\lambda \cap \{x_n < \lambda\})$.

Case 1: If $x \in \partial D$. Then $v(x_\lambda) = 0$ and $x_\lambda \in \partial D$. Since D is a ball and $\lambda > 0$, we get that $x = x_\lambda \in \partial D$. Since v is C^1, we get that there exists $\tau_i \in ((x^i)_n, 2\lambda_i - (x^i)_n)$ such that
$$v(x^i) - v((x^i)_{\lambda_i}) = \partial_n v((x')^i, \tau_i) \times 2((x^i)_n - \lambda_i).$$
Letting $i \to +\infty$, using that $(x^i)_n < \lambda_i$ and (15.31), we get that $\partial_n v(x) \geq 0$. On the other hand, we have that
$$\partial_n v(x) = \frac{\partial v}{\partial \nu}(x) \cdot (\nu(x)|\vec{e}_n) = \frac{\lambda}{|x + \vec{e}_1/2|} \frac{\partial v}{\partial \nu}(x) < 0.$$
A contradiction with (15.25).

Case 2: If $x \in D$. Since $v(x_\lambda) = v(x)$, we then get that $x_\lambda \in D$. Since $x \in \partial(D_\lambda \cap \{x_n < \lambda\})$, we then get that $x \in D \cap \{x_n = \lambda\}$. With the same argument as in the preceding step, we get that $\partial_n v(x) \geq 0$. On the other hand, with (15.30), we get that $2\partial_n v(x) = \partial_n w(x) < 0$. A contradiction.

This proves that $\lambda = 0$ in either one of the two cases considered above. It now follows from the definition (15.29) of λ that $v(x', x_n) \geq v(x', -x_n)$ for all $x \in D$ such that $x_n \leq 0$. With the same technique, we get the reverse inequality, and then, we get that $v(x', x_n) = v(x', -x_n)$ for all $x = (x', x_n) \in D$. In other words, v is symmetric with respect to the hyperplane $\{x_n = 0\}$. The same analysis holds for any hyperplane containing $\vec{e}_1$. Coming back to the initial function u, this complete the proof of Theorem 15.3.1.

15.4. The Sobolev-Hardy-Rellich inequalities

The same type of interpolation between the Sobolev and the Hardy-Rellich inequalities yield the following.

THEOREM 15.4.1. *(Sobolev-Hardy-Rellich inequality)* Assume $n > 4$ and $0 < s < 4$. Setting $2^{**}(s) := \dfrac{2(n-s)}{n-4}$, then for any domain Ω in $\mathbb{R}^n$, there exists a constant $C > 0$ such that

(15.32) $\quad \left(\int_\Omega \dfrac{|u|^{2^{**}}}{|x|^s} dx\right)^{\frac{1}{2^{**}}} \leq C \left(\int_\Omega |\Delta u|^2 dx\right)^{\frac{1}{2}}$ for all $u \in H_0^2(\Omega)$.

If Ω is bounded, then the inequality holds with $2^{**}(s)$ replaced by any q, $2 \leq q \leq 2^{**}(s)$.

Proof: Again, for $s = 0$, this is just Sobolev's inequality and for $s = 4$, it is the Hardy-Rellich inequality. Now, for $0 < s < 4$ and $2 < q < 2^{**}(s)$ we have

$$\int_\Omega \frac{|u|^q}{|x|^s} \, dx = \int_\Omega \left(\frac{|u|^2}{|x|^4} \, dx\right)^{\frac{s}{4}} |u|^{q-\frac{s}{2}} \, dx$$

$$\leq \left(\int_\Omega \frac{|u|^2}{|x|^4} \, dx\right)^{\frac{s}{4}} \left(\int_\Omega |u|^{(q-\frac{s}{2})\frac{4}{4-s}}\right)^{\frac{4-s}{4}} \, dx.$$

We use Hardy-Rellich's inequality for the first term, while for the second term we note that if $q \leq 2^{**}(s) = \frac{2(n-s)}{n-4}$, then $(q - \frac{s}{2})\frac{4}{4-s} \leq 2^{**}(0) = \frac{2n}{n-4}$. If Ω is bounded then the Sobolev inequality applies and we get that

$$\int_\Omega \frac{|u|^q}{|x|^s} \, dx \leq \left(\int_\Omega |\Delta u|^2 \, dx\right)^{\frac{s}{4}} \left(\int_\Omega |\Delta u|^2 \, dx\right)^{(q-\frac{s}{2})\frac{4}{4-s}\frac{4-s}{4}\frac{1}{2}} = \left(\int_\Omega |\Delta u|^2 \, dx\right)^{\frac{q}{2}}.$$

We now consider the best constant in (15.32), i.e.,

$$(15.33) \qquad \nu_s(\Omega) := \inf\left\{\frac{\int_\Omega |\Delta u|^2 \, dx}{\left(\int_\Omega \frac{u^q}{|x|^s}\right)^{\frac{2}{q}} \, dx}; u \in H_0^2(\Omega) \setminus \{0\}\right\},$$

where $q = 2^{**}(s) = \frac{2(n-s)}{n-4}$.

When $s = 0$, one can prove as in the second order case, that $\mu_0(\Omega)$ is independent of the domain Ω, and that it is only attained when $\Omega = \mathbb{R}^n$. Moreover, the extremals are given by the following functions:

$$(15.34) \qquad V_\epsilon(x) = k(\epsilon)\left(\epsilon + |x|^2\right)^{\frac{4-n}{2}},$$

where $k(\epsilon) := \epsilon^{\frac{n-4}{2}}(n(n-2)(n+2)(n-4))^{\frac{n-4}{8}}$. They then satisfy the corresponding Euler-Lagrange equation on $\mathbb{R}^n$,

$$(15.35) \qquad \Delta^2 u = u^{\frac{n+4}{n-4}} \text{ and } u > 0 \quad \text{on } \mathbb{R}^n.$$

However, none of these results are known when $0 < s < 4$, and in particular no explicit solution is known for the following equation.

$$(15.36) \qquad \Delta^2 u = \frac{u^{2^{**}(s)-1}}{|x|^s} \text{ and } u > 0 \text{ on } \mathbb{R}^n.$$

We have however the following consequence of the Pohozaev identity, which implies in particular that $\mu_s(\Omega)$ is never attained on a bounded star-shaped domain.

PROPOSITION 15.4.1. *If Ω is a bounded domain in $\mathbb{R}^n$ that is star-shaped around 0, then the equation*

$$(15.37) \qquad \begin{cases} \Delta^2 u = \mu \frac{u^{q-1}}{|x|^s} + \lambda |u|^{r-1} u & \text{in } \Omega \\ |\nabla u| = u = 0 & \text{on } \partial\Omega, \end{cases}$$

has no non-trivial solution, whenever $r = 2^{**}(0) = \frac{2n}{n-4}$ and $q = 2^{**}(s) = \frac{2(n-s)}{n-4}$ with $0 < s < 4$.

Proof: First we show that

$$(15.38) \quad (\frac{n}{2} - 2) \int_\Omega |\Delta u|^2 + \frac{1}{2} \int_{\partial\Omega} |\Delta u|^2 x.\nu ds = \lambda \frac{n}{r} \int_\Omega |u|^r + \mu \frac{n-s}{q} \int_\Omega \frac{|u|^q}{|x|^s}.$$

For that, we multiply both sides of the equation by $x \cdot \nabla u$ and integrate over Ω to get

$$\begin{aligned} R.H.S. &= \lambda \int_\Omega x \cdot \nabla u |u|^{r-2} u + \mu \int_\Omega \frac{|u|^{q-2} u}{|x|^s} x \cdot \nabla u \\ &= \frac{\lambda}{r} \int_\Omega x \cdot \nabla |u|^r + \frac{\mu}{q} \int_\Omega \frac{x \cdot \nabla |u|^q}{|x|^s} \\ &= -\lambda \frac{n}{r} \int_\Omega |u|^r - \mu \frac{(n-s)}{q} \int_\Omega \frac{|u|^q}{|x|^s}, \end{aligned}$$

and

$$\begin{aligned} L.H.S. &= \int_\Omega \Delta^2 u \, x \cdot \nabla u = \int_\Omega \Delta u \Delta(x \cdot \nabla u) - \int_{\partial\Omega} \Delta u \partial_\nu (x \cdot \nabla u) \\ &=: A_1 + A_2. \end{aligned}$$

Since $\Delta(x \cdot \nabla u) = 2\Delta u + x \cdot \nabla \Delta u$, we can rewrite A_1 as

$$A_1 = 2 \int_\Omega |\Delta u|^2 + \frac{1}{2} \int_\Omega x \cdot \nabla |\Delta u|^2.$$

Applying Green's formula we simplify A_1 to

$$A_1 = (2 - \frac{n}{2}) \int_\Omega |\Delta u|^2 + \frac{1}{2} \int_{\partial\Omega} |\Delta u|^2 x \cdot \nu dS.$$

On the other hand, since $u = |\nabla u| = 0$ on $\partial\Omega$, we have $u_i = |\nabla u| \nu_i$ and $u_{i,j} = \Delta u \, \nu_i \nu_j$ on $\partial\Omega$. Therefore,

$$A_2 = -\int_{\partial\Omega} |\Delta u|^2 dS.$$

Finally,

$$L.H.S. = (2 - \frac{n}{2}) \int_\Omega |\Delta u|^2 - \frac{1}{2} \int_{\partial\Omega} |\Delta u|^2 x \cdot \nu dS.$$

This completes the proof of (15.38).

Now, multiply equation (15.37) by u and integrate to get

$$(15.39) \quad \int_\Omega |\Delta u|^2 \, dx = \lambda \int_\Omega |u|^r \, dx + \mu \int_\Omega \frac{|u|^q}{|x|^s} \, dx.$$

By combining the previous identity with (15.38) we obtain

$$\frac{1}{2} \int_{\partial\Omega} |\Delta u|^2 x.\nu ds = \lambda(\frac{n}{r} - \frac{n-4}{2}) \int_\Omega |u|^r \, dx + \mu(\frac{n-s}{q} - \frac{n-4}{2}) \int_\Omega \frac{|u|^q}{|x|^s} \, dx.$$

This clearly yields the claim of the proposition. □

15.5. Further comments and remarks

The Caffarelli-Kohn-Nirenberg inequalities were first establised in [**76**]. The change of variable 15.5 reduces most questions related to these inequalities (best constants, extremals, etc) to those concerned with the Hardy-Sobolev inequality. The extremals for the latter on $\mathbb{R}^n$ were obtained in [**155**] and probably earlier. A good account on Schwarz symmetrization can be found in Lieb and Loss [**203**]. The Bliss lemma can be found in [**28**] or [**51**]).

Just like in the Hardy-Rellich inequalities (Chapter 6), the symmetry breaking phenomenon appears for special values of the parameters in the Caffarelli-Kohn-Nirenberg inequalities on $\mathbb{R}^n$. We treat here a simple case where standard symmetrization techniques apply, and we refer to the founding work of Felli and Schneider [**145**], which has motivated other important works by Dolbeault-Esteban-Tarantello-Tertikas [**117**] and Dolbeault-Esteban-Loss-Tarantello [**118**] and [**119**].

The symmetry of the extremals in half-space was first proved by Ghoussoub-Robert in [**158**]. The method of moving plane used in that proof originated in Alexandrov [**19**] and was developed further by Serrin [**246**], Gidas-Ni-Nirenberg [**167, 168**], and Caffarelli-Gidas-Spruck [**75**].

Just like in the case of the Hardy inequality (and those of Hardy-Sobolev studied in the next chapter), one can again study the case when $0 \in \partial\Omega$ and in particular, conical domains. The best constants are then different from when $0 \in \Omega$ and their attainability will again depend on the domain. See for example the recent preprint by Caldiroli and Musina [**78**] for some very partial results. That the best constant in the critical Hardy-Rellich-Sobolev inequality is never attained on bounded star-shaped domains was verified by M. Fazly [**135**].

Open problem (17): Is there an explicit formula for the extremals of the Hardy-Sobolev inequality on $\mathbb{R}^n_+$, i.e., an explicit solution for

$$(15.40) \quad \begin{cases} \Delta v + \dfrac{v^{2^*(s)-1}}{|x|^s} = 0 & \text{in } \mathbb{R}^n_- \\ v > 0 & \text{on } \mathbb{R}^n_-, \\ v = 0 & \text{on } \partial\mathbb{R}^n_- . \end{cases}$$

Recall that the answer is affirmative if half-space is replaced by the whole of $\mathbb{R}^n$.

Open problem (18): Is there an explicit formula for the extremals of the Hardy-Rellich-Sobolev inequality in $\mathbb{R}^n$, i.e., an explicit solution for

$$(15.41) \quad \begin{cases} \Delta^2 v = \dfrac{v^{2^{**}(s)-1}}{|x|^s} & \text{in } \mathbb{R}^n \\ v > 0 & \text{on } \mathbb{R}^n . \end{cases}$$

We have seen that the answer is affirmative when $s = 0$ as the extremals are given by (15.34). See also some related work of Lieb on doubly weighted Hardy-Littlewood-Sobolev inequality [**202**].

CHAPTER 16

Domain Curvature and Best Constants in the Hardy-Sobolev Inequalities

This chapter addresses the question of attainability of the best constant $\mu_s(\Omega)$ in the Hardy-Sobolev inequality on a smooth domain Ω of $\mathbb{R}^n$, when 0 is on the boundary $\partial\Omega$. This question is closely related to the geometry of $\partial\Omega$, since in dimension $n \geq 3$, the negativity of the mean curvature of $\partial\Omega$ at 0 is sufficient to ensure the attainability of $\mu_s(\Omega)$. The proof, which relies on a fine analysis of the asymptotic behaviour of appropriate minimizing sequences, will only be given in dimension $n \geq 4$. The result holds true also in dimension 3 but the more involved proof is omitted and can be viewed in [**160**].

16.1. From the subcritical to the critical case in the Hardy-Sobolev inequalities

We consider again the attainability of the best constant $\mu_s(\Omega)$ in the critical Hardy-Sobolev inequality, that is

$$(16.1) \qquad \mu_s(\Omega) = \inf\left\{ \frac{\int_\Omega |\nabla u|^2\, dx}{\left(\int_\Omega \frac{|u|^{2^*(s)}}{|x|^s}\, dx\right)^{\frac{2}{2^*(s)}}}; u \in H_0^1(\Omega) \setminus \{0\}\right\},$$

where Ω is a smooth domain of $\mathbb{R}^n$, $n \geq 3$ $s \in [0,2]$ and $2^*(s) = \frac{2(n-s)}{n-2}$. Throughout this chapter, $2^*(s)$ will be denoted by $2^\star$. We also consider the ground state solutions in $H_0^1(\Omega) \cap C^1(\overline{\Omega})$ for the corresponding Euler-Lagrange equation satisfied by any potential extremal,

$$(16.2) \qquad \begin{cases} -\Delta u = \dfrac{u^{2^\star - 1}}{|x|^s} & \text{in } \Omega \\ u > 0 & \text{in } \Omega \\ u = 0 & \text{on } \partial\Omega. \end{cases}$$

The following theorem is the main result of this chapter. It stands in contrast to the previous chapter, where it is shown that if 0 belongs to the interior of a domain Ω, then $\mu_s(\Omega) = \mu_s(\mathbb{R}^n)$ for any $0 < s < 2$ and that $\mu_s(\Omega)$ is never attained unless $\text{cap}(\mathbb{R}^n \setminus \Omega) = 0$.

THEOREM 16.1.1. *Let Ω be a smooth bounded oriented domain of $\mathbb{R}^n$ where $n \geq 4$, such that $0 \in \partial\Omega$ and assume $s \in (0,2)$. If the mean curvature of $\partial\Omega$ at 0 is negative, then the infimum $\mu_s(\Omega)$ in (16.1) is achieved. In addition, the set of minimizers of (16.1) is pre-compact in the $H_0^1(\Omega)$–topology.*

The attainability result will be obtained via a fine study of the asymptotic behaviour of solutions to the corresponding subcritical PDE's. For that we first consider for any $\varepsilon \in (0, 2^\star - 2)$, the infimum

$$(16.3) \qquad \mu_{s,\varepsilon}(\Omega) := \inf_{u \in H_0^1(\Omega) \setminus \{0\}} \frac{\int_\Omega |\nabla u|^2 \, dx}{\left(\int_\Omega \frac{|u|^{2^\star - \varepsilon}}{|x|^s} \, dx \right)^{\frac{2}{2^\star - \varepsilon}}},$$

which –as we shall prove– is achieved by a function $u_\varepsilon \in H_0^1(\Omega)$, $u_\varepsilon > 0$ in Ω in $C^1(\overline{\Omega}) \cap C^2(\overline{\Omega} \setminus \{0\})$ that satisfies the system

$$(16.4) \qquad \begin{cases} -\Delta u_\varepsilon = \frac{u_\varepsilon^{2^\star - 1 - \varepsilon}}{|x|^s} & \text{in } \Omega \\ u_\varepsilon > 0 & \text{in } \Omega \\ \int_\Omega \frac{|u_\varepsilon|^{2^\star - \varepsilon}}{|x|^s} \, dx = (\mu_{s,\varepsilon}(\Omega))^{\frac{2^\star - \varepsilon}{2^\star - 2 - \varepsilon}}. \end{cases}$$

The sequence (u_ε) can eventually develop a singularity at zero as we approach the critical exponent $2^\star$ (i.e., when $\varepsilon \to 0$). In the next sections, we shall describe the way these sequences may blow up, which makes for an interesting analysis in its own right. The following theorem –established in Section 16.4– explicitly shows how the curvature assumption prevents such blow-up and restores compactness.

THEOREM 16.1.2. *Let Ω be a smooth bounded oriented domain of $\mathbb{R}^n$ where $n \geq 4$, and assuming that u_ε converges weakly to zero (i.e. when blow-up occurs), then there exists a solution v for the equation*

$$(16.5) \qquad \begin{cases} -\Delta v = \frac{v^{2^\star - 1}}{|x|^s} & \text{in } \mathbb{R}^n_- \\ v > 0 & \text{in } \mathbb{R}^n_- \\ v = 0 & \text{on } \partial \mathbb{R}^n_-, \end{cases}$$

such that

$$(16.6) \qquad \int_{\mathbb{R}^n_-} |\nabla v|^2 \, dx = \mu_s(\Omega)^{\frac{2^\star}{2^\star - 2}} = \mu_s(\mathbb{R}^n_-)^{\frac{2^\star}{2^\star - 2}},$$

while -modulo passing to a subsequence- we have

$$(16.7) \qquad \lim_{\varepsilon \to 0} \varepsilon \left(\max_\Omega u_\varepsilon \right)^{\frac{2}{n-2}} = \frac{(n-s) \int_{\partial \mathbb{R}^n_-} |x|^2 |\nabla v|^2 \, dx}{n(n-2)^2 \mu_s(\mathbb{R}^n_-)^{\frac{n-s}{2-s}}} \cdot H(0),$$

where $H(0)$ is the mean curvature of the oriented boundary $\partial \Omega$ at 0.

We also consider the case where the equations involve a linear term $a(x) \in C^1(\overline{\Omega})$ such that

$$(16.8) \qquad -\Delta + a \text{ is coercive in } \Omega,$$

that is, there exists $c_0 > 0$ such that for all $\varphi \in C_c^1(\Omega)$,

$$(16.9) \qquad \int_\Omega (|\nabla \varphi|^2 + a \varphi^2) \, dx \geq c_0 \int_\Omega \varphi^2 \, dx.$$

THEOREM 16.1.3. *Let Ω be a smooth bounded oriented domain of $\mathbb{R}^n$ where $n \geq 4$, such that $0 \in \partial \Omega$. Assume $s \in (0, 2)$ and consider a $C^1(\overline{\Omega})$-function a such that the operator $-\Delta + a$ is coercive in Ω. If the mean curvature of $\partial \Omega$ at 0 is negative, then there exists a solution $u \in H_0^1(\Omega) \cap C^1(\overline{\Omega})$ for*

$$\begin{cases} -\Delta u + au = \frac{u^{2^\star - 1}}{|x|^s} & \text{in } \Omega \\ u > 0 & \text{in } \Omega \\ u = 0 & \text{on } \partial \Omega. \end{cases}$$

16.1. FROM THE SUBCRITICAL TO THE CRITICAL CASE

We first deal with the subcritical case.

PROPOSITION 16.1.1. *Let Ω be a smooth bounded domain of $\mathbb{R}^n$, $n \geq 3$ and $s \in (0,2)$. Let $a \in C^1(\overline{\Omega})$ such that $-\Delta + a$ is coercive. Then for any $p \in (2, 2^*)$, the infimum*

$$(16.10) \qquad \mu_{s,p}(a,\Omega) := \inf_{u \in H_0^1(\Omega) \setminus \{0\}} \frac{\int_\Omega (|\nabla u|^2 + au^2)\, dx}{\left(\int_\Omega \frac{|u|^p}{|x|^s}\, dx \right)^{\frac{2}{p}}},$$

is achieved by a positive function $u \in H_0^1(\Omega)$. Moreover, $u \in C^1(\overline{\Omega}) \cap C^2(\overline{\Omega} \setminus \{0\})$ and can be assumed to satisfy the system

$$(16.11) \qquad \begin{cases} -\Delta u + au = \frac{u^{p-1}}{|x|^s} & \text{in } \Omega \\ u > 0 & \text{in } \Omega \\ \int_\Omega \frac{|u|^p}{|x|^s}\, dx = (\mu_{s,p}(a,\Omega))^{\frac{p}{p-2}}. \end{cases}$$

Proof: We claim that there exists a minimizer for $\mu_{s,p}(a, \Omega)$. Indeed, let $(u_k)_{k \in \mathbb{N}} \in H_0^1(\Omega)$ be a minimizing sequence for $\mu_{s,p}(a, \Omega)$ such that

$$\int_\Omega \frac{|u_k|^p}{|x|^s}\, dx = 1 \text{ and } \mu_{s,p}(a, \Omega) = \int_\Omega (|\nabla u_k|^2 + au_k^2)\, dx + o(1)$$

where $\lim_{k \to +\infty} o(1) = 0$. Since $\|u_k\|_{H_0^1(\Omega)} = O(1)$ when $k \to +\infty$, there exists $\tilde{u} \in H_0^1(\Omega)$ such that, up to a subsequence, $u_k \rightharpoonup \tilde{u}$ weakly in $H_0^1(\Omega)$ when $k \to +\infty$ and $\lim_{k \to +\infty} u_k(x) = \tilde{u}(x)$ a.e. in Ω. Let $\theta_k = u_k - \tilde{u} \in H_0^1(\Omega)$. As easily checked, we have that

$$(16.12) \qquad \mu_{s,p}(a, \Omega) = \int_\Omega (|\nabla \tilde{u}|^2 + a\tilde{u}^2)\, dx + \int_\Omega |\nabla \theta_k|^2\, dx + o(1),$$

where $\lim_{k \to +\infty} o(1) = 0$. Let $\eta \in C_c^\infty(\mathbb{R})$ such that $\eta(x) = 1$ for all $x \in [-1, 1]$. Let $A > 0$. With Lebesgue's theorem, we have that

$$\left| \int_\Omega \frac{|u_k|^p}{|x|^s}\, dx - \int_\Omega \frac{|\tilde{u}|^p}{|x|^s}\, dx \right|$$
$$= \left| \int_\Omega \left(\eta\left(\frac{u_k}{A}\right) \frac{|u_k|^p}{|x|^s} - \eta\left(\frac{\tilde{u}}{A}\right) \frac{|\tilde{u}|^p}{|x|^s} \right) dx \right|$$
$$+ \int_\Omega \left| 1 - \eta\left(\frac{u_k}{A}\right) \right| \frac{|u_k|^p}{|x|^s}\, dx + \int_\Omega \left| 1 - \eta\left(\frac{\tilde{u}}{A}\right) \right| \frac{|\tilde{u}|^p}{|x|^s}\, dx$$
$$\leq o(1) + \frac{1}{A^\varepsilon} \int_\Omega \left| 1 - \eta\left(\frac{u_k}{A}\right) \right| \frac{|u_k|^{2^*}}{|x|^s}\, dx + \frac{1}{A^\varepsilon} \int_\Omega \left| 1 - \eta\left(\frac{\tilde{u}}{A}\right) \right| \frac{|\tilde{u}|^{2^*}}{|x|^s}\, dx$$
$$\leq o(1) + \frac{1}{A^\varepsilon} \int_\Omega \frac{|u_k|^{2^*}}{|x|^s}\, dx + \frac{1}{A^\varepsilon} \int_\Omega \frac{|\tilde{u}|^{2^*}}{|x|^s}\, dx$$
$$\leq o(1) + \frac{1}{A^\varepsilon} \mu_s(\Omega)^{-\frac{2^*}{2}} \left(\|u_k\|_{H_0^1(\Omega)}^{2^*} + \|\tilde{u}\|_{H_0^1(\Omega)}^{2^*} \right)$$

where $\lim_{k \to +\infty} o(1) = 0$. Letting $k \to +\infty$, and then $A \to +\infty$, we get that

$$\lim_{k \to +\infty} \int_\Omega \frac{|u_k|^p}{|x|^s}\, dx = \int_\Omega \frac{|\tilde{u}|^p}{|x|^s}\, dx.$$

It then follows that $\int_\Omega \frac{|\tilde{u}|^p}{|x|^s} dx = 1$. With the definition of $\mu_{s,p}(a,\Omega)$, we then get that

$$\mu_{s,p}(a,\Omega) \leq \int_\Omega (|\nabla \tilde{u}|^2 + a\tilde{u}^2)\, dx.$$

With (16.12), we then get that $\lim_{k \to +\infty} \theta_k = 0$ in $H_0^1(\Omega)$. As a consequence, $\mu_{s,p}(a,\Omega)$ is attained by $\tilde{u}$. This proves the claim.

Up to replacing $\tilde{u}$ by $|\tilde{u}|$, we can assume that $\tilde{u} \geq 0$. We let

$$\mu = \mu_{s,p}(a,\Omega)^{\frac{1}{p-2}} \tilde{u}.$$

As easily checked, $u \geq 0$ is also a minimizer for $\mu_{s,p}(a,\Omega)$. It satisfies

$$-\Delta u + au = \frac{u^{p-1}}{|x|^s} \text{ in } \Omega.$$

Moreover, it follows from the appendix and standard elliptic theory that $u \in C^1(\overline{\Omega}) \cap C^2(\overline{\Omega} \setminus \{0\})$. Since $(-\Delta + a)u \geq 0$ in Ω and $u \not\equiv 0$, it follows from the strong comparison principle that $u > 0$ in Ω. □

The following chart around a given point $x_0 \in \partial\Omega$ will be useful throughout this chapter. Since $\partial\Omega$ is smooth and $x_0 \in \partial\Omega$, there exist U, V open subsets of $\mathbb{R}^n$, there exists I an open interval of $\mathbb{R}$, there exists U' an open subset of $\mathbb{R}^{n-1}$ such that $0 \in U = I \times U'$ and $x_0 \in V$. There exist $\varphi \in C^\infty(U,V)$ and $\varphi_0 \in C^\infty(U')$ such that

(16.13)
- (i) $\varphi : U \to V$ is a C^∞-diffeomorphism
- (ii) $\varphi(0) = x_0$
- (iii) $D_0\varphi = Id_{\mathbb{R}^n}$
- (iv) $\varphi(U \cap \{x_1 < 0\}) = \varphi(U) \cap \Omega$ and $\varphi(U \cap \{x_1 = 0\}) = \varphi(U) \cap \partial\Omega$.
- (v) $\varphi(x_1, y) = x_0 + (x_1 + \varphi_0(y), y)$ for all $(x_1, y) \in I \times U' = U$
- (vi) $\varphi_0(0) = 0$ and $\nabla\varphi_0(0) = 0$.

Here $D_x\varphi$ denotes the differential of φ at x.

We now consider for each $\varepsilon \in (0, 2^\star - 2)$, the subcritical exponent $p_\varepsilon = 2^\star - \epsilon$, and use the following notation

(16.14) $\quad \mu_s(a,\Omega) := \mu_{s,2^\star}(a,\Omega) \quad$ and $\quad \mu_{s,\varepsilon}(a,\Omega) := \mu_{s,p_\varepsilon}(a,\Omega),$

in such a way that $\mu_s(\Omega) = \mu_s(0,\Omega)$.

PROPOSITION 16.1.2. *Let Ω be a smooth bounded domain of $\mathbb{R}^n$, $n \geq 3$, such that $0 \in \partial\Omega$ and consider $s \in (0,2)$. Let $a \in C^0(\overline{\Omega})$ and a family $(a_\varepsilon)_{\varepsilon>0} \in C^1(\mathcal{U})$ such that $\lim_{\varepsilon \to 0} a_\varepsilon = a$ in $C^1_{loc}(\mathcal{U})$ where $\mathcal{U}$ is a neighborhood of $\overline{\Omega}$. Then,*
1. $\mu_s(a,\Omega) \leq \mu_s(\mathbb{R}^n_-)$.
2. $\lim_{\varepsilon \to 0} \mu_{s,\varepsilon}(a_\varepsilon,\Omega) = \mu_s(a,\Omega)$.
3. *If u_ε is a minimizer for $\mu_{s,\varepsilon}(a_\varepsilon,\Omega)$, then $(u_\varepsilon)_\varepsilon$ is bounded in $H_0^1(\Omega)$ and any non-zero weak limit v of $(u_\varepsilon)_\varepsilon$ is a minimiser for $\mu_s(a,\Omega)$.*

16.1. FROM THE SUBCRITICAL TO THE CRITICAL CASE

Proof: (1) To prove the upper bound for $\mu_{s,a}(\Omega)$, we let $\alpha > 0$ and $u \in C_c^\infty(\mathbb{R}_-^n) \setminus \{0\}$ such that

$$\frac{\int_{\mathbb{R}_-^n} |\nabla u|^2 \, dx}{\left(\int_{\mathbb{R}_-^n} \frac{|u|^{2^\star}}{|x|^s} \, dx\right)^{\frac{2}{2^\star}}} \leq \mu_s(\mathbb{R}_-^n) + \alpha.$$

Taking $x_0 = 0$ in (16.13), we define

$$u(x) = \varepsilon^{-\frac{n-2}{2}} u\left(\frac{\varphi^{-1}(x)}{\varepsilon}\right)$$

for all $x \in \Omega$ and all $\varepsilon > 0$. As easily checked, for $\varepsilon > 0$ small enough, we have that $u \in C_c^\infty(\Omega)$. With a change of variable, we get that

$$\int_\Omega \frac{|u_\varepsilon|^{2^\star}}{|x|^s} \, dx = \int_{\mathbb{R}^n} \frac{|u(y)|^{2^\star}}{\left|\frac{\varphi(\varepsilon y)}{\varepsilon}\right|^s} \cdot |\mathrm{Jac}(\varepsilon y)| \, dy.$$

Since u is compactly supported, we get from (iii) of (16.13) and Lebesgue's convergence theorem that

$$\lim_{\varepsilon \to 0} \int_\Omega \frac{|u_\varepsilon|^{2^\star}}{|x|^s} \, dx = \int_{\mathbb{R}_-^n} \frac{|u|^{2^\star}}{|x|^s} \, dx.$$

On the other hand, we have that

$$\int_\Omega (|\nabla u_\varepsilon|^2 + a u_\varepsilon^2) \, dx = \int_{\mathbb{R}_-^n} (|\nabla u|_{g_\varepsilon}^2 + \varepsilon^2 a \circ \varphi(\varepsilon x) u^2) \cdot \sqrt{|g_\varepsilon|} \, dx,$$

where $(g_\varepsilon(x))_{ij} = (\partial_i \varphi(\varepsilon x), \partial_j \varphi(\varepsilon x))$, and $|g_\varepsilon| = \det(g_\varepsilon)$. From (iii) of (16.13) and Lebesgue's convergence theorem, we get that

$$\lim_{\varepsilon \to 0} \int_\Omega (|\nabla u_\varepsilon|^2 + a u_\varepsilon^2) \, dx = \int_{\mathbb{R}_-^n} |\nabla u|^2 \, dx.$$

As a consequence, we get that

$$\mu_s(a, \Omega) \leq \frac{\int_\Omega (|\nabla u_\varepsilon|^2 + a u_\varepsilon^2) \, dx}{\left(\int_\Omega \frac{|u_\varepsilon|^{2^\star}}{|x|^s} \, dx\right)^{\frac{2}{2^\star}}} = \frac{\int_{\mathbb{R}_-^n} |\nabla u|^2 \, dx}{\left(\int_{\mathbb{R}_-^n} \frac{|u|^{2^\star}}{|x|^s} \, dx\right)^{\frac{2}{2^\star}}} + o(1) \leq \mu_s(\mathbb{R}_-^n) + \alpha + o(1)$$

where $\lim_{\varepsilon \to 0} o(1) = 0$. Letting $\varepsilon \to 0$ and $\alpha \to 0$ yields the conclusion of (1).

(2) In order to show that

$$\lim_{\varepsilon \to 0} \mu_{s,a_\varepsilon}^\varepsilon(\Omega) = \mu_{s,a}(\Omega),$$

we let $\alpha > 0$ and $u \in C_c^\infty(\Omega) \setminus \{0\}$ such that

$$\frac{\int_\Omega (|\nabla u|^2 + a u^2) \, dx}{\left(\int_\Omega \frac{|u|^{2^\star}}{|x|^s} \, dx\right)^{\frac{2}{2^\star}}} \leq \mu_{s,a}(\Omega) + \alpha.$$

We have that

$$\lim_{\varepsilon \to 0} \frac{\int_\Omega (|\nabla u|^2 + a_\varepsilon u^2) \, dx}{\left(\int_\Omega \frac{|u|^{2^\star - \varepsilon}}{|x|^s} \, dx\right)^{\frac{2}{2^\star - \varepsilon}}} = \frac{\int_\Omega (|\nabla u|^2 + a u^2) \, dx}{\left(\int_\Omega \frac{|u|^{2^\star}}{|x|^s} \, dx\right)^{\frac{2}{2^\star}}} \leq \mu_{s,a}(\Omega) + \alpha.$$

Letting $\varepsilon \to 0$ and $\alpha \to 0$, we get that
(16.15)
$$\limsup_{\varepsilon \to 0} \mu^\varepsilon_{s,a_\varepsilon}(\Omega) \le \mu_{s,a}(\Omega).$$

We now let $v \in C_c^\infty(\Omega) \setminus \{0\}$. It follows from Hölder's inequality that

$$\left(\int_\Omega \frac{|v|^{2^*-\varepsilon}}{|x|^s} dx\right)^{\frac{2}{2^*-\varepsilon}} \le \left(\int_\Omega \frac{dx}{|x|^s}\right)^{\frac{2\varepsilon}{2^* \cdot (2^*-\varepsilon)}} \left(\int_\Omega \frac{|v|^{2^*}}{|x|^s} dx\right)^{\frac{2}{2^*}}$$

and then

$$\frac{\int_\Omega (|\nabla v|^2 + av^2) dx}{\left(\int_\Omega \frac{|v|^{2^*}}{|x|^s} dx\right)^{\frac{2}{2^*}}} \le \left(\int_\Omega \frac{dx}{|x|^s}\right)^{\frac{2\varepsilon}{2^* \cdot (2^*-\varepsilon)}} \cdot \frac{\int_\Omega (|\nabla v|^2 + a_\varepsilon v^2) dx}{\left(\int_\Omega \frac{|v|^{2^*-\varepsilon}}{|x|^s} dx\right)^{\frac{2}{2^*-\varepsilon}}}$$

$$+ \frac{\int_\Omega (a - a_\varepsilon) v^2 \, dx}{\left(\int_\Omega \frac{|v|^{2^*-\varepsilon}}{|x|^s} dx\right)^{\frac{2}{2^*-\varepsilon}}}$$

for $\varepsilon > 0$ small. Here, we have used that $-\Delta + a_\varepsilon$ is coercive on Ω for $\varepsilon > 0$ small, which is a consequence of (16.8). Taking the infimum, using Hölder's inequality and the fact that a_ε converges uniformly to a, we get that

(16.16)
$$\mu_{s,a}(\Omega) \le (1 + o(1)) \mu^\varepsilon_{s,a_\varepsilon}(\Omega),$$

where $\lim_{\varepsilon \to 0} o(1) = 0$. Claim (2) of the proposition then follows from (16.15) and (16.16).

(3) We now prove that, when it is nonzero, the weak limit of the u_ε's is a minimizer for $\mu_s(a, \Omega)$. It is clear from Proposition 16.1.1, the uniform convergence of a_ε and (16.8) that

$$\|u_\varepsilon\|_{H^1_0(\Omega)} = O(1)$$

when $\varepsilon \to 0$. Then there exists $u_0 \in H^1_0(\Omega)$ such that, up to a subsequence, $u_\varepsilon \rightharpoonup u_0$ weakly in $H^1_0(\Omega)$ when $\varepsilon \to 0$. We assume that $u_0 \not\equiv 0$. It then follows from the definition of $\mu_{s,a}(\Omega)$ that

$$\frac{\int_\Omega (|\nabla u_0|^2 + a u_0^2) dx}{\left(\int_\Omega \frac{|u_0|^{2^*}}{|x|^s} dx\right)^{\frac{2}{2^*}}} \ge \mu_{s,a}(\Omega).$$

Testing the weak inequality $-\Delta u_\varepsilon + a_\varepsilon u_\varepsilon = \frac{u_\varepsilon^{2^*-1-\varepsilon}}{|x|^s}$ on u_0 and letting $\varepsilon \to 0$, we get that

$$\int_\Omega (|\nabla u_0|^2 + a u_0^2) dx = \int_\Omega \frac{|u_0|^{2^*}}{|x|^s} dx.$$

We then obtain that

$$\int_\Omega \frac{|u_0|^{2^*}}{|x|^s} dx \ge \mu_{s,a}(\Omega)^{\frac{2^*}{2^*-2}}.$$

Since $u_\varepsilon \rightharpoonup u_0$ when $\varepsilon \to 0$, we get from the definition of u_ε in Proposition 16.1.1 and Step 4.2 that

$$\int_\Omega \frac{|u_0|^{2^*}}{|x|^s} dx \le \liminf_{\varepsilon \to 0} \int_\Omega \frac{|u_\varepsilon|^{2^*-\varepsilon}}{|x|^s} dx = \mu_{s,a}(\Omega)^{\frac{2^*}{2^*-2}}.$$

Consequently, we get that

(16.17)
$$\int_\Omega (|\nabla u_0|^2 + a u_0^2) dx = \int_\Omega \frac{|u_0|^{2^*}}{|x|^s} dx = \mu_{s,a}(\Omega)^{\frac{2^*}{2^*-2}}.$$

Since $\mu_{s,a_\varepsilon}^\varepsilon(\Omega)^{\frac{2^\star-\varepsilon}{2^\star-2-\varepsilon}} = \int_\Omega(|\nabla u_\varepsilon|^2 + a_\varepsilon u_\varepsilon^2)\,dx$, we get from the definition of u_ε in Proposition 16.1.1 that

$$(16.18) \quad \mu_{s,a}(\Omega)^{\frac{2^\star}{2^\star-2}} = \int_\Omega (|\nabla u_0|^2 + au_0^2)\,dx + \int_\Omega |\nabla(u_\varepsilon - u_0)|^2\,dx + o(1)$$

with $\lim_{\varepsilon\to 0} o(1) = 0$. It follows from (16.17) and (16.18) that $\lim_{\varepsilon\to 0} u_\varepsilon = u_0$ in $H_0^1(\Omega)$. As easily checked, u_0 is a minimizer for $\mu_{s,a}(\Omega)$. □

16.2. Preliminary blow-up analysis

From now on, we let Ω be a smooth bounded domain of $\mathbb{R}^n$, $n \geq 3$, such that $0 \in \partial\Omega$. We let $s \in (0,2)$. For any $\varepsilon > 0$, we let $r_\varepsilon \in [0, 2^\star - 2)$ such that

$$(16.19) \quad \lim_{\varepsilon\to 0} r_\varepsilon = 0.$$

We consider $a \in C^1(\overline{\Omega})$ and a family $(a_\varepsilon)_{\varepsilon>0} \in C^1(\overline{\Omega})$ such that (16.8) holds and $a_\varepsilon \to a$ uniformly around Ω. For any $\varepsilon > 0$, we consider $u_\varepsilon \in H_0^1(\Omega) \cap C^2(\overline{\Omega}\setminus\{0\})$ a solution to the system

$$(16.20) \quad \begin{cases} -\Delta u_\varepsilon + a_\varepsilon u_\varepsilon = \frac{u_\varepsilon^{2^\star-1-r_\varepsilon}}{|x|^s} & \text{in } \mathcal{D}'(\Omega) \\ u_\varepsilon > 0 & \text{in } \Omega \end{cases}$$

for all $\varepsilon > 0$. We assume that u_ε is of minimal energy type, that is

$$(16.21) \quad \int_\Omega \frac{|u_\varepsilon|^{2^\star-r_\varepsilon}}{|x|^s}\,dx = \mu_s(\Omega)^{\frac{2^\star}{2^\star-2}} + o(1),$$

where $\lim_{\varepsilon\to 0} o(1) = 0$. We have already seen that

$$(16.22) \quad \|u_\varepsilon\|_{H_0^1(\Omega)} = O(1) \text{ when } \varepsilon \to 0.$$

We now assume that blow-up occurs, that is

$$(16.23) \quad u_\varepsilon \rightharpoonup 0 \text{ weakly in } H_0^1(\Omega) \text{ when } \varepsilon \to 0.$$

Such a family arises naturally when $u_0 \equiv 0$ in Proposition 16.1.2. In the remainder of this section, we shall describe precisely the behaviour of the u_ε's.

Proposition 16.5.1 below gives that $u_\varepsilon \in C^0(\overline{\Omega})$. We let $x_\varepsilon \in \Omega$ and $\mu_\varepsilon, k_\varepsilon > 0$ such that

$$(16.24) \quad \max_\Omega u_\varepsilon = u_\varepsilon(x_\varepsilon) = \mu_\varepsilon^{-\frac{n-2}{2}} \text{ and } k_\varepsilon := \mu_\varepsilon^{1-\frac{r_\varepsilon}{2^\star-2}}.$$

We let $\varphi : U \to V$ a local chart as in (16.13) with $x_0 = 0$, where U, V are open neighborhoods of 0. For any $\varepsilon > 0$ and any $x \in \frac{U}{k_\varepsilon} \cap \{x_1 \leq 0\}$, we define the maximum rescaling of u_ε as follows

$$(16.25) \quad v_\varepsilon(x) := \frac{u_\varepsilon \circ \varphi(k_\varepsilon x)}{u_\varepsilon(x_\varepsilon)},$$

where $x_\varepsilon, k_\varepsilon$ are as in (16.24). As easily checked, for any $\eta \in C_c^\infty(\mathbb{R}^n)$, we have that $\eta v_\varepsilon \in H_0^1(\mathbb{R}_-^n)$. In this section, we prove the following proposition:

PROPOSITION 16.2.1. *Let Ω be a smooth bounded domain of $\mathbb{R}^n$, $n \geq 3$ and $s \in (0,2)$. Consider $(r_\varepsilon)_{\varepsilon>0}$ such that $r_\varepsilon \in [0, 2^\star - 2)$ for all $\varepsilon > 0$ and let $(u_\varepsilon)_{\varepsilon>0} \in H_0^1(\Omega)$ be such that (16.8), (16.20), (16.21) and (16.23) hold. Also let v_ε be as in (16.25). Then,*

(1) There exists $v \in H_0^1(\mathbb{R}_-^n) \setminus \{0\}$ such that for any $\eta \in C_c^\infty(\mathbb{R}^n)$,

(16.26) $$\eta v_\varepsilon \rightharpoonup \eta v \text{ in } H_0^1(\mathbb{R}_-^n) \text{ when } \varepsilon \to 0.$$

(2) Moreover, v verifies that

(16.27) $$-\Delta v = \frac{v^{2^*-1}}{|x|^s} \text{ in } \mathbb{R}_-^n,$$

and

(16.28) $$\int_{\mathbb{R}_-^n} |\nabla v|^2 \, dx = \mu_s(\Omega)^{\frac{2^*}{2^*-2}} = \mu_s(\mathbb{R}_-^n)^{\frac{2^*}{2^*-2}}.$$

(3) In addition, there exists $\theta \in (0,1)$ such that $v \in C^{1,\theta}(\overline{\mathbb{R}_-^n})$ and

(16.29) $$v_\varepsilon \to v \text{ in } C_{loc}^{1,\theta}(\overline{\mathbb{R}_-^n}) \text{ when } \varepsilon \to 0.$$

(4) Moreover, we have that

(16.30) $$\lim_{\varepsilon \to 0} \mu_\varepsilon^{r_\varepsilon} = 1.$$

PROOF. We first claim that

(16.31) $$\mu_\varepsilon = o(1) \text{ when } \varepsilon \to 0.$$

We proceed by contradiction and assume that $\lim_{\varepsilon \to 0} \mu_\varepsilon \neq 0$. In this case, up to a subsequence, there exists $C > 0$ such that $u_\varepsilon(x) \leq C$ for all $x \in \Omega$ and all $\varepsilon > 0$. Since (16.23) hold, it follows from standard elliptic theory that $\lim_{\varepsilon \to 0} u_\varepsilon = 0$ in $C^0(\overline{\Omega})$. A contradiction to (16.21). This proves (16.31).

We now claim that

(16.32) $$|x_\varepsilon| = O(k_\varepsilon) \text{ when } \varepsilon \to 0.$$

We proceed again by contradiction and assume that

(16.33) $$\lim_{\varepsilon \to 0} \frac{|x_\varepsilon|}{k_\varepsilon} = +\infty.$$

For any $\varepsilon > 0$, we let

(16.34) $$\beta_\varepsilon = |x_\varepsilon|^{\frac{s}{2}} u_\varepsilon(x_\varepsilon)^{\frac{2+r_\varepsilon-2^*}{2}} = |x_\varepsilon|^{\frac{s}{2}} k_\varepsilon^{\frac{2-s}{2}}.$$

It follows from the definition (16.34) of β_ε and (16.33) that

(16.35) $$\lim_{\varepsilon \to 0} \beta_\varepsilon = 0, \; \lim_{\varepsilon \to 0} \frac{\beta_\varepsilon}{k_\varepsilon} = +\infty \text{ and } \lim_{\varepsilon \to 0} \frac{\beta_\varepsilon}{|x_\varepsilon|} = 0.$$

We now consider the following two cases:

Case (1): Assume that there exists $\rho > 0$ such that

(16.36) $$\frac{d(x_\varepsilon, \partial\Omega)}{\beta_\varepsilon} \geq 2\rho \text{ for all } \varepsilon > 0.$$

In this case we define $x \in B_{2\rho}(0)$ and $\varepsilon > 0$, the function $\bar{v}_\varepsilon(x) := \frac{u_\varepsilon(x_\varepsilon + \beta_\varepsilon x)}{u_\varepsilon(x_\varepsilon)}$. Note that this is well defined since $x_\varepsilon + \beta_\varepsilon x \in \Omega$ for all $x \in B_{2\rho}(0)$. As easily checked, we have that $\bar{v}_\varepsilon$ is a weak solution of

$$-\Delta \bar{v}_\varepsilon + \beta_\varepsilon^2 a_\varepsilon(x_\varepsilon + \beta_\varepsilon x) \bar{v}_\varepsilon = \frac{\bar{v}_\varepsilon^{2^*-1-r_\varepsilon}}{\left|\frac{x_\varepsilon}{|x_\varepsilon|} + \frac{\beta_\varepsilon}{|x_\varepsilon|} \cdot x\right|^s} \text{ in } B_{2\rho}(0).$$

Since (16.35) holds, we have that
$$-\Delta \bar{v}_\varepsilon + \beta_\varepsilon^2 a_\varepsilon(x_\varepsilon + \beta_\varepsilon x)\bar{v}_\varepsilon = (1+o(1))\bar{v}_\varepsilon^{2^*-1-r_\varepsilon}$$
weakly in $B_{2\rho}(0)$, where $\lim_{\varepsilon \to 0} o(1) = 0$ in $C^0_{loc}(B_{2\rho}(0))$. Since $0 \leq \bar{v}_\varepsilon(x) \leq \bar{v}_\varepsilon(0) = 1$ for all $x \in B_{2\rho}(0)$, it follows from standard elliptic theory that there exists $v \in C^1(B_{2\rho}(0))$ such that $v \geq 0$ and
$$\bar{v}_\varepsilon \to \bar{v} \text{ in } C^1_{loc}(B_{2\rho}(0)) \text{ when } \varepsilon \to 0.$$

In particular,
(16.37) $$\bar{v}(0) = \lim_{\varepsilon \to 0} \bar{v}_\varepsilon(0) = 1.$$

With a change of variables and the definition (16.34) of β_ε, we get that
$$\int_{\Omega \cap B_{\rho\beta_\varepsilon}(x_\varepsilon)} \frac{u_\varepsilon^{2^*-r_\varepsilon}}{|x|^s} dx = \frac{u_\varepsilon(x_\varepsilon)^{2^*-r_\varepsilon} \beta_\varepsilon^n}{|x_\varepsilon|^s} \int_{B_\rho(0)} \frac{\bar{v}_\varepsilon^{2^*-r_\varepsilon}}{\left|\frac{x_\varepsilon}{|x_\varepsilon|} + \frac{\beta_\varepsilon}{|x_\varepsilon|} \cdot x\right|^s} dx$$
$$\geq \left(\frac{\beta_\varepsilon}{k_\varepsilon}\right)^{n-2} \int_{B_\rho(0)} \frac{\bar{v}_\varepsilon^{2^*-r_\varepsilon}}{\left|\frac{x_\varepsilon}{|x_\varepsilon|} + \frac{\beta_\varepsilon}{|x_\varepsilon|} x\right|^s} dx.$$

Using (16.21), (16.35) and passing to the limit $\varepsilon \to 0$ (note that $\mu_\varepsilon^{-1} \geq 1$ for $\varepsilon > 0$ small), we get that
$$\int_{B_\rho(0)} \bar{v}^{2^*} dx = 0,$$
and then $\bar{v} \equiv 0$ in $B_\rho(0)$, which contradicts (16.37). It follows that (16.33) does not hold, and this proves that (16.32) holds in Case (1).

Case (2): We now assume that, up to a subsequence,
(16.38) $$\lim_{\varepsilon \to 0} \frac{d(x_\varepsilon, \partial\Omega)}{\beta_\varepsilon} = 0.$$

In this case, we have necessarily that $\lim_{\varepsilon \to 0} x_\varepsilon = x_0 \in \partial\Omega$. Since $x_0 \in \partial\Omega$, we let $\varphi : U \to V$ as in (16.13), where U, V are open neighborhoods of 0 and x_0 respectively. We let $\tilde{u}_\varepsilon = u_\varepsilon \circ \varphi$, which is defined on $U \cap \{x_1 \leq 0\}$. For any $i, j = 1, ..., n$, we let $g_{ij} = (\partial_i \varphi, \partial_j \varphi)$, where $(\cdot, \cdot)$ denotes the Euclidean scalar product on $\mathbb{R}^n$, and we consider g as a metric on $\mathbb{R}^n$. We let $\Delta_g = \text{div}_g(\nabla)$ the Laplace-Beltrami operator with respect to the metric g. In our basis, we have that
$$-\Delta_g = -g^{ij}\left(\partial_{ij} - \Gamma_{ij}^k \partial_k\right),$$
where $g^{ij} = (g^{-1})_{ij}$ are the coordinates of the inverse of the tensor g and the Γ_{ij}^k's are the Christoffel symbols of the metric g. As easily checked, we have that $\tilde{u}_\varepsilon$ is a weak solution of
$$-\Delta_g \tilde{u}_\varepsilon + a_\varepsilon \circ \varphi(x) \cdot \tilde{u}_\varepsilon = \frac{\tilde{u}_\varepsilon^{2^*-1-r_\varepsilon}}{|\varphi(x)|^s}$$
in $U \cap \{x_1 < 0\}$. We let $z_\varepsilon \in \partial\Omega$ such that
(16.39) $$|z_\varepsilon - x_\varepsilon| = d(x_\varepsilon, \partial\Omega),$$
and consider $\tilde{x}_\varepsilon, \tilde{z}_\varepsilon \in U$ such that
(16.40) $$\varphi(\tilde{x}_\varepsilon) = x_\varepsilon \text{ and } \varphi(\tilde{z}_\varepsilon) = z_\varepsilon.$$

It follows from the properties (16.13) of φ that

(16.41) $$\lim_{\varepsilon \to 0} \tilde{x}_\varepsilon = \lim_{\varepsilon \to 0} \tilde{z}_\varepsilon = 0, \ (\tilde{x}_\varepsilon)_1 < 0 \text{ and } (\tilde{z}_\varepsilon)_1 = 0.$$

Finally, we let
$$\tilde{v}_\varepsilon(x) := \frac{\tilde{u}_\varepsilon(\tilde{z}_\varepsilon + \beta_\varepsilon x)}{\tilde{u}_\varepsilon(\tilde{x}_\varepsilon)} \text{ for all } x \in \frac{U - \tilde{z}_\varepsilon}{\beta_\varepsilon} \cap \{x_1 < 0\}.$$

From (16.41), we get that $\tilde{v}_\varepsilon$ is defined on $B_R(0) \cap \{x_1 < 0\}$ for all $R > 0$, as long as ε is small enough. The function $\tilde{v}_\varepsilon$ is a weak solution of

$$-\Delta_{\tilde{g}_\varepsilon} \tilde{v}_\varepsilon + \beta_\varepsilon^2 a_\varepsilon \circ \varphi(\tilde{z}_\varepsilon + \beta_\varepsilon x) \tilde{v}_\varepsilon = \frac{\tilde{v}_\varepsilon^{2^*-1-r_\varepsilon}}{\left|\frac{\varphi(\tilde{z}_\varepsilon + \beta_\varepsilon x)}{|x_\varepsilon|}\right|^s}$$

in $B_R(0) \cap \{x_1 < 0\}$. In this expression, $\tilde{g}_\varepsilon = g(\tilde{z}_\varepsilon + \beta_\varepsilon x)$ and $-\Delta_{\tilde{g}_\varepsilon}$ is the Laplace-Beltrami operator with respect to the metric $\tilde{g}_\varepsilon$. From (16.38), (16.39) and (16.40), we get that

$$\varphi(\tilde{z}_\varepsilon + \beta_\varepsilon x) = x_\varepsilon + O_R(1) \beta_\varepsilon,$$

for all $x \in B_R(0) \cap \{x_1 \leq 0\}$ and all $\varepsilon > 0$, where there exists $C_R > 0$ such that $|O_R(1)| \leq C_R$ for all $x \in B_R(0) \cap \{x_1 < 0\}$. With (16.35), we then get that

$$\lim_{\varepsilon \to 0} \frac{|\varphi(\tilde{z}_\varepsilon + \beta_\varepsilon x)|}{|x_\varepsilon|} = 1$$

in $C^0(B_R(0) \cap \{x_1 \leq 0\})$. It then follows that $\tilde{v}_\varepsilon$ is a weak solution for

$$-\Delta_{\tilde{g}_\varepsilon} \tilde{v}_\varepsilon + \beta_\varepsilon^2 a_\varepsilon \circ \varphi(\tilde{z}_\varepsilon + \beta_\varepsilon x) \tilde{v}_\varepsilon = (1 + o(1)) \tilde{v}_\varepsilon^{2^*-1-r_\varepsilon}$$

in $B_R(0) \cap \{x_1 < 0\}$, where $\lim_{\varepsilon \to 0} o(1) = 0$ in $C^0(B_R(0) \cap \{x_1 \leq 0\})$. Since $\tilde{v}_\varepsilon$ vanishes on $B_R(0) \cap \{x_1 = 0\}$ (in the sense of the trace) and that $0 \leq \tilde{v}_\varepsilon \leq 1$, it follows from standard elliptic theory that there exists $\tilde{v} \in C^1(B_R(0) \cap \{x_1 \leq 0\})$ such that

$$\lim_{\varepsilon \to 0} \tilde{v}_\varepsilon = \tilde{v} \text{ in } C^0(B_{\frac{R}{2}}(0) \cap \{x_1 \leq 0\}).$$

In particular,

(16.42) $$\tilde{v} \equiv 0 \text{ on } B_{\frac{R}{2}}(0) \cap \{x_1 = 0\}.$$

Moreover, it follows from (16.39) and (16.40) that

$$\tilde{v}_\varepsilon \left(\frac{\tilde{x}_\varepsilon - \tilde{z}_\varepsilon}{\beta_\varepsilon}\right) = 1 \text{ and } \lim_{\varepsilon \to 0} \frac{\tilde{x}_\varepsilon - \tilde{z}_\varepsilon}{\beta_\varepsilon} = 0.$$

In particular, $\tilde{v}(0) = 1$, which contradicts (16.42). It follows that (16.33) does not hold, and this proves (16.32) in Case (2)

Note now that a consequence of (16.32) is that $\lim_{\varepsilon \to 0} x_\varepsilon = 0 \in \partial\Omega$. We therefore let $\varphi : U \to V$ as in (16.13) be a local chart of $\partial\Omega$ with $x_0 = 0$ (in other words, $\varphi(0) = 0$), where U, V are open neighborhoods of 0. We write $x_\varepsilon = \varphi(x_{1,\varepsilon}, z_\varepsilon)$, where $x_{1,\varepsilon} < 0$ and $z_\varepsilon \in \mathbb{R}^{n-1}$ are such that $(x_{1,\varepsilon}, z_\varepsilon) \in U$.

We now claim that when $\varepsilon \to 0$,

(16.43) $$d(x_\varepsilon, \partial\Omega) = (1 + o(1))|x_{1,\varepsilon}| = O(k_\varepsilon) \text{ and } z_\varepsilon = O(k_\varepsilon),$$

Indeed, with (16.32), we get that

(16.44) $$d(x_\varepsilon, \partial\Omega) \leq |x_\varepsilon| = O(k_\varepsilon)$$

when $\varepsilon \to 0$. We first remark that
$$d(x_\varepsilon, \partial\Omega) \le d(x_\varepsilon, \varphi(0, z_\varepsilon)) = |x_{1,\varepsilon}|.$$
We let $a_\varepsilon \in \text{span}(\vec{e}_2, ..., \vec{e}_n)$ and $Y_\varepsilon = \varphi(0, a_\varepsilon) \in \partial\Omega$ such that $d(x_\varepsilon, \partial\Omega) = |x_\varepsilon - Y_\varepsilon|$. Since $d(x_\varepsilon, \partial\Omega) \le |x_{1,\varepsilon}|$, we get that
$$z_\varepsilon - a_\varepsilon = O(|x_{1,\varepsilon}|),$$
when $\varepsilon \to 0$. Since $\nabla\varphi_0(0) = 0$ (where φ_0 is as in (16.13)), we get that
$$\varphi_0(z_\varepsilon) = \varphi_0(a_\varepsilon) + o(|z_\varepsilon - a_\varepsilon|) = \varphi_0(a_\varepsilon) + o(|x_{1,\varepsilon}|)$$
when $\varepsilon \to 0$. Moreover,
$$\begin{aligned} d(x_\varepsilon, \partial\Omega) &= |x_\varepsilon - Y_\varepsilon| \\ &= |(x_{1,\varepsilon} + \varphi_0(z_\varepsilon) - \varphi_0(a_\varepsilon), z_\varepsilon - a_\varepsilon)| \\ &= |(x_{1,\varepsilon} + o(|x_{1,\varepsilon}|), z_\varepsilon - a_\varepsilon)| \le |x_{1,\varepsilon}| \end{aligned}$$
when $\varepsilon \to 0$. It then follows that $z_\varepsilon - a_\varepsilon = o(|x_{1,\varepsilon}|)$ and $d(x_\varepsilon, \partial\Omega) = (1+o(1))|x_{1,\varepsilon}|$ when $\varepsilon \to 0$. This last result, coupled with (16.32) and (16.44) prove (16.43).

We now let
$$(16.45) \qquad \lambda_\varepsilon := -\frac{x_{1,\varepsilon}}{k_\varepsilon} > 0 \text{ and } \theta_\varepsilon := \frac{z_\varepsilon}{k_\varepsilon}.$$
It follows from (16.43) that there exist $\lambda_0 \ge 0$ and $\theta_0 \in \mathbb{R}^{n-1}$ such that
$$(16.46) \qquad \lim_{\varepsilon \to 0} \lambda_\varepsilon = \lambda_0 \text{ and } \lim_{\varepsilon \to 0} \theta_\varepsilon = \theta_0.$$
For any $\varepsilon > 0$ and any $x \in \frac{U}{k_\varepsilon} \cap \{x_1 \le 0\}$, we let (as in (16.25))
$$(16.47) \qquad v_\varepsilon(x) := \frac{u_\varepsilon \circ \varphi(k_\varepsilon x)}{u_\varepsilon(x_\varepsilon)},$$
where $\varphi : U \to V$ is defined in (16.13) (with $x_0 = 0$) and $k_\varepsilon, x_\varepsilon$ are as in (16.24). As easily checked, for any $\eta \in C_c^\infty(\mathbb{R}^n)$, we have that $\eta v_\varepsilon \in H_0^1(\mathbb{R}^n_-)$ for all $\varepsilon > 0$.

We now proceed to prove Proposition 16.2.1. First, we claim that for any $\eta \in C_c^\infty(\mathbb{R}^n)$, there exists $v_\eta \in H_0^1(\mathbb{R}^n_-)$ such that, up to a subsequence,
$$\eta v_\varepsilon \rightharpoonup v_\eta \text{ weakly in } H_0^1(\mathbb{R}^n_-).$$
Indeed, as easily checked, we have that
$$\nabla(\eta v_\varepsilon)(x) = v_\varepsilon \nabla\eta + \frac{k_\varepsilon}{u_\varepsilon(x_\varepsilon)} \eta \cdot D_{(k_\varepsilon x)}\varphi[(\nabla u_\varepsilon)(\varphi(k_\varepsilon x))],$$
for all $\varepsilon > 0$ and all $x \in \mathbb{R}^n_-$. In this expression, $D_x\varphi$ is the differential of the function φ at x. It is standard that for any $\alpha > 0$, there exists $C_\alpha > 0$ such that
$$(x+y)^2 \le C_\alpha x^2 + (1+\alpha) \cdot y^2 \text{ for all } x, y > 0.$$
With this inequality, we get that
$$\int_{\mathbb{R}^n_-} |\nabla(\eta v_\varepsilon)|^2 \, dx \le C_\alpha \int_{\mathbb{R}^n_-} |\nabla\eta|^2 v_\varepsilon^2 \, dx$$
$$+ (1+\alpha) \int_{\mathbb{R}^n_-} \eta^2 \frac{k_\varepsilon^2}{u_\varepsilon(x_\varepsilon)^2} \cdot |D_{(k_\varepsilon x)}\varphi[(\nabla u_\varepsilon)(\varphi(k_\varepsilon x))]|^2 \, dx.$$

Since $D_0\varphi = Id_{\mathbb{R}^n}$, we get that with Hölder's inequality and a change of variables that

$$\int_{\mathbb{R}^n_-} |\nabla(\eta v_\varepsilon)|^2 \, dx \le C_\alpha \int_{\mathbb{R}^n_-} |\nabla \eta|^2 v_\varepsilon^2 \, dx$$

$$+ (1+\alpha) \cdot (1+O(k_\varepsilon)) \int_{\mathbb{R}^n_-} \eta^2 \frac{k_\varepsilon^2}{u_\varepsilon(x_\varepsilon)^2} \cdot |\nabla u_\varepsilon|^2(\varphi(k_\varepsilon x)) \, dx$$

$$\le c_\alpha \|\nabla \eta\|_n^2 \cdot \|v_\varepsilon\|^2_{L^{\frac{2n}{n-2}}(\operatorname{Supp}\nabla\eta)}$$

(16.48) $$+ (1+\alpha) \cdot (1+O(k_\varepsilon)) \cdot \mu_\varepsilon^{\frac{r_\varepsilon(n-2)}{2^*-2}} \int_\Omega |\nabla u_\varepsilon|^2 \, dx.$$

With another change of variables, we get that

$$\int_{\mathbb{R}^n_-} |\nabla(\eta v_\varepsilon)|^2 \, dx \le C_\alpha \cdot \mu_\varepsilon^{\frac{(n-2)r_\varepsilon}{2^*-2}} \|\nabla\eta\|_n^2 \cdot \|u_\varepsilon\|^2_{L^{\frac{2n}{n-2}}(\Omega)}$$

(16.49) $$+ (1+\alpha) \cdot (1+O(k_\varepsilon)) \cdot \mu_\varepsilon^{\frac{r_\varepsilon(n-2)}{2^*-2}} \int_\Omega |\nabla u_\varepsilon|^2 \, dx.$$

With (16.22), Sobolev's inequality and since $\mu_\varepsilon^{r_\varepsilon} \le 1$ for all $\varepsilon > 0$ small enough, we get with (16.49) that

$$\|\eta v_\varepsilon\|_{H_0^1(\mathbb{R}^n_-)} = O(1) \text{ when } \varepsilon \to 0.$$

It then follows that there exists $v_\eta \in H_0^1(\mathbb{R}^n_-)$ such that, up to a subsequence, $\eta v_\varepsilon \rightharpoonup v_\eta$ weakly in $H_0^1(\mathbb{R}^n_-)$ when $\varepsilon \to 0$.

We now show item (1) of Proposition 16.2.1, that is there exists $v \in H_0^1(\mathbb{R}^n_-)$ such that for any $\eta \in C_c^\infty(\mathbb{R}^n)$, we have, up to a subsequence,

$$\eta v_\varepsilon \rightharpoonup \eta v \text{ weakly in } H_0^1(\mathbb{R}^n_-) \text{ as } \varepsilon \to 0.$$

Indeed, we let $\eta_1 \in C_c^\infty(\mathbb{R}^n)$ such that $\eta_1 \equiv 1$ in $B_1(0)$ and $\eta_1 \equiv 0$ in $\mathbb{R}^n \setminus B_2(0)$. For any $R > 0$, we let $\eta_R(x) = \eta_1(\frac{x}{R})$ for all $x \in \mathbb{R}^n$. With a diagonal argument, we can assume that, up to a subsequence, for any $R > 0$, there exists $v_R \in H_0^1(\mathbb{R}^n_-)$ such that $\eta_R v_\varepsilon \rightharpoonup v_R$ weakly in $H_0^1(\mathbb{R}^n_-)$ when $\varepsilon \to 0$, and that $(\eta_R v_\varepsilon)(x) \to v_R(x)$ when $\varepsilon \to 0$ for a.e. $x \in \mathbb{R}^n_-$. Letting $\varepsilon \to 0$ in (16.49), with (16.22), Sobolev's inequality and since $\mu_\varepsilon^{r_\varepsilon} \le 1$ for all $\varepsilon > 0$ small enough, we get that there exists a constant $C > 0$ independent of R such that

$$\int_{\mathbb{R}^n_-} |\nabla v_R|^2 \, dx \le C_\alpha \|\nabla \eta_R\|_n^2 \cdot C + (1+\alpha) \cdot C \text{ for all } R > 0.$$

Since $\|\nabla \eta_R\|_n^2 = \|\nabla \eta_1\|_n^2$ for all $R > 0$, we get that there exists $C > 0$ independent of R such that $\int_{\mathbb{R}^n_-} |\nabla v_R|^2 \, dx \le C$, for all $R > 0$. It then follows that there exists $v \in H_0^1(\mathbb{R}^n_-)$ such that $v_R \rightharpoonup v$ weakly in $H_0^1(\mathbb{R}^n_-)$ when $R \to +\infty$ and $v_R(x) \to v(x)$ when $R \to +\infty$ for a.e. $x \in \mathbb{R}^n_-$. As easily checked, we then obtain that $v_\eta = \eta v$.

We now show that

$$v \not\equiv 0.$$

We proceed as in Case (2) of the proof of (16.32). We let $(\tilde{g}_\varepsilon)_{ij} = (\partial_i \varphi(k_\varepsilon x), \partial_j \varphi(k_\varepsilon x))$, where $(\cdot, \cdot)$ denotes the Euclidean scalar product on $\mathbb{R}^n$ and consider $\tilde{g}_\varepsilon$ as a metric on $\mathbb{R}^n$. We let

$$-\Delta_{\tilde{g}_\varepsilon} = -\tilde{g}_\varepsilon^{ij} \left(\partial_{ij} - \Gamma_{ij}^k(\tilde{g}_\varepsilon) \partial_k \right),$$

where $\tilde{g}_\varepsilon^{ij} := (\tilde{g}_\varepsilon^{-1})_{ij}$ are the coordinates of the inverse of the tensor $\tilde{g}_\varepsilon$ and the $\Gamma_{ij}^k(\tilde{g}_\varepsilon)$'s are the Christoffel symbols of the metric $\tilde{g}_\varepsilon$. With a change of variable and the definition (16.47), equation (16.20) can be rewritten as

$$(16.50) \quad -\Delta_{\tilde{g}_\varepsilon}(\eta_R v_\varepsilon) + k_\varepsilon^2 a_\varepsilon \circ \varphi(k_\varepsilon x) \eta_R v_\varepsilon = \frac{(\eta_R v_\varepsilon)^{2^*-1-r_\varepsilon}}{\left|\frac{\varphi(k_\varepsilon x)}{k_\varepsilon}\right|^s} \quad \text{in } B_R(0) \cap \{x_1 < 0\},$$

for all $\varepsilon > 0$. With (16.24), (16.47) and since $s \in (0,2)$, we get that $0 \le v_\varepsilon \le 1$ and that there exists $p > \frac{n}{2}$ such that the RHS of (16.50) is bounded in L^p when $\varepsilon \to 0$. It follows from standard elliptic theory that there exists $\alpha > 0$ such that

$$\|\eta_R v_\varepsilon\|_{C^{0,\alpha}(B_{R/2}(0) \cap \{x_1 \le 0\})} = O(1) \text{ when } \varepsilon \to 0.$$

It then follows from Ascoli's theorem that for any $\alpha' \in (0, \alpha)$, $v_R \in C^{0,\alpha'}(B_{R/2}(0) \cap \{x_1 \le 0\})$ and that, up to a subsequence,

$$(16.51) \quad \lim_{\varepsilon \to 0} \eta_R v_\varepsilon = v_R \text{ in } C^{0,\alpha'}(B_{R/4}(0) \cap \{x_1 \le 0\}).$$

From (16.47) and (16.45), we have that $(\eta_R v_\varepsilon)(-\lambda_\varepsilon, \theta_\varepsilon) = 1$ for all $\varepsilon > 0$ and $R > 0$ large enough. Passing to the limit $\varepsilon \to 0$ in this last equality, and using (16.51) and (16.46), we get that $v_R(-\lambda_0, \theta_0, 0) = 1$ for $R > 0$ large enough. With the same type of arguments, we get that $v \in C^{0,\alpha}(\{x_1 \le 0\})$ and that $\lim_{R \to +\infty} v_R = v$ in $C_{loc}^{0,\alpha}(\{x_1 \le 0\})$. Since $\eta_R v = v_R$, we get that $v(-\lambda_0, \theta_0) = 1$. In particular, $v \not\equiv 0$ and $\lambda_0 > 0$.

We now show claim (3) of Proposition 16.2.1, that is there exists $\theta \in (0,1)$ such that $v \in C^{1,\theta}(\overline{\mathbb{R}_-^n})$ and $v_\varepsilon \to v$ in $C_{loc}^{1,\theta}(\overline{\mathbb{R}_-^n})$ when $\varepsilon \to 0$. Indeed, it follows from the previous step that there exists $\alpha > 0$ such that for all $R > 0$, there exists $C(R) > 0$ such that

$$\|v_\varepsilon\|_{C^{0,\alpha}(B_R(0) \cap \{x_1 \le 0\})} \le C(R).$$

Let now

$$\alpha_0 := \sup\{\alpha \in (0,1)/ \forall R > 0, \exists C(R) > 0 \text{ s.t. } \|v_\varepsilon\|_{C^{0,\alpha}(B_R(0) \cap \{x_1 \le 0\})} \le C(R)\}.$$

For a given $\alpha \in (0, \alpha_0)$ and $R > 0$, we associate to each $\tilde{R} > R$, a constant $C(\tilde{R}) > 0$ such that

$$(16.52) \quad \|v_\varepsilon\|_{C^{0,\alpha}(B_{\tilde{R}}(0) \cap \{x_1 \le 0\})} \le C(\tilde{R}).$$

Since $v_\varepsilon \equiv 0$ on $\partial \mathbb{R}_-^n$, we get with (16.52) that

$$(16.53) \quad |v_\varepsilon(x)| = |v_\varepsilon(x) - v_\varepsilon(x - (x_1, 0))| \le C(\tilde{R})|x_1|^\alpha$$

for all $B_{\tilde{R}}(0) \cap \{x_1 < 0\}$ and all $\varepsilon > 0$. It then follows from the properties of φ (see (16.13) with $x_0 = 0$) that

$$0 \le f_\varepsilon(x) := \frac{(\eta v_\varepsilon)^{2^*-1-r_\varepsilon}}{\left|\frac{\varphi(k_\varepsilon x)}{k_\varepsilon}\right|^s} \le \frac{C}{|x|^{s-(2^*-1-r_\varepsilon)\alpha}}$$

for all $\varepsilon > 0$ and all $x \in B_{\tilde{R}}(0) \cap \{x_1 < 0\}$. With the properties (16.13), we get that for any $\tilde{R} > 0$ and any $p > 1$, we have that

$$\int_{B_{\tilde{R}}(0) \cap \{x_1 < 0\}} \frac{dx}{\left|\frac{\varphi(k_\varepsilon x)}{k_\varepsilon}\right|^p} \le C \int_{B_{\tilde{R}}(0)} \frac{dx}{|x|^p} \text{ for all } \varepsilon > 0.$$

(Note that the RHS can be infinite.) Using the same strategy as in the proof of Proposition 16.5.1, we get that there exists $\theta \in (0,1)$ such that $v \in C^{1,\theta}(\overline{\mathbb{R}^n_-})$ and $v_\varepsilon \to v$ in $C^{1,\theta}_{loc}(\overline{\mathbb{R}^n_-})$, when $\varepsilon \to 0$. We omit the proof and refer to the proof of Proposition 16.5.1 below for the details.

We now establish claim (2) of Proposition 16.2.1, i.e., that v verifies the equation

$$(16.54) \qquad -\Delta v = \frac{v^{2^*-1}}{|x|^s} \text{ in } \mathbb{R}^n_-,$$

and that $\int_{\mathbb{R}^n_-} |\nabla v|^2\, dx = \mu_s(a,\Omega)^{\frac{2^*}{2^*-2}} = \mu_s(\mathbb{R}^n_-)^{\frac{2^*}{2^*-2}}$. Indeed, passing first to the weak limit as $\varepsilon \to 0$ and then to the weak limit as $R \to +\infty$ in (16.50), we get that $-\Delta v = \frac{v^{2^*-1}}{|x|^s}$ in $\mathbb{R}^n_-$. Testing this equality with $v \in H^1_0(\mathbb{R}^n_-)\setminus\{0\}$ and using the optimal Hardy-Sobolev inequality (16.1), we get that

$$(16.55) \qquad \left(\int_{\mathbb{R}^n_-} |\nabla v|^2\, dx\right)^{\frac{2^*-2}{2^*}} = \frac{\int_{\mathbb{R}^n_-} |\nabla v|^2\, dx}{\left(\int_{\mathbb{R}^n_-} \frac{v^{2^*}}{|x|^s}\, dx\right)^{\frac{2}{2^*}}} \geq \mu_s(\mathbb{R}^n_-).$$

We then obtain that

$$(16.56) \qquad \int_{\mathbb{R}^n_-} |\nabla v|^2\, dx \geq \mu_s(\mathbb{R}^n_-)^{\frac{2^*}{2^*-2}}.$$

Since $0 \leq v_\varepsilon \leq 1$, it follows from Lebesgue's theorem that when $\varepsilon \to 0$, $v_\varepsilon \to v$ strongly in $L^{\frac{2n}{n-2}}_{loc}(\mathbb{R}^n_- \cap \{x_1 \leq 0\})$. Passing to the weak limit in (16.48) and using (16.21), we get that

$$(16.57) \qquad \begin{aligned}\int_{\mathbb{R}^n_-} |\nabla v_R|^2\, dx &\leq C_\alpha \|\nabla \eta_R\|_n^2 \cdot \|v\|^2_{L^{\frac{2n}{n-2}}(B_{2R}(0)\setminus B_R(0))} \\ &\quad + (1+\alpha)\cdot(\lim_{\varepsilon\to 0}\mu_\varepsilon^{\frac{r_\varepsilon(n-2)}{2^*-2}})\mu_{s,a}(\Omega)^{\frac{2^*}{2^*-2}}\end{aligned}$$

for all $R > 0$. Since $v \in H^1_0(\mathbb{R}^n_-)$, it follows from Sobolev's theorem that $v \in L^{\frac{2n}{n-2}}(\mathbb{R}^n_-)$. Since $\|\nabla \eta_R\|_n^2 = \|\nabla \eta_1\|_n^2$ is independent of $R > 0$ and $v \in L^{\frac{2n}{n-2}}(\mathbb{R}^n_-)$, letting $R \to +\infty$ in (16.57), we get that

$$(16.58) \qquad \int_{\mathbb{R}^n_-} |\nabla v|^2\, dx \leq (1+\alpha)\cdot(\lim_{\varepsilon\to 0}\mu_\varepsilon^{\frac{r_\varepsilon(n-2)}{2^*-2}})\mu_{s,a}(\Omega)^{\frac{2^*}{2^*-2}}.$$

Since $\alpha > 0$ is arbitrary and $\mu_\varepsilon \leq 1$, we get from (16.56), (16.58), Proposition 16.1.2 and (16.55) that

$$\int_{\mathbb{R}^n_-} |\nabla v|^2\, dx = \mu_s(a,\Omega)^{\frac{2^*}{2^*-2}} = \mu_s(\mathbb{R}^n_-)^{\frac{2^*}{2^*-2}},$$

and that

$$\lim_{\varepsilon \to 0} \mu_\varepsilon(\Omega)^{r_\varepsilon} = 1.$$

This finishes the proof of Proposition 16.2.1. □

We shall later need the following lemma.

LEMMA 16.2.1. *Under the hypothesis of Proposition 16.2.1, we have that*

$$\lim_{R \to +\infty} \lim_{\varepsilon \to 0} \int_{\Omega \setminus B_{Rk_\varepsilon}(0)} \frac{u_\varepsilon^{2^\star - r_\varepsilon}}{|x|^s} \, dx = 0. \tag{16.59}$$

PROOF. Since $D_0\varphi = Id_{\mathbb{R}^n}$ and $\varphi(0) = 0$, we have that $\varphi\left(B_{\frac{R}{2}k_\varepsilon}(0)\right) \subset B_{Rk_\varepsilon}(0)$, for all $R > 0$ and $\varepsilon > 0$ small enough. With a change of variable and using (16.21), we get that

$$\int_{\Omega \setminus B_{Rk_\varepsilon}(0)} \frac{u_\varepsilon^{2^\star - r_\varepsilon}}{|x|^s} \, dx \leq \int_{\Omega \setminus \varphi\left(B_{\frac{R}{2}k_\varepsilon}(0)\right)} \frac{u_\varepsilon^{2^\star - r_\varepsilon}}{|x|^s} \, dx$$

$$\leq \int_{\Omega} \frac{u_\varepsilon^{2^\star - r_\varepsilon}}{|x|^s} \, dx - \int_{\varphi\left(B_{\frac{R}{2}k_\varepsilon}(0)\right)} \frac{u_\varepsilon^{2^\star - r_\varepsilon}}{|x|^s} \, dx$$

$$\leq \mu_s(a, \Omega)^{\frac{2^\star}{2^\star - 2}} + o(1)$$

$$- \mu_\varepsilon(\Omega)^{-r_\varepsilon \frac{(n-2)^2}{2(2-s)}} (1 + o(1)) \int_{B_{\frac{R}{2}}(0)} \frac{v_\varepsilon^{2^\star - r_\varepsilon}}{|x|^s} \, dx.$$

Letting $\varepsilon \to 0$ and then $R \to +\infty$, we get with (16.30) and Proposition 16.2.1 that

$$\lim_{R \to +\infty} \lim_{\varepsilon \to 0} \int_{\Omega \setminus B_{Rk_\varepsilon}(0)} \frac{u_\varepsilon^{2^\star - r_\varepsilon}}{|x|^s} \, dx \leq \mu_s(a, \Omega)^{\frac{2^\star}{2^\star - 2}} - \lim_{R \to +\infty} \int_{B_{\frac{R}{2}}(0)} \frac{v^{2^\star}}{|x|^s} \, dx$$

$$\leq \mu_s(a, \Omega)^{\frac{2^\star}{2^\star - 2}} - \int_{\mathbb{R}^n_-} \frac{v^{2^\star}}{|x|^s} \, dx = 0.$$

This last inequality yields (16.59). □

16.3. Refined blow-up analysis and strong pointwise estimates

The objective of this section is the proof of the following strong pointwise estimate.

PROPOSITION 16.3.1. *Let Ω be a smooth bounded domain of $\mathbb{R}^n$, $n \geq 3$ and $s \in (0,2)$. We let $(r_\varepsilon)_{\varepsilon > 0}$ such that $r_\varepsilon \in [0, 2^\star - 2)$ for all $\varepsilon > 0$ and (16.19) holds. We consider $(u_\varepsilon)_{\varepsilon > 0} \in H_0^1(\Omega)$ such that (16.8), (16.20), (16.21) and (16.23) hold. We let μ_ε as in (16.24). Then, there exists $C > 0$ such that for all $\varepsilon > 0$ and all $x \in \Omega$, we have*

$$u_\varepsilon(x) \leq C \cdot \left(\frac{\mu_\varepsilon}{\mu_\varepsilon^2 + |x|^2}\right)^{\frac{n-2}{2}}. \tag{16.60}$$

PROOF. We first claim that there exists $C > 0$ such that for all $\varepsilon > 0$ and all $x \in \Omega$,

$$|x|^{\frac{n-2}{2}} u_\varepsilon(x)^{1 - \frac{r_\varepsilon}{2^\star - 2}} \leq C. \tag{16.61}$$

Indeed by contradiction, assume there are $y_\varepsilon \in \Omega$ such that

$$|y_\varepsilon|^{\frac{n-2}{2}} u_\varepsilon(y_\varepsilon)^{1 - \frac{r_\varepsilon}{2^\star - 2}} = \sup_{x \in \Omega} |x|^{\frac{n-2}{2}} u_\varepsilon(x)^{1 - \frac{r_\varepsilon}{2^\star - 2}} \to +\infty \tag{16.62}$$

when $\varepsilon \to 0$. We let

$$\nu_\varepsilon := u_\varepsilon(y_\varepsilon)^{-\frac{2}{n-2}} \text{ and } \ell_\varepsilon := \nu_\varepsilon^{1 - \frac{r_\varepsilon}{2^\star - 2}} \tag{16.63}$$

for all $\varepsilon > 0$. It follows from (16.62) and (16.63) that

(16.64) $$\lim_{\varepsilon \to 0} \nu_\varepsilon = 0 \quad \text{and} \quad \lim_{\varepsilon \to 0} \frac{|y_\varepsilon|}{\ell_\varepsilon} = +\infty.$$

From (16.24) and (16.30) we have that

(16.65) $$\lim_{\varepsilon \to 0} \nu_\varepsilon^{r_\varepsilon} = 1.$$

For all $\varepsilon > 0$, set

(16.66) $$\gamma_\varepsilon^2 := |y_\varepsilon|^s |u_\varepsilon(y_\varepsilon)|^{-(2^*-2-p_\varepsilon)}.$$

It follows from (16.64) that

(16.67) $$\lim_{\varepsilon \to 0} \frac{\gamma_\varepsilon}{|y_\varepsilon|} = 0.$$

We again consider two cases:

Case (1): Assume that, up to a subsequence, there exists $\rho > 0$ such that for all $\varepsilon > 0$,

(16.68) $$\frac{d(y_\varepsilon, \partial\Omega)}{\gamma_\varepsilon} \geq 3\rho.$$

For any $x \in B_{2\rho}(0)$ and any $\varepsilon > 0$, we let

(16.69) $$w_\varepsilon(x) := \nu_\varepsilon^{\frac{n-2}{2}} u_\varepsilon(y_\varepsilon + \gamma_\varepsilon x).$$

Note that w_ε is well defined thanks to (16.68). From (16.62) and (16.66), we get that

$$\left| \frac{y_\varepsilon}{|y_\varepsilon|} + \frac{\gamma_\varepsilon}{|y_\varepsilon|} x \right|^{\frac{n-2}{2}} w_\varepsilon(x)^{1 - \frac{r_\varepsilon}{2^*-2}} \leq 1.$$

In particular, with (16.64), there exists $C_0 > 0$ such that for all $x \in B_{2\rho}(0)$ and all $\varepsilon > 0$,

(16.70) $$0 \leq w_\varepsilon(x) \leq C_0.$$

From (16.20), we get that for all $x \in B_{2\rho}(0)$ and all $\varepsilon > 0$,

$$-\Delta w_\varepsilon + \gamma_\varepsilon^2 a_\varepsilon(y_\varepsilon + \gamma_\varepsilon x) w_\varepsilon = \frac{w_\varepsilon^{2^*-1-r_\varepsilon}}{\left| \frac{y_\varepsilon}{|y_\varepsilon|} + \frac{\gamma_\varepsilon}{|y_\varepsilon|} x \right|^s}.$$

Since (16.64) and (16.70) hold, it follows from standard elliptic theory that there exists $w \in C^1(B_{2\rho}(0))$ such that $w \geq 0$ and

(16.71) $$\lim_{\varepsilon \to 0} w_\varepsilon = w \text{ in } C^1_{loc}(B_{2\rho}(0)).$$

It follows from (16.69) that $w(0) = 1$. With a change of variable, we get that

(16.72) $$\int_{B_{\rho\gamma_\varepsilon}(y_\varepsilon)} \frac{u_\varepsilon(x)^{2^*-r_\varepsilon}}{|x|^s} dx = \frac{\gamma_\varepsilon^n u_\varepsilon(y_\varepsilon)^{2^*-r_\varepsilon}}{|y_\varepsilon|^s} \int_{B_\rho(0)} \frac{w_\varepsilon(x)^{2^*-r_\varepsilon}}{\left| \frac{y_\varepsilon}{|y_\varepsilon|} + \frac{\gamma_\varepsilon}{|y_\varepsilon|} \cdot x \right|^s} dx.$$

With (16.66), (16.65), (16.64) and (16.63), we then get that

$$\frac{\gamma_\varepsilon^n u_\varepsilon(y_\varepsilon)^{2^*-r_\varepsilon}}{|y_\varepsilon|^s} = (1+o(1)) \cdot \left(\frac{|y_\varepsilon|}{\ell_\varepsilon} \right)^{\frac{s(n-2)}{2}} \to +\infty \quad \text{as } \varepsilon \to 0.$$

With (16.72), (16.71) and (16.21), we get that $\int_{B_\rho(0)} w^{2^*} dx = 0$, and therefore $w \equiv 0$, which is a contradiction since $w(0) = 1$. This ends Case (1)

Case (2): Now assume that

$$\lim_{\varepsilon \to 0} \frac{d(y_\varepsilon, \partial\Omega)}{\gamma_\varepsilon} = 0. \tag{16.73}$$

It then follows that there exists $y_0 \in \partial\Omega$ such that $\lim_{\varepsilon \to 0} y_\varepsilon = y_0$. Since the latter is on $\partial\Omega$, which is smooth, we let $\varphi : U \to V$ as in (16.13) with $x_0 = y_0$ and where U, V are open neighborhoods of 0 and y_0 respectively. We let $\tilde{u}_\varepsilon = u_\varepsilon \circ \varphi$, which is defined on $U \cap \{x_1 \leq 0\}$. For any $i, j = 1, ..., n$, we let $g_{ij} = (\partial_i \varphi, \partial_j \varphi)$, where $(\cdot, \cdot)$ denotes the Euclidean scalar product on $\mathbb{R}^n$, and we consider g as a metric on $\mathbb{R}^n$. We let $\Delta_g = div_g(\nabla)$ the Laplace-Beltrami operator with respect to the metric g. We write

$$-\Delta_g = -g^{ij}\left(\partial_{ij} - \Gamma^k_{ij}\partial_k\right),$$

where $g^{ij} = (g^{-1})_{ij}$ are the coordinates of the inverse of the tensor g and the Γ^k_{ij}'s are the Christoffel symbols of the metric g. As easily checked, we have that $\tilde{u}_\varepsilon$ is a weak solution for

$$-\Delta_g \tilde{u}_\varepsilon + a_\varepsilon \circ \varphi(x) \cdot \tilde{u}_\varepsilon = \frac{\tilde{u}_\varepsilon^{2^*-1-r_\varepsilon}}{|\varphi(x)|^s}$$

in $U \cap \{x_1 < 0\}$. We let $z_\varepsilon \in \partial\Omega$ such that

$$|z_\varepsilon - y_\varepsilon| = d(y_\varepsilon, \partial\Omega), \tag{16.74}$$

and $\tilde{y}_\varepsilon, \tilde{z}_\varepsilon \in U$ such that

$$\varphi(\tilde{y}_\varepsilon) = y_\varepsilon \text{ and } \varphi(\tilde{z}_\varepsilon) = z_\varepsilon. \tag{16.75}$$

It follows from the properties of φ that

$$\lim_{\varepsilon \to 0} \tilde{y}_\varepsilon = \lim_{\varepsilon \to 0} \tilde{z}_\varepsilon = 0, \ (\tilde{y}_\varepsilon)_1 < 0 \text{ and } (\tilde{z}_\varepsilon)_1 = 0. \tag{16.76}$$

Finally, we let $\tilde{w}_\varepsilon(x) := \frac{\tilde{u}_\varepsilon(\tilde{z}_\varepsilon + \gamma_\varepsilon x)}{\tilde{u}_\varepsilon(\tilde{y}_\varepsilon)}$ for all $x \in \frac{U-\tilde{z}_\varepsilon}{\gamma_\varepsilon} \cap \{x_1 < 0\}$. From (16.76), we get that $\tilde{w}_\varepsilon$ is defined on $B_R(0) \cap \{x_1 < 0\}$ for all $R > 0$, as soon as ε is small enough. The function $\tilde{w}_\varepsilon$ verifies

$$-\Delta_{\tilde{g}_\varepsilon} \tilde{w}_\varepsilon + \gamma_\varepsilon^2 a_\varepsilon \circ \varphi(\tilde{z}_\varepsilon + \gamma_\varepsilon x)\tilde{w}_\varepsilon = \frac{\tilde{w}_\varepsilon^{2^*-1-r_\varepsilon}}{\left|\frac{\varphi(\tilde{z}_\varepsilon + \gamma_\varepsilon x)}{|y_\varepsilon|}\right|^s}$$

weakly in $B_R(0) \cap \{x_1 < 0\}$. In this expression, $\tilde{g}_\varepsilon = g(\tilde{z}_\varepsilon + \gamma_\varepsilon x)$ and $-\Delta_{\tilde{g}_\varepsilon}$ is the Laplace-Beltrami operator with respect to the metric $\tilde{g}_\varepsilon$. From (16.73), (16.74) and (16.75), we get that $\varphi(\tilde{z}_\varepsilon + \gamma_\varepsilon x) = y_\varepsilon + O_R(1)\gamma_\varepsilon$, for all $x \in B_R(0) \cap \{x_1 \leq 0\}$ and all $\varepsilon > 0$, where there exists $C_R > 0$ such that $|O_R(1)| \leq C_R$ for all $x \in B_R(0) \cap \{x_1 < 0\}$. From (16.67), we then get that $\lim_{\varepsilon \to 0} \frac{|\varphi(\tilde{z}_\varepsilon + \gamma_\varepsilon x)|}{|y_\varepsilon|} = 1$ in $C^0(B_R(0) \cap \{x_1 \leq 0\})$. It then follows that

$$-\Delta_{\tilde{g}_\varepsilon} \tilde{w}_\varepsilon + \gamma_\varepsilon^2 a_\varepsilon \circ \varphi(\tilde{z}_\varepsilon + \gamma_\varepsilon x)\tilde{w}_\varepsilon = (1 + o(1))\tilde{w}_\varepsilon^{2^*-1-r_\varepsilon}$$

weakly in $B_R(0) \cap \{x_1 < 0\}$, where $\lim_{\varepsilon \to 0} o(1) = 0$ in $C^0(B_R(0) \cap \{x_1 \leq 0\})$. Since $\tilde{w}_\varepsilon$ vanishes on $B_R(0) \cap \{x_1 = 0\}$ (in the sense of the trace) and that $0 \leq \tilde{w}_\varepsilon \leq 2$ (see for instance the proof of (16.70)), it follows from standard elliptic theory that there exists $\tilde{w} \in C^1(B_R(0) \cap \{x_1 \leq 0\})$ such that $\lim_{\varepsilon \to 0} \tilde{w}_\varepsilon = \tilde{w}$ in $C^0(B_{\frac{R}{2}}(0) \cap \{x_1 \leq 0\})$. In particular,

(16.77) $$\tilde{w} \equiv 0 \text{ on } B_{\frac{R}{2}}(0) \cap \{x_1 = 0\}.$$

Moreover, it follows from (16.73), (16.74) and (16.75) that

$$\tilde{w}_\varepsilon\left(\frac{\tilde{y}_\varepsilon - \tilde{z}_\varepsilon}{\gamma_\varepsilon}\right) = 1 \text{ and } \lim_{\varepsilon \to 0} \frac{\tilde{y}_\varepsilon - \tilde{z}_\varepsilon}{\gamma_\varepsilon} = 0.$$

In particular, $\tilde{w}(0) = 1$, which contradicts (16.77). This ends the proof of (16.61) also in Case (2).

REMARK 16.3.1. It follows from (16.20), (16.23), (16.61) and standard elliptic theory that

(16.78) $$\lim_{\varepsilon \to 0} u_\varepsilon = 0 \text{ in } C^2_{loc}(\overline{\Omega} \setminus \{0\}).$$

We now show the following slight improvement of (16.61):

(16.79) $$\lim_{R \to +\infty} \lim_{\varepsilon \to 0} \sup_{x \in \Omega \setminus B_{Rk_\varepsilon}(0)} |x|^{\frac{n-2}{2}} u_\varepsilon(x)^{1-\frac{r_\varepsilon}{2^\star - 2}} = 0.$$

Again by contradiction, assume there exists $\varepsilon_0 > 0$ and a family $(y_\varepsilon)_{\varepsilon > 0} \in \Omega$ such that

(16.80) $$|y_\varepsilon|^{\frac{n-2}{2}} u_\varepsilon(y_\varepsilon)^{1-\frac{r_\varepsilon}{2^\star - 2}} \geq \varepsilon_0 \text{ and } \lim_{\varepsilon \to 0} \frac{|y_\varepsilon|}{k_\varepsilon} = +\infty.$$

We let

(16.81) $$\nu_\varepsilon := u_\varepsilon(y_\varepsilon)^{-\frac{2}{n-2}} \text{ and } \gamma_\varepsilon := \nu_\varepsilon^{1-\frac{r_\varepsilon}{2^\star - 2}}.$$

It follows from (16.78), (16.61), (16.80) and (16.81) that there exists $\rho_0 \in \mathbb{R}$ such that

(16.82) $$\lim_{\varepsilon \to 0} y_\varepsilon = 0, \lim_{\varepsilon \to 0} \nu_\varepsilon = 0 \text{ and } \lim_{\varepsilon \to 0} \frac{|y_\varepsilon|}{\gamma_\varepsilon} = \rho_0 > 0.$$

Note that it follows from (16.24) and (16.30) that

(16.83) $$\lim_{\varepsilon \to 0} \nu_\varepsilon^{r_\varepsilon} = 1.$$

We let $\varphi : U \to V$ as in (16.13) with $x_0 = 0$ and where U, V are open neighborhoods of 0. For any $x \in \frac{U}{\gamma_\varepsilon} \cap \{x_1 < 0\}$, we let

(16.84) $$\overline{w}_\varepsilon(x) := \nu_\varepsilon^{\frac{n-2}{2}} u_\varepsilon \circ \varphi(\gamma_\varepsilon x).$$

It follows from (16.61) and the properties (16.13) of φ that there exists $C > 0$ such that for all $x \in \frac{U}{\gamma_\varepsilon} \cap \{x_1 < 0\}$ and all $\varepsilon > 0$, we have

(16.85) $$|x|^{\frac{n-2}{2}} \overline{w}_\varepsilon(x)^{1-\frac{r_\varepsilon}{2^\star - 2}} \leq C.$$

Consider again the metric $(\bar{g}_\varepsilon)_{ij} = (\partial_i \varphi, \partial_j \varphi)(\gamma_\varepsilon x)$ for $i,j = 1, ..., n$. From (16.20), we get that for all $\varepsilon > 0$ and all x in $\frac{U}{\gamma_\varepsilon} \cap \{x_1 < 0\}$,

(16.86) $$-\Delta_{\bar{g}_\varepsilon} \overline{w}_\varepsilon + \gamma_\varepsilon^2 a_\varepsilon \circ \varphi(\gamma_\varepsilon x) \overline{w}_\varepsilon = \frac{\overline{w}_\varepsilon^{2^\star - 1 - r_\varepsilon}}{\left|\frac{\varphi(\gamma_\varepsilon x)}{\gamma_\varepsilon}\right|^s}.$$

Moreover, $\overline{w}_\varepsilon$ vanishes on $\frac{U}{\gamma_\varepsilon} \cap \{x_1 = 0\}$. It then follows from (16.85), (16.86) and standard elliptic theory that there exists $\overline{w} \in C^0(\mathbb{R}^n_- \cap \{x_1 = 0\}) \setminus \{0\}$ such

16.3. REFINED BLOW-UP ANALYSIS AND STRONG POINTWISE ESTIMATES

that $\overline{w} \geq 0$ and $\lim_{\varepsilon \to 0} \overline{w}_\varepsilon = \overline{w}$ in $C^0(\mathbb{R}^n_- \cap \{x_1 = 0\}) \setminus \{0\})$. We now write $y_\varepsilon = \varphi(\gamma_\varepsilon \tilde{y}_\varepsilon)$. It follows from (16.82) that $\lim_{\varepsilon \to 0} = y_0 \neq 0$. As a consequence, $\overline{w}(y_0) = \lim_{\varepsilon \to 0} \overline{w}_\varepsilon(\tilde{y}_\varepsilon) = 1$, and therefore $\overline{w} \not\equiv 0$.

We let now $0 < \delta < R$, and with a change of variable, we have that

$$(16.87) \quad \lim_{\varepsilon \to 0} \int_{\varphi(B_{R\gamma_\varepsilon}(0)) \setminus \varphi(B_{\delta\gamma_\varepsilon}(0))} \frac{u_\varepsilon(x)^{2^* - r_\varepsilon}}{|x|^s} dx = \int_{B_R(0) \setminus B_\delta(0)} \frac{\overline{w}(x)^{2^*}}{|x|^s} dx.$$

From (16.80), we get that for any $\rho > 0$, $B_{\rho k_\varepsilon}(0) \cap (\varphi(B_{R\gamma_\varepsilon}(0)) \setminus \varphi(B_{\delta\gamma_\varepsilon}(0))) = \emptyset$ for all $\varepsilon > 0$ small enough, and up to a subsequence. It then follows from (16.59) that

$$\lim_{\varepsilon \to 0} \int_{\varphi(B_{R\gamma_\varepsilon}(0)) \setminus \varphi(B_{\delta\gamma_\varepsilon}(0))} \frac{u_\varepsilon(x)^{2^* - r_\varepsilon}}{|x|^s} dx = 0.$$

This equality and (16.87) yield $\int_{B_R(0) \setminus B_\delta(0)} \frac{\overline{w}^{2^*}}{|x|^s} dx = 0$ for all $R > \delta > 0$. We then get that $\overline{w} \equiv 0$, which contradicts $\overline{w}(y_0) = 1$. This complete the proof of (16.61).

We now prove a first approximation of (16.60). More precisely, we claim that for any $\alpha \in (0, n-2)$, there exists $C_\alpha > 0$ such that for all $\varepsilon > 0$ and all $x \in \Omega$,

$$(16.88) \quad |x|^\alpha \mu_\varepsilon^{\frac{n-2}{2} - \alpha} u_\varepsilon(x) \leq C_\alpha.$$

Indeed, since $-\Delta + a$ is coercive on Ω and $(a_\varepsilon)_{\varepsilon>0}$ converges uniformly to a, there exists U_0 an open subset of $\mathbb{R}^n$ such that $\overline{\Omega} \subset\subset U_0$, as well as $\alpha_0 > 0$ and $\lambda > 0$ such that

$$(16.89) \quad \int_{U_0} \left(|\nabla \varphi|^2 + (a_\varepsilon - 2\alpha_0)\varphi^2\right) dx \geq \lambda \int_{U_0} \varphi^2 dx$$

for all $\varphi \in C_c^1(U_0)$ and all $\varepsilon > 0$. In other words, the family of the operators $-\Delta + a_\varepsilon - \alpha_0$ is uniformly coercive in a neighborhood of $\overline{\Omega}$. We let $G_\varepsilon \in C^2(U_0 \times U_0 \setminus \{(x,x)/x \in U_0\})$ be the Green's function for $-\Delta + a_\varepsilon - \alpha_0$ with Dirichlet condition in U_0. G_ε satisfies

$$(16.90) \quad -\Delta G_\varepsilon(x, \cdot) + (a_\varepsilon - \alpha_0) G_\varepsilon(x, \cdot) = \delta_x$$

weakly on U. Since $0 \in U$, that there exists $C > 0$ such that for all $\varepsilon > 0$ and all $x \in \overline{U} \setminus \{0\}$,

$$(16.91) \quad 0 < G_\varepsilon(0, x) \leq C \cdot |x|^{2-n}.$$

More precisely, there exists $\delta_0 > 0$ and $C_0 > 0$ such that for all $\varepsilon > 0$ and all $x \in B_{\delta_0}(0) \setminus \{0\}$,

$$(16.92) \quad G_\varepsilon(0, x) \geq C_0 \cdot |x|^{2-n} \quad \text{and} \quad \frac{|\nabla G_\varepsilon(0, x)|}{|x|^{n-2}} \geq \frac{C_0}{|x|}.$$

We consider the operator

$$L_\varepsilon = -\Delta + \left(a_\varepsilon - \frac{u_\varepsilon^{2^* - 2 - r_\varepsilon}}{|x|^s}\right),$$

and claim that there exist $\nu_0 \in (0,1)$ and $R_1 > 0$ such that for any $\nu \in (0, \nu_0)$ and any $R > R_1$, we have for all $x \in \Omega \setminus B_{Rk_\varepsilon}(0)$ and for all $\varepsilon > 0$ sufficiently small,

$$(16.93) \quad L_\varepsilon G_\varepsilon^{1-\nu} > 0.$$

Indeed, we let $\nu_0 \in (0,1)$ such that for any $\nu \in (0,\nu_0)$, we have for all $\varepsilon > 0$ and all $x \in \Omega$,

(16.94) $$\nu \cdot (a_\varepsilon(x) - \alpha_0) \geq -\frac{1}{2}\alpha_0.$$

By (16.90), we get that for all $x \in \Omega \setminus \{0\}$ and all $\varepsilon > 0$,

(16.95) $$\frac{L_\varepsilon G_\varepsilon^{1-\nu}}{G_\varepsilon^{1-\nu}}(x) = \alpha_0 + \nu \cdot (a_\varepsilon(x) - \alpha_0) + \nu \cdot (1-\nu) \cdot \frac{|\nabla G_\varepsilon|^2}{G_\varepsilon^2}(x) - \frac{u_\varepsilon(x)^{2^*-2-r_\varepsilon}}{|x|^s}.$$

It follows from the pointwise estimate (16.79) that there exists $R_1 > 0$ such that for any $R > R_1$, we have for all $\varepsilon > 0$ and all $x \in \Omega \setminus B_{Rk_\varepsilon}(0)$,

(16.96) $$|x|^{2-s} u_\varepsilon(x)^{2^*-2-r_\varepsilon} \leq \frac{1}{2}\nu(1-\nu)C_0^2.$$

Here, $C_0 > 0$ is as in (16.92). We are now in position to prove (16.93). We let $\nu \in (0,\nu_0)$ and $R > R_1$. We first let $x \in \Omega$ such that $|x| \geq \delta_0$. It follows from (16.95) and (16.94) that

$$\frac{L_\varepsilon G_\varepsilon^{1-\nu}}{G_\varepsilon^{1-\nu}}(x) \geq \frac{\alpha_0}{2} - \frac{u_\varepsilon(x)^{2^*-2-r_\varepsilon}}{\delta_0^s}$$

for all $\varepsilon > 0$. Inequality (16.93) then follows with this inequality and (16.78). This proves (16.93) when $|x| \geq \delta_0$.

We now let $x \in B_{\delta_0}(0) \setminus B_{Rk_\varepsilon}(0)$. It follows from (16.95), (16.92) and (16.96) that

$$\frac{L_\varepsilon G_\varepsilon^{1-\nu}}{G_\varepsilon^{1-\nu}}(x) \geq \frac{\alpha_0}{2} + \frac{\nu \cdot (1-\nu) \cdot C_0^2}{|x|^2} - \frac{\nu \cdot (1-\nu) \cdot C_0^2}{2 \cdot |x|^2} > 0,$$

which clearly proves (16.93) in the remaining case when $x \in B_{\delta_0}(0) \setminus B_{Rk_\varepsilon}(0)$.

We let $R < R_1$ and $\nu \in (0,\nu_0)$. We now claim that there exists $C(R) > 0$ such that

(16.97) $$\left\{ \begin{array}{ll} L_\varepsilon\left(C(R)\mu_\varepsilon^{\frac{n-2}{2}-\nu(n-2)}G_\varepsilon(0,\cdot)^{1-\nu}\right) > L_\varepsilon u_\varepsilon & \text{in } \Omega \setminus B_{Rk_\varepsilon}(0) \\ C(R)\mu_\varepsilon^{\frac{n-2}{2}-\nu(n-2)}G_\varepsilon(0,\cdot)^{1-\nu} > u_\varepsilon & \text{on } \partial\Omega \setminus B_{Rk_\varepsilon}(0) \end{array} \right\}$$

Indeed, the first inequality is trivial since $L_\varepsilon u_\varepsilon = 0$ and (16.93) holds. Concerning the second inequality, we get from Definition (16.24) of μ_ε, the limit (16.30) and (16.92) that for all $x \in \Omega \cap \partial B_{Rk_\varepsilon}(0)$,

$$\frac{u_\varepsilon(x)}{\mu_\varepsilon^{\frac{n-2}{2}-\nu(n-2)} G_\varepsilon(0,x)^{1-\nu}} \leq C_0^{\nu-1} \cdot \mu_\varepsilon^{-(n-2)(1-\nu)} \cdot |x|^{(n-2)(1-\nu)}$$

$$\leq 2 \cdot C_0^{1-\nu} \cdot R^{(n-2)(1-\nu)} := C(R).$$

The inequalities (16.97) are proved.

Since $G_\varepsilon(0,x)^{1-\nu} > 0$ in $\overline{\Omega \cap \partial B_{Rk_\varepsilon}(0)}$ and $L_\varepsilon G_\varepsilon(0,x)^{1-\nu} > 0$ in $\Omega \cap \partial B_{Rk_\varepsilon}(0)$, it follows that L_ε verifies the comparison principle (see for example [**49**]). It then follows from (16.97) that

$$u_\varepsilon(x) \leq C(R)\mu_\varepsilon^{\frac{n-2}{2}-\nu(n-2)} G_\varepsilon(0,x)^{1-\nu}$$

for all $x \in \Omega \setminus \overline{B}_{Rk_\varepsilon}(0)$. By (16.91), we get that there exists $C'(R) > 0$ such that

$$u_\varepsilon(x) \leq C'(R)\mu_\varepsilon^{\frac{n-2}{2}-\nu(n-2)} |x|^{2-n+\nu(n-2)}$$

for all $x \in \Omega \setminus \overline{B}_{Rk_\varepsilon}(0)$. Up to taking a larger $C'(R)$, it follows from (16.24) that this inequality holds on the whole set Ω. Taking $\alpha = (n-2)\cdot(1-\nu)$, we get (16.88) for α close enough to $n-2$. As easily checked, this implies the inequality for all $\alpha \in (0, n-2)$. This ends the proof of (16.88).

We are in position to prove Proposition 16.3.1. For all $\varepsilon > 0$, we let $y_\varepsilon \in \Omega$ such that
$$\max_{x \in \Omega} |x|^{n-2} u_\varepsilon(x_\varepsilon) u_\varepsilon(x) = |y_\varepsilon|^{n-2} u_\varepsilon(x_\varepsilon) u_\varepsilon(y_\varepsilon).$$

Clearly, Proposition 16.3.1 is equivalent to proving that

(16.98) $$|y_\varepsilon|^{n-2} u_\varepsilon(x_\varepsilon) u_\varepsilon(y_\varepsilon) = O(1) \text{ as } \varepsilon \to 0.$$

We shall again consider two different cases:

Case 1): We assume first that $|y_\varepsilon| = O(k_\varepsilon)$ as $\varepsilon \to 0$. We then get with (16.24) that
$$|y_\varepsilon|^{n-2} u_\varepsilon(x_\varepsilon) u_\varepsilon(y_\varepsilon) = O(1) \text{ as } \varepsilon \to 0,$$
which proves (16.98) in that case.

Case 2): We now assume that

(16.99) $$\lim_{\varepsilon \to 0} \frac{|y_\varepsilon|}{k_\varepsilon} = +\infty.$$

As in the beginning of the last step, we choose U_0 such that $\overline{\Omega} \subset\subset U_0$ such that $-\Delta + a_\varepsilon$ is coercive on U_0. We let H_ε be the Green's function for $-\Delta + a_\varepsilon$ on U_0 with Dirichlet boundary condition. It follows from Green's representation formula and standard estimates on the Green's function that

(16.100) $$u_\varepsilon(x) \leq \int_\Omega H_\varepsilon(x,y) \cdot \frac{u_\varepsilon(y)^{2^*-1-r_\varepsilon}}{|y|^s} dy \leq C \int_\Omega |x-y|^{2-n} \cdot \frac{u_\varepsilon(y)^{2^*-1-r_\varepsilon}}{|y|^s} dy$$

for all $x \in \Omega$. We now let
$$\hat{v}_\varepsilon(x) = \mu_\varepsilon^{-\frac{n-2}{2}} u_\varepsilon(k_\varepsilon x)$$

for all $x \in k_\varepsilon^{-1}\Omega$ and all $\varepsilon > 0$. It follows from Proposition 16.2.1 and (16.88) that for any $\alpha \in (0, n-2)$, there exists $C_\alpha > 0$ such that for all $x \in k_\varepsilon^{-1}\Omega$ and all $\varepsilon > 0$,

(16.101) $$\hat{v}_\varepsilon(x) \leq \frac{C_\alpha}{1+|x|^\alpha}.$$

It follows from (16.100) and a change of variable that

(16.102) $$\begin{aligned}\mu_\varepsilon^{-\frac{n-2}{2}} u_\varepsilon(y_\varepsilon) &\leq C \int_{k_\varepsilon^{-1}\Omega} |y_\varepsilon - k_\varepsilon y|^{2-n} \frac{\hat{v}_\varepsilon(y)^{2^*-1-r_\varepsilon}}{|y|^s} dx \\ &\leq C \int_{k_\varepsilon^{-1}\Omega \cap \{|y_\varepsilon - k_\varepsilon y| \geq \frac{|y_\varepsilon|}{2}\}} \frac{1}{|y_\varepsilon - k_\varepsilon y|^{n-2}} \cdot \frac{\hat{v}_\varepsilon(y)^{2^*-1-r_\varepsilon}}{|y|^s} dx \\ &\quad + C \int_{k_\varepsilon^{-1}\Omega \cap \{|y_\varepsilon - k_\varepsilon y| < \frac{|y_\varepsilon|}{2}\}} \frac{1}{|y_\varepsilon - k_\varepsilon y|^{n-2}} \cdot \frac{\hat{v}_\varepsilon(y)^{2^*-1-r_\varepsilon}}{|y|^s} dx.\end{aligned}$$

We estimate the two integrals of the RHS separately. With (16.101), we get that for all $\varepsilon > 0$ small and α close enough to $n-2$,

$$\int_{k_\varepsilon^{-1}\Omega \cap \{|y_\varepsilon - k_\varepsilon y| \geq \frac{|y_\varepsilon|}{2}\}} \frac{1}{|y_\varepsilon - k_\varepsilon y|^{n-2}} \cdot \frac{\hat{v}_\varepsilon(y)^{2^*-1-r_\varepsilon}}{|y|^s} dx$$

$$\leq C \cdot |y_\varepsilon|^{2-n} \int_{k_\varepsilon^{-1}\Omega} \frac{1}{|y|^s(1+|y|^{\alpha \cdot (2^*-1-r_\varepsilon)})} dy$$

(16.103) $$\leq C \cdot |y_\varepsilon|^{2-n}.$$

On the other hand, by (16.101), we get that

$$\int_{k_\varepsilon^{-1}\Omega \cap \{|y_\varepsilon - k_\varepsilon y| \leq \frac{|y_\varepsilon|}{2}\}} \frac{1}{|y_\varepsilon - k_\varepsilon y|^{n-2}} \cdot \frac{\hat{v}_\varepsilon(y)^{2^*-1-r_\varepsilon}}{|y|^s} dx$$

$$\leq C \int_{k_\varepsilon^{-1}\Omega \cap \{|y_\varepsilon - k_\varepsilon y| \leq \frac{|y_\varepsilon|}{2}\}} \frac{1}{|y_\varepsilon - k_\varepsilon y|^{n-2}} \cdot \frac{1}{|y|^{\alpha(2^*-1-r_\varepsilon)+s}} dx$$

$$\leq \frac{C \cdot k_\varepsilon^{\alpha(2^*-1-r_\varepsilon)+s}}{|y_\varepsilon|^{\alpha(2^*-1-r_\varepsilon)+s}} \int_{k_\varepsilon^{-1}\Omega \cap \{|y_\varepsilon - k_\varepsilon y| \leq \frac{|y_\varepsilon|}{2}\}} \frac{1}{|y_\varepsilon - k_\varepsilon y|^{n-2}} dy$$

$$\leq \frac{C \cdot k_\varepsilon^{\alpha(2^*-1-r_\varepsilon)+s}}{|y_\varepsilon|^{\alpha(2^*-1-r_\varepsilon)+s}} \cdot \frac{|y_\varepsilon|^2}{|k_\varepsilon|^n}$$

$$\leq C|y_\varepsilon|^{2-n} \cdot \left(\frac{k_\varepsilon}{|y_\varepsilon|}\right)^{(2^*-1-r_\varepsilon)\alpha+s-n}.$$

Since $\lim_{\alpha \to n-2}\lim_{\varepsilon \to 0}(2^*-1-r_\varepsilon)\alpha + s - n = 2 - s > 0$, we get from (16.99) and α close enough to $n-2$ that

(16.104) $$\int_{k_\varepsilon^{-1}\Omega \cap \{|y_\varepsilon - k_\varepsilon y| \geq \frac{|y_\varepsilon|}{2}\}} \frac{1}{|y_\varepsilon - k_\varepsilon y|^{n-2}} \cdot \frac{\hat{v}_\varepsilon(y)^{2^*-1-r_\varepsilon}}{|y|^s} dx = o\left(|y_\varepsilon|^{2-n}\right),$$

when $\varepsilon \to 0$. Plugging (16.103) and (16.104) into (16.102), we finally obtain that

$$\mu_\varepsilon^{-\frac{n-2}{2}} u_\varepsilon(y_\varepsilon) = O\left(|y_\varepsilon|^{2-n}\right) \text{ as } \varepsilon \to 0,$$

which proves (16.98) holds in Case 2) and we are done with (16.98).

As easily checked, (16.60) and then Proposition 16.3.1 follow from (16.98) and (16.24). This completes the proof of Proposition 16.3.1. □

We shall now use Proposition 16.3.1, to derive pointwise estimates for v_ε. This is the object of the following proposition.

PROPOSITION 16.3.2. *Assume that the hypothesis of Proposition* 16.3.1 *are satisfied. Then there exists $C > 0$ such that for all $\varepsilon > 0$ and all $x \in \frac{U}{k_\varepsilon} \cap \{x_1 < 0\}$,*

(16.105) $$v_\varepsilon(x) \leq \frac{C}{(1+|x|^2)^{\frac{n-2}{2}}} \text{ and } |\nabla v_\varepsilon(x)| \leq \frac{C}{(1+|x|^2)^{\frac{n-1}{2}}},$$

where v_ε was defined in (16.47) and U is as in (16.13) with $x_0 = 0$.

PROOF. The first inequality of the proposition is an immediate consequence of estimate (16.60) and the definition (16.47) of v_ε. Concerning the second inequality, we proceed by contradiction and assume that there exists a family $(y_\varepsilon)_{\varepsilon>0}$ such that $y_\varepsilon \in U$ for all $\varepsilon \to 0$ and such that

$$\lim_{\varepsilon \to 0} \left(1 + \left|\frac{y_\varepsilon}{k_\varepsilon}\right|\right)^{n-1} \left|\nabla v_\varepsilon\left(\frac{y_\varepsilon}{k_\varepsilon}\right)\right| = +\infty. \tag{16.106}$$

We consider three different cases:

Case 1): Assume that $y_\varepsilon \not\to 0$ when $\varepsilon \to 0$. It follows from the pointwise estimate (16.60) that for any $\delta > 0$, there exists $C(\delta) > 0$ such that for all $x \in \overline{\Omega} \setminus B_\delta(x_0)$ and all $\varepsilon > 0$, we have

$$u_\varepsilon(x) \le C(\delta) \mu_\varepsilon^{\frac{n-2}{2}}.$$

We then get that

$$-\Delta(\mu_\varepsilon^{\frac{2-n}{2}} u_\varepsilon) + a_\varepsilon \cdot (\mu_\varepsilon^{\frac{2-n}{2}} u_\varepsilon) = \mu_\varepsilon^{\frac{n-2}{2}(2^*-2-r_\varepsilon)} \frac{(\mu_\varepsilon^{\frac{2-n}{2}} u_\varepsilon)^{2^*-1-r_\varepsilon}}{|x|^s}$$

weakly in $\Omega \setminus \bar{B}_\delta(x_0)$. It then follows from standard elliptic theory that

$$\|\mu_\varepsilon^{\frac{2-n}{2}} u_\varepsilon\|_{C^2(\overline{\Omega} \setminus B_{3\delta}(x_0))} = O(1) \text{ as } \varepsilon \to 0. \tag{16.107}$$

Since $y_\varepsilon \not\to 0$, there exists $\delta > 0$ such that, up to a subsequence, $|y_\varepsilon| \ge 4\delta$ for $\varepsilon > 0$. It follows from (16.107) that $\nabla u_\varepsilon(\varphi(y_\varepsilon)) = O(\mu_\varepsilon^{\frac{n-2}{2}})$ when $\varepsilon \to 0$. A contradiction with (16.106). This proves the Proposition in Case 1).

Case 2): We now assume that

$$\lim_{\varepsilon \to 0} y_\varepsilon = 0 \text{ and } \lim_{\varepsilon \to 0} \frac{|y_\varepsilon|}{k_\varepsilon} = +\infty. \tag{16.108}$$

We let φ as in (16.13) with $x_0 = 0$ and define

$$h_\varepsilon(x) := \frac{|y_\varepsilon|^{n-2}}{k_\varepsilon^{\frac{n-2}{2}}} u_\varepsilon \circ \varphi(|y_\varepsilon|x)$$

for all $x \in \frac{U}{|y_\varepsilon|} \cap \{x_1 \le 0\}$. It follows from (16.60) and (16.30) that there exists $C > 0$ such that for all $x \in \frac{U}{|y_\varepsilon|} \cap \{x_1 \le 0\}$, $x \ne 0$,

$$h_\varepsilon(x) \le C \cdot |x|^{2-n}. \tag{16.109}$$

We let $\Delta_{\bar{g}_\varepsilon} = \bar{g}_\varepsilon^{ij}\left(\partial_{ij} - \Gamma_{ij}^k(\bar{g}_\varepsilon)\partial_k\right)$ be the Laplace-Beltrami operator for the metric $(\bar{g}_\varepsilon)_{ij} = (\partial_i\varphi, \partial_j\varphi)(k_\varepsilon x)$ where again, the $\bar{g}_\varepsilon^{ij} = (\bar{g}_\varepsilon^{-1})_{ij}$ are the coordinates of the inverse of the tensor $\bar{g}_\varepsilon$ and the $\Gamma_{ij}^k(\bar{g}_\varepsilon)$ are the Christoffel symbols associated to the metric $\bar{g}_\varepsilon$. After a change of variables, (16.20) can be rewriten as

$$-\Delta_{\bar{g}_\varepsilon} h_\varepsilon + |y_\varepsilon|^2 a_\varepsilon(\varphi(|y_\varepsilon|x)) h_\varepsilon = k_\varepsilon^{r_\varepsilon \frac{n-2}{2}} \left(\frac{k_\varepsilon}{|y_\varepsilon|}\right)^{2-s-r_\varepsilon(n-2)} \frac{h_\varepsilon^{2^*-1-r_\varepsilon}}{\left|\frac{\varphi(|y_\varepsilon|x)}{|y_\varepsilon|}\right|^s}$$

weakly in $\frac{U}{|y_\varepsilon|} \cap \{x_1 < 0\}$. Since (16.30), (16.108) and (16.109) hold and since $s \in (0,2)$, there exists $p > \frac{n}{2}$ such that

$$-\Delta_{\bar{g}_\varepsilon} h_\varepsilon + |y_\varepsilon|^2 a_\varepsilon(\varphi(|y_\varepsilon|x)) h_\varepsilon = f_\varepsilon \text{ in } \frac{U}{|y_\varepsilon|} \cap \{x_1 < 0\},$$

where $f_\varepsilon \in L^p_{loc}(\frac{U}{|y_\varepsilon|} \cap \{x_1 \leq 0\} \setminus \{0\})$ uniformly wrt $\varepsilon \to 0$. Since $h_\varepsilon \equiv 0$ on $\frac{U}{|y_\varepsilon|} \cap \{x_1 = 0\}$ and (16.109) holds, it follows from standard elliptic theory that for any $\delta_1 > \delta_2 > 0$, there exists $C'(\delta_1, \delta_2) > 0$ such that for all $\varepsilon > 0$,

$$\|h_\varepsilon\|_{C^1((B_{\delta_1}(0) \setminus B_{\delta_2}(0)) \cap \{x_1 \leq 0\})} \leq C'(\delta_1, \delta_2).$$

It then follows that

$$\left|\nabla h_\varepsilon\left(\frac{y_\varepsilon}{|y_\varepsilon|}\right)\right| = O(1) \text{ as } \varepsilon \to 0.$$

Coming back to the definitions of h_ε and v_ε, we get a contradiction from (16.106), which proves the Proposition in Case 2).

Case 3): We finally assume that

$$|y_\varepsilon| = O(k_\varepsilon) \text{ as } \varepsilon \to 0.$$

In this case, we have from Proposition 16.2.1 that $\left|\nabla v_\varepsilon\left(\frac{y_\varepsilon}{k_\varepsilon}\right)\right| = O(1)$ when $\varepsilon \to 0$, which is in contradiction with (16.106). This proves the claim in Case 3, and completes the proof of Proposition 16.3.2. □

We end this section with the following corollary.

COROLLARY 16.3.1. *Let $(u_\varepsilon)_{\varepsilon>0}$ as in the hypothesis of Proposition 16.3.1. Then, there exists $H \in C^1(\overline{\Omega} \setminus \{0\})$ such that*

$$u_\varepsilon(x_\varepsilon) u_\varepsilon \to H \text{ in } C^1_{loc}(\overline{\Omega} \setminus \{0\}) \text{ as } \varepsilon \to 0.$$

Proof: We let $H_\varepsilon(x) := u_\varepsilon(x_\varepsilon) u_\varepsilon(x)$ for all $x \in \Omega$ and all $\varepsilon > 0$. It follows from Proposition 16.3.1 that for any open subset U such that $\overline{U} \subset \overline{\Omega} \setminus \{0\}$, there exists $C(U) > 0$ such that $|H_\varepsilon(x)| \leq C(U)$ for all $x \in U$ and all $\varepsilon > 0$. Equation (16.20) can be rewritten as

$$-\Delta H_\varepsilon + a_\varepsilon H_\varepsilon = u_\varepsilon(x_\varepsilon)^{2+r_\varepsilon - 2^\star} \frac{H_\varepsilon^{2^\star - 1 - r_\varepsilon}}{|x|^s} \text{ in } \Omega.$$

The conclusion of the Corollary is then a consequence of standard elliptic theory.□

16.4. Pohozaev identity and proof of attainability

We start by proving the following estimate.

PROPOSITION 16.4.1. *Let Ω be a smooth bounded domain of $\mathbb{R}^n$, $n \geq 4$ and $s \in (0,2)$. We let $(r_\varepsilon)_{\varepsilon>0}$ be in $[0, 2^\star - 2)$ such that (16.19) holds. We consider $(u_\varepsilon)_{\varepsilon>0} \in H_0^1(\Omega)$ such that (16.8), (16.20), (16.21) and (16.23) hold. We let μ_ε be as in (16.24) and v as in Proposition 16.2.1. Then, we have*

$$(16.110) \qquad \lim_{\varepsilon \to 0} \frac{r_\varepsilon}{\mu_\varepsilon} = \frac{(n-s) \int_{\partial \mathbb{R}^n_-} |x|^2 |\nabla v|^2 \, dx}{n(n-2)^2 \mu_s(\mathbb{R}^n_-)^{\frac{n-s}{2-s}}} \cdot H(0),$$

where $H(0)$ is the mean curvature of the oriented boundary $\partial \Omega$ at 0.

Proof: First we write a Pohozaev-type identity for u_ε. Note that we have from Proposition 16.5.1 below that $u_\varepsilon \in C^1(\overline{\Omega})$ and that $-\Delta u_\varepsilon \in L^p(\Omega)$ for all $p \in (1, \frac{n}{s})$. In the sequel, we denote by $\nu(x)$ the outward normal vector at $x \in \partial\Omega$

16.4. POHOZAEV IDENTITY AND PROOF OF ATTAINABILITY

of the oriented hypersurface $\partial\Omega$ (oriented as the boundary of Ω). Integrating by parts, we get that

$$\int_\Omega x^i \partial_i u_\varepsilon \cdot -\Delta u_\varepsilon \, dx$$

$$= -\int_{\partial\Omega} x^i \partial_i u_\varepsilon \partial_\nu u_\varepsilon \, d\sigma + \int_\Omega \partial_j(x^i \partial_i u_\varepsilon) \partial_j u_\varepsilon \, dx$$

$$= -\int_{\partial\Omega} x^i \partial_i u_\varepsilon \partial_\nu u_\varepsilon \, d\sigma + \int_\Omega |\nabla u_\varepsilon|^2 \, dx + \int_\Omega x^i \partial_i \frac{|\nabla u_\varepsilon|^2}{2} \, dx$$

$$= \left(1 - \frac{n}{2}\right) \int_\Omega |\nabla u_\varepsilon|^2 \, dx + \int_{\partial\Omega} \left((x,\nu) \frac{|\nabla u_\varepsilon|^2}{2} - x^i \partial_i u_\varepsilon \partial_\nu u_\varepsilon\right) d\sigma$$

$$= \left(1 - \frac{n}{2}\right) \left(\int_{\partial\Omega} u_\varepsilon \partial_\nu u_\varepsilon \, d\sigma + \int_\Omega u_\varepsilon \cdot -\Delta u_\varepsilon \, dx\right)$$

$$+ \int_{\partial\Omega} \left((x,\nu) \frac{|\nabla u_\varepsilon|^2}{2} - x^i \partial_i u_\varepsilon \partial_\nu u_\varepsilon\right) d\sigma.$$

Using the equation (16.20) in the RHS, we get that

$$\int_\Omega x^i \partial_i u_\varepsilon \cdot -\Delta u_\varepsilon \, dx = \left(1 - \frac{n}{2}\right) \left(\int_\Omega \frac{u_\varepsilon^{2^\star - r_\varepsilon}}{|x|^s} \, dx - \int_\Omega a_\varepsilon u_\varepsilon^2 \, dx\right)$$

(16.111)
$$+ \int_{\partial\Omega} \left(\left(1 - \frac{n}{2}\right) u_\varepsilon \partial_\nu u_\varepsilon + (x,\nu) \frac{|\nabla u_\varepsilon|^2}{2} - x^i \partial_i u_\varepsilon \partial_\nu u_\varepsilon\right) d\sigma.$$

On the other hand, using the equation (16.20) satisfied by u_ε, we get that

$$\int_\Omega x^i \partial_i u_\varepsilon \cdot -\Delta u_\varepsilon \, dx$$

$$= \int_\Omega x^i \partial_i u_\varepsilon \frac{u_\varepsilon^{2^\star - 1 - \varepsilon}}{|x|^s} \, dx - \int_\Omega x^i \partial_i u_\varepsilon a_\varepsilon u_\varepsilon \, dx$$

$$= \int_\Omega x^i |x|^{-s} \partial_i \left(\frac{u_\varepsilon^{2^\star - r_\varepsilon}}{2^\star - r_\varepsilon}\right) dx - \int_\Omega x^i \partial_i u_\varepsilon a_\varepsilon u_\varepsilon \, dx$$

(16.112)
$$= -\int_\Omega \partial_i(x^i |x|^{-s}) \frac{u_\varepsilon^{2^\star - r_\varepsilon}}{2^\star - r_\varepsilon} \, dx + \int_{\partial\Omega} \frac{(x,\nu)}{2^\star - r_\varepsilon} \cdot \frac{u_\varepsilon^{2^\star - r_\varepsilon}}{|x|^s} \, d\sigma - \int_\Omega x^i \partial_i u_\varepsilon a_\varepsilon u_\varepsilon \, dx$$

$$= -\int_\Omega \frac{n-s}{|x|^s} \cdot \frac{u_\varepsilon^{2^\star - r_\varepsilon}}{2^\star - r_\varepsilon} \, dx + \frac{1}{2} \int_\Omega (na_\varepsilon + x^i \partial_i a_\varepsilon) u_\varepsilon^2 \, dx$$

$$+ \int_{\partial\Omega} \frac{(x,\nu)}{2^\star - r_\varepsilon} \cdot \frac{u_\varepsilon^{2^\star - r_\varepsilon}}{|x|^s} \, d\sigma - \int_{\partial\Omega} \frac{(x,\nu)}{2} a_\varepsilon u_\varepsilon^2 \, d\sigma.$$

Plugging together (16.111) and (16.112), we get that

(16.113)
$$\left(\frac{n-2}{2} - \frac{n-s}{2^\star - r_\varepsilon}\right) \int_\Omega \frac{u_\varepsilon^{2^\star - r_\varepsilon}}{|x|^s} \, dx + \int_\Omega \left(a_\varepsilon + \frac{(x, \nabla a_\varepsilon)}{2}\right) u_\varepsilon^2 \, dx$$

$$= \int_{\partial\Omega} \left(-\frac{n-2}{2} u_\varepsilon \partial_\nu u_\varepsilon + (x,\nu) \frac{|\nabla u_\varepsilon|^2}{2}\right.$$

(16.114)
$$\left. - x^i \partial_i u_\varepsilon \partial_\nu u_\varepsilon - \frac{(x,\nu)}{2^\star - r_\varepsilon} \cdot \frac{u_\varepsilon^{2^\star - r_\varepsilon}}{|x|^s}\right) d\sigma + \int_{\partial\Omega} \frac{(x,\nu)}{2} a_\varepsilon u_\varepsilon^2 \, dx$$

for all $\varepsilon > 0$. Since $u_\varepsilon \equiv 0$ on $\partial\Omega$, we get that

(16.115) $\quad \dfrac{(n-2)r_\varepsilon}{2\cdot(2^\star - r_\varepsilon)} \displaystyle\int_\Omega \dfrac{u_\varepsilon^{2^\star - r_\varepsilon}}{|x|^s}\,dx - \int_\Omega \left(a_\varepsilon + \dfrac{(x,\nabla a_\varepsilon)}{2}\right) u_\varepsilon^2\,dx = \dfrac{1}{2}\int_{\partial\Omega}(x,\nu)|\nabla u_\varepsilon|^2\,d\sigma.$

We now deal with the RHS of (16.115). We take φ as in (16.13) with $x_0 = 0$. With the pointwise limit of Corollary 16.3.1, we get that

$$\int_{\partial\Omega}(x,\nu)|\nabla u_\varepsilon|^2\,d\sigma = \int_{\partial\Omega\cap\varphi(U)}(x,\nu)|\nabla u_\varepsilon|^2\,d\sigma + o(\mu_\varepsilon) \text{ when } \varepsilon \to 0$$

as soon as $n \geq 4$. With a change of variable, we get that

(16.116) $\quad \displaystyle\int_{\partial\Omega}(x,\nu)|\nabla u_\varepsilon|^2\,d\sigma = (1+o(1))\cdot \int_{D_\varepsilon}\left(\dfrac{\varphi(k_\varepsilon x)}{k_\varepsilon}, \nu\circ\varphi(k_\varepsilon x)\right)|\nabla v_\varepsilon|^2_{\tilde g_\varepsilon}\sqrt{|\tilde g_\varepsilon|}\,dx$
$\qquad\qquad + o(\mu_\varepsilon^{n-2}),$

where the metric $\tilde g_\varepsilon$ is such that $(\tilde g_\varepsilon)_{ij} = (\partial_i\varphi, \partial_j\varphi)(k_\varepsilon x)$ for all $i,j = 2,...,n$, $|\tilde g_\varepsilon| = \det(\tilde g_\varepsilon)$ and $D_\varepsilon = \dfrac{U}{k_\varepsilon}\cap\{x_1 = 0\}$.

Using the expression of φ (see (16.13)), we have

$$\nu(\varphi(x)) = \dfrac{(1, -\partial_2\varphi_0(x), ..., -\partial_n\varphi_0(x))}{\sqrt{1+\sum_{i=2}^n(\partial_i\varphi_0(x))^2}}$$

for all $x \in U \cap \{x_1 = 0\}$. We then get that

$$(\nu\circ\varphi(x), \vec X) = (1+O(|x|^2))\cdot\left(X^1 - \sum_{i=2}^n X^i\partial_i\varphi_0(x)\right)$$

for all $x \in U \cap \{x_1 = 0\}$ and all $\vec X \in \mathbb{R}^n$. In this expression $O(1)$ is bounded for $x \in U \cap \{x_1 = 0\}$ and $\vec X \in \mathbb{R}^n$. With the expression of φ (see (16.13)), we get that

$(\varphi(k_\varepsilon x), \nu\circ\varphi(k_\varepsilon x)) = (1+O(k_\varepsilon^2|x|^2))\left(\varphi_0(k_\varepsilon x) - k_\varepsilon \sum_{i=2}^n x^i \partial_i\varphi_0(k_\varepsilon x)\right)$

(16.117) $\qquad\qquad = (1+O(k_\varepsilon^2|x|^2))\cdot\left(-\dfrac{1}{2}k_\varepsilon^2\partial_{ij}\varphi(0)x^ix^j + O(1)(k_\varepsilon^3|x|^3)\right)$

for $\varepsilon > 0$ and $x \in \dfrac{U}{k_\varepsilon}\cap\{x_1 = 0\}$. Plugging (16.117) into (16.116), using the estimates of Proposition 16.3.1, Lebesgue's convergence theorem and letting $\varepsilon \to 0$, we get that

(16.118) $\quad \displaystyle\int_{\partial\Omega}(x,\nu)|\nabla u_\varepsilon|^2\,d\sigma = \left(-\dfrac{1}{2}\int_{\partial\mathbb{R}^n_-}\partial_{ij}\varphi_0(0)x^ix^j|\nabla v|^2\,dx + o(1)\right)\cdot k_\varepsilon$

when $n \geq 4$ and where $\lim_{\varepsilon\to 0} o(1) = 0$.

Note that we have from Proposition 16.3.1 that if $n \geq 4$, then

(16.119) $\qquad\qquad \displaystyle\int_\Omega u_\varepsilon^2\,dx = o(\mu_\varepsilon) \text{ when } \varepsilon \to 0.$

Plugging (16.118) into (16.115), using (16.21) and (16.119), we get that

(16.120) $\quad \left(\dfrac{n-2}{2\cdot 2^\star}\mu_s(\mathbb{R}^n_-)^{\frac{n-s}{2-s}} + o(1)\right)r_\varepsilon = \left(-\dfrac{1}{4}\int_{\partial\mathbb{R}^n_-}\partial_{ij}\varphi_0(0)x^ix^j|\nabla v|^2\,dx + o(1)\right)\cdot\mu_\varepsilon$

where $\lim_{\varepsilon\to 0} o(1) = 0$. By (16.120), we get that

$$\lim_{\varepsilon\to 0}\dfrac{n-2}{2\cdot 2^\star}\mu_s(\mathbb{R}^n_-)^{\frac{n-s}{2-s}}\cdot\dfrac{r_\varepsilon}{\mu_\varepsilon} = -\dfrac{1}{4}\int_{\partial\mathbb{R}^n_-}\partial_{ij}\varphi_0(0)x^ix^j|\nabla v|^2\,dx.$$

16.4. POHOZAEV IDENTITY AND PROOF OF ATTAINABILITY

We now consider the second fundamental form associated to $\partial\Omega$, namely $II_p(x,y) = (d\nu_p x, y)$ for all $p \in \partial\Omega$ and all $x, y \in T_p\partial\Omega$ (recall that ν is the outward normal vector at the hypersurface $\partial\Omega$). In the canonical basis of $\partial\mathbb{R}_-^n = T_0\partial\Omega$, the matrix of the bilinear form II_0 is $-D_0^2\varphi_0$, where $D_0^2\varphi_0$ is the Hessian matrix of φ_0 at 0. With this remark and (16.120), we get that

$$(16.121) \qquad \lim_{\varepsilon \to 0} \frac{r_\varepsilon}{\mu_\varepsilon} = \frac{(n-s)}{(n-2)^2} \mu_s(\mathbb{R}_-^n)^{-\frac{n-s}{2-s}} \cdot \int_{\partial\mathbb{R}_-^n} II_0(x,x)|\nabla v|^2 \, dx.$$

Since $v \geq 0$, that $v \in C^2(\overline{\mathbb{R}_-^n})$ and v verifies (16.54), it follows from the strong maximum principle that $v > 0$ in $\mathbb{R}_-^n$. Moreover, it follows from the definition (16.47) and the pointwise estimate (16.60) that there exists $C > 0$ such that $v(x) \leq \frac{C}{(1+|x|^2)^{\frac{n-2}{2}}}$ for all $x \in \mathbb{R}_-^n$. We let $\tilde{v}(x) := |x|^{2-n} v\left(\frac{x}{|x|^2}\right)$ be the Kelvin transform of v. As easily checked, $\tilde{v} \in C^2(\overline{\mathbb{R}_-^n} \setminus \{0\})$ and verifies

$$-\Delta \tilde{v} = \frac{\tilde{v}^{2^\star - 1}}{|x|^s} \quad \text{and} \quad \tilde{v}(x) \leq \frac{C}{(1+|x|^2)^{\frac{n-2}{2}}} \quad \text{for all } x \in \mathbb{R}_-^n.$$

Since $\tilde{v}$ vanishes on $\partial\mathbb{R}_-^n$, it then follows from standard elliptic theory that $\tilde{v} \in C^1(\overline{\mathbb{R}_-^n})$ and then, that there exists $C > 0$ such that $\tilde{v}(x) \leq C|x|$ for all $x \in B_1(0) \cap \mathbb{R}_-^n$. Coming back to the function v, we get that tehre exists $C > 0$ such that

$$v(x) \leq \frac{C}{(1+|x|^2)^{\frac{n-1}{2}}} \quad \text{for all } x \in \mathbb{R}_-^n.$$

It follows from Proposition 15.3.1 that there exists $w \in C^2(\mathbb{R}_-^\star \times \mathbb{R})$ such that $v(x_1, x') = w(x_1, |x'|)$ for all $(x_1, x') \in \mathbb{R}_-^\star \times \mathbb{R}^{n-1}$. In particular, $|\nabla v|(0, x')$ is radially symmetrical with respect to $x' \in \partial\mathbb{R}_-^n$. Since we have chosen a chart φ that is Euclidean at 0, we get that

$$\int_{\partial\mathbb{R}_-^n} II_0(x,x)|\nabla v|^2 \, dx = \frac{\sum_{i=2}^n (II_0)^{ii}}{n} \int_{\partial\mathbb{R}_-^n} |x|^2 |\nabla v|^2 \, dx$$

$$= \frac{H(0)}{n} \int_{\partial\mathbb{R}_-^n} |x|^2 |\nabla v|^2 \, dx.$$

Note that we have used here that in the chart φ defined in (16.13), the matrix of the first fundamental form at 0 is the identity. Plugging this last inequality in (16.121), we get that

$$(16.122) \qquad \lim_{\varepsilon \to 0} \frac{r_\varepsilon}{\mu_\varepsilon} = \frac{(n-s) \int_{\partial\mathbb{R}_-^n} |x|^2 |\nabla v|^2 \, dx}{n(n-2)^2 \mu_s(\mathbb{R}_-^n)^{\frac{n-s}{2-s}}} \cdot H(0).$$

$\square$

We are now in position to prove Theorems 16.1.1 and 16.1.3.

Proof of Theorem 16.1.1: We proceed by contradiction and assume that there are no extremals for (16.1). It follows from Propositions 16.1.1 that there exists $u_\varepsilon \in H_0^1(\Omega)$ such that (16.20), (16.21) and (16.23) hold with $a_\varepsilon \equiv 0$ and $r_\varepsilon = \varepsilon$. Since $0 < s < 2$, then (16.122) holds with $r_\varepsilon = \varepsilon$ when $n \geq 4$. We then get that $H(0) \geq 0$. A contradiction with the assumptions of Theorem 16.1.1. This proves the first point of Theorem 16.1.1 when $n \geq 4$.

Concerning the compactness, we note that any sequence of minimizers of (16.1) satisfies (16.20) and (16.21) with $r_\varepsilon \equiv 0$ and $a \equiv 0$. If the sequence of minimizers blows up, we get from (16.122) that $H(0) = 0$, which contradicts our initial assumption. It follows then that the minimizing sequence does not blow up, and therefore, from standard elliptic theory, that it converges in $H_0^1(\Omega)$. This proves Theorem 16.1.1 when $n \geq 4$.

The proof of Theorem 16.1.3 is quite similar by taking this time $a_\varepsilon \equiv a$ and $r_\varepsilon = \varepsilon$.

16.5. Appendix: Regularity of weak solutions

In this section, we prove the following regularity result that was used repeatedly in the previous sections.

PROPOSITION 16.5.1. *Let Ω be a smooth bounded domain of $\mathbb{R}^n$, $n \geq 3$. We let $s \in (0,2)$ and $a \in C^0(\overline{\Omega})$. We let $\varepsilon \in [0, 2^\star - 2)$ and consider $u \in H_0^1(\Omega)$ a weak solution of*

$$-\Delta u + au = \frac{|u|^{2^\star - 2 - \varepsilon} u}{|x|^s} \quad \text{in } \Omega.$$

Then there exists $\theta \in (0,1)$ such that $u \in C^{1,\theta}(\overline{\Omega})$.

Proof: We follow a strategy developed by Trudinger. Let $\beta \geq 1$, and $L > 0$. We consider

$$G_L(t) = \begin{cases} |t|^{\beta-1} t & \text{if } |t| \leq L \\ \beta L^{\beta-1}(t - L) + L^\beta & \text{if } t \geq L \\ \beta L^{\beta-1}(t + L) - L^\beta & \text{if } t \leq -L \end{cases}$$

and

$$H_L(t) = \begin{cases} |t|^{\frac{\beta-1}{2}} t & \text{if } |t| \leq L \\ \frac{\beta+1}{2} L^{\frac{\beta-1}{2}}(t - L) + L^{\frac{\beta+1}{2}} & \text{if } t \geq L \\ \frac{\beta+1}{2} L^{\frac{\beta-1}{2}}(t + L) - L^{\frac{\beta+1}{2}} & \text{if } t \leq -L. \end{cases}$$

As easily checked, we have for all $t \in \mathbb{R}$ and all $L > 0$,

$$0 \leq t G_L(t) \leq H_L(t)^2 \quad \text{and} \quad G_L'(t) = \frac{4\beta}{(\beta+1)^2}(H_L'(t))^2.$$

For $\eta \in C_c^\infty(\mathbb{R}^n)$, it is easy to check that both $\eta^2 G_L(u)$ and $\eta H_L(u)$ are in $H_0^1(\Omega)$. With the equation verified by u, we get that

(16.123) $$\int_\Omega \nabla u \nabla(\eta^2 G_L(u))\, dx = \int_\Omega \frac{|u|^{2^\star - 2 - \varepsilon}}{|x|^s} \eta^2 u G_L(u)\, dx - \int_\Omega a \eta^2 u G_L(u)\, dx.$$

We let $J_L(t) = \int_0^t G_L(\tau)\, d\tau$ for all $t \in \mathbb{R}$. Integrating by parts, we get that

(16.124)
$$\int_\Omega \nabla u \nabla(\eta^2 G_L(u))\, dx$$
$$= \int_\Omega \eta^2 G_L'(u) |\nabla u|^2\, dx + \int_\Omega \nabla \eta^2 \nabla J_L(u)\, dx$$
$$= \frac{4\beta}{(\beta+1)^2} \int_\Omega \eta^2 |\nabla H_L(u)|^2\, dx + \int_\Omega (-\Delta \eta^2) J_L(u)\, dx$$
$$= \frac{4\beta}{(\beta+1)^2} \int_\Omega |\nabla(\eta H_L(u))|^2\, dx + \frac{4\beta}{(\beta+1)^2} \int_\Omega \eta - \Delta \eta |H_L(u)|^2\, dx$$
$$\quad + \int_\Omega (-\Delta \eta^2) J_L(u)\, dx.$$

16.5. APPENDIX: REGULARITY OF WEAK SOLUTIONS

On the other hand, with Hölder's inequality and the definition of $\mu_s(\mathbb{R}^n)$, we have that

$$\int_\Omega \left(\frac{|u|^{2^*-2-\varepsilon}}{|x|^s} - a \right) \cdot \eta^2 u G_L(u) \, dx$$

$$\leq \int_\Omega \left(|a| + \frac{|u|^{2^*-2-\varepsilon}}{|x|^s} \right) \cdot (\eta H_L(u))^2 \, dx$$

(16.125)
$$\leq \left(\int_{\Omega \cap \mathrm{Supp}\,\eta} \frac{(|a| \cdot |x|^s + |u|^{2^*-2-r_\varepsilon})^{\frac{2^*-\varepsilon}{2^*-2-\varepsilon}}}{|x|^s} \right)^{1-\frac{2}{2^*-\varepsilon}}$$

$$\times \left(\int_\Omega \frac{|\eta H_L(u)|^{2^*}}{|x|^s} \right)^{\frac{2}{2^*}} \times \left(\int_\Omega \frac{dx}{|x|^s} \right)^{\frac{2\varepsilon}{2^* \cdot (2^*-\varepsilon)}}$$

$$\leq \alpha \cdot \int_\Omega |\nabla(\eta H_L(u))|^2 \, dx$$

where

$$\alpha := $$
$$\left(\int_{\Omega \cap \mathrm{Supp}\,\eta} \frac{(|a| \cdot |x|^s + |u|^{2^*-2-r_\varepsilon})^{\frac{2^*-\varepsilon}{2^*-2-\varepsilon}}}{|x|^s} \, dx \right)^{1-\frac{2}{2^*-\varepsilon}} \mu_s(\mathbb{R}^n)^{-1} \left(\int_\Omega \frac{dx}{|x|^s} \right)^{\frac{2\varepsilon}{2^* \cdot (2^*-\varepsilon)}}$$

Plugging (16.124) and (16.125) into (16.123), we get that

(16.126)
$$A \cdot \int_\Omega |\nabla(\eta H_L(u))|^2 \, dx \leq \frac{4\beta}{(\beta+1)^2} \int_\Omega |\eta - \Delta \eta| |H_L(u)|^2 \, dx$$
$$+ \int_\Omega |-\Delta(\eta^2) J_L(u)| \, dx,$$

where

$$A := \frac{4\beta}{(\beta+1)^2}$$
$$- \left(\int_{\Omega \cap \mathrm{Supp}\,\eta} \frac{(|a| \cdot |x|^s + |u|^{2^*-2-r_\varepsilon})^{\frac{2^*-\varepsilon}{2^*-2-\varepsilon}}}{|x|^s} \, dx \right)^{1-\frac{2}{2^*-\varepsilon}} \mu_s(\mathbb{R}^n)^{-1} \left(\int_\Omega \frac{dx}{|x|^s} \right)^{\frac{2\varepsilon}{2^* \cdot (2^*-\varepsilon)}}$$

We let $p_0 = \sup\{p \geq 1/\ u \in L^p(\Omega)\}$. It follows from Sobolev's embedding theorem that $p_0 \geq \frac{2n}{n-2}$. We claim that

$$p_0 = +\infty.$$

We proceed by contradiction and assume that $p_0 < \infty$. It follows from the definition of p_0 that $u \in L^p(\Omega)$ for any $p \in (2, p_0)$. Fix such a p and let $\beta = p - 1 > 1$. For any $x \in \overline{\Omega}$, we let $\delta_x > 0$ be small enough so that

(16.127)
$$\mu_s(\mathbb{R}^n)^{-1} \left(\int_{\Omega \cap B_{2\delta_x}(x)} \frac{(|a| \cdot |x|^s + |u|^{2^*-2-r_\varepsilon})^{\frac{2^*-\varepsilon}{2^*-2-\varepsilon}}}{|x|^s} \, dx \right)^{1-\frac{2}{2^*-\varepsilon}} \left(\int_\Omega \frac{dx}{|x|^s} \right)^{\frac{2\varepsilon}{2^* \cdot (2^*-\varepsilon)}}$$
$$\leq \frac{2\beta}{(\beta+1)^2}.$$

Since $\overline{\Omega}$ is compact, we get that there exists $x_1, ..., x_N \in \overline{\Omega}$ such that $\overline{\Omega} \subset \bigcup_{i=1}^{N} B_{\delta_{x_i}}(x_i)$. We fix $i \in \{1, ..., N\}$ and let $\eta \in C^\infty(B_{2\delta_{x_i}}(x_i))$ such that $\eta(x) = 1$ for all $x \in B_{\delta_{x_i}}(x_i)$. We then get from (16.126) and (16.127) that

$$\frac{2\beta}{(\beta+1)^2} \int_\Omega |\nabla(\eta H_L(u))|^2 \, dx \leq \frac{4\beta}{(\beta+1)^2} \int_\Omega |\eta - \Delta\eta| |H_L(u)|^2 \, dx$$
(16.128)
$$+ \int_\Omega |-\Delta\eta^2| \cdot |J_L(u) \, dx.$$

From the Sobolev inequality, there exists $K(n,2) > 0$ that depends only on n such that for all $f \in H_0^1(\mathbb{R}^n)$,

(16.129)
$$\left(\int_{\mathbb{R}^n} |f|^{\frac{2n}{n-2}} \, dx \right)^{\frac{n-2}{n}} \leq K(n,2) \int_{\mathbb{R}^n} |\nabla f|^2 \, dx.$$

It follows from (16.128) and (16.129) that for all $L > 0$,

$$\frac{2\beta}{(\beta+1)^2} K(n,2)^{-1} \left(\int_\Omega |\eta H_L(u)|^{\frac{2n}{n-2}} \, dx \right)^{\frac{n-2}{n}} \leq \frac{4\beta}{(\beta+1)^2} \int_\Omega |\eta - \Delta\eta| |H_L(u)|^2 \, dx$$
$$+ \int_\Omega |-\Delta\eta^2| \cdot |J_L(u)| \, dx.$$

As easily checked, there exists $C_0 > 0$ such that $|J_L(t)| \leq C_0 \cdot |t|^{\beta+1}$ for all $t \in \mathbb{R}$ and all $L > 0$. Since $u \in L^{\beta+1}(\Omega)$, we get that there exists a constant $C = C(\eta, u, \beta, \Omega)$ independant of L such that

$$\int_{\Omega \cap B_{\delta_{x_i}}(x_i)} |H_L(u)|^{\frac{2n}{n-2}} \, dx \leq \int_\Omega |\eta H_L(u))|^{\frac{2n}{n-2}} \, dx \leq C$$

for all $L > 0$. Letting $L \to +\infty$, we get that

$$\int_{\Omega \cap B_{\delta_{x_i}}(x_i)} |u|^{\frac{n}{n-2}(\beta+1)} \, dx < +\infty,$$

for all $i = 1...N$. We then get that $u \in L^{\frac{n}{n-2}(\beta+1)}(\Omega) = L^{\frac{n}{n-2}p}(\Omega)$. And then, $\frac{n}{n-2}p \leq p_0$ for all $p \in (2, p_0)$. Letting $p \to p_0$, we get a contradiction. It follows that $p_0 = +\infty$ and that $u \in L^p(\Omega)$ for all $p \geq 1$.

In the next step we show that $u \in C^{0,\alpha}(\overline{\Omega})$ for all $\alpha \in (0,1)$. Indeed, it follows from the previous step and the assumption $0 < s < 2$ that there exists $p > \frac{n}{2}$ such that $f_\varepsilon := \frac{|u|^{2^*-2-\varepsilon} u}{|x|^s} - au \in L^p(\Omega)$. It follows from standard elliptic theory that, in this case, $u \in C^{0,\alpha}(\overline{\Omega})$ for all $\alpha \in (0, \min\{2-s, 1\})$. We let

$$\alpha_0 = \sup\{\alpha \in (0,1) / u \in C^{0,\alpha}(\overline{\Omega})\}.$$

For any $\alpha \in (0, \alpha_0)$, we have $u \in C^{0,\alpha}(\overline{\Omega})$, and since $u(0) = 0$, we have that

(16.130)
$$|u(x)| \leq |u(x) - u(0)| \leq C|x|^\alpha.$$

We then get from (16.130) that for all $x \in \Omega$,

$$|f_\varepsilon(x)| = \left| \frac{|u(x)|^{2^*-1-\varepsilon} u}{|x|^s} - au \right| \leq \frac{C}{|x|^{s-(2^*-1-\varepsilon)\alpha}}.$$

We shall show that $\alpha_0 = 1$, and for that we distinguish 2 cases:

Case 1): $s - (2^* - 1 - \varepsilon)\alpha_0 \leq 0$. In this case, we have for any $p > 1$, and up to taking α close enough to α_0, that $f_\varepsilon \in L^p(\Omega)$. Since $-\Delta u + au = f_\varepsilon$ and $u \in H_0^1(\Omega)$, it follows from standard elliptic theory that there exist exists $\theta \in (0,1)$ such that $u \in C^{1,\theta}(\overline{\Omega})$ and therefore $\alpha_0 = 1$.

Case 2): $s - (2^\star - 1 - \varepsilon)\alpha_0 > 0$. In this case, we have for any $p < \frac{n}{s-(2^\star-1-\varepsilon)\alpha_0}$, and up to taking α close enough to α_0, that $f_\varepsilon \in L^p(\Omega)$.

We distinguish the following 3 subcases:

Case 2.a): $s - (2^\star - 1 - \varepsilon)\alpha_0 < 1$, in which case –and up to taking α close enough to α_0– there exists $p > n$ such that $f_\varepsilon \in L^p(\Omega)$. Since $-\Delta u = f_\varepsilon$ and $u \in H_0^1(\Omega)$, it follows from standard elliptic theory that there exist exists $\theta \in (0,1)$ such that $u \in C^{1,\theta}(\overline{\Omega})$, hence again $\alpha_0 = 1$.

Case 2.b): $s - (2^\star - 1 - \varepsilon)\alpha_0 = 1$, in which case we have for any $p < n$ – and up to taking α close enough to α_0– that $f_\varepsilon \in L^p(\Omega)$. Since $-\Delta u + au = f_\varepsilon$ and $u \in H_0^1(\Omega)$, it follows from standard elliptic theory that $u \in C^{0,\tilde{\alpha}}(\overline{\Omega})$ for all $\tilde{\alpha} \in (0,1)$, and hence $\alpha_0 = 1$.

Case 2.c): $s - (2^\star - 1 - \varepsilon)\alpha_0 > 1$. It then follows from standard elliptic theory that $u \in C^{0,\tilde{\alpha}}(\overline{\Omega})$ for all $\tilde{\alpha} \leq 2 - (s - (2^\star - 1 - \varepsilon)\alpha_0)$. From the definition of α_0, we get that $\alpha_0 \geq 2 - (s - (2^\star - 1 - \varepsilon)\alpha_0)$, and then $0 \geq 2 - s + (2^\star - 2 - \varepsilon)\alpha_0 > 0$, which is a contradiction since $s < 2$ and $\varepsilon < 2^\star - 2$. This proves that this case does not occur, and we are back to the other cases. where $\alpha_0 = 1$.

We are now ready to prove that there exists $\theta \in (0,1)$ such that $u \in C^{1,\theta}(\overline{\Omega})$. We proceed again as in the previous step and consider any $\alpha \in (0, \alpha_0) = (0,1)$. We then get that for all $x \in \Omega$, that $|f_\varepsilon(x)| = \left|\frac{|u(x)|^{2^\star - 1 - \varepsilon} u}{|x|^s}\right| \leq \frac{C}{|x|^{s-(2^\star-1-\varepsilon)\alpha}}$.

We distinguish 2 cases:

Case 1): $s - (2^\star - 1 - \varepsilon) \leq 0$, in which case we have for any $p > 1$ –up to taking α close enough to $\alpha_0 = 1$– that $f_\varepsilon \in L^p(\Omega)$. Since $-\Delta u + au = f_\varepsilon$ and $u \in H_0^1(\Omega)$, it follows from standard elliptic theory that there exist exists $\theta \in (0,1)$ such that $u \in C^{1,\theta}(\overline{\Omega})$.

Case 2): $s - (2^\star - 1 - \varepsilon) > 0$, in which case we have for any $p < \frac{n}{s-(2^\star-1-\varepsilon)}$ –up to taking α close enough to $\alpha_0 = 1$– we get that $f_\varepsilon \in L^p(\Omega)$. As easily checked, $1 - (s - (2^\star - 1 - \varepsilon)) = 2 - s + (2^\star - 1 - \varepsilon) - 1 > 2^\star - 2 - \varepsilon$, and we therefore get that $f_\varepsilon \in L^p(\Omega)$ for some $p > n$. Since $-\Delta u + au = f_\varepsilon$ and $u \in H_0^1(\Omega)$, it follows from standard elliptic theory that there exists $\theta \in (0,1)$ such that $u \in C^{1,\theta}(\overline{\Omega})$, and we are done. □

16.6. Further comments

Egnell [**127**] was first to realize that when 0 belongs to the boundary of the domain, then things may be different in terms of the attainability of the best constant in the Hardy-Sobolev inequality. He considers open cones of the form $C = \{x \in \mathbb{R}^n; x = r\theta, \theta \in \Sigma \text{ and } r > 0\}$ where the base Σ is a connected domain of the unit sphere S^{n-1} of $\mathbb{R}^n$ and shows that $\mu_s(C)$ is then attained for $0 < s < 2$ even when $\bar{C} \neq \mathbb{R}^n$.

Ghoussoub-Kang considered the smooth case in [**157**] and showed that in dimension $n \geq 4$, the negativity of all principal curvatures at 0 –which is essentially a condition of *"strict concavity"* at 0– leads to attainability of the best constant for problems with Dirichlet boundary conditions, while the Neumann problems required the positivity of the mean curvature at 0. On the other hand, standard

Pohozaev type arguments show non-attainability in the cases where Ω is convex or star-shaped at 0.

Eventually, Ghoussoub-Robert [**158**] showed that the negativity of the mean curvature at 0 is sufficient for $n \geq 4$. It is their proof that we include here. While the proof of Kang-Ghoussoub relies on a Brezis-Nirenberg type argument, the one given here uses a much more powerful blow-up technique. In [**159**], Ghoussoub and Robert carry the analysis further and proved the result also in dimension $n = 3$, but the more involved proof is omitted and can be viewed in that paper, where the existence of an infinite number of sign changing solutions for (16.2) is also established under slightly stronger conditions on the curvature. In [**160**], they tackle similar questions for various critical equations involving a whole affine subspace of singularities on the boundary.

The study of blow-up solutions in certain nonlinear elliptic equations was initiated by Atkinson-Peletier [**31**] (see also Brézis-Peletier [**64**]). In the Riemannian context, such asymptotics were first studied by Schoen [**248**] and Hebey-Vaugon [**184**]. The techniques of blow-up have been developed in a general context by Druet, Hebey and Robert [**125**]. They turned out to be very powerful tools for the study of best constant problems in Sobolev inequalities, see for instance Druet [**122**], Hebey-Vaugon [**184**], [**185**] and Robert [**244**]). We also mention the work of Han [**176**], Hebey [**183**], Druet-Robert [**126**] and Robert [**243**]) on the asymptotics for solutions to nonlinear pde's, the 3−dimensional conjecture of Brézis solved by Druet [**123**] and the intricate compactness issues in the Riemannian context (see for instance Schoen [**248**] and Druet [**124**]).

One can also study the best constant in the Hardy-Rellich-Sobolev inequality and whether it can be attained in the case when $0 \in \partial\Omega$, in particular in conical domains. See for example the recent preprint by Caldiroli-Musina [**78**] for some very partial results.

Hsia-Lin-Wadade [**210**] refined the Brezis-Nirenberg type argument used by Ghoussoub-Kang [**157**] to tackle the more general Euler-Lagrange equations corresponding to the critical Caffarelli-Kohn-Nirenberg inequalities. This leads to the following question.

Open problem (19): Establish Theorem 16.1.3 in the case where the potential a is not necessarily C^1 around 0 but still in such a way that $-\Delta + a$ is still coercive. The key example being $a(x) = -\frac{\mu}{|x|^2}$ where $0 < \mu < \frac{(n-2)^2}{4}$.

In view of Corollary 15.1.1, such an extension will then yield that the negative mean curvature condition at $0 \in \partial\Omega$ also ensures that the best constant in the Caffarelli-Kohn-Nirenberg inequality

$$(16.131) \qquad \left(\int_\Omega |x|^{-bq}|u|^q\right)^{\frac{2}{q}} dx \leq C \int_\Omega |x|^{-2a}|\nabla u|^2 dx,$$

is also attained whenever $-\infty < a < \frac{n-2}{2}$, $0 \leq b - a \leq 1$ and $q = \frac{2n}{n-2+2(b-a)}$.

Part 6

Aubin-Moser-Onofri Inequalities

CHAPTER 17

Log-Sobolev Inequalities on the Real Line

We consider the functional

$$I_\alpha(g) = \frac{\alpha}{2}\int_{-1}^{1}(1-x^2)|g'(x)|^2\,dx + \int_{-1}^{1}g(x)\,dx - \log\frac{1}{2}\int_{-1}^{1}e^{2g(x)}\,dx$$

on the space $H^1(-1,1)$ of functions in $L^2(-1,1)$ such that $\int_{-1}^{1}(1-x^2)|g'(x)|^2 dx < \infty$. If I_α is restricted to the manifold $\mathcal{G} = \left\{g \in H^1(-1,1); \int_{-1}^{1}e^{2g(x)}x\,dx = 0\right\}$, then the following holds:

- If $\alpha \geq \frac{1}{2}$, then $\inf_{g \in \mathcal{G}} I_\alpha(g) = 0$.
- if $\alpha < \frac{1}{2}$, then $\inf_{g \in \mathcal{G}} I_\alpha(g) = -\infty$.

We also show, that if u^* denotes the Legendre transform of a function u, then the functional

$$\Phi(u) = \int_{-1}^{1}u(x)\,dx - \log\left(\frac{1}{2}\int_{-\infty}^{+\infty}e^{-2u^*(x)}\,dx\right)$$

is convex on the cone $\mathcal{W}$ of all bounded convex functions u on $(-1,1)$, and that

$$\inf_{u \in \mathcal{W}} \Phi(u) = \log(\frac{4}{\pi}).$$

Both inequalities will play a key role in the next two chapters, which address the Moser-Trudinger and the Moser-Onofri-Aubin inequalities on the two-dimensional sphere $\mathbb{S}^2$.

17.1. One-dimensional version of the Moser-Aubin inequality

Consider the functional

$$I_\alpha(g) = \frac{\alpha}{2}\int_{-1}^{1}(1-x^2)|g'(x)|^2\,dx + \int_{-1}^{1}g(x)\,dx - \log\frac{1}{2}\int_{-1}^{1}e^{2g(x)}\,dx$$

on the space $H^1(-1,1)$ of functions in $L^2(-1,1)$ such that

$$\|g\|_{H^1} = (\int_{-1}^{1}|g(x)|^2 dx)^{1/2} + (\int_{-1}^{1}(1-x^2)|g'(x)|^2 dx)^{1/2} < \infty.$$

We start by showing that for small α, the functional is not bounded below even when restricted to the manifold

$$\mathcal{G} = \left\{g \in H^1(-1,1); \int_{-1}^{1}e^{2g(x)}x\,dx = 0\right\}.$$

LEMMA 17.1.1. *If $\alpha < 1/2$, then I_α is not bounded below on the manifold $\mathcal{G}$.*

Proof: Consider the trial functions

(17.1) $$g(x) = \begin{cases} c\log(1-x) & \text{for } 0 < x < 1-\epsilon \\ c\log(\epsilon) & \text{for } 1-\epsilon < x < 1 \end{cases}$$

extended as even functions to the whole interval $(-1, 1)$. It is clear that such functions belong to $\mathcal{G}$ and a straightforward calculation shows that for small ϵ

$$I_\alpha(g) = 2\alpha c^2 |\log(\epsilon)| - \log\left(-\frac{\epsilon^{2c+1}}{2c+1} + \frac{1}{2c+1} + \epsilon^{2c+1}\right) + O(1).$$

If $2c + 1 < 0$, this becomes

$$I_\alpha(g) = p(c)|\log(\epsilon)| + O(1),$$

where $p(c) = 2\alpha c^2 + 2c + 1$. Now suppose $\alpha < 1/2$. Then the discriminant of $p(c)$, namely $4 - 8\alpha$, is positive. Hence $p(c)$ has real roots and must be negative for some value of c. For this value of c, $2c + 1 < p(c) < 0$, so $I_\alpha(g)$ tends to $-\infty$ as ϵ becomes small. □

On the other hand, we have the following result.

PROPOSITION 17.1.1. *If $\alpha > \frac{1}{2}$, then the functional I_α is bounded below and is coercive on the manifold $\mathcal{G}$.*

The coercivity will follow immediately from the following Lemma.

LEMMA 17.1.2. *Suppose g is a function on $(-1,1)$ such that $g(0) = 0$, $\|g\|^2 = \int_{-1}^{1} (1-x^2)|g'(x)|^2\, dx < \infty$, and $\int_{-1}^{1} e^{2g(x)} x\, dx = 0$. Then, for all $0 < \beta < 1$, we have*

$$\int_{-1}^{1} e^{2g(x)} dx \leq C_\beta e^{\|g\|^2/4\beta^2}. \tag{17.2}$$

Proof: Introduce the following notation

$$\|g\|_+^2 = \int_0^1 (1-x^2)|g'(x)|^2 dx \quad \text{and} \quad \|g\|_-^2 = \int_{-1}^0 (1-x^2)|g'(x)|^2 dx.$$

We first establish the following simple claim. If $\|g\| < \infty$ and $g(0) = 0$ then

$$|g(x)| \leq \begin{cases} \|g\|_+ \operatorname{arctanh}^{\frac{1}{2}}(|x|) & \text{if } 0 \leq x < 1 \\ \|g\|_- \operatorname{arctanh}^{\frac{1}{2}}(|x|) & \text{if } -1 \leq x < 0. \end{cases} \tag{17.3}$$

Indeed, for $0 \leq x < 1$ we have

$$|g(x)| = \left|\int_0^x g'(y)\,dy\right| \leq \int_0^x \frac{1}{\sqrt{1-y^2}} \sqrt{1-y^2}|g'(y)|\,dy$$

$$\leq \left[\int_0^x \frac{1}{1-y^2} dy\right]^{1/2} \|g\|_+ = \|g\|_+ \operatorname{arctanh}^{\frac{1}{2}}(|x|).$$

The proof for $-1 < x \leq 0$ is similar.

To establish the lemma, we first note that since $\|g\|_+^2 + \|g\|_-^2 = \|g\|^2$, we have $\|g\|_+^2 = \|g\|^2/2 - \delta$ and $\|g\|_-^2 = \|g\|^2/2 + \delta$ for some δ with $-\|g\|^2/2 < \delta < \|g\|^2/2$. For convenience, assume $\delta \geq 0$. Now,

$$\operatorname{arctanh}(|x|) = \frac{1}{2}\log\frac{1+|x|}{1-|x|} \leq \frac{1}{2}\log\frac{2}{1-|x|}.$$

This inequality and (17.3) imply that for $0 \leq x < 1$

$$\begin{aligned}
e^{2g(x)} &\leq \exp\left(2\|g\|_+ \operatorname{arctanh}^{\frac{1}{2}}(|x|)\right) \\
&\leq \exp\left(\sqrt{2}\|g\|_+ \left[\log 2 + \log\frac{1}{1-x}\right]^{1/2}\right) \\
&= \exp\left(2\frac{\|g\|_+}{\sqrt{2}\beta}\beta\left[\log 2 + \log\frac{1}{1-|x|}\right]^{1/2}\right) \\
&\leq \exp\left(\frac{\|g\|_+^2}{2\beta^2} + \beta^2\left[\log 2 + \log\frac{1}{1-|x|}\right]\right) \\
&= \exp\left(\frac{\|g\|^2}{4\beta^2} - \frac{\delta}{2\beta^2} + \beta^2\left[\log 2 + \log\frac{1}{1-|x|}\right]\right) \\
&\leq 2\exp\left(\frac{\|g\|^2}{4\beta^2}\right)\lambda^{-1}(1-|x|)^{-\beta^2},
\end{aligned}$$

that is

(17.4) $$e^{2g(x)} \leq 2\exp\left(\frac{\|g\|^2}{4\beta^2}\right)\lambda^{-1}(1-|x|)^{-\beta^2},$$

where $\lambda = \exp\left(\frac{\delta}{2\beta^2}\right)$. Similarly, for $-1 \leq x < 0$,

(17.5) $$e^{2g(x)} \leq 2\exp\left(\frac{\|g\|^2}{4\beta^2}\right)\lambda(1-|x|)^{-\beta^2}.$$

Now, since $\int_{-1}^{1} e^{2g(x)} x\, dx = 0$, we have

$$\int_{-1}^{1} e^{2g(x)}\, dx = \int_{-1}^{1} e^{2g(x)}(1+\lambda x)\, dx \leq \int_{-1}^{1} e^{2g(x)}(1+\lambda x)_+\, dx,$$

where $(1+\lambda x)_+ = \max\{1+\lambda x, 0\}$. Thus, using (17.4) and (17.3), we have that for any $0 < \beta < 1$

$$\begin{aligned}
&\int_{-1}^{1} e^{2g(x)}\, dx \\
&\leq 2\exp\left(\frac{\|g\|^2}{4\beta^2}\right)\left(\frac{1}{\lambda}\int_0^1 (1-|x|)^{-\beta^2}(1+\lambda x)_+\, dx + \lambda \int_{-1}^0 (1-|x|)^{-\beta^2}(1+\lambda x)_+\, dx\right) \\
&= 2\exp\left(\frac{\|g\|^2}{4\beta^2}\right)\left(\frac{1}{\lambda}\int_0^1 (1-x)^{-\beta^2}(1+\lambda x)\, dx + \lambda \int_{-1/\lambda}^0 (1+x)^{-\beta^2}(1+\lambda x)\, dx\right).
\end{aligned}$$

The proof is completed by recalling that $\lambda \geq 1$ and noting that for constants depending only on β we have

$$\int_0^1 (1-x)^{-\beta^2}(1+\lambda x)\, dx \leq C_1 + C_2\lambda,$$

and

$$\begin{aligned}
\int_{-1/\lambda}^0 (1+x)^{-\beta^2}(1+\lambda x)\, dx &\leq \int_{-1/\lambda}^0 (1+x)^{-\beta^2}\, dx \\
&= (1-\beta^2)^{-1}(1-(1-\lambda^{-1})^{1-\beta^2}) \\
&\leq C_3/\lambda.
\end{aligned}$$

It is now clear that by taking β in such a way that $\frac{1}{2\alpha} < \beta^2 < 1$, the above lemma combined with the Sobolev embedding yields that for any $g \in \mathcal{G}$,

$$I_\alpha(g) \geq \gamma_1 \|g\|^2 - \gamma_2 \|g\| - \gamma_3$$

where γ_1, γ_2 and γ_3 are positive constants. That is I_α is coercive on $\mathcal{G}$. □

17.2. The Euler-Lagrange equation and the case $\alpha \geq \frac{2}{3}$

First, we analyze the critical points of the functional I_α restricted to $\mathcal{G}$.

PROPOSITION 17.2.1. *If $0 < \alpha < 1$, then any critical point g of I_α restricted to $\mathcal{G}$ satisfies the following differential equation*

$$(17.6) \qquad \alpha \frac{d}{dx}(1-x^2)\frac{d}{dx}g - 1 + \frac{2}{\lambda}e^{2g} = 0,$$

where $\lambda = \int_{-1}^{1} e^{2g} dx$.

Proof: Indeed, any critical point g of I_α restricted to $\mathcal{G}$ satisfies the following Euler-Lagrange equation:

$$(17.7) \qquad \alpha \frac{d}{dx}(1-x^2)\frac{d}{dx}g - 1 + \left(\frac{2}{\lambda} + \mu x\right)e^{2g} = 0,$$

where μ is a Lagrange multiplier and $\lambda = \int_{-1}^{1} e^{2g} dx$. We need to show that $\mu = 0$. For that, we multiply equation (17.6) by the function $f(x) = (1-x^2)g'(x) - \frac{x}{\alpha}$ and integrate between -1 and 1. After a series of integration by parts and using that $g \in \mathcal{G}$, we obtain the following:

The first term becomes

$$\alpha \int_{-1}^{1} f(x)\frac{d}{dx}((1-x^2)g'(x))\,dx = -\alpha \int_{-1}^{1} f'(x)(1-x^2)g'(x)\,dx$$

$$= -\alpha \int_{-1}^{1} [(-2xg' + (1-x^2)g'' - \frac{1}{\alpha}](1-x^2)g'(x)\,dx$$

$$= \alpha \int_{-1}^{1} 2x(1-x^2)(g')^2\,dx - \alpha \int_{-1}^{1} (1-x^2)^2 g' g''\,dx$$

$$+ \int_{-1}^{1} (1-x^2)g'\,dx$$

$$= \alpha \int_{-1}^{1} 2x(1-x^2)(g')^2\,dx - \frac{\alpha}{2}\int_{-1}^{1} (1-x^2)^2 d((g')^2)$$

$$+ \int_{-1}^{1} (1-x^2)g'\,dx$$

$$= \int_{-1}^{1} (1-x^2)g'\,dx.$$

The second term is

$$(17.8) \qquad -\int_{-1}^{1} f(x)\,dx = -\int_{-1}^{1}(1-x^2)g'\,dx$$

17.2. THE EULER-LAGRANGE EQUATION AND THE CASE $\alpha \geq \frac{2}{3}$

while the fact that $\int_{-1}^{1} xe^{2g}\, dx = 0$ implies that the third term is

$$\int_{-1}^{1} f(x)(\frac{2}{\lambda} + \mu x)e^{2g}\, dx$$

$$= \frac{2}{\lambda}\int_{-1}^{1}(1-x^2)g'e^{2g}\, dx + \mu\int_{-1}^{1} x(1-x^2)g'e^{2g}\, dx - \frac{\mu}{\alpha}\int_{-1}^{1} x^2 e^{2g}\, dx$$

$$= \frac{2}{\lambda}\int_{-1}^{1} xe^{2g}\, dx + \frac{\mu}{2}\int_{-1}^{1} x(1-x^2)d(e^{2g}) - \frac{\mu}{\alpha}\int_{-1}^{1} x^2 e^{2g}\, dx$$

$$= -\frac{\mu}{2}\int_{-1}^{1}(1-3x^2)e^{2g}\, dx - \frac{\mu}{\alpha}\int_{-1}^{1} x^2 e^{2g}\, dx$$

$$= \frac{\mu}{2}\int_{-1}^{1} e^{2g}[(3-\frac{2}{\alpha})x^2 - 1]\, dx.$$

Finally, by adding the 3 terms, we get that $\mu L_\alpha = 0$ where

$$L_\alpha = \frac{1}{2}\int_{-1}^{1} e^{2g}[(3-\frac{2}{\alpha})x^2 - 1]\, dx.$$

But it is easy to see that $L_\alpha < 0$ when $0 < \alpha < 1$. Hence μ is necessarily zero. □

Now, we show the following

PROPOSITION 17.2.2. *Suppose $g \in H^1(-1,1)$ satisfies the differential equation*

(17.9) $$\alpha\frac{d}{dx}(1-x^2)\frac{d}{dx}g - 1 + \frac{2}{\lambda}e^{2g} = 0,$$

where $\lambda = \int_{-1}^{1} e^{2g}dx$. If $\alpha \neq 1$, then the function $G(x) = (1-x^2)g'(x)$ belongs to $H^1(-1,1)$ and satisfies

(17.10) $$\alpha G' - 1 + \frac{2}{\lambda}e^{2g} = 0,$$

and

(17.11) $$\begin{cases} (1-x^2)G'' + \frac{2}{\alpha}G - 2GG' = 0 \\ G(-1) = G(1) = 0 \text{ and } \int_{-1}^{1} G(x)\, dx = 0. \end{cases}$$

Proof: Let $G(x) = (1-x^2)g'(x)$ and note that since $g \in H^1$, we necessarily have $G(1) = G(-1) = 0$ and that $G \in L^2(-1,1)$. By substituting in equation (17.6), we get

(17.12) $$\alpha G'(x) - 1 + \frac{2}{\lambda}e^{2g(x)} = 0 \text{ for } x \in (-1,1).$$

It follows that $G \in H^1(-1,1)$ and by differentiating, we also get

(17.13) $$\alpha G''(x) + \frac{4}{\lambda}e^{2g(x)}g'(x) = 0 \text{ for } x \in (-1,1).$$

Multiply (17.13) by $1 - x^2$ and use (17.12) to get

(17.14) $$\alpha(1-x^2)G''(x) + 2(1-\alpha G'(x))G(x) = 0 \text{ for } x \in (-1,1).$$

or

(17.15) $$(1-x^2)G''(x) + \frac{2}{\alpha}G(x) - 2G'(x)G(x) = 0 \text{ for } x \in (-1,1).$$

Integrate (17.15) between -1 and 1 to obtain:

$$(17.16) \qquad \int_{-1}^{1}(1-x^2)G''(x)\,dx + \frac{2}{\alpha}\int_{-1}^{1} G(x)\,dx - \int_{-1}^{1} 2G(x)G'(x)\,dx = 0.$$

Note that the last term is equal to $\int_{-1}^{1} d(G^2) = 0$, while the first term is equal—modulo two integration by parts— to

$$(17.17) \qquad \int_{-1}^{1}(1-x^2)G''(x)\,dx = \int_{-1}^{1}(1-x^2)\,d(G')$$

$$= (1-x^2)G'(x)|_{-1}^{1} - \int_{-1}^{1} -2xG'(x)\,dx$$

$$= (1-x^2)G'(x)|_{-1}^{1} - 2\int_{-1}^{1} G(x)\,dx.$$

Combine (17.16) and (17.17) to get

$$\left(\frac{2}{\alpha}-2\right)\int_{-1}^{1} G(x)\,dx = -(1-x^2)G'(x)|_{-1}^{1}.$$

On the other hand, and back to (17.12) we have, in view of (17.2)) that

$$\alpha(1-x^2)G'(x)|_{-1}^{1} = (1-x^2)|_{-1}^{1} - \frac{2}{\lambda}(1-x^2)e^{2g}|_{-1}^{1} = 0,$$

which, since $\alpha \neq 1$, implies that $\int_{-1}^{1} G(x)\,dx = 0$. $\qquad \square$

COROLLARY 17.2.1. *Assume $\alpha \geq \frac{2}{3}$ and $\alpha \neq 1$, then the following hold:*
(1) *The only solutions of the equation (17.9) are the constant functions.*
(2) *In particular, the only critical points of the functional I_α restricted to $\mathcal{G}$ are the constant functions, and therefore $\inf_{g \in \mathcal{G}} I_{\frac{2}{3}}(g) = 0$.*

Proof: With the notation of the previous proposition, (17.11) means that G is orthogonal to the first eigenspace of the operator $\frac{d}{dx}((1-x^2)\frac{d}{dx})$ on $H^1(-1,1)$. Since the second eigenvalue is 2, we have

$$(17.18) \qquad 2\int_{-1}^{1}|G(x)|^2 \leq \int_{-1}^{1}(1-x^2)|G'(x)|^2\,dx.$$

On the other hand, if we multiply (17.15) above by G and integrate by parts twice, we get

$$(17.19) \qquad \int_{-1}^{1}(1-x^2)|G'(x)|^2\,dx = \left(\frac{2}{\alpha}-1\right)\int_{-1}^{1}|G(x)|^2.$$

So, by comparing (17.19) and (17.18), we get that either $\alpha \leq 2/3$ or that $G \equiv 0$ and hence g is a constant. $\qquad \square$

17.3. The optimal bound in the one-dimensional Aubin-Moser-Onofri inequality

The next theorem improves considerably the above result and provides the optimal bound.

THEOREM 17.3.1. *If $\alpha \geq \frac{1}{2}$, then the only critical points of the functional I_α restricted to $\mathcal{G}$ are constant functions, and therefore $\inf_{g \in \mathcal{G}} I_{\frac{1}{2}}(g) = 0$.*

17.3. THE OPTIMAL BOUND

To this end, we need some notation and some basic facts about Legendre's polynomials. Let $P_n(x)$ be the n-th Legendre polynomial, i.e., P_n satisfies
$$((1-x^2)P_n')' + \lambda_n P_n = 0, \lambda_n = n(n+1), \quad n = 0, 1, ...$$
Note that $P_0 = 1, P_1 = x, P_2 = \frac{1}{2}(3x^2 - 1), ...$ Moreover (see [1])

(17.20) $$|P_n'(x)| \leq \frac{1}{2}\lambda_n, \int_{-1}^{1} P_n^2 = \frac{2}{2n+1}.$$

Let now g be a solution of (17.6), and set $G = (1-x^2)g'$ as above. Write
$$G(x) = \beta x + a_2 \frac{1}{2}(3x^2 - 1) + \sum_{k=3}^{\infty} a_k P_k(x),$$
$G_2 = \sum_{k=3}^{\infty} a_k P_k(x)$, and $b_k^2 = a_k^2 \int_{-1}^{1} P_k^2$, $k \geq 2$.

We first derive some equalities:

(17.21) $$\int_{-1}^{1} (1-x^2)(G')^2 = (\frac{2}{\alpha} - 1)\int_{-1}^{1} G^2,$$

(17.22) $$\int_{-1}^{1} P_1 G = \frac{2}{3}\beta,$$

(17.23) $$\int_{-1}^{1} (1-x^2)\frac{e^{2g}}{\lambda} = \frac{2}{3}(1 - \alpha\beta),$$

(17.24) $$\int_{-1}^{1} P_k G = -\frac{2}{\alpha\lambda_k}\int_{-1}^{1}(1-x^2)P_k'\frac{e^{2g}}{\lambda}, k \geq 2,$$

(17.25) $$\int_{-1}^{1} G^2 = (6 - \frac{2}{\alpha})\frac{2}{3}\beta,$$

(17.26) $$\frac{2}{3}\beta(4\beta + (7 - \frac{2}{\alpha})(\frac{2}{\alpha} - 6)) = \int_{-1}^{1}(1-x^2)(G_2')^2 - 6\int_{-1}^{1} G_2^2,$$

(17.27) $$\int_{-1}^{1}(1-x^2)(G_2')^2 - 6\int_{-1}^{1} G_2^2 = \sum_{k=3}^{\infty}(\lambda_k - 6)b_k^2.$$

Proofs of 17.21-17.27: (17.21) was established in (17.11). The relation (17.22) follows by definition. Multiplying (17.6) by $\int_{-1}^{x} P_k(s)ds, k \geq 1$ and integrating over $[-1,1]$ we obtain (17.23) and (17.24). Multiplying (17.15) by x and integrating from -1 to 1 we obtain (17.25). To show (17.26), we just need to use (17.21), (17.25) and the definition of G_2. The equality (17.27) follows from definition. □

LEMMA 17.3.2. *If $\beta = 0$, then $G = 0$.*

Proof: Indeed, $\beta = 0$ means that G is orthogonal to the second eigenspace of the operator $\frac{d}{dx}((1-x^2)\frac{d}{dx})$ on $H^1(-1,1)$. Since the third eigenvalue is equal to 6, we have

(17.28) $$6\int_{-1}^{1}|G(x)|^2 \leq \int_{-1}^{1}(1-x^2)|G'(x)|^2 \, dx.$$

On the other hand, if we multiply (17.15) above by G and integrate by parts twice, we get

(17.29) $$\int_{-1}^{1}(1-x^2)|G'(x)|^2 \, dx = (\frac{2}{\alpha} - 1)\int_{-1}^{1}|G(x)|^2.$$

So, by comparing (17.29) and (17.28), we get that either $\alpha \leq 2/7$ or that $G \equiv 0$ and hence g is constant. $\square$

The rest of the proof consists of showing that $\beta = 0$. The strategy is to show that if $\beta \neq 0$, then
$$\beta = \frac{1}{\alpha},$$
which will lead to a contradiction.

Assuming that $\beta \neq 0$, we shall derive the following inequalities. From (17.23) we have

(17.30) $$\frac{1}{\alpha} - \beta > 0.$$

By definition we have
$$b_k^2 = a_k^2 \int_{-1}^1 P_k^2 = \frac{(\int_{-1}^1 GP_k)^2}{\int_{-1}^1 P_k^2}$$
$$\leq \frac{2k+1}{2}(\frac{2}{\alpha\lambda_k}\int_{-1}^1 (1-x^2)|P_k'|\frac{e^{2g}}{\lambda})^2$$
$$\leq \frac{2k+1}{2}(\frac{2}{\alpha\lambda_k}\frac{\lambda_k}{2}\frac{2}{3}(1-\alpha\beta))^2.$$

Hence we obtain

(17.31) $$b_k^2 \leq \frac{2(2k+1)}{9}(\frac{1}{\alpha} - \beta)^2, \quad k \geq 2.$$

Similarly we obtain

(17.32) $$\frac{3}{5}|a_2| \leq \frac{1}{\alpha} - \beta.$$

From (17.26) and since again $\beta > 0$,
$$4\beta + (7 - \frac{2}{\alpha})(\frac{2}{\alpha} - 6) \geq 0.$$

Since $\alpha \geq 0.5$, we have

(17.33) $$\beta \geq \frac{1}{4}(7 - \frac{2}{\alpha})(6 - \frac{2}{\alpha}) \geq 1.5.$$

From (17.26) and (17.30), we have
$$\frac{4}{\alpha} + (7 - \frac{2}{\alpha})(\frac{2}{\alpha} - 6) \geq 0.$$
which implies that
$$\alpha \leq 0.537.$$

From (17.26) we have
$$\frac{2}{3}\beta(4\beta + (7 - \frac{2}{\alpha})(\frac{2}{\alpha} - 6)) = \int_{-1}^1 (1-x^2)(G_2')^2 - 6\int_{-1}^1 G_2^2$$
$$\geq \frac{1}{2}\int_{-1}^1 (1-x^2)(G_2')^2$$
$$\geq \frac{1}{2}[\int_{-1}^1 (1-x^2)(G')^2 - \frac{4}{3}\beta^2 - \frac{12}{5}a_2^2]$$
$$\geq \frac{1}{2}[(\frac{2}{\alpha} - 1)(6 - \frac{2}{\alpha})\frac{2}{3}\beta - \frac{4}{3}\beta^2 - \frac{12}{5}a_2^2].$$

Hence we obtain

$$\frac{2}{3}\beta[\frac{5}{\alpha}+(7-\frac{2}{\alpha})(\frac{2}{\alpha}-6)-\frac{1}{2}(\frac{2}{\alpha}-1)(6-\frac{2}{\alpha})] \geq \frac{10}{3}\beta(\frac{1}{\alpha}-\beta)-\frac{6}{5}a_2^2$$
$$\geq \frac{10}{3}\beta(\frac{1}{\alpha}-\beta)-\frac{6}{5}\times\frac{25}{9}(\frac{1}{\alpha}-\beta)^2$$
(17.34)
$$\geq \frac{10}{3}(2\beta-\frac{1}{\alpha})(\frac{1}{\alpha}-\beta).$$

Since $(\frac{1}{\alpha}-\beta)\geq 0$ and $2\beta-\frac{1}{\alpha}\geq 0$, we conclude that (since $\beta > 0$)

(17.35) $$\frac{5}{\alpha}+(7-\frac{2}{\alpha})(\frac{2}{\alpha}-6)-\frac{1}{2}(\frac{2}{\alpha}-1)(6-\frac{2}{\alpha})\geq 0.$$

which implies, by a simple computation, that

(17.36) $$\alpha \leq 0.52.$$

Moreover, since $\alpha \geq 0.5$ and $\beta < 1.5$, we obtain from (17.34) that

(17.37) $$\frac{1}{\alpha}-\beta \leq \frac{\beta}{5(2\beta-\frac{1}{\alpha})} \leq \frac{\beta}{5}.$$

To obtain better estimates, we fix an integer $n \geq 3$. We have by (17.26) and (17.27)

$$\frac{2}{3}\beta(4\beta+(7-\frac{2}{\alpha})(\frac{2}{\alpha}-6)) = \sum_{k=3}^{\infty}(\lambda_k-6)b_k^2$$
$$= \sum_{k=3}^{n}(\lambda_k-6)b_k^2 + \sum_{k=n+1}^{\infty}(\lambda_k-6)b_k^2$$
$$\geq \sum_{k=3}^{n}(\lambda_k-6)b_k^2 + \frac{\lambda_{n+1}-6}{\lambda_{n+1}}\sum_{k=n+1}^{\infty}\lambda_k b_k^2$$
$$= \sum_{k=3}^{n}(\lambda_k-6)b_k^2 + \frac{\lambda_{n+1}-6}{\lambda_{n+1}}(\frac{2}{3}\beta(\frac{2}{\alpha}-1)(6-\frac{2}{\alpha})-\frac{4}{3}\beta^2-\frac{12}{5}a_2^2-\sum_{k=3}^{n}\lambda_k b_k^2)$$
$$= \sum_{k=3}^{n}(\lambda_k-6-\frac{\lambda_{n+1}-6}{\lambda_{n+1}}\lambda_k)b_k^2 + \frac{\lambda_{n+1}-6}{\lambda_{n+1}}(\frac{2}{3}\beta(\frac{2}{\alpha}-1)(6-\frac{2}{\alpha})-\frac{4}{3}\beta^2-\frac{12}{5}a_2^2)$$
$$= \sum_{k=3}^{n}6\frac{\lambda_k-\lambda_{n+1}}{\lambda_{n+1}}b_k^2-\frac{12}{5}a_2^2\frac{\lambda_{n+1}-6}{\lambda_{n+1}}+\frac{\lambda_{n+1}-6}{\lambda_{n+1}}(\frac{2}{3}\beta(\frac{2}{\alpha}-1)(6-\frac{2}{\alpha})-\frac{4}{3}\beta^2).$$

Hence we have

$$\frac{2}{3}\beta(4\beta+(7-\frac{2}{\alpha})(\frac{2}{\alpha}-6))-\frac{\lambda_{n+1}-6}{\lambda_{n+1}}(\frac{2}{3}\beta(\frac{2}{\alpha}-1)(6-\frac{2}{\alpha})-\frac{4}{3}\beta^2)$$

(17.38) $$\geq \sum_{k=3}^{n}6\frac{\lambda_k-\lambda_{n+1}}{\lambda_{n+1}}b_k^2-\frac{12}{5}a_2^2\frac{\lambda_{n+1}-6}{\lambda_{n+1}}.$$

After some simple computations, the left hand of (17.38) equals to

$$12\beta(\frac{1}{\alpha}-2)+\frac{4\beta}{\lambda_{n+1}}[(\frac{2}{\alpha}-1)(6-\frac{2}{\alpha})-\frac{2}{\alpha}]-4\beta(1-\frac{2}{\lambda_{n+1}})(\frac{1}{\alpha}-\beta).$$

Thus we have by (17.31), (17.32) and (17.38)

(17.39)
$$12\beta(\frac{1}{\alpha} - 2) + \frac{4\beta}{\lambda_{n+1}}[(\frac{2}{\alpha} - 1)(6 - \frac{2}{\alpha}) - \frac{2}{\alpha}]$$

$$\geq 4\beta(1 - \frac{2}{\lambda_{n+1}})(\frac{1}{\alpha} - \beta) - \frac{12}{5}a_2^2\frac{\lambda_{n+1} - 6}{\lambda_{n+1}} + 6\sum_{k=3}^{n}\frac{\lambda_k - \lambda_{n+1}}{\lambda_{n+1}}\frac{2(2k+1)}{9}(\frac{1}{\alpha} - \beta)^2$$

$$\geq 4\beta(1 - \frac{2}{\lambda_{n+1}})(\frac{1}{\alpha} - \beta) - \frac{20}{3}\frac{\lambda_{n+1} - 6}{\lambda_{n+1}}(\frac{1}{\alpha} - \beta)^2 - \frac{4}{3}\sum_{k=3}^{n}\frac{\lambda_{n+1} - \lambda_k}{\lambda_{n+1}}(2k+1)(\frac{1}{\alpha} - \beta)^2$$

$$\geq [4\beta(1 - \frac{2}{\lambda_{n+1}}) - \frac{20}{3}\frac{\lambda_{n+1} - 6}{\lambda_{n+1}}(\frac{1}{\alpha} - \beta) - \frac{4}{3}c_n(\frac{1}{\alpha} - \beta)](\frac{1}{\alpha} - \beta),$$

where
$$c_n = \sum_{k=3}^{n}\frac{\lambda_{n+1} - \lambda_k}{\lambda_{n+1}}(2k+1).$$

Since $1/2 < \alpha \leq 1$ and $\lambda_n > 2$ for $n \geq 1$, we have

$$12\beta(\frac{1}{\alpha} - 2) + \frac{4\beta}{\lambda_{n+1}}[(\frac{2}{\alpha} - 1)(6 - \frac{2}{\alpha}) - \frac{2}{\alpha}] - \frac{8\beta}{\lambda_{n+1}}$$

$$= 4\beta(\frac{1}{\alpha} - 2)[3 - \frac{4}{\lambda_{n+1}}(\frac{1}{\alpha} - 1)] \leq 0.$$

Thus the left hand side of (17.39) satisfies

(17.40)
$$LHS\ of\ (17.39) \leq \frac{8\beta}{\lambda_{n+1}}.$$

We now claim

(17.41)
$$\frac{1}{\alpha} - \beta \leq \frac{4}{\lambda_n}, \qquad \forall n \geq 4.$$

By (17.40), we just need to show that the right hand side of (17.39) satisfies

(17.42)
$$RHS\ of\ (17.39) \geq 2\beta(\frac{1}{\alpha} - \beta).$$

We prove it by induction, and we start with $n = 4$. To this end, we iterate the inequality (17.39). Note that the right hand side of (17.39) with $n = 3$ equals

$$[4\beta(1 - \frac{2}{20}) - \frac{20}{3}\frac{20-6}{20}(\frac{1}{\alpha} - \beta) - \frac{4}{3}\frac{20-12}{20} \times 7(\frac{1}{\alpha} - \beta)](\frac{1}{\alpha} - \beta)$$

$$\geq [4\beta\frac{9}{10} - \frac{14}{3}(\frac{1}{\alpha} - \beta) - \frac{56}{15}(\frac{1}{\alpha} - \beta)](\frac{1}{\alpha} - \beta)$$

$$\geq [3.6\beta - \frac{126}{15}(\frac{1}{\alpha} - \beta)](\frac{1}{\alpha} - \beta)$$

$$\geq [3.6\beta - \frac{126}{15}\frac{\beta}{5}](\frac{1}{\alpha} - \beta) \qquad (by\ (2.14))$$

(17.43)
$$\geq 1.92\beta(\frac{1}{\alpha} - \beta).$$

By using (17.40) and (17.39) again, we obtain

(17.44)
$$\frac{1}{\alpha} - \beta \leq \frac{8}{20}\frac{1}{1.92} < 0.25.$$

Similarly, by using (17.44), we have

$$\text{RHS of (17.39)} \geq [3.6\beta - \frac{126}{15} \times 0.25](\frac{1}{\alpha} - \beta) \quad \text{(by (17.44))}$$

$$\geq 2\beta(\frac{1}{\alpha} - \beta) \quad \text{(since } \beta > 1.5 \text{ by (17.37))}$$

Thus (17.42) holds for $n = 4$ and hence (17.41) holds for $n = 4$.

Let us now assume that

$$\frac{1}{\alpha} - \beta \leq \frac{4}{\lambda_k}, k = n \geq 4.$$

We observe that for $n \geq 4$

$$c_n = \sum_{k=3}^{n}(2k+1) - \frac{1}{\lambda_{n+1}}\sum_{k=3}^{n}\lambda_k(2k+1)$$

$$= \sum_{k=3}^{n}(2k+1) - \frac{1}{\lambda_{n+1}}\sum_{k=3}^{n}k(k+1)(2k+1)$$

$$= \frac{1}{2}\lambda_{n+1} - 9 + \frac{36}{\lambda_{n+1}}.$$

Hence we have by (17.38)

$$12\beta(\frac{1}{\alpha} - 2) + \frac{4\beta}{\lambda_{n+1}}[(\frac{2}{\alpha} - 1)(6 - \frac{2}{\alpha}) - \frac{2}{\alpha}] \geq$$

(17.45) $[4\beta(1 - \frac{2}{\lambda_{n+1}}) - (\frac{20}{3}\frac{\lambda_{n+1} - 6}{\lambda_{n+1}} + \frac{4}{3}(\frac{1}{2}\lambda_{n+1} - 9 + \frac{36}{\lambda_{n+1}}))(\frac{1}{\alpha} - \beta)](\frac{1}{\alpha} - \beta).$

The right hand of (17.45) satisfies

$$\text{RHS of (17.45)} \geq [4\beta(1 - \frac{2}{\lambda_{n+1}}) + \frac{64}{3}\frac{1}{\lambda_n} - \frac{32}{\lambda_n\lambda_{n+1}} - \frac{8}{3}\frac{\lambda_{n+1}}{\lambda_n}](\frac{1}{\alpha} - \beta).$$

To show (17.42), we only need to show

$$\beta(1 - \frac{4}{\lambda_{n+1}}) \geq -\frac{32}{3}\frac{1}{\lambda_n} + \frac{16}{\lambda_n\lambda_{n+1}} + \frac{4}{3}\frac{\lambda_{n+1}}{\lambda_n},$$

or

$$\beta \geq \frac{4}{3} \cdot \frac{\lambda_{n+1}^2 - 8\lambda_{n+1} + 12}{\lambda_n(\lambda_{n+1} - 4)}.$$

In view of the inductive assumption, it suffices to show

$$\frac{1}{\alpha} \geq \frac{4}{3} \cdot \frac{\lambda_{n+1}}{\lambda_n} \cdot \frac{\lambda_{n+1} - 5}{\lambda_{n+1} - 4}.$$

Because of (17.36), it is easy to verify that the above inequality holds for $n \geq 4$.

In conclusion, we have obtained (17.41). Finally we can finish the proof by letting $n \to +\infty$ in (17.41) to obtain that $\frac{1}{\alpha} - \beta = 0$, which is a contradiction to (17.30). This implies that $\beta = 0$ and therfore $G \equiv 0$. Hence $g' \equiv 0$, and g is identically constant.

17.4. Ghigi's inequality for convex bounded functions on the line

For a convex function $u : \mathbb{R}^n \to (-\infty, \infty]$, we denote by u^* its Legendre transform, which is defined by the formula

(17.46) $$u^*(y) = \sup \{xy - u(x) \ ; \ x \in \mathbb{R}\}.$$

If u is defined only on a subset $\Omega \subset \mathbb{R}^n$, one first extends it to all of $\mathbb{R}^n$ by putting it equal to $+\infty$ on $\mathbb{R}^n \setminus \Omega$, then applies the formula above to define its Legendre transform.

Consider now the function

(17.47) $$u_0(x) = \log\left(\frac{1 + e^{2x}}{2e^x}\right),$$

and denote by $\mathcal{V}$ the space of smooth functions on the real line such that

(17.48) $$\begin{cases} u = u_0 + a & \text{for } x \ll 0 \\ u = u_0 + b & \text{for } x \gg 0, \end{cases}$$

where here, a and b are constants depending on the function u.

We now prove the first of two inequalities that will be used in the next chapter.

THEOREM 17.4.1. *Consider the functional*

(17.49) $$\Psi(u) = \int_{-\infty}^{+\infty} \left(xu'(x) - u(x)\right) u''(x) \, dx.$$

Ψ *is then a well-defined functional on* $\mathcal{V}$, *and satisfies*

(17.50) $$\Psi(u) \geq \int_{-1}^{1} u^*(y) \, dy \text{ for all } u \in \mathcal{V}.$$

Moreover, equality holds if u is strictly convex.

We shall need the following observations.

LEMMA 17.4.2. *The following statements hold:*
  a) *If u_1 and u_2 are functions on $\mathbb{R}^n$, then $\|u_1^* - u_2^*\|_\infty \leq \|u_1 - u_2\|_\infty$.*
  b) *The Legendre transform of u_0 is given by*
  $$u_0^*(y) = \frac{1}{2}(1+y) \log(1+y) + \frac{1}{2}(1-y) \log(1-y),$$
  which is a continuous function on $[-1, 1]$.
  c) *If $u \in \mathcal{V}$, then $u^* \in L^\infty(-1, 1)$.*
  d) *If $u \in \mathcal{V}$ and $y \in (-1, 1)$, then the supremum in the definition of u^* is attained at some point x such that $u'(x) = y$.*

Proof: (a) From the definition (17.46)

$$\begin{aligned} u_1^*(y) &= \sup_{x \in \mathbb{R}^n} \left(xy - u_1(x)\right) \\ &\leq \sup_{x \in \mathbb{R}^n} \left(xy - u_2(x)\right) + \sup_{x \in \mathbb{R}^n} \left(u_2(x) - u_1(x)\right) \\ &\leq u_2^*(y) + \|u_1 - u_2\|_\infty. \end{aligned}$$

Interchanging u_1 and u_2 and taking the sup in y one gets the result. Note that this still holds when the functions attain infinite values. Indeed, if the set where they are infinite is not the same for both, clearly $\|u_1 - u_2\|_\infty = \infty$ and there is nothing to prove. While if they are both finite on the same set $\Omega \subset \mathbb{R}^n$, it suffices to compute the suprema above on the set Ω. (We use the convention $\infty - \infty = 0$.) (b) is an elementary computation, and (c) follows immediately from (a) and (b). To prove (d), let $|y| < 1$ in such a way that $xy - u(x) = x(y-1) + (x - u(x))$. Now $x - u(x) = x - u_0(x) + (u_0(x) - u(x))$ is bounded for $x > 0$. On the other hand as $x \to +\infty$, $x(y-1)$ tends to $-\infty$. Therefore $\lim_{x \to +\infty}(xy - u(x)) = -\infty$ and similarly for $x \to -\infty$. Hence the supremum is attained at some point $\bar{x}$. But $z(x) = yx - u(x)$ is a smooth function of x, therefore $z'(\bar{x}) = y - u'(\bar{x}) = 0$. □

Proof of Theorem 17.4.1: To show that Ψ is well-defined, it is enough to prove that for any $u \in \mathcal{V}$, we have that $u'' \in L^1$ and $xu' - u \in L^\infty$. Note first that the case of $u = u_0$ is immediate since

$$(17.51) \qquad u_0'(x) = \frac{e^{2x} - 1}{e^{2x} + 1} \qquad u_0''(x) = \frac{4e^{2x}}{(e^{2x} + 1)^2}.$$

The case of a general $u \in \mathcal{V}$ follows directly from (17.48).

Assume now that $w(x) = xu'(x) - u(x) \geq -M$ for some $M \in \mathbb{R}$, and put $\bar{u}(x) = u(x) - M$. Then $\bar{u}' = u'$ and $\bar{w}(x) = x\bar{u}'(x) - \bar{u}(x) = w + M \geq 0$. Moreover $\bar{u}^* = u^* - M$. Since $\lim_{x \to \pm\infty} u'(x) = \lim_{x \to \pm\infty} u_0'(x) = \pm 1$, $\int u'' = 2$, $\Psi(u) = \Psi(\bar{u}) - 2M$ and $\int_{-1}^{1} \bar{u}^* = \int_{-1}^{1} u^* - 2M$. It therefore suffices to prove (17.50) for $u = \bar{u}$. Put $f = \bar{u}' : \mathbb{R} \to \mathbb{R}$. From the coarea formula ([**166**] p. 82, Theorem 2) and since $\bar{w} \geq 0$ one gets

$$\Psi(\bar{u}) = \int_{-\infty}^{+\infty} \bar{w}(x) f'(x)\,dx = \int_{-\infty}^{+\infty}\left[\sum_{f^{-1}(y)} \bar{w}(x)\right]dy \geq \int_{-1}^{1}\left[\sum_{f^{-1}(y)} \bar{w}(x)\right]dy.$$

Now again from $\bar{w} \geq 0$ and (d) of Lemma 17.4.2, it follows that $\sum_{f^{-1}(y)} \bar{w}(x) \geq \bar{u}^*(y)$, whence the result. Finally, if u is strictly convex then $u^*(u'(x)) = xu'(x) - u(x) = w(x)$, and it suffices to make the substitution $y = u'(x)$ to prove the equality in (17.50). □

THEOREM 17.4.3. *Let $\mathcal{W}$ denote the space of bounded convex functions on $(-1,1)$. The functional $\Phi : \mathcal{W} \to \mathbb{R}$ defined by*

$$(17.52) \qquad \Phi(u) = \int_{-1}^{1} u(y)\,dy - \log\left(\frac{1}{2}\int_{-\infty}^{+\infty} e^{-2u^*(x)}\,dx\right)$$

is then finitely-valued convex and bounded below on $\mathcal{W}$.

We start with the following elementary property of Legendre transforms.

LEMMA 17.4.4. *Let u_1, u_2 be functions on a convex subset $\Omega \subset \mathbb{R}^n$. For $\lambda \in [0,1]$ put $u = \lambda u_1 + (1-\lambda)u_2$ and denote by u_1^*, u_2^*, u^* the Legendre transforms of u_1, u_2 and u respectively. Then for any $x, y \in \Omega$*

$$(17.53) \qquad u^*(\lambda x + (1-\lambda)y) \leq \lambda u_1^*(x) + (1-\lambda)u_2^*(y).$$

Proof: It is enough to apply (17.46):

$$\begin{aligned}
\lambda u_1^*(x) + (1-\lambda)u_2^*(y) &= \lambda \sup_{\xi \in \Omega}\{x \cdot \xi - u_1(\xi)\} + (1-\lambda)\sup_{\eta \in \Omega}\{y \cdot \eta - u_2(\eta)\} \\
&\geq \sup_{\xi \in \Omega}\{\lambda(x \cdot \xi - u_1(\xi)) + (1-\lambda)(y \cdot \xi - u_2(\xi))\} \\
&= \sup_{\xi \in \Omega}\{(\lambda x + (1-\lambda)y) \cdot \xi - u(\xi)\} \\
&= u^*(\lambda x + (1-\lambda)y).
\end{aligned}$$

$\square$

We shall need the Prékopa-Leindler inequality in the following form.

LEMMA 17.4.5. *Let φ, ψ and μ be nonnegative measurable functions on $[0, \infty)$ such that for all $x, y \in [0, \infty), \lambda \in [0, 1]$, we have*

(17.54) $$\mu(x^\lambda y^{1-\lambda}) \geq \varphi(x)^\lambda \psi(y)^{1-\lambda}.$$

Then

(17.55) $$\int_0^\infty \mu \geq \left(\int_0^\infty \varphi\right)^\lambda \left(\int_0^\infty \psi\right)^{1-\lambda}.$$

Proof: The Prékopa-Leindler inequality states that if f, g and m are nonnegative measurable functions on $\mathbb{R}^n$, such that for all $x, y \in \mathbb{R}^n, \lambda \in [0, 1]$, we have

(17.56) $$m(\lambda x + (1-\lambda)y) \geq f(x)^\lambda g(y)^{1-\lambda},$$

then

(17.57) $$\int_{\mathbb{R}^n} m \geq \left(\int_{\mathbb{R}^n} f\right)^\lambda \left(\int_{\mathbb{R}^n} g\right)^{1-\lambda}.$$

We shall only need it in dimension one, where it follows immediately from the arithmetic-geometric mean inequality. Indeed, by homogeneity, we may assume that $\int_{\mathbb{R}} f(x)dx = \int_{\mathbb{R}} g(x)dx = 1$ and – modulo an approximation – that f and g are continuous with strictly positive values. Define $x, y : [0, 1] \to \mathbb{R}$ by

$$\int_{-\infty}^{x(t)} f(q)dq = t, \quad \int_{-\infty}^{y(t)} g(q)dq = t.$$

Therefore x and y are increasing and differentiable and

$$x'(t)f(x(t)) = y'(t)g(y(t)) = 1.$$

Set $z(t) = \theta x(t) + (1-\theta)y(t), t \in [0, 1]$. By the arithmetic-geometric mean inequality, for every t,

$$z'(t) = \theta x'(t) + (1-\theta)y'(t) \geq (x'(t))^\theta (y'(t))^{1-\theta}.$$

Since z is injective, we get from the hypothesis (17.56) on m and the above inequality,

$$\begin{aligned}
\int_{\mathbb{R}} m\,dx &\geq \int_0^1 m(z(t))z'(t)dt \\
&\geq \int_0^1 f(x(t))^\theta g(y(t))^{1-\theta} (x'(t))^\theta (y'(t))^{1-\theta} dt \\
&= \int_0^1 [f(x(t))x'(t)]^\theta [g(y(t))y'(t)]^{1-\theta} dt \\
&= 1,
\end{aligned}$$

17.4. GHIGI'S INEQUALITY FOR CONVEX BOUNDED FUNCTIONS ON THE LINE

and we are done with (17.57).

Put now $f(x) = \varphi(e^x)e^x$, $g(x) = \psi(e^x)e^x$ and $m(x) = \mu(e^x)e^x$. Then f, g, m satisfy (17.56). To get the result, apply (17.55) and use the change of variables in order to check that the integrals in (17.57) coincide with the ones in (17.55). □

Proof of Theorem 17.4.3: To show that Φ is finite on $\mathcal{W}$, consider $w \in \mathcal{W}$ and note that $||w - u_0^*||_\infty < \infty$ since both w and u_0^* are bounded. Lemma 17.4.2 (a) then yields that $||w^* - u_0||_\infty < \infty$ as well, and therefore the integral inside the logarithm in (17.52) converges, which means that $\Phi(w)$ is well-defined for $w \in \mathcal{W}$.

Note that $\mathcal{W}$ is a convex subset of $C^0(-1,1)$, so it makes sense to talk about convexity of the functional Φ. To prove it, we see that the first term of Φ in (17.52) is linear so convex. It is now enough to check that the second is concave. For that, let $u_1, u_2 \in \mathcal{W}$, $\lambda \in [0,1]$ and put $u = \lambda u_1 + (1-\lambda) u_2$. It follows from Lemma 17.4.4 that
$$e^{-2u^*(\lambda x + (1-\lambda)y)} \geq e^{-2\lambda u_1^*(x) - 2(1-\lambda)u_2^*(y)} = \left(e^{-2u_1^*(x)}\right)^\lambda \left(e^{-2u_2^*(y)}\right)^{(1-\lambda)}.$$

Applying Lemma 17.4.5 we get
$$\left(\int_{-\infty}^\infty e^{-2u^*}\right) \geq \left(\int_{-\infty}^\infty e^{-2u_1^*}\right)^\lambda \left(\int_{-\infty}^\infty e^{-2u_2^*}\right)^{(1-\lambda)}.$$

Hence
$$\log\left(\frac{1}{2} \int_{-\infty}^\infty e^{-2u^*(x)}\, dx\right) \geq \lambda \log\left(\frac{1}{2} \int_{-\infty}^\infty e^{-2u_1^*(x)}\, dx\right)$$
$$+ (1-\lambda) \log\left(\frac{1}{2} \int_{-\infty}^\infty e^{-2u_2^*(x)}\, dx\right).$$

Therefore, the second term in (17.52) is concave and Φ is convex.

If now $u \in \mathcal{W}$, then $w(y) = u(-y)$ is also in $\mathcal{W}$ and $w^*(x) = u^*(-x)$, hence $\Phi(w) = \Phi(u)$. From the convexity of Φ it follows that if $\bar{u} = (u+w)/2$, then
$$\Phi(\bar{u}) \leq \frac{\Phi(u) + \Phi(w)}{2} = \Phi(u).$$

To compute the infimum of Φ, we can therefore restrict it to even functions in $\mathcal{W}$. For such a function $u \in \mathcal{W}$, we have
$$\Phi(u) = 2 \int_0^1 u(y)\, dy - \log\left(\int_0^\infty e^{-2u^*(x)}\, dx\right).$$

Using Jensen inequality

(17.58)
$$e^{-\Phi(u)} = \exp\left(-2 \int_0^1 u(y)\, dy\right) \int_0^\infty e^{-2u^*(x)}\, dx$$
$$\leq \int_0^1 e^{-2u(y)}\, dy \int_0^\infty e^{-2u^*(x)}\, dx.$$

So it is enough to show that for some constant C and for any even function $u \in \mathcal{W}$ we have

(17.59)
$$\int_0^1 e^{-2u(y)}\, dy \int_0^\infty e^{-2u^*(x)}\, dx \leq C.$$

For that, put
$$\psi(x) = e^{-2u^*(x)}, \quad \mu(t) = e^{-t^2},$$

and
$$\varphi(y) = \begin{cases} e^{-2u(y)} & y \in [0,1] \\ 0 & y \in (1,\infty). \end{cases}$$

Since $u(y) + u^*(x) \geq xy$, we have $\sqrt{\varphi(y)\psi(x)} \leq \mu(\sqrt{xy})$, that is (17.54) with $\lambda = 1/2$. Using Lemma 17.4.5 (i.e. the Prékopa-Leindler inequality) we conclude that

$$\sqrt{\left(\int_0^\infty f\right)\left(\int_0^\infty g\right)} \leq \int_0^\infty e^{-t^2}\,dt = \frac{\sqrt{\pi}}{2}.$$

Taking the square we get (17.59) with $C = \pi/4$. This concludes the proof of the theorem. □

17.5. Further comments

This chapter is motivated by the axially symmetric case of the Moser and the Moser-Onofri-Aubin inequality on the 2-dimensional sphere. The general case will be tackled in the next two chapters, and the proofs there will make use of the results of this chapter. The fact that once I_α is restricted to

$$\mathcal{G} = \left\{g \in H^1(-1,1); \int_{-1}^1 e^{2g(x)} x\,dx = 0\right\},$$

then it is non-negative for $\alpha \geq 2/3$ (Corollary 17.2.1) was first established by Feldman, Froese, Ghoussoub and Gui [140]. Actually, they showed it for $\alpha \geq 16/25$. It was eventually proved for all $\alpha \geq 1/2$ by Gui and Wei [172] and independently by Lin [206]. Theorems 17.4.1 and 17.4.3 are due to Ghigi [154] in his proof of the Moser inequality of $\mathbb{S}^2$.

Note that, when applied to axially symmetric functions, the Moser-Onofri theorem established in the next chapter yields the following one-dimensional result:

$$(17.60) \quad \begin{cases} \text{If } \alpha \geq 1, & \text{then } \inf_{g \in H^1} I_\alpha(g) = 0. \\ \text{If } \alpha < 1, & \text{then } \inf_{g \in H^1} I_\alpha(g) = -\infty. \end{cases}$$

Theorem 17.4.3 and more specifically inequality (17.59) are a particular case of the following more general inequality established by K. Ball: If $\varphi : \mathbb{R}^n \to \mathbb{R}$ is an even measurable function such that $0 < \int e^{-\varphi}\,dx < \infty$, then,

$$(17.61) \quad \int_{\mathbb{R}^n} e^{-\varphi}\,dx \int_{\mathbb{R}^n} e^{-\varphi}\,dx \leq \left(\int_{\mathbb{R}^n} e^{-\frac{|x|^2}{2}}\,dx\right)^2,$$

with equality if and only if φ is a positive definite quadratic form a.e. See Klartag [194]) for a proof based on Caffarelli's mass transport theorem [74] mentioned in Chapter 12.

Open problem (20): Find a direct 1-dimensional proof of (17.60) without going through the results of Chapter 18. Specifically, does (17.60) follow directly from Theorem 17.4.3?

CHAPTER 18

Trudinger-Moser-Onofri Inequality on $\mathbb{S}^2$

Let $\mathbb{S}^2$ be the 2-dimensional sphere and let J_α denote the functional on the Sobolev space $H^1(\mathbb{S}^2)$ defined by

$$J_\alpha(u) = \alpha \int_{\mathbb{S}^2} |\nabla u|^2 \frac{dV_0}{4\pi} + 2 \int_{\mathbb{S}^2} u \frac{dV_0}{4\pi} - \log\left(\int_{\mathbb{S}^2} e^{2u} \frac{dV_0}{4\pi}\right),$$

where dV_0 denotes Lebesgue measure $dV_0 := \sin\theta\, d\theta \wedge d\varphi$ on the unit sphere. We establish the following Trudinger-Moser-Onofri inequality stating that

(1) If $\alpha \geq 1$, then $\inf_{u \in H^1(\mathbb{S}^2)} J_\alpha(u) = 0$.
(2) If $\alpha < 1$, then $\inf_{u \in H^1(\mathbb{S}^2)} J_\alpha(u) = -\infty$.

18.1. The Trudinger-Moser inequality on $\mathbb{S}^2$

Consider the 2-dimensional unit sphere $\mathbb{S}^2$ equipped with the standard metric g_0, whose constant Gaussian curvature is $K_0 = 1$ and with corresponding volume form dV_0 such that $\int_{\mathbb{S}^2} dV_0 = 4\pi$. Trudinger established that for some constants C, μ and ν, we have for every $u \in H^1(\mathbb{S}^2)$,

(18.1) $$\int_{\mathbb{S}^2} e^u dV_0 \leq C \exp\left[\mu \int_{\mathbb{S}^2} |\nabla u|^2 dV_0 + \nu \int_{\mathbb{S}^2} u^2 dV_0\right].$$

Later, Moser improved the above inequality by showing the following remarkable result.

For every $\beta \leq 4\pi$, there exists a constant $C(\beta)$ such that for any $u \in H^1(\mathbb{S}^2)$, we have

(18.2) $$\int_{\mathbb{S}^2} \exp\left(\frac{\beta(u-\bar{u})^2}{\int_{\mathbb{S}^2} |\nabla u|^2 dV_0}\right) \leq C(\beta),$$

where $\bar{u} := \frac{1}{4\pi} \int_{\mathbb{S}^2} u\, dV_0$.

Moreover 4π is the best constant in the sense that if $\beta > 4\pi$, then

(18.3) $$\sup_{u \in H^1(\mathbb{S}^2)} \int_{\mathbb{S}^2} \exp\left(\frac{\beta(u-\bar{u})^2}{\int_{\mathbb{S}^2} |\nabla u|^2 dV_0}\right) = +\infty.$$

Note that by using the inequality $2ab \leq a^2 + b^2$, one can write

$$2(u-\bar{u}) \leq \frac{\beta(u-\bar{u})^2}{\int_{\mathbb{S}^2} |\nabla u|^2 dV_0} + \frac{1}{\beta} \int_{\mathbb{S}^2} |\nabla u|^2 dV_0,$$

and Moser's inequality then yields that for any $\beta \leq 4\pi$,

(18.4) $$\frac{1}{4\pi} \int_{\mathbb{S}^2} e^{2u} dV_0 \leq C \exp\left(\frac{1}{\beta} \int_{\mathbb{S}^2} |\nabla u|^2 dV_0 + 2 \int_{\mathbb{S}^2} u \frac{dV_0}{4\pi}\right).$$

In other words, for any $\alpha \geq 1$, the functional

$$J_\alpha(u) = \alpha \int_{\mathbb{S}^2} |\nabla u|^2 \frac{dV_0}{4\pi} + 2 \int_{\mathbb{S}^2} u \frac{dV_0}{4\pi} - \log \int_{\mathbb{S}^2} e^{2u} \frac{dV_0}{4\pi}$$

is bounded below on the Sobolev space $H^1(\mathbb{S}^2)$, that is

(18.5) $\qquad C_\alpha = \inf\left\{J_\alpha(u); u \in H^1(\mathbb{S}^2)\right\} > -\infty$ for $\alpha \geq 1,$

while

(18.6) $\qquad\qquad\qquad C_\alpha = -\infty$ when $\alpha < 1.$

Note that we always have $C_\alpha \leq 0$. Eventually, Onofri showed that $C_1 = 0$.

This chapter is devoted to the proof of these two results. Note that by setting
$$I_\alpha(u) = J_\alpha(-\frac{u}{2}),$$
the Moser result can be formulated in the following way.

THEOREM 18.1.1. *If $\alpha \geq 1$, then the functional I_α defined on $H^1(\mathbb{S}^2)$ by*

(18.7) $\qquad I_\alpha(v) = \dfrac{\alpha}{16\pi}\displaystyle\int_{\mathbb{S}^2} |\nabla v|^2 dV_0 - \dfrac{1}{4\pi}\int_{\mathbb{S}^2} v\, dV_0 - \log\left(\dfrac{1}{4\pi}\int_{\mathbb{S}^2} e^{-v} dV_0\right),$

is bounded below on $H^1(\mathbb{S}^2)$.

We first apply symmetrization to reduce the question to a one-dimensional problem.

LEMMA 18.1.2. *Let $\mathcal{D}$ denote the space of functions on the sphere that are constant on parallel circles and that are constant near the poles. Then*
$$\inf_{C^\infty(\mathbb{S}^2)} I_1 = \inf_{\mathcal{D}} I_1$$

Proof: *Spherical symmetrization* is a process that associates to a smooth function φ on $\mathbb{S}^2$ a function $\varphi^\#$, which is constant on the parallel circles, in such a way that

(18.8) $\qquad \displaystyle\int_{\mathbb{S}^2} f(\varphi^\#) = \int_{\mathbb{S}^2} f(\varphi)$ and $\displaystyle\int_{\mathbb{S}^2} |\nabla \varphi^\#|^2 \leq \int_{\mathbb{S}^2} |\nabla \varphi|^2,$

where f is any continuous function on the real line. For details, see Baernstein [34, Corollary 3 p. 60]. One then gets that $I(\varphi^\#) \leq I(\varphi)$. A density argument based on the continuity of I in the H^1-norm shows that one can further reduce to $\mathcal{D}$.

Denote now by (θ, y) the usual coordinates on $\mathbb{S}^2$, namely $\theta \in (-\pi/2, \pi/2)$ is the longitude, that is the signed distance from equator, and y is latitude, that we consider as a periodic (geodesic) parameter on the equator itself. The metric and the volume form are then given by

(18.9) $\qquad\qquad g = d\theta^2 + \cos^2\theta\, dy^2 \qquad \omega = \cos\theta\, d\theta \wedge dy.$

Consider the one-dimensional function

(18.10) $\qquad\qquad\qquad x = \log\tan\left(\dfrac{\theta}{2} + \dfrac{\pi}{4}\right)$

and use $(x, y) \in \mathbb{R} \times \mathbb{R}$ as coordinates on $\mathbb{S}^2 \setminus \{\text{poles}\}$. ($z = x + \mathbb{I}y$ being a complex parameter on $\mathcal{C}^* \subset \mathbb{P}^1(\mathcal{C}) = \mathbb{S}^2$.)

Set again

(18.11) $\qquad\qquad\qquad u_0(x) = \log\left(\dfrac{1 + e^{2x}}{2e^x}\right),$

18.1. THE TRUDINGER-MOSER INEQUALITY ON $\mathbb{S}^2$

and define as –in Chapter 17– the space $\mathcal{V}$ of smooth functions on the real line such that

(18.12) $$\begin{cases} u = u_0 + a & \text{for } x \ll 0 \\ u = u_0 + b & \text{for } x \gg 0, \end{cases}$$

where a and b are constants depending on the function u. Also recall from the last chapter that the functional

(18.13) $$\Psi(u) = \int_{-\infty}^{+\infty} (xu'(x) - u(x))u''(x)\, dx$$

is well defined on $\mathcal{V}$ and that

$$\Psi(u) \geq \int_{-1}^{1} u^*(y)dy \quad \text{for all } u \in \mathcal{V}.$$

If now $\varphi \in \mathcal{D}$, then it does not depend on y, and it is clear that the function

(18.14) $$u(x) := u_0(x) + \frac{\varphi(x)}{2}$$

belongs to $\mathcal{V}$. We shall consistently use in the sequel the fact that

(18.15) $$u_0 + \mathcal{D} \subset \mathcal{V},$$

and the following expression of $I_1(\varphi)$ when $\varphi \in \mathcal{D}$ in terms of $u \in \mathcal{V}$.

For $\varphi \in C^\infty(\mathbb{S}^2)$ set

$$B(\varphi) = \frac{1}{16\pi} \int_{\mathbb{S}^2} |\nabla \varphi|^2 dV_0 - \frac{1}{4\pi} \int_{\mathbb{S}^2} \varphi dV_0 \quad \text{and} \quad A(\varphi) = \log\left(\frac{1}{4\pi}\int_{\mathbb{S}^2} e^{-\varphi} dV_0\right),$$

Clearly $I_1 = B - A$.

Also recall from Theorem 17.4.3 that the functional

(18.16) $$\Phi(u) = \int_{-1}^{1} u(y)\, dy - \log\left(\frac{1}{2} \int_{-\infty}^{+\infty} e^{-2u^*(x)}\, dx\right)$$

is a well defined convex and bounded below on the cone $\mathcal{W}$ of bounded convex functions on $(-1, 1)$.

PROPOSITION 18.1.1. For $\varphi \in \mathcal{D}$ and $u = u_0 + \frac{\varphi}{2}$, we have

(18.17) $$B(\varphi) = \Psi(u) - \Psi(u_0)$$

(18.18) $$A(\varphi) = \log\left(\frac{1}{2}\int_{-\infty}^{+\infty} e^{-2u(x)}\, dx\right)$$

(18.19) $$I_1(\varphi) \geq \Phi(u^*) - \Psi(u_0) = \Phi(u_0^* \star (\varphi/2)^*) - \Psi(u_0).$$

Proof: We already know that Ψ is well-defined on $\mathcal{V}$, hence on $u_0 + \mathcal{D}$. To prove the formula for B, fix $\varphi \in \mathcal{D}$ and note that $\theta = 2\arctan e^x - \frac{\pi}{2}$, $\theta' = \frac{2e^x}{1+e^{2x}}$, $|\frac{\partial}{\partial \theta}|^2 = 1$, and an easy computation yields that

$$\nabla \varphi = \frac{\partial \varphi}{\partial \theta}\frac{\partial}{\partial \theta} = \varphi' \frac{dx}{d\theta}\frac{\partial}{\partial \theta}$$

$$|\nabla \varphi|^2 = \frac{(\varphi')^2}{(\theta')^2} = \frac{(\varphi')^2}{u_0''}$$

$$|\nabla \varphi|^2 dV_0 = (\varphi')^2 dx \wedge dy.$$

It follows that $\frac{1}{16\pi}\int_{\mathbb{S}^2} |\nabla \varphi|^2 dV_0 = \frac{1}{8}\int_{-\infty}^{+\infty} (\varphi')^2\, dx$.

Since now φ' has compact support we can integrate by parts:

$$\frac{1}{16\pi}\int_{\mathbb{S}^2}|\nabla\varphi|^2 dV_0 = \frac{1}{8}\int_{-\infty}^{+\infty}(\varphi')^2 = -\frac{1}{8}\int_{-\infty}^{+\infty}\varphi\varphi''$$

$$= -\frac{1}{2}\int_{-\infty}^{+\infty}uu'' + \frac{1}{2}\int_{-\infty}^{+\infty}uu_0'' + \frac{1}{2}\int_{-\infty}^{+\infty}u_0 u'' - \frac{1}{2}\int_{-\infty}^{+\infty}u_0 u_0''.$$

On the other hand

$$\frac{1}{4\pi}\int_{\mathbb{S}^2}\varphi dV_0 = \frac{1}{2}\int_{-\infty}^{+\infty}\varphi u_0'' = \int_{-\infty}^{+\infty}uu_0'' - \int_{-\infty}^{+\infty}u_0 u_0''.$$

Hence

$$B(\varphi) = -\frac{1}{2}\int_{-\infty}^{+\infty}uu'' + \frac{1}{2}\int_{-\infty}^{+\infty}u_0 u_0'' + \frac{1}{2}\int_{-\infty}^{+\infty}(u_0 u'' - uu_0'').$$

The last integral contains some asymptotic information. Indeed, for $R \gg 0$ integration by parts gives

$$\int_{-R}^{R}(u_0 u'' - uu_0'') = \Big[u_0 u' - uu_0'\Big]_{-R}^{R}$$
$$= u_0(R)u_0'(R) - (u_0(R)+b)u_0'(R)$$
$$\quad - u_0(-R)u_0'(-R) + (u_0(-R)+a)u_0'(-R).$$

Letting R tend to ∞ we get $\int_{-\infty}^{+\infty}(u_0 u'' - uu_0'') = -(a+b)$, so that

(18.20) $$B(\varphi) = -\frac{1}{2}\int_{-\infty}^{+\infty}uu'' + \frac{1}{2}\int_{-\infty}^{+\infty}u_0 u_0'' - \frac{1}{2}(a+b).$$

(Here a and b are as in (18.12) so they depend on u.) On the other hand

$$\int_{-R}^{R}xu'u'' = \Big[(xu')u'\Big]_{-R}^{R} - \int_{-\infty}^{+\infty}(u' + xu'')u'$$
$$= \frac{1}{2}\Big[x(u')^2\Big]_{-R}^{R} - \frac{1}{2}\int_{-\infty}^{+\infty}(u')^2$$
$$= \frac{1}{2}\Big[x(u')^2\Big]_{-R}^{R} - \frac{1}{2}\Big[uu'\Big]_{-R}^{R} + \frac{1}{2}\int_{-R}^{R}uu''.$$

If $R \gg 0$, then

$$\int_{-R}^{R}xu'u'' - \int_{-R}^{R}xu_0'u_0'' = = \frac{1}{2}\int_{-R}^{R}uu'' - \frac{1}{2}\int_{-R}^{R}u_0 u_0'' - \frac{1}{2}\Big[uu' - u_0 u_0'\Big]_{-R}^{R}$$

and again

$$\Big[uu' - u_0 u_0'\Big]_{-R}^{R} = -(a+b),$$

so that

$$\int_{-\infty}^{+\infty}xu'u'' - \int_{-\infty}^{+\infty}xu_0 u_0'' = \frac{1}{2}\int_{-\infty}^{+\infty}uu'' - \frac{1}{2}\int_{-\infty}^{+\infty}u_0 u_0'' + \frac{1}{2}(a+b)$$

and

$$\begin{aligned}E(u)-E(u_0) &= \int_{-\infty}^{+\infty} xu'u'' - \int_{-\infty}^{+\infty} xu_0 u_0'' - \int_{-\infty}^{+\infty} uu'' + \int_{-\infty}^{+\infty} u_0 u_0'' \\ &= -\frac{1}{2}\int_{-\infty}^{+\infty} uu'' + \frac{1}{2}\int_{-\infty}^{+\infty} u_0 u_0'' + \frac{1}{2}(a+b) \\ &= B(\varphi).\end{aligned}$$

This proves (18.17). To prove (18.18) observe that $u_0'' = e^{-2u_0}$. Therefore

$$\frac{1}{4\pi}\int_{\mathbb{S}^2} e^{-\varphi} dV_0 = \frac{1}{4\pi}\int_{\mathbb{S}^2} e^{-\varphi} e^{-2u_0}\, dx \wedge dy = \frac{1}{2}\int_{-\infty}^{+\infty} e^{-2u(x)}\, dx$$

which proves (18.18). $\square$

Finally, using (18.17), (18.18), the fact that $u^{**} \leq u$ and Theorem 17.4.1, we obtain

$$\begin{aligned}I_1(\varphi) &= B(\varphi) - A(\varphi) \\ &= \Psi(u) - \Psi(u_0) - \log\left(\frac{1}{2}\int_{-\infty}^{+\infty} e^{-2u(x)}\, dx\right) \\ &= \Psi(u) - \Psi(u_0) - \log\left(\frac{1}{2}\int_{-\infty}^{+\infty} e^{-2u^{**}(x)}\, dx\right) \\ &= \Psi(u) - \Psi(u_0) + \Phi(u^*) - \int_{-1}^{1} u^*(y)\, dy \\ &\geq -\Psi(u_0) + \Phi(u^{**}).\end{aligned}$$

PROPOSITION 18.1.2. *The following holds*

$$\inf_{\mathcal{D}} I_1 \geq \log(\frac{4}{\pi}) - E(u_0).$$

Proof: If $\varphi \in \mathcal{D}$, then $u \in \mathcal{V}$ and u^* is bounded by Lemma 17.4.2 (a), so $u^* \in \mathcal{W}$. By Theorem 17.4.3, Φ is well defined convex and bounded below on $\mathcal{W}$ by the constant $\log(\frac{4}{\pi})$. Using (18.19), it follows that

$$\inf_{\mathcal{D}} I_1 \geq \inf_{\mathcal{W}} \Phi - E(u_0) = \log(\frac{4}{\pi}) - E(u_0).$$

18.2. The optimal Moser-Onofri inequality

We shall now establish the exact lower bounds on I_α.

THEOREM 18.2.1. *For $\alpha \geq 0$, consider the functional I_α defined for any $u \in H^1(\mathbb{S}^2)$ by*

$$(18.21) \quad I_\alpha(v) = \frac{\alpha}{16\pi}\int_{\mathbb{S}^2} |\nabla v|^2 dV_0 - \frac{1}{4\pi}\int_{\mathbb{S}^2} v\, dV_0 - \log\left(\frac{1}{4\pi}\int_{\mathbb{S}^2} e^{-v} dV_0\right).$$

(1) *If $\alpha \geq 1$, then* $\inf_{v \in H^1(\mathbb{S}^2)} I_\alpha(v) = 0$.
(2) *If $\alpha < 1$, then* $\inf_{v \in H^1(\mathbb{S}^2)} I_\alpha(v) = -\infty$.

Proof: By Theorem 18.1.1, I_α is bounded below on $H^1(\mathbb{S}^2)$ for every $\alpha > 1$. Recalling that $J_\alpha(u) = I_\alpha(-2u)$, it follows that for every $\epsilon > 0$, the functional $J_{1+\epsilon}$ is coercive which means that every minimizing sequence is bounded in $H^1(\mathbb{S}^2)$.

Since $J_{1+\varepsilon}$ is weakly lower semi-continuous on $H^1(\mathbb{S}^2)$, it follows that the infimum $C_{1+\varepsilon}$ is attained by $u_\varepsilon \in H^1(\mathbb{S}^2)$. Since u_ε is a critical point for $J_{1+\varepsilon}$, it satisfies

$$(18.22) \qquad (1+\varepsilon)\Delta u_\varepsilon + \frac{e^{2u_\varepsilon}}{\int_{\mathbb{S}^2} e^{2u_\varepsilon} \frac{dV_0}{4\pi}} = 1.$$

We shall show that $u_\varepsilon = 0$.

By symmetrization, we may suppose u_ε axially symmetric about the N-S axis. So let's consider the functional in the axially symmetric case. Let θ and φ denote the usual angular coordinates on the sphere, and define $x = \cos\theta$. The standard metric is given by $ds^2 = (1-x^2)^{-1}dx^2 + (1-x^2)d\varphi^2$. Thus, the normalized measure $\frac{dV_0}{4\pi}$ is given by $(4\pi)^{-1}dxd\varphi$, while

$$\int_{\mathbb{S}^2} |\nabla u|^2 \frac{dV_0}{4\pi} = (4\pi)^{-1} \int_{-1}^{1} \int_0^{2\pi} [(1-x^2)\left|\frac{\partial u}{\partial x}\right|^2 + (1-x^2)^{-1}\left|\frac{\partial u}{\partial \varphi}\right|^2] dx d\varphi.$$

Axially symmetric functions only depend on x, and for such functions, the functional $J_\alpha(w) = I_\alpha(-2w)$ is then given by

$$J_\alpha(w) = \frac{\alpha}{2} \int_{-1}^{1} (1-x^2)|w'(x)|^2 \, dx + \int_{-1}^{1} w(x) \, dx - \log \frac{1}{2} \int_{-1}^{1} e^{2w(x)} \, dx.$$

We shall still denote by J_α the restriction to the axially symmetric functions, equivalently the space $H^1(-1,1)$ consisting of those functions in $L^2(-1,1)$ such that

$$\|w\|_{H^1} = \left(\int_{-1}^{1} (1-x^2)|w'(x)|^2 dx\right)^{1/2} < \infty.$$

Note now that u_ε satisfies the one dimensional equation

$$(18.23) \qquad (1+\varepsilon)\frac{d}{dx}(1-x^2)\frac{d}{dx}u_\varepsilon - 1 + \frac{2}{\lambda}e^{2u_\varepsilon} = 0,$$

where $\lambda = \int_{-1}^{1} e^{2u_\varepsilon} dx$. It follows from Corollary 17.2.1 that u_ε is a constant, which clearly implies that the infimum $C_{1+\varepsilon} = I_{1+\varepsilon}(u_\varepsilon) = 0$, and hence $C_1 = 0$.

To establish 2) we first prove that if $\beta > 4\pi$, then

$$(18.24) \qquad \sup_{u \in H^1(\mathbb{S}^2)} \int_{\mathbb{S}^2} \exp\left(\frac{\beta(u-\bar{u})^2}{\int_{\mathbb{S}^2} |\nabla u|^2 dV_0}\right) dV_0 = +\infty,$$

where $\bar{u} := \frac{1}{4\pi} \int_{\mathbb{S}^2} u \, dV_0$. To show this, it is enough to construct a sequence $\{\varphi_n\}$ such that

$$(18.25) \qquad \int_{\mathbb{S}^2} |\nabla \varphi_n|^2 dV_0 = 1, \quad \int_{\mathbb{S}^2} \varphi_n dV_0 = 0, \quad \text{and} \quad \int_{\mathbb{S}^2} \exp(\beta\varphi_n^2) dV_0 \to \infty.$$

Introduce the longitude φ and latitude θ on $\mathbb{S}^2$ so that the canonical metric is given by

$$ds^2 = d\theta^2 + \cos^2\theta d\varphi^2, \quad |\theta| > \frac{\pi}{2},$$

and where $\theta = \pm\frac{\pi}{2}$ correspond to the two poles. Let $\varphi(\theta)$ be a radially symmetric function and define the variables t, $w(t)$, and $\rho(t)$ by

$$e^{t/2} := \tan\left(\frac{\theta}{2} + \frac{\pi}{4}\right), \quad w(t) := \sqrt{4\pi}\varphi(\theta), \quad \text{and} \quad \rho(t) = \frac{1}{e^t + e^{-t} + 2}.$$

18.2. THE OPTIMAL MOSER-ONOFRI INEQUALITY

In order to show (18.25), it is enough to construct a sequence $\{w_n(t)\}$ such that for any $\beta > 4\pi$,

(18.26) $$\int_{-\infty}^{\infty} |w_n'(t)|^2 dt = 1, \quad \int_{-\infty}^{\infty} w_n(t)\rho(t)dt = 0,$$

and

$$\int_{-\infty}^{\infty} \exp(\frac{\beta}{4\pi} w_n^2(t))\rho(t)dt \to \infty \text{ as } n \to +\infty.$$

Define $z_n(t)$ by

$$z_n(t) = \begin{cases} 0 & \text{if } t < 0 \\ \frac{t}{\sqrt{n}} & \text{if } 0 \leq t \leq n \\ \sqrt{n}, & \text{if } n \leq t. \end{cases}$$

Clearly $\int_{-\infty}^{\infty} (z_n'(t))^2 dt = 1$, and $\bar{z}_n := \int_{-\infty}^{\infty} z_n(t)\rho(t)dt < \infty$. Now let $w_n = z_n - \bar{z}_n$. We have

$$\int_0^{\infty} \exp(\frac{\beta}{4\pi} w_n^2(t) - t)dt \geq \int_n^{\infty} \exp(\frac{\beta}{4\pi}(\sqrt{n} - \bar{z}_n)^2 - t)dt$$

$$= \exp(\frac{\beta}{4\pi}(\sqrt{n} - \bar{z}_n)^2 - n).$$

Since $\beta > 4\pi$, we have $\int_0^{\infty} \exp(\frac{\beta}{4\pi} w_n^2(t))\rho(t)dt \to \infty$ as $n \to \infty$. Thus w_n satisfies (18.26) and consequently (18.24) follows.

Let now $\alpha < 1$. In order to show that the functional

$$I_\alpha(v) = \frac{\alpha}{16\pi} \int_{\mathbb{S}^2} |\nabla v|^2 dV_0 + \frac{1}{4\pi} \int_{\mathbb{S}^2} v dV_0 - \ln(\frac{1}{4\pi} \int_{\mathbb{S}^2} e^u dV_0)$$

is unbounded below on $H^1(\mathbb{S}^2)$, we fix $\beta := \frac{4\pi}{\alpha} > \beta' > 4\pi$ and let φ_n be the sequence in (18.25). Set $a_n = 2M_n\beta$, where M_n is the maximum of φ_n. Since

$$\beta'\varphi_n^2 - a_n\varphi_n + \frac{1}{4\beta}a_n^2 = \beta'(\varphi_n - \frac{a_n}{2\beta'})^2 + \frac{a_n^2}{4}(\frac{1}{\beta} - \frac{1}{\beta'}),$$

we have that

$$a_n\varphi_n - \frac{1}{4\beta}a_n^2 \geq \beta'\varphi_n^2 \quad \text{on the set } \{p \in \mathbb{S}^2 : \varphi_n(p) = M_n\}.$$

It follows that

$$\int_{\mathbb{S}^2} \exp(a_n\varphi_n - \frac{1}{4\beta}a_n^2)dV_0 \geq \int_{\{\varphi_n = M_n\}} \exp(\beta'\varphi_n^2)dV_0$$

(18.27) $$= \int_n^{\infty} \exp(\frac{\beta'}{4\pi} w_n^2(t))\rho(t)dt$$

$$\geq \frac{1}{4} \int_n^{\infty} \exp(\frac{\beta'}{4\pi} w_n^2(t) - t)dt.$$

Since $\beta' > 4\pi$, the right hand side of the (18.27) tends to infinity as $n \to \infty$ and consequently

(18.28) $$\lim_{n \to \infty} \int_{\mathbb{S}^2} \exp(a_n\varphi_n - \frac{1}{4\beta}a_n^2)dV_0 = \infty.$$

We then have
$$\begin{aligned}I_\alpha(a_n\varphi_n) &= \frac{\alpha a_n^2}{16\pi} - \log(\frac{1}{4\pi}\int_{\mathbb{S}^2} e^{a_n\varphi_n}dV_0) \\ &= \frac{a_n^2}{4\beta} - \log(\int_{\mathbb{S}^2} e^{a_n\varphi_n}dV_0) + \ln(4\pi) \\ &= -\log(\int_{\mathbb{S}^2}\exp(a_n\varphi_n - \frac{1}{4\beta}a_n^2)dV_0) + \log(4\pi).\end{aligned}$$

Finally, we get from (18.28) that $\lim_{n\to\infty} I_\alpha(a_n\varphi_n) = -\infty$, and we are done. □

The Moser-Onofri theorem applied to axially symmetric functions yields then the following interesting one-dimensional inequality – already mentioned in Chapter 17.

COROLLARY 18.2.1. *Consider the functional*
$$J_\alpha(g) = \frac{\alpha}{2}\int_{-1}^{1}(1-x^2)|g'(x)|^2\,dx + \int_{-1}^{1}g(x)\,dx - \log\frac{1}{2}\int_{-1}^{1}e^{2g(x)}\,dx$$
on the space $H^1(-1,1)$.
 (1) *If* $\alpha \geq 1$, *then* $\inf_{g\in H^1} I_\alpha(g) = 0$.
 (2) *If* $\alpha < 1$, *then* $\inf_{g\in H^1} I_\alpha(g) = -\infty$.

18.3. Conformal invariance of J_1 and its applications

Let g_0 be the standard metric on $\mathbb{S}^2$, and let g be a conformal metric, i.e., $g = e^{2u}g_0$ with Gaussian curvature K, then,

(18.29) $$\Delta u + K(x)e^{2u} = 1 \quad \text{on } \mathbb{S}^2,$$

where $\Delta = \Delta_{g_0}$ is the original Laplacian associated to the standard metric g_0. The solutions of (18.29) are the critical points of the functional

(18.30) $$F_K(u) = \int_{\mathbb{S}^2}|\nabla u|^2\,d\omega + 2\int_{\mathbb{S}^2}u\,d\omega - \log\int_{\mathbb{S}^2}K(x)e^{2u}\,d\omega,$$

on $H^1(\mathbb{S}^2)$, where $d\omega = \frac{dV_0}{4\pi}$.

For any $P \in \mathbb{S}^2$ and $t \geq 1$, use stereographic coordinates with P at infinity and denote $\psi(z) = \psi_{P,t}(z) = tz$, where
$$\mathbf{x} \equiv z = \cot(\frac{\theta}{2})e^{i\varphi} = \frac{x_1 + ix_2}{1-x_3}.$$

These are the conformal transformation on $\mathbb{S}^2$, which can also be identified with fractional linear transformations of the form $\psi(z) = \frac{\alpha z + \beta}{\gamma z + \delta}$ (in $SL(2,\mathbb{C})$), with $\alpha\delta - \beta\gamma = 1$. Note that

(18.31) $$\frac{1}{2}\log\det|d\psi(z)| = 2\log\frac{1+|z|^2}{|\alpha z + \beta|^2 + |\gamma z + \delta|^2}.$$

For any $u \in H^1(\mathbb{S}^2)$, we denote by $T_\psi u$ its transform by the conformal transformation ψ, that is $T_\psi(u)(x) := u(\psi(x))$ for all $x \in \mathbb{S}^2$.

Basic geometric considerations yield that (K,u) satisfy (18.29) if and only if $(K\circ\psi, T_\psi(u))$ satisfies (18.29) for any conformal transformation ψ. It follows that all the solutions of the equation

(18.32) $$\Delta u + e^{2u} = 1 \quad \text{on } \mathbb{S}^2$$

(which corresponds to gaussian curvature $K = 1$) are of the form $u = \frac{1}{2}\log det|d\psi|$, where ψ is a conformal transformation. More generally, if K is a constant c, then $u = \frac{1}{2}\log det|d\psi| + \log c$ is another non-trivial solution of (18.29) with $F_c(u) = \ln c$.

THEOREM 18.3.1. *With the above notation, the following assertions hold:*
  (1) *J_1 is conformally invariant, that is $J_1(T_\varphi(u)) = J_1(u)$ for any $u \in H^1(\mathbb{S}^2)$ and any conformal transformation ψ.*
  (2) *For any $u \in H^1(\mathbb{S}^2)$, there exists a conformal transformation ψ such that $T_\psi u \in \mathcal{M}$, where $\mathcal{M}$ is the submanifold of $H^1(\mathbb{S}^2)$ defined by*

$$\text{(18.33)} \qquad \mathcal{M} := \{u \in H^1(\mathbb{S}^2); \int_{\mathbb{S}^2} e^u \mathbf{x}\, dw = 0\}, \text{ where } \mathbf{x} = (x_1, x_2, x_3) \in \mathbb{S}^2.$$

  (3) *Consequently,*

$$\text{(18.34)} \qquad \inf_{u \in \mathcal{M}} J_1(u) = \inf_{u \in H^1(\mathbb{S}^2)} J_1(u) = 0,$$

and the infimum is attained at the constant functions $u_\psi := \frac{1}{2}\log det|d\psi|$, where ψ is a conformal transformation of $\mathbb{S}^2$.

Proof: It suffices to prove (2), which will be obtained by Brouwer's fixed point theorem. Indeed, given $u \in H^1(\mathbb{S}^2)$, we need to show that the map

$$\psi \in SL(2,\mathbb{C}) \to \int_{\mathbb{S}^2} e^{T_\psi u} \mathbf{x}\, dw \in \mathbb{R}^3$$

contains 0 in its range. For that, we note that the set of conformal transformations can be paramatrized by the unit ball $B^3 = \mathbb{S}^2 \times [1, +\infty)/\mathbb{S}^2 \times \{1\}$. Also note that for a transformation $\psi := \psi_{P,t}$, we can write the change of variable

$$\int_{\mathbb{S}^2} e^{T_\psi u} x_j\, dw = \int_{\mathbb{S}^2} e^u (x_j \circ \psi_{P,t}^{-1})\, dw.$$

It follows that for a fixed $u \in H^1(\mathbb{S}^2)$, the center of mass map

$$C : \psi := \psi_{P,t} \to \frac{\int_{\mathbb{S}^2} e^u (x_j \circ \psi_{P,t}^{-1})\, dw}{\int_{\mathbb{S}^2} e^u\, dw},$$

may be considered as a map from B^3 to B^3 with a continuous extension to the boundary map $P \in \mathbb{S}^2 \to -P \in \mathbb{S}^2$. It follows from Brouwer's fixed point theorem that there exists $P \in \mathbb{S}^2$ and $t \in [1, +\infty)$ such that $\frac{\int_{\mathbb{S}^2} e^u (x_j \circ \psi_{P,t}^{-1})\, dw}{\int_{\mathbb{S}^2} e^u\, dw} = 0$. □

One application of the Moser-Onofri inequality is an isoperimetric statement for the determinant of Laplacians established by Osgood-Philips-Sarnak. Indeed, for any conformal metric g on $\mathbb{S}^2$ one can associates its Laplace-Beltrami operator Δ_g and its eigenvalues $(\lambda_i^g)_i$. One needs to define the notion of a determinant for Δ_g that can reflect the expression $det\Delta_g = \Pi_i \lambda_i^g$. This is usually expressed and justified via the following formula

$$\text{(18.35)} \qquad det\Delta_g = e^{-\xi'(0)} \quad \text{where} \quad \xi(s) = \Sigma_{k>0} \frac{1}{\lambda_k^s} \quad \text{for } Re(s) > 1.$$

One then have the following formula established by Ray-Singer-Polyakov.

THEOREM 18.3.2. *If g is any conformal metric of the form $g = e^{2u}g_0$ and $\int_{\mathbb{S}^2} e^{2u} d\omega = 1$ (equal volume), then*

$$\text{(18.36)} \qquad \log \frac{det\Delta_g}{det\Delta_{g_0}} = -\frac{1}{12\pi} \int_{\mathbb{S}^2} (2u + |\nabla u|^2)\, d\omega.$$

The following result of Osgood-Philips-Sarnak is now an immediate application of the optimal Moser-Onofri inequality.

COROLLARY 18.3.1. *Among all conformal metrics on $\mathbb{S}^2$ with equal volume, $\log \det \Delta_{g_0}$ is maximum, where g_0 is the standard metric.*

There are also connections with the three-dimensional steady-state Navier-Stokes equations

$$-\Delta u + (u \cdot \nabla)u + \nabla p = 0 \text{ on } \mathbb{R}^3 \tag{18.37}$$
$$\text{div} u = 0, \tag{18.38}$$

which has a non-trivial scaling symmetry $u(x) \to \lambda u(\lambda x)$ and it is therefore natural to try to find solutions which are invariant under this scaling. Explicit examples of such solutions were first calculated by L.D. Landau in 1944. The question is whether there are other solutions (besides Landau's explicit solutions) that are invariant under such a scaling. Recently, Sverak (2009) answered this question by showing the following:

THEOREM 18.3.3. *The only (-1)-homogeneous solutions of the stationary Navier-Stokes equation on $\mathbb{R}^3$ are the Landau solutions.*

Sverak's proof consists of noting first that any (-1)-homogeneous solution can be written as $u(x) = \mathbf{v}(\mathbf{x}) + f(x)\mathbf{x}$ (where $\mathbf{v}(\mathbf{x})$ is tangent to $\mathbb{S}^2$), and that u is a solution of (18.37) if and only if $\mathbf{v} = \nabla \varphi$, where

$$\Delta \varphi + 2e^\varphi = 2 \text{ on } \mathbb{S}^2.$$

Finally, recall that the only solutions of this last equation are given by $\varphi = \frac{1}{2} \log \det |d\psi|$, where ψ is a conformal transformation.

18.4. Further comments

It was N. Trudinger [265] who first proved an exponential case of the Sobolev embedding theorem on bounded domains of $\mathbb{R}^2$. Later, Moser improved the inequality by finding the best constant and by establishing the corresponding result on the 2-dimensional sphere, which is what concerns us in this chapter. Other proofs of –or rather variations on– the Trudinger-Moser inequality were given by Adachi-Tanaka [3], Carlson-Chang [82], Flucher [146], McLeod-Peletier [218], Chang and Yang [86, 87], among others.

The proof given here of the fact that J_1 is bounded below is quite recent and is due to Alessandro Ghigi [154]. In [229], Onofri used the conformal invariance of J_1 to prove that its infimum is actually equal to zero. Other proofs were also given by Hong [187] and by Osgood-Phillips-Sarnak [230], who used it to prove an isoperimetric inequality for the determinant of Laplacians via the Ray-Singer-Polyakov formula.

Note that Onofri's inequality on the sphere $\mathbb{S}^2$ can be re-written as

$$\int_{\mathbb{S}^2} e^{2u - 2\int_{\mathbb{S}^2} u\, d\sigma}\, d\sigma \leq e^{\|\nabla u\|^2_{L^2(\mathbb{S}^2, d\sigma)}}, \tag{18.39}$$

for all $u \in \mathcal{E} = \{u \in L^1(\mathbb{S}^2, d\sigma) \mid |\nabla u| \in L^2(\mathbb{S}^2, d\sigma)\}$, where $d\sigma$ denotes the measure induced by Lebesgue's measure on $\mathbb{R}^3 \supset \mathbb{S}^2$, normalized so that $\int_{\mathbb{S}^2} d\sigma = 1$. Using the stereographic projection from $\mathbb{S}^2$ onto $\mathbb{R}^2$, one sees that (18.39) is equivalent to the following inequality on $\mathbb{R}^2$:

$$\int_{\mathbb{R}^2} e^{v-\int_{\mathbb{R}^2} v\, d\mu}\, d\mu \le e^{\frac{1}{16\pi}\|\nabla v\|^2_{L^2(\mathbb{R}^2, dx)}},$$

for all $v \in \mathcal{D} = \{v \in L^1(\mathbb{R}^2, d\mu) : |\nabla v| \in L^2(\mathbb{R}^2, dx)\}$ where $d\mu$ denotes the probability measure $d\mu = \frac{dx}{\pi(1+|x|^2)^2}$. Recently, Dolbeault-Esteban-Tarantello [**119**] extended the above (Euclidean) Moser-Trudinger inequality to the family of probability measures $d\mu_\alpha = \frac{\alpha+1}{\pi} \frac{|x|^{2\alpha}\, dx}{(1+|x|^{2(\alpha+1)})^2}$ by showing that the weighted inequality

(18.40) $$\int_{\mathbb{R}^2} e^{v-\int_{\mathbb{R}^2} v\, d\mu_\alpha}\, d\mu_\alpha \le e^{\frac{1}{16\pi(\alpha+1)}\|\nabla v\|^2_{L^2(\mathbb{R}^2, dx)}}$$

holds for $\alpha > -1$ as long as v is in the subspace of

$$\mathcal{E}_\alpha = \left\{v \in L^1(\mathbb{R}^2, d\mu_\alpha) : |\nabla v| \in L^2(\mathbb{R}^2, dx)\right\}$$

consisting of radially symmetric about the origin. On the other hand, they show that without symmetry assumption, inequality (18.40) holds in $\mathcal{E}_\alpha$ if and only if $\alpha \in (-1, 0]$. They also use the above information to investigate possible symmetry breaking phenomena for extremal functions of the Caffarelli-Kohn-Nirenberg inequalities in two space dimensions.

Note that Ghigi's proof of the Trudinger-Moser inequality presented above relies on a one-dimensional inequality (Theorem 17.4.3), which itself follows from the Prépoka-Leindler inequality (Lemma 17.4.5). It is well known that this last inequality is closely related to geometric inequalities that can be obtained from optimal mass transportation.

Open problem (21): Find a direct proof of the Moser-Trudinger inequality on $\mathbb{S}^2$ and its best constant, by using optimal mass transport on the 2-dimensional sphere.

CHAPTER 19

Optimal Aubin-Moser-Onofri Inequality on $\mathbb{S}^2$

We consider again the functional

$$J_\alpha(u) = \alpha \int_{\mathbb{S}^2} |\nabla u|^2 \frac{dV_0}{4\pi} + 2 \int_{\mathbb{S}^2} u \frac{dV_0}{4\pi} - \ln \int_{\mathbb{S}^2} e^{2u} \frac{dV_0}{4\pi}$$

on the Sobolev space $H^1(\mathbb{S}^2)$, where $\mathbb{S}^2$ is the 2-dimensional unit sphere equipped with the standard metric g_0 and the corresponding volume form $dV_0 := \sin\theta\, d\theta \wedge d\varphi$. We show here that the Moser-Onofri inequality can be improved when J_α is restricted to the submanifold

$$\mathcal{M} = \{u \in H^1(\mathbb{S}^2); \int_{\mathbb{S}^2} e^{2u} \mathbf{x}\, dV_0 = 0\}.$$

In particular,

(1) If $\alpha \geq \frac{2}{3}$, then $\inf_{u \in \mathcal{M}} J_\alpha(u) = 0$.

(2) If $\alpha < \frac{1}{2}$, then $\inf_{u \in \mathcal{M}} J_\alpha(u) = -\infty$.

The question whether (1) remains true for $1/2 \leq \alpha < 2/3$ is still an open problem.

19.1. The Aubin inequality

Let $\mathbb{S}^2$ be the 2-dimensional unit sphere with the standard metric g_0 with the corresponding volume form $dV_0 := \sin\theta\, d\theta \wedge d\varphi$ in such a way that $\int_{\mathbb{S}^2} dV_0 = 4\pi$, and consider the manifold

$$\mathcal{M} = \{u \in H^1(\mathbb{S}^2); \int_{\mathbb{S}^2} e^{2u} \mathbf{x}\, dV_0 = 0\}.$$

This section is devoted to the following result of T. Aubin.

THEOREM 19.1.1. *If $\alpha > \frac{1}{2}$, then there exists a constant C_α such that the following holds for any $u \in \mathcal{M}$,*

$$(19.1) \quad \frac{1}{4\pi} \int_{\mathbb{S}^2} e^{2u}\, dV_0 \leq C_\alpha \exp\left(\alpha \int_{\mathbb{S}^2} |\nabla u|^2 \frac{dV_0}{4\pi} + 2 \int_{\mathbb{S}^2} u \frac{dV_0}{4\pi} \right).$$

Proof: Without loss of generality we may assume $\int_{\mathbb{S}^2} u\, dV_0 = 0$. It is clearly equivalent to show that for $\alpha > \frac{1}{32\pi}$ and $u \in \mathcal{M} \cap C^\infty(\mathbb{S}^2)$ the following inequality holds

$$(19.2) \quad \int_{\mathbb{S}^2} e^u\, dV_0 \leq C_\alpha \exp\left(\alpha \int_{\mathbb{S}^2} |\nabla u|^2\, dV_0 \right),$$

for some C_α independent of u. Define

$$K_i^+ := \mathbb{S}^2 \setminus \{(x_1, x_2, x_3) \in \mathbb{S}^2 : x_i > 0\}, \quad K_i^- := \mathbb{S}^2 \setminus \{(x_1, x_2, x_3) \in \mathbb{S}^2 : x_i < 0\},$$

$$\Omega_i^+ := \{x \in \mathbb{S}^2 : x_i \geq \frac{1}{3}\}, \quad \text{and} \quad \Omega_i^- := \{x \in \mathbb{S}^2 : x_i \leq -\frac{1}{3}\}.$$

Let $0 \leq h_i^+, h_i^-, g_i^+, g_i^- \leq 1$ be C^1 functions such that

$$h_i^+(\Omega_i^+) = h_i^-(\Omega_i^-) = g_i^+(K_i^+) = g_i^-(K_i^-) = \{1\},$$

$\mathrm{supp}(h_i^+) \cap \mathrm{supp}(g_i^-) = \emptyset$, and $\mathrm{supp}(h_i^-) \cap \mathrm{supp}(g_i^+) = \emptyset$.

For any $\eta > 0$, there exists $a > 0$ such that the set $\Omega_a := \{x \in \mathbb{S}^2 : u(x) \geq a\}$ has measure η, i.e., $\int_{\Omega_a} dV_0 = \eta$. Define $u_a = \max\{u - a, 0\}$.

Since $|x_1| + |x_2| + |x_3| \geq 1$ for any $(x_1, x_2, x_3) \in \mathbb{S}^2$, at least one of the functions $f_i^+ = \max\{0, x_i\}$ or $f_i^- = \max\{0, -x_i\}$, $1 \leq i \leq 3$, is greater than $\frac{1}{3}$. Thus $\bigcup_{i=1}^3 (\Omega_i^+ \cup \Omega_i^-) = \mathbb{S}^2$ and

$$\int_{\mathbb{S}^2} e^{u_a} dV_0 \leq \sum_{i=1}^3 \left(\int_{\Omega_i^+} e^{u_a} dV_0 + \int_{\Omega_i^-} e^{u_a} dV_0 \right).$$

There exists $t_0 \in \{1, 2, 3\}$ such that

$$\int_{\Omega_{t_0}^-} e^{u_a} dV_0 \geq \int_{\Omega_i^+} e^{u_a} dV_0 \quad \text{and} \quad \int_{\Omega_{t_0}^-} e^{u_a} dV_0 \geq \int_{\Omega_i^-} e^{u_a} dV_0,$$

for all $1 \leq i \leq 3$. If now

(19.3) $$\|\nabla(u_a h_{t_0}^-)\|_{L^2(\mathbb{S}^2)} \leq \|\nabla(u_a g_{t_0}^+)\|_{L^2(\mathbb{S}^2)},$$

then using Moser's inequality we have

$$\int_{\mathbb{S}^2} e^u dV_0 \leq 6e^a \int_{\Omega_{t_0}^-} e^{u_a} dV_0 \leq 6e^a \int_{\mathbb{S}^2} e^{u_a h_{t_0}^-} dV_0$$

$$\leq 6e^a C(\alpha) \exp\left(2(\alpha + \frac{1}{32\pi}) \|\nabla(u_a h_{t_0}^-)\|^2_{L^2(\mathbb{S}^2)} + \frac{\int_{\mathbb{S}^2} u h_{t_0}^- dV_0}{4\pi} \right).$$

Since

$$2\|\nabla(u_a h_{t_0}^-)\|^2_{L^2(\mathbb{S}^2)} \leq \|\nabla(u_a h_{t_0}^-)\|^2_{L^2(\mathbb{S}^2)} + \|\nabla(u_a g_{t_0}^+)\|^2_{L^2(\mathbb{S}^2)},$$

there exist constant β and γ such that

$$2\|\nabla(u_a h_{t_0}^-)\|^2_{L^2(\mathbb{S}^2)} \leq \|\nabla u_a\|^2_{L^2(\mathbb{S}^2)} + \beta \|\nabla u_a\|_{L^2(\mathbb{S}^2)} \|u_a\|_{L^2(\mathbb{S}^2)} + \gamma \|u_a\|^2_{L^2(\mathbb{S}^2)},$$

for all $u \in H^1(\mathbb{S}^2) \cap C^\infty(\mathbb{S}^2)$ and $a > 0$. Now choose $\epsilon_1 > 0$ small enough so that

$$(\alpha + \frac{1}{32\pi})(1 + \beta\epsilon_1) > \alpha > \frac{1}{32\pi}.$$

By Young's inequality, there exists $M > 0$ such that

$$2\|\nabla(u_a h_{t_0}^-)\|^2_2 \leq (1 + \beta\epsilon_1) \|\nabla u_a\|^2_{L^2(\mathbb{S}^2)} + (\gamma + M\beta) \|u_a\|^2_{L^2(\mathbb{S}^2)},$$

hence,

(19.4) $$\int_{\mathbb{S}^2} e^u dV_0$$

$$\leq 6e^a C(\alpha) \exp\left((\alpha + \frac{1}{32\pi})[(1 + \beta\epsilon_1)\|\nabla u_a\|^2_2 + (\gamma + M\beta)\|u_a\|^2_2] + \int_{\mathbb{S}^2} |u_a| \frac{dV_0}{4\pi} \right)$$

$$\leq 6e^a C(\alpha) \exp\left((\alpha + \frac{1}{32\pi})[(1 + \beta\epsilon_1)\|\nabla u_a\|^2_2 + (\frac{1}{2\sqrt{\pi\alpha}} + \gamma + M\beta)\|u_a\|^2_2] + 1 \right),$$

where we have applied Hölder inequality to obtain the last inequality and all constants appearing above are independent of u and a.

Now we relate a and the norms of u_a to η and the norms of u. By Hölder's and Sobolev's inequalities, we have for some $D > 0$,

$$\|u_a\|^2_{L^2(\mathbb{S}^2)} \leq \|u_a\|^2_{L^4(\mathbb{S}^2)} \eta^{\frac{1}{2}} \leq D \|\nabla u_a\|^2_{L^2(\mathbb{S}^2)} \eta^{\frac{1}{2}} \leq D \|\nabla u\|^2_{L^2(\mathbb{S}^2)} \eta^{\frac{1}{2}},$$

Since $\int_{\mathbb{S}^2} u \, dV_0 = 0$,
$$\int_{\mathbb{S}^2} u^+ dV_0 = \int_{\mathbb{S}^2} u^- dV_0 = \frac{1}{2}\int_{\mathbb{S}^2} u \, dV_0 \leq C_1 \|\nabla u\|_{L^2(\mathbb{S}^2)}.$$

Therefore $\|u_a\|_{L^1(\mathbb{S}^2)} \leq C_1 \|\nabla u\|_{L^2(\mathbb{S}^2)}$. Consequently, for every $\rho > 0$ there exists $C(\rho)$ such that
$$a \leq \frac{\|u_a\|_{L^1(\mathbb{S}^2)}}{\eta} \leq \frac{C_1}{\eta}\|\nabla u\|_{L^2(\mathbb{S}^2)} \leq \frac{\rho}{\eta}\|\nabla u\|^2_{L^2(\mathbb{S}^2)} + \frac{C(\rho)}{\eta}.$$

In particular,
$$a \leq \frac{C_1}{\eta}\|\nabla u\|_{L^2(\mathbb{S}^2)} \leq \eta\|\nabla u\|^2_{L^2(\mathbb{S}^2)} + \frac{C(\eta)}{\eta}.$$

Therefore, we have
$$\int_{\mathbb{S}^2} e^u dV_0 \leq 6 e^{(\frac{C(\eta)}{\eta}+1)} C(\alpha) \exp\left(((\alpha + \frac{1}{32\pi})[(1+\beta\epsilon_1) + \mu D\eta^{\frac{1}{2}}] + \eta)\|\nabla u\|^2_{L^2(\mathbb{S}^2)}\right),$$

where $\mu := \frac{1}{\sqrt{2\pi\alpha}} + \gamma + M\beta$. If $\eta > 0$ is small enough ($\eta < \eta_1$) such that
$$(\alpha + \frac{1}{32\pi})[(1+\beta\epsilon_1) + \mu D\eta^{\frac{1}{2}}] + \eta < \alpha,$$

then (19.2) holds.

On the other hand, if $\|\nabla(u_a h^-_{t_0})\|_{L^2(\mathbb{S}^2)} > \|\nabla(u_a g^+_{t_0})\|_{L^2(\mathbb{S}^2)}$, then
$$\int_{\Omega_{t_0}} e^{u_a} dV_0 \leq 3\int_{K^-_{t_0}} -x_{t_0} e^{u_a} dV_0 = 3\int_{K^-_{t_0}} x_{t_0} e^{u_a} dV_0 \leq 3\int_{K^+_{t_0}} e^u dV_0,$$

and we obtain again by Moser's inequality that
$$\int_{K^+_{t_0}} x_{t_0} e^{u_a} dV_0 \leq \int_{\mathbb{S}^2} e^{u_a g^+_{t_0}} dV_0$$
$$\leq C(\alpha) \exp\left(2(\alpha + \frac{1}{32\pi})\|\nabla(u_a g^+_{t_0})\|^2_{L^2(\mathbb{S}^2)} + \frac{\int_{\mathbb{S}^2} u_a g^+_{t_0} dV_0}{4\pi}\right).$$

Since
$$2\|\nabla(u_a g^+_{t_0})\|_{L^2(\mathbb{S}^2)} \leq \|\nabla(u_a g^+_{t_0})\|_{L^2(\mathbb{S}^2)} + \|\nabla(u_a h^-_{t_0})\|_{L^2(\mathbb{S}^2)},$$

one can similarly show that there exists $\eta_2 > 0$ such that if $\eta < \eta_2$, then inequality (19.2) holds.

Finally, if we let $\eta < \min\{\eta_1, \eta_2\}$, then the validity of the inequality (19.2) holds in either case and the proof is now complete. $\square$

19.2. Towards an optimal Aubin-Moser-Onofri inequality on $\mathbb{S}^2$

Another way of formulating (19.1.1) is to say that for any $\alpha > \frac{1}{2}$, the functional
$$J_\alpha(u) = \alpha \int_{\mathbb{S}^2} |\nabla u|^2 \frac{dV_0}{4\pi} + 2\int_{\mathbb{S}^2} u \frac{dV_0}{4\pi} - \ln \int_{\mathbb{S}^2} e^{2u} \frac{dV_0}{4\pi}$$

is bounded below on the submanifold $\mathcal{M}$. In other words, for any $\alpha > \frac{1}{2}$, we have
(19.5) $$C_\alpha := \inf\{J_\alpha(u); u \in \mathcal{M}\} > -\infty.$$

The rest of the chapter will deal with the computation of this lower bound.

THEOREM 19.2.1. *Consider the functional J_α on the manifold*

$$\mathcal{M} = \{u \in H^1(\mathbb{S}^2); \int_{\mathbb{S}^2} e^u \mathbf{x} dV_0 = 0\}.$$

(1) *If $\alpha \geq \frac{2}{3}$, then $C_\alpha = \inf_{u \in \mathcal{M}} J_\alpha(u) = 0$.*
(2) *If $\alpha < \frac{1}{2}$, then $C_\alpha = \inf_{u \in \mathcal{M}} J_\alpha(u) = -\infty$.*

Proof: We start by proving (2) and for that we consider the functions

(19.6) $$g_{c,\epsilon}(\theta, \varphi) = \begin{cases} c \log(1 - \cos\theta) & \text{for } 0 < \cos\theta < 1 - \epsilon \\ c \log(\epsilon) & \text{for } 1 - \epsilon < \cos\theta < 1, \end{cases}$$

with θ and φ denoting the usual angular coordinates on the sphere. Extend these functions to the whole sphere as even functions of $\cos\theta$ on the whole interval $(-1, 1)$. It is clear that such functions are axially symmetric and that they belong to $\mathcal{M}$.

When the function u only depends on $x = \cos(\theta)$, the functional J_α is then given by

$$J_\alpha(u) = \frac{\alpha}{2} \int_{-1}^{1} (1 - x^2)|u'(x)|^2 \, dx + \int_{-1}^{1} u(x) \, dx - \log \frac{1}{2} \int_{-1}^{1} e^{2u(x)} \, dx.$$

Now use the computation at the beginning of Section 17.1 to show that for $\alpha < 1/2$, there exists $c < -\frac{1}{2}$ so that $J_\alpha(g_{c,\epsilon})$ tends to $-\infty$ as ϵ becomes small.

In order to establish Part (1) of Theorem 19.2.1, we shall need the following lemmas. The first is a tricky result of Kazdan-Warner.

LEMMA 19.2.2. *If v is a solution of the equation*

(19.7) $$\Delta v = c - he^v \quad \text{on } \mathbb{S}^2,$$

then

(19.8) $$\int_{\mathbb{S}^2} e^v \nabla h \cdot \nabla F dV_0 = (2 - c) \int_{\mathbb{S}^2} e^v h F dV_0,$$

for all spherical harmonics F of degree 1.

Proof: First we observe the following identity which holds for any pair of smooth functions u, F on any given Riemannian manifold.

(19.9) $2\Delta u(\nabla F \cdot \nabla u) = \nabla(2(\nabla F \cdot \nabla u)\nabla u - |\nabla u|^2 \nabla F) - (2H_F - (\Delta F)g)(\nabla u, \nabla u).$

Here g denotes the metric tensor and H_F is the Hessian or second covariant derivative of F. (In Euclidean space, the matrix of the symmetric bilinear form H_F with respect to the canonical bases is just the matrix of second partial derivatives of F). In the notation of tensor calculus this identity becomes

(19.10) $$2u^j_{;j}(F^i u_i) = 2((F^i u_i)u^j)_{;j} - ((u^i u_i)F^j)_{;j} - 2F^i_{;j} u_i u^j + F^j_{;j} u^i u_i,$$

which can be readily verified. Now we restrict our attention to $\mathbb{S}^2$ with its standard metric. If F is a first order spherical harmonic (i.e. the restriction to $\mathbb{S}^2$ of a linear function in R^3) then

(19.11) $$\Delta F = -2F \quad \text{and} \quad 2H_F - (\Delta F)g = 0.$$

For such an F, identity (19.9) becomes

(19.12) $$\Delta u(\nabla F \cdot \nabla u) \sim 0,$$

where we use $\sim$ to denote equality modulo terms which are divergences. We now apply (19.12) to derive an integrability condition for solutions u of the equation $\Delta u = c - he^u$. Replacing Δu in (19.12) by $c - he^u$ we obtain

$$c\nabla u \cdot \nabla F \sim he^u \nabla u \cdot \nabla F. \tag{19.13}$$

Consider the left hand side of (19.13). Taking a derivative off of F and placing it on ∇u, again substituting $c - he^u$ for Δu, and observing that $c^2 F$ is a divergence since F is a spherical harmonic, we get that

$$c\nabla u \cdot \nabla F \sim -cF\Delta u = -cF(c - he^u) \sim chFe^u. \tag{19.14}$$

On the right hand side of (19.14), observe that $e^u \nabla u = \nabla e^u$, remove the derivative from e^u, place it on $h\nabla F$, and use the fact $\Delta F = -2F$ to obtain

$$he^u \nabla u \cdot \nabla F \sim -e^u \nabla(h\nabla F) = 2he^u F - e^u \nabla h \cdot \nabla F. \tag{19.15}$$

Combining (19.15) and (19.14) with (19.13) we see that

$$e^u \nabla h \cdot \nabla F \sim (2-c)hFe^u. \tag{19.16}$$

Integrating both sides of (19.16) over $\mathbb{S}^2$ completes the proof of the lemma.

LEMMA 19.2.3. *If $0 < \alpha \leq 1$, then any critical point $u \in H^1(\mathbb{S}^2)$ for the functional J_α restricted on the manifold $\mathcal{M}$, satisfies the equation*

$$\alpha \Delta u + e^{2u} = 1 \quad on \ \mathbb{S}^2. \tag{19.17}$$

Proof: Accounting for the Lagrange multipliers, the Euler-Lagrange equation for the critical point u of J_α restricted to $\mathcal{M}$ is

$$\alpha \Delta u + e^{2u} = 1 + \sum_{j=1}^{3} a_j x_j e^{2u} \quad \text{on } \mathbb{S}^2 \tag{19.18}$$

for some constants a_j ($j = 1, 2, 3$). We shall now show that if $\alpha \leq 1$, then $a_j = 0$ for $j = 1, 2, 3$. Apply Lemma 19.2.2 with $v = 2u$, $c = 2/\alpha$, $h = (2/\alpha)(1 - \sum_{i=1}^{3} a_i x_i)$ to obtain for each $j = 1, 2, 3$,

$$-\frac{2}{\alpha} \int_{\mathbb{S}^2} e^{2u} \left(\sum_{i=1}^{3} a_i x_i\right) \cdot \nabla x_j dV_0 = (2 - \frac{2}{\alpha}) \cdot \frac{2}{a} \int_{\mathbb{S}^2} e^{2u} \left(1 - \sum_{i=1}^{3} a_i x_i\right) x_j dV_0$$
$$= -(2 - \frac{2}{\alpha}) \cdot \frac{2}{a} \int_{\mathbb{S}^2} e^{2u} \left(\sum_{i=1}^{3} a_i x_i\right) x_j dV_0. \tag{19.19}$$

Multiplying (19.19) by a_j and suming over $j = l, 2, 3$, we get

$$-\frac{2}{\alpha} \int_{\mathbb{S}^2} e^{2u} |\nabla(\sum_{i=1}^{3} a_i x_i)|^2 dV_0 = -(2 - \frac{2}{\alpha}) \cdot \frac{2}{a} \int_{\mathbb{S}^2} e^{2u} |\sum_{i=1}^{3} a_i x_i|^2 dV_0. \tag{19.20}$$

When $\alpha < 1$, the left hand side of (19.20) is always negative while the right hand side is always positive (or zero when $\alpha = 1$) unless $\sum_{i=1}^{3} a_i x_i \equiv 0$, i.e. $a_j = 0$ for all $i = 1, 2, 3$, which finishes the proof of the lemma. $\square$

We shall also need the following lemma, which is a consequence of an inequality due to Bol, and whose proof will be given in the last section of this chapter.

LEMMA 19.2.4. *Let Ω be a simply connected domain in $\mathbb{R}^2$, and suppose $g \in C^2(\Omega)$ satisfies*

$$\begin{cases} \Delta g + e^g > 0 & \text{in } \Omega \text{ and} \\ \int_\Omega e^g dy \leq 8\pi. \end{cases}$$

Consider an open set $\omega \subset \Omega$ such that $\lambda_{1,g}(\omega) \leq 0$, where $\lambda_{1,g}(\omega)$ is the first eigenvalue of the operator $\Delta + e^g$ on $H_0^1(\omega)$. Then, we necessarily have that

$$\int_\omega e^g dy > 4\pi. \tag{19.21}$$

End of proof of Theorem 19.2.1: First, notice that Theorem 19.1.1 yields that for any $\alpha > \frac{1}{2}$, J_α is coercive on $\mathcal{M}$. Since J_α is weakly lower semi-continuous and $\mathcal{M}$ is weakly closed, the infimum C_α of J_α on $\mathcal{M}$ is attained at a function $u \in \mathcal{M}$. It follows from Lemma 19.2.3 that u satisfies the equation

$$\alpha \Delta u + (e^{2u} - 1) = 0 \quad \text{on } \mathbb{S}^2, \tag{19.22}$$

We shall now prove Theorem 19.2.1 by showing that if $\alpha \geq \frac{2}{3}$, then any solution of (19.22) is necessarily axially symmetric and is therefore identically zero by Theorem 17.2.1.

Suppose u is a solution of (19.22), and let ξ_0 be a critical point of u. Without loss of generality, we may assume $\xi_0 = (0, 0, -1)$. By using the stereographic projection $\Pi : \mathbb{S}^2 \to \mathbb{R}^2$ with respect to the North pole $N = (0, 0, 1)$, that is

$$\Pi(x) := \left(\frac{x_1}{1 - x_3}, \frac{x_2}{1 - x_3} \right),$$

we can associate to u, the function $\tilde{u}$ on $\mathbb{R}^2$,

$$\tilde{u}(y) := u(\Pi^{-1}(y)) \quad \text{for } y \in \mathbb{R}^2.$$

which is then a solution for

$$\Delta \tilde{u} + \frac{1}{\alpha} J(y) \left(e^{2\tilde{u}} - 1 \right) = 0 \quad \text{in } \mathbb{R}^2,$$

where $J(y) := \left(\frac{2}{1+|y|^2} \right)^2$ is the Jacobian of Π. By letting

$$v(y) := 2\tilde{u}(y) + \frac{2}{\alpha} \log \left((1+|y|^2)^{-2} \right) + \log\left(\frac{64}{\alpha}\right) \quad \text{for } y \in \mathbb{R}^2, \tag{19.23}$$

we have that v satisfies

$$\Delta v + (1 + |y|^2)^l e^v = 0 \quad \text{in } \mathbb{R}^2, \tag{19.24}$$

where $l = 2(\frac{1}{\alpha} - 1)$, as well as

$$\int_{\mathbb{R}^2} (1 + |y|^2)^l e^v dy = \frac{8\pi}{\alpha}, \tag{19.25}$$

and

$$\nabla v(0) = 0. \tag{19.26}$$

Consider now the function $\varphi(y) := y_2 \frac{\partial v}{\partial y_1} - y_1 \frac{\partial v}{\partial y_2}$. It is bounded by (19.23) and satisfies

$$\Delta \varphi + (1 + |y|^2)^l e^v \varphi = 0 \quad \text{in } \mathbb{R}^2, \tag{19.27}$$

19.2. TOWARDS AN OPTIMAL AUBIN-MOSER-ONOFRI INEQUALITY ON S^2

If now $\varphi \not\equiv 0$, then by (19.26), we can write φ as

$$\varphi(y) = Q(y) + \text{higher order terms} \quad \text{for } |y| \ll 1,$$

where $Q(y)$ is a quadratic polynomial of degree m with $m \geq 2$, that is also a harmonic function, i.e., $\Delta Q = 0$. Thus, the nodal line $\{y \mid \varphi(y) = 0\}$ divides a small neighborhood of the origin into at least four regions. Let γ_i, $i = 1, 2, 3, 4$, be four branches of nodal line of φ emanating from the origin. If γ_i does not intersect with γ_j, $i \neq j$, then $\mathbb{R}^2 \setminus \bigcup_{i=1}^{4} \gamma_i$ contains at least four simply-connected components. See Figure 1 below. If γ_i intersects with some γ_j, then $\mathbb{R}^2 \setminus \bigcup_{i=1}^{4} \gamma_i$ contains at least three simply-connected components. See Figure 2.

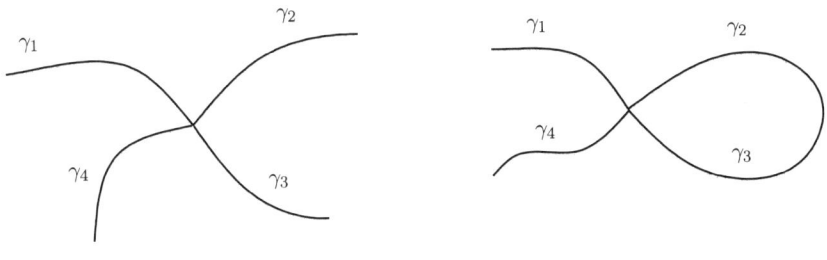

Fig.1   Fig.2

If there are more branches of nodal line of φ issuing from the origin, then $\mathbb{R}^2 \setminus \{\varphi = 0\}$ is divided into more components of simply-connected domains. Therefore, we conclude that $\mathbb{R}^2$ is divided by the nodal line $\{y \mid \varphi(y) = 0\}$ into at least 3 regions, i.e.,

$$(19.28) \qquad \mathbb{R}^2 \setminus \{y \mid \varphi(y) = 0\} = \bigcup_{j=1}^{3} \Omega_j.$$

In each component Ω_j, the first eigenvalue of $\Delta + (1 + |y|^2)^l e^v$ being equal to 0. Let now

$$(19.29) \qquad g := \log\left((1 + |y|^2)^l e^v\right).$$

and note that

$$(19.30) \qquad \Delta g + e^g > 0 \quad \text{in } \mathbb{R}^2.$$

Now we are in position to complete the proof of Theorem 19.2.1. Indeed, Lemma 19.2.4 applied to g in (19.29) above and to each Ω_j, $j = 1, 2, 3$, yields

$$\int_{\Omega_j} e^g \, dy = \int_{\Omega_j} (1 + |y|^2)^l e^v \, dy > 4\pi.$$

It follows that

$$\frac{8\pi}{\alpha} = \int_{\mathbb{R}^2} (1 + |y|^2)^l e^v \, dy = \sum_{j=1}^{3} \int_{\Omega_j} (1 + |y|^2)^l e^v \, dy > 12\pi,$$

which is a contradiction if we had assumed that $\alpha > \frac{2}{3}$. It follows that $\varphi \equiv 0$, i.e., v, and therefore u is axially symmetric. By Theorem 17.2.1, we can conclude that

u is constant, which is necessarily equal to 0. It follows that $C_\alpha = J_\alpha(u) = 0$ for all $\alpha > \frac{2}{3}$, hence (1) of Theorem 19.2.1 is proved. □

REMARK 19.2.5. If we further assume that the antipodal of ξ_0 is also a critical point of u, then $\mathbb{R}^2 \setminus \{y \mid \varphi(y) = 0\} = \bigcup_{j=1}^{m} \Omega_j$, where $m \geq 4$. Lemma 19.2.4 then yields

$$\frac{8\pi}{\alpha} = \int_{\mathbb{R}^2} (1+|y|^2)^l e^v dy \geq \sum_{j=1}^{m} \int_{\Omega_j} (1+|y|^2)^l e^v dy > 4m\pi \geq 16\pi,$$

which is a contradiction whenever $\alpha > \frac{1}{2}$. By Theorem 17.2.1, we have again that $u \equiv 0$. For example, if u is even on $\mathbb{S}^2$ (i.e., $u(z) = u(-z)$ for all $z \in \mathbb{S}^2$), then Theorem 19.2.1.1) holds for $\alpha > \frac{1}{2}$.

REMARK 19.2.6. Theorem 19.2.1.1) actually holds for $\alpha \geq \frac{2}{3} - \epsilon_0$ for some $\epsilon_0 > 0$. Indeed, it suffices to show that for α smaller but close to $\frac{2}{3}$, the functional J_α is still non-negative. Assuming not, then there exists a sequence of $\{\alpha_k\}_k$ such that $\frac{1}{2} < \alpha_k < \frac{2}{3}$, $\lim_k \alpha_k = \frac{2}{3}$ and $\inf_\mathcal{M} J_{\alpha_k} < 0$. Since J_α is coercive for each $\alpha > \frac{1}{2}$, a standard compactness argument yields the existence of a minimizer $u_k \in \mathcal{M}$ for J_{α_k}. Moreover, $\|u_k\|_{H^1} < C$ for some positive constant independent of k. Modulo extracting a subsequence, u_k then converges weakly to some u_0 in $\mathcal{M}$ as $k \to \infty$, and u_0 is necessarily a minimizer for $I_{\frac{2}{3}}$ in $\mathcal{M}$. By the above result, $u_0 \equiv 0$. Now, we claim that u_k actually converges strongly in H^1 to $u_0 \equiv 0$. Indeed, the Euler-Lagrange equations are

$$(19.31) \qquad \alpha_k \Delta u_k - 1 + \frac{1}{\lambda_k} e^{2u_k} = 0,$$

where $\lambda_k = \int_{\mathbb{S}^2} e^{2u_k} dw < C$ for some positive constant C. Multiplying (19.31) by u_k and integrating over $\mathbb{S}^2$, we obtain

$$(19.32) \qquad \alpha_k \int_{\mathbb{S}^2} |\nabla u_k|^2 dw + \int_{\mathbb{S}^2} u_k(x) \, dw = \frac{1}{\lambda_k} \int_{\mathbb{S}^2} e^{2u_k(x)} u_k(x) \, dw.$$

Applying Aubin's inequality for u_k and using that $\|u_k\|_{H^1} < C$, we get that $\int_{\mathbb{S}^2} e^{2u_k} dw$ is also uniformly bounded. This combined with Hölder's inequality and the fact that u_k converges strongly to 0 in L^2 yields that $\int_{\mathbb{S}^2} e^{2u_k} u_k \, dw \to 0$. Use now (19.32) to conclude that $\|u_k\|_{H^1} \to 0$ as $k \to \infty$.

Now, write $u = v + o(\|u\|)$ for $\|u\|$ small, where v belongs to the tangent space of the submanifold $\mathcal{M}$ at $u_0 \equiv 0$ in $H^1(\mathbb{S}^2)$. Since $\int_{\mathbb{S}^2} v\mathbf{x} \, dw = 0$, we can calculate the second variation of J_α in $\mathcal{M}$ at $u_0 \equiv 0$ and get the following estimate around 0

$$J_\alpha(u) = \alpha \int_{\mathbb{S}^2} |\nabla v|^2 dw - 2 \int_{\mathbb{S}^2} |v|^2 dw + o(\|u\|^2).$$

Note that the eigenvalues of the Laplacian on $\mathbb{S}^2$ corresponding to the eigenspace generated by x_1, x_2, x_3 are $\lambda_2 = \lambda_3 = \lambda_4 = 2$, while $\lambda_5 = 6$. Since v is orthogonal to $\mathbf{x}$, we have $\int_{\mathbb{S}^2} |\nabla v|^2 dw \geq 6 \int_{\mathbb{S}^2} |v|^2 dw$, and therefore

$$J_\alpha(u) \geq (\alpha - \frac{1}{3}) \|\nabla u\|_2^2 + o(\|\nabla u\|^2).$$

Taking $\alpha = \alpha_k$ and $u = u_k$ for k large enough, we get that $J_{\alpha_k}(u_k) \geq 0$, which clearly contradicts our initial assumption on u_k.

19.3. Bol's isoperimetric inequality

We shall now establish the following result which was crucial to the proof of Section 19.2.

PROPOSITION 19.3.1. Let Ω be a simply-connected bounded domain of $\mathbb{R}^2$ and $v \in C^2(\bar{\Omega})$ such that $-\Delta v < e^v$ in $\bar{\Omega}$. Assume there exits $\varphi \in C^1(\bar{\Omega})$ such that

(19.33) $\quad\quad\quad \Delta\varphi + e^v\varphi = 0$ in Ω, $\quad \varphi = 0$ on $\partial\Omega$, and $\varphi \not\equiv 0$.

Then

(19.34) $$\int_\Omega e^v\, dx > 4\pi.$$

The proposition relies on the following isoperimetric inequality of Bol.

LEMMA 19.3.1. (Bol's inequality) Let Ω be a simply-connected domain of $\mathbb{R}^2$ and assume u is a function in $C^2(\mathbb{R}^2) \cap C^0(\bar{\Omega})$ that satisfies

(19.35) $$-\Delta u \leq e^u \quad on\ \Omega.$$

Then

(19.36) $$(\int_{\partial\Omega} e^{\frac{u}{2}}\, ds)^2 \geq \frac{1}{2}(\int_\Omega e^u\, dx)(8\pi - \int_\Omega e^u\, dx).$$

Proof: Let h be the harmonic extension of the restriction of u on $\partial\Omega$, that is,

$$\begin{aligned} -\Delta h &= 0 \text{ in } \Omega \\ h &= u \text{ on } \partial\Omega. \end{aligned}$$

For each $\omega \subset \Omega$ with sufficiently smooth boundary, the following inequality holds,

(19.37) $$(\int_{\partial\omega} e^{h/2} ds)^2 \geq 4\pi \int_\omega e^h dx.$$

Indeed, consider an analytic function $g = g(z)$ in Ω such that $|g'|^2 = e^h$. We then have

$$\int_{\partial\omega} e^{h/2} ds = \int_{\partial\omega} |g'|\, ds \text{ and } \int_\omega e^h dx = \int_\omega |g'|^2\, dx.$$

Therefore (19.37) is nothing but the isoperimetric inequality for the flat Riemannian surface $g(\omega)$. Now set $v = e^u e^{-h}$, which solves

$$\begin{aligned} -\Delta \log v &\leq ve^h \text{ in } \Omega \\ v &= 1 \text{ on } \partial\Omega, \end{aligned}$$

and define two right continuous and strictly increasing functions $k(t)$ and $\mu(t)$ as follows:

(19.38) $$k(t) = \int_{\{v>t\}} ve^h dx,$$

and

(19.39) $$\mu(t) = \int_{\{v>t\}} e^h dx.$$

We shall derive a differential inequality satisfied by the functions $k(t)$ and $\mu(t)$. The co-area formula yields

(19.40) $\quad -k'(t) = \displaystyle\int_{\{v=t\}} \frac{ve^h}{|\nabla v|} ds = t\int_{\{v=t\}} \frac{e^h}{|\nabla v|} ds = -t\mu'(t) \quad$ for a.e. t.

On the other hand, by Green's formula and the Sard's lemma we have,

(19.41) $$\int_{\{v>t\}} (-\Delta \log v) dx = \int_{\{v=t\}} \frac{|\nabla v|}{v} = \frac{1}{t} \int_{\{v=t\}} |\nabla v| ds \quad \text{for a.e. } t > 1.$$

Hence, (19.38) yields

(19.42) $$\frac{1}{t} \int_{\{v=t\}} |\nabla v| ds \le \int_{\{v>t\}} v e^h dx = k(t) \quad \text{for a.e. } t > 1.$$

By Schwarz inequality and (19.37) we get

$$\begin{aligned}
-k(t)k'(t) &\ge \frac{1}{t} \int_{\{v=t\}} |\nabla v| ds \cdot t \int_{\{v=t\}} \frac{e^h}{|\nabla v|} ds \\
&\ge (\int_{\{v=t\}} e^{\frac{h}{2}} ds)^2 \\
&\ge 4\pi \int_{\{v>t\}} e^h dx \\
&= 4\pi \mu(t) \quad \text{for a.e. } t > 1.
\end{aligned}$$

Therefore

(19.43) $$\frac{d}{dt}\{t\mu(t) - k(t) + \frac{1}{8\pi} k^2(t)\} = \mu(t) + \frac{1}{4\pi} k(t) k'(t) \le 0 \quad \text{for a.e. } t > 1.$$

Note that $k(t)$ is increasing and right continuous and the function

(19.44) $$g(t) := k(t) - \mu(t) t = \int_{\{v>t\}} (v-t) e^h dx$$

is continuous. Indeed, it is clear that $\lim_{\epsilon \to 0} g(t+\epsilon) = g(t)$. On the other hand,

$$\lim_{\epsilon \to 0} (g(t) - g(t-\epsilon)) = \int_{\{v=t\}} (v-t) e^h dx = 0.$$

It then follows from (19.43) that

(19.45) $$\left[t\mu(t) - k(t) + \frac{1}{8\pi} k^2(t) \right]_1^\infty = -\left\{ \mu(1) - k(1) + \frac{1}{8\pi} k^2(1) \right\} \le 0.$$

On the other hand, we have

$$\begin{aligned}
k(1) - \mu(1) = g(1) &= \int_{v>1} (v-1) e^h dx \\
&\ge \int_\Omega (v-1) e^h dx \\
&= \int_\Omega e^u dx - \int_\Omega e^h dx
\end{aligned}$$

as well as

$$k(1) \le \int_\Omega e^u dx.$$

It then follows from (19.45) that

(19.46) $$(\int_\Omega e^u)(1 - \frac{1}{8\pi} \int_\Omega e^u) \le \int_\Omega e^h dx.$$

Combining the above inequality with (19.37), we obtain (19.36). □

We shall need Bol's isoperimetric inequality in the following form.

LEMMA 19.3.2. *Let Ω be a simply-connected domain in $\mathbb{R}^2$ and suppose $v \in C^2(\bar{\Omega})$ satisfies*

(19.47) $$-\Delta v < e^v \quad \text{in } \bar{\Omega}.$$

Then, for any compact $\omega \subseteq \Omega$ of class C^1, the following inequality holds:

(19.48) $$(\int_{\partial \omega} e^{v/2})^2 > \frac{1}{2}(\int_\omega e^v)(8\pi - \int_\omega e^v).$$

Proof: By (19.47) we can find $\epsilon > 0$ such that $-\Delta v(x) \leq (1-\epsilon)e^{v(x)}$ for all $x \in \Omega$. Therefore the function $\tilde{v} := v + \log(1-\epsilon)$ satisfies assumptions (19.35). Hence, by applying Bol's inequality to the function $\tilde{v}$ we get

$$(\int_{\partial \omega} e^{v/2})^2 \geq \frac{1}{2}(\int_\omega e^v)(8\pi - (1-\epsilon)\int_\omega e^v),$$

from which (19.48) follows. □

Proof of Proposition 19.3.1: Proceeding by contradiction, we shall prove that if v is such that $\int_\Omega e^v \, dx \leq 4\pi$, then any $\varphi \in C^1(\bar{\Omega})$ with $\varphi \not\equiv 0$ satisfies the following Faber-Krahn type inequality

(19.49) $$\frac{\int_\Omega |\nabla \varphi|^2}{\int_\Omega e^v \varphi^2} > 1.$$

Note that since φ is continuous up to the boundary, the following property on the upper level sets holds:

$$\Omega_t := \{|\varphi| > t\} \subset\subset \Omega, \quad \forall t \geq 0.$$

Set $U(x) = -2\log(1 + \frac{|x|^2}{8})$ which satisfies $\Delta U + e^U = 0$ in $\mathbb{R}^2$. Note that this function realizes the equality in (19.48) when ω is a ball centered at the origin. We shall make a rearrangement of the function $|\varphi|$ with respect to the measure $e^U dx$ and $e^u dx$. To this end, define first the balls Ω^* and Ω_t^* centered at the origin as follows:

$$\int_{\Omega^*} e^{U(x)} dx = \int_\Omega e^{v(x)}, \quad \int_{\Omega_t^*} e^{U(x)} dx = \int_{\Omega_t} e^{v(x)} dx.$$

The balls Ω_t^* can be seen as geodesic balls on the two-dimensional sphere having the same measure as the set $\{|\varphi| > t\}$ endowed with the measure $e^v dx$. Define the symmetrization $\varphi^* : \Omega^* \to \mathbb{R}$ of the function $|\varphi|$ by $\varphi^*(x) = \sup\{t \in \mathbb{R} : x \in \Omega_t^*\}$. We obtain this way an equimeasurable rearrangement with respect to the measure $e^U dx$ and $e^v dx$, i.e.

(19.50) $$\int_{\{\varphi^* > t\}} e^U = \int_{\Omega_t} e^v, \quad \forall t > 0.$$

In particular, we have (Cavalieri's principle):

(19.51) $$\int_{\Omega^*} e^U |\varphi^*|^2 = \int_\Omega e^v |\varphi|^2.$$

We now prove that such a re-arrangement necessarily decreases the Dirichlet integral. Indeed, by applying the co-area formula, the Schwarz inequality, and then

Bol's inequality, we get:

$$
\begin{aligned}
-\frac{d}{dt}\int_{\Omega_t}|\nabla|\varphi||^2 &= \int_{\{||\varphi|=t|\}}|\nabla\varphi| \\
&\geq \left(\int_{\{|\varphi|=t\}}e^{v/2}\right)^2\left(\int_{\{\varphi=t|\}}\frac{e^v}{|\nabla\varphi|}\right)^{-1} \\
&= \left(\int_{\{|\varphi|=t\}}e^{v/2}\right)^2\left(-\frac{d}{dt}\int_{\Omega_t}e^v\right)^{-1} \\
&> \frac{1}{2}\left(\int_{\Omega_t}e^v\right)\left(8\pi-\int_{\Omega_t}e^v\right)\left(-\frac{d}{dt}\int_{\Omega_t}e^v\right)^{-1} \\
&> \frac{1}{2}\left(\int_{\Omega_t^*}e^v\right)\left(8\pi-\int_{\Omega_t^*}e^v\right)\left(-\frac{d}{dt}\int_{\Omega_t^*}e^v\right)^{-1}
\end{aligned}
$$

for almost every $t \geq 0$. Furthermore, since e^U realizes the equality in (19.48) on each ball ω, one checks easily that

(19.52) $\quad -\dfrac{d}{dt}\displaystyle\int_{\Omega_t^*}|\nabla\varphi^*|^2 = \dfrac{1}{2}\left(\displaystyle\int_{\Omega_t^*}e^v\right)\left(8\pi-\displaystyle\int_{\Omega_t^*}e^v\right)\left(-\dfrac{d}{dt}\displaystyle\int_{\Omega_t^*}e^v\right)^{-1}.$

Hence,

(19.53) $\quad -\dfrac{d}{dt}\displaystyle\int_{\Omega_t}|\nabla\varphi|^2 > -\dfrac{d}{dt}\displaystyle\int_{\Omega_t^*}|\nabla\varphi^*|^2, \quad$ a.e. $t \geq 0.$

By integrating the above inequality with respect to t, we obtain

(19.54) $\quad \displaystyle\int_{\Omega}|\nabla\varphi|^2 > \displaystyle\int_{\Omega^*}|\nabla\varphi^*|^2.$

We then deduce that

$$\frac{\int_\Omega|\nabla\varphi|^2}{\int_\Omega e^v|\varphi|^2} > \frac{\int_\Omega|\nabla\varphi^*|^2}{\int_\Omega e^v|\varphi^*|^2} \geq \lambda_1(e^U,\Omega^*),$$

where we set for $B \subset\subset \mathbb{R}^2$:

$$\lambda_1(e^U,B) = \inf\left\{\frac{\int_\Omega|\nabla\xi|^2}{\int_\Omega e^U\xi^2} : \xi \in H_0^1(B),\ \xi \neq 0\right\}.$$

Now we claim that under the assumption, $\int_\Omega e^v \leq 4\pi$, we necessarily have that $\lambda_1(e^U,\Omega^*) \geq 1$. Indeed, a straightforward computation shows that $\psi(r) = \frac{8-r^2}{8+r^2}$ solves

$$-\Delta\psi = e^U\psi, \quad \psi > 0 \ \text{in}\ B_{\sqrt{8}}, \quad \psi \in H_0^1(B_{\sqrt{8}}),$$

where $B_{\sqrt{8}}$ denotes the ball $B(0,\sqrt{8})$, and therefore, $\lambda_1(e^U,B_{\sqrt{8}}) = 1$.

On the other hand we know that $\int_\Omega e^U \leq 4\pi$, and so by explicit calculations, one can show that $\Omega^* \subseteq B_{\sqrt{8}}$, from which we deduce that

$$\lambda_1(e^U,\Omega^*) \geq \lambda_1(e^U,B_{\sqrt{8}}) = 1,$$

and we are done. $\square$

REMARK 19.3.3. We note that Lemma 19.2.4 follows immediately from Proposition 19.3.1 whenever ω is bounded. We now show that it can still be applied even when ω is unbounded. Indeed, for simplicity, we shall assume that for some $\beta \geq 2$, we have
$$g(y) = -\beta \log |y| + O(1) \quad \text{at } \infty.$$
We shall also assume that the corresponding null-eigenfunction φ in ω, i.e.,
$$\begin{cases} \Delta \varphi + e^g \varphi = 0 & \text{in } \omega, \\ \varphi|_{\partial \omega} = 0, \end{cases}$$
is bounded in $\overline{\omega}$. Indeed, both of these conditions are satisfied in the setting of the proof of Theorem 19.2.1 above. Without loss of generality, we may also assume that $0 \notin \overline{\omega}$. Now set
$$\hat{g}(x) = g(\frac{x}{|x|^2}) - 2\log|x| \quad \text{and} \quad \hat{\varphi}(x) = \varphi(\frac{x}{|x|^2}) \quad \text{for } x \in \omega^* = \{y = \frac{x}{|x|^2};\ x \in \omega\}.$$
Since $\beta \geq 2$, $e^{\hat{g}}$ is a Hölder function at $0 \in \overline{\omega^*}$, and $\hat{g}$ and $\hat{\varphi}$ satisfy
$$\Delta \hat{g} + e^{\hat{g}} > 0 \quad \text{in } \omega^* \setminus \{0\} \quad \text{and} \quad \Delta \hat{\varphi} + e^{\hat{g}} \hat{\varphi} = 0 \quad \text{in } \omega^*.$$
By the boundedness of $\hat{\varphi}$, $\hat{\varphi}$ is continuous on $\overline{\omega^*}$. If $0 \in \omega^*$, then by noting that $\hat{g}$ satisfies $\Delta \hat{g} + e^{\hat{g}} \geq (\beta - 2)\delta_0$, where δ_0 is the Dirac measure at 0 and $\beta - 2 \geq 0$, we can then apply a version of Proposition 19.3.1, where $\hat{g}$ can have a singularity (see [**60**]) to deduce that
$$\int_{\omega^*} e^{\hat{g}(x)} dx = \int_{\omega} e^{g(x)} dx \geq 4\pi.$$
We note that in the proof of Theorem 19.2.1, we have that φ is bounded on all of $\mathbb{R}^2$.

19.4. Further comments

The idea of gaining coercivity by restricting energy functionals to submanifolds of H^1 that are orthogonal to first eigenfunctions of the Laplace-Beltrami operator is due to Aubin and Cherrier. A more general version than Theorem 19.1.1 was proved by Aubin [**30**] in any compact Riemannian manifold. In particular, he showed that if one restricts J_α to the class of $\mathcal{M}$ of functions g for which e^{2g} has centre of mass equal to 0, then J_α is bounded below by a non-positive constant C_α for $\alpha \geq \frac{1}{2}$. In their work on Nirenberg's prescribing Gaussian curvature problem on $\mathbb{S}^2$, A. Chang and P. Yang [**86**] and [**87**] showed that $C_\alpha = 0$ for α close enough to 1. This led them to the conjecture that $C_\alpha = 0$ for $\alpha \geq \frac{1}{2}$. In the axially symmetric case, the result was established for $\alpha \geq 16/25 - \epsilon$ by Feldman-Froese-Ghoussoub-Gui in [**140**] and for $\alpha \geq \frac{1}{2}$ by Gui and Wei [**172**]. The general case was established recently by Ghoussoub-Lin [**165**] but only for $\alpha \geq 2/3$. The question on whether this still holds for $\alpha \geq \frac{1}{2}$ is still open. A proof of Bol's inequality is given by Bandle [**38**] when u is real analytic. Suzuki [**255**] refined Bandle's argument to show that Bol's inequality holds for functions in $C^2(\mathbb{R}^2) \cap C^0(\bar{\Omega})$. Inequality (19.37) is essentially due to Nehari [**228**].

Open problem (22): Show that Part 1 of Theorem 19.2.1 holds for $\alpha = \frac{1}{2}$.

The proof of Theorem 19.2.1 connects the conjecture of Chang-Yang to an equally interesting Liouville type theorem on $\mathbb{R}^2$.

Open problem (23): Is the function

(19.55) $$v^*(y) = -2(\ell+2)\log(1+|y|^2) + \log(4(\ell+2)) \quad \text{for } y \in \mathbb{R}^2,$$

the only solution of the equation

(19.56) $$\Delta v + (1+|y|^2)^\ell e^v = 0 \quad \text{in } \mathbb{R}^2,$$

such that

(19.57) $$\frac{1}{2\pi}\int_{\mathbb{R}^2}(1+|y|^2)^\ell e^v dy = 2(\ell+2).$$

It is indeed the case in the radially symmetric case provided $\ell \leq 2$. We refer to [**206**] for more details about this extremely interesting conjecture.

Bibliography

[1] M. Abramowitz, I. A. Stegun, *Handbook of Mathematical Functions with Formulas, Graphs, and Mathematical Tables*, Washington, National Bureau of Standards Applied Mathematics, 1964.

[2] B. Abdellaoui, E. Colorado, I. Peral, *Some improved Caffarelli-Kohn-Nirenberg inequalities*, Calc. Var. Partial Differential Equations, 23 (2005) 327–345.

[3] S. Adachi, K. Tanaka, *Trudinger type inequalities in $\mathbb{R}^N$ and their best exponents*, Proc. Amer. Math. Soc. 128 (2000) 2051-2057.

[4] R. A. Adams, *Sobolev spaces*, Academic Press, New-york-San Francisco-London (1975).

[5] Adimurthi, N. Chaudhuri, N. Ramaswamy, *An improved Hardy Sobolev inequality and its applications*, Proc. Amer. Math. Soc. 130 (2002) 489-505.

[6] Adimurthi, S. Filippas, A. Tertikas, *On the best constant of Hardy-Sobolev inequalities*, Nonlinear Anal., 70 (2009) 2826–2833.

[7] Adimurthi, Maria J. Esteban, *An improved Hardy-Sobolev inequality in $W^{1,p}$ and its application to Schrödinger operators*, NoDEA Nonlinear Differential Equations Appl. 12 no. 2 (2005) 243-263.

[8] Adimurthi, M. Grossi, S. Santra, *Optimal Hardy-Rellich inequalities, maximum principles and related eigenvalue problems*, J. Funct. Anal. 240 (2006) 36-83.

[9] Adimurthi, A. Sekar, *Role of the fundamental solution in Hardy-Sobolev-type inequalities*, Proceedings of the Royal Society of Edinburgh, 136A (2006) 1111-1130.

[10] Adimurthi; K. Sandeep, *A singular Moser-Trudinger embedding and its applications.* NoDEA Nonlinear Differential Equations Appl. 13, no. 5-6 (2007) 585–603.

[11] Adimurthi; Yang, Y., *An interpolation of Hardy inequality and Trudinger-Moser inequality in $\mathbb{R}^N$ and its applications*, Int. Math. Res. Not. IMRN, no. 13 (2010) 2394–2426.

[12] R. P. Agarwal, M. Bohner, W-T. Li, *Nonoscillation and oscillation: theory for functional differential equations*, Dekker, New York (1995).

[13] A. Aghajani, A. Moradifam, *Oscillations of solutions of second-order nonlinear differential equations of Euler type*, J. Math. Anal. Appl, 326 (2007) 1076-1089.

[14] M. Agueh, *Existence of solutions to degenerate parabolic equations via the Monge-Kantorovich theory*, Adv. Differential Equations., Vol 10, No 3 (2005) 309-360.

[15] M. Agueh, *Sharp Gagliardo-Nirenberg Inequalities and Mass Transport Theory*, Journal of Dynamics and Differential Equations (2006) DOI: 10.1007/s10884-006-9039-9

[16] M. Agueh, *Gagliardo-Nirenberg inequalities involving the gradient L^2-norm*, C. R. Math. Acad. Sci. Paris, 346 (2008) 757–762.

[17] M. Agueh, N. Ghoussoub, X. S. Kang, *Geometric inequalities via a general comparison principle for interacting gases*, Geom. And Funct. Anal., Vol 14, 1 (2004) 215-244.

[18] Allegretto, W., *Nonoscillation theory of elliptic equations of order $2n$*, Pacific J. Math. **64** (1976) 1–16.

[19] A. D. Alexandrov, *Uniqueness theorems for surfaces in the large*, Amer. Math. Soc. Transl. 21 (1962) 412-416.

[20] A. Alvino, V. Ferone, and G. Trombetti, *On the best constant in a Hardy-Sobolev inequality*, Appl. Anal., 85 (2006) 171–180.

[21] L.A. Ambrosio, N. Gigli, G. Savare, *Gradient flows in metric spaces and in the Wasserstein space of probability measures*, Lecture Notes in Mathematics, Birkhäuser, 2005.

[22] C. Ané, S. Blachère, D. Chafai, P. Fougères, I. Gentil, F. Malrieu, C. Roberto, and G. Scheffer, *Sur les inégalités de Sobolev logarithmiques*, vol. 10 of Panoramas et Synthèses, Société Mathématique de France, Paris, 2000. With a preface by D. Bakry and M. Ledoux.

[23] T. Aoki, *Calcul exponentiel des opérateurs microdifferentiels d'ordre infini. I*, Ann. Inst. Fourier (Grenoble) **33** (1983) 227-250.

[24] G. Arioli, F. Gazzola, H.-C. Grunau, E. Mitidieri, *A semilinear fourth order elliptic problem with exponential nonlinearity*, SIAM J. Math. Anal. 36 No. 4 (2005) 1226-1258.

[25] A. Arnold, J.-P. Bartier, and J. Dolbeault, *Interpolation between logarithmic Sobolev and Poincaré inequalities*, Communications in Mathematical Sciences, 5 (2007) 971–979.

[26] A. Arnold, P. Markowich, G. Toscani, and A. Unterreiter, *On convex Sobolev inequalities and the rate of convergence to equilibrium for Fokker-Planck type equations*, Comm. Partial Differential Equations, 26 (2001) 43–100.

[27] T. Aubin, *Problèmes isopérimétriques et espaces de Sobolev*, J. Differential Geometry, 11 (1976) 573–598.

[28] T. Aubin, *Nonlinear Analysis on Manifolds. Monge-Ampère Equations*, Springer-Verlag, 1982.

[29] T. Aubin, *Some nonlinear problems in Riemannian geometry*, Springer Monographs in Mathematics, Springer-Verlag, Berlin, 1998.

[30] T. Aubin, *Meilleures constantes dans des théorèmes d'inclusion de Sobolev at un théorème de Fredholm non linéaire pour la transformation conforme de la courbure scalaire*, J. Funct. Anal., 32 (1979) 149-179.

[31] F.V. Atkinson, L.A. Peletier, *Elliptic equations with nearly critical growth*, J. Diff. Equ., 70 (1987) 349-365.

[32] F.G. Avkhadiev, K.J. Wirths, *Unified Poincare and Hardy inequalities with sharp constants for convex domains*, Math. Mech. 87, No. 8-9 (2007) 632-642.

[33] M. Badiale and G. Tarantello, *A Sobolev-Hardy inequality with applications to a nonlinear elliptic equation arising in astrophysics*, Arch. Ration. Mech. Anal., 163 (2002) 259–293.

[34] A. Baernstein, *A unified approach to symmetrization*, Partial differential equations of elliptic type, Cortona, 1992, p. 47–91. Cambridge Univ. Press, Cambridge, 1994.

[35] D. Bakry and M. Émery, *Hypercontractivité de semi-groupes de diffusion*, C. R. Acad. Sci. Paris Sér. I Math., 299 (1984) 775–778.

[36] D. Bakry, M. Emery, *Diffusions hypercontractives*, In Sém. Prob. XIX, LNM, 1123, Springer (1985) 177-206.

[37] K. Ball, *An elementary introduction to modern convex geometry*. In *Flavors of geometry*, pages 1–58. Cambridge Univ. Press, Cambridge, 1997.

[38] C. Bandle, *On a Differential Inequality and Its Applications to Geometry*, Math Z., Vol. 147 (1976) 253-261.

[39] C. Bandle, *Isoperimetric Inequalities and aplications*, Pitman, London, 1980.

[40] B. Barbatis, S. Filippas, A. Tertikas, *Series expansions for L^p Hardy inequalities*, Indiana Univ. Math. J. 52, no. 1 (2003) 171-190.

[41] W. Beckner, *Weighted inequalities and Stein-Weiss potentials*, Forum Math. 20 (2008), 587-606.

[42] W. Beckner, *Geometric asymptotics and the logarithmic Sobolev inequality*, Forum Math. 11, No. 1 (1999) 105-137.

[43] W. Beckner, *Sharp Sobolev inequalities on the sphere and the Moser-Trudinger inequality*, Ann. of Math. (2), 138 (1993) 213–242.

[44] R. D. Benguria, R. L. Frank, and M. Loss, *The sharp constant in the Hardy-Sobolev-Maz'ya inequality in the three dimensional upper half-space*, Math. Res. Lett., 15 (2008) 613–622.

[45] R. D. Benguria and M. Loss, *Connection between the Lieb-Thirring conjecture for Schrödinger operators and an isoperimetric problem for ovals on the plane*, in Partial differential equations and inverse problems, vol. 362 of Contemp. Math., Amer. Math. Soc., Providence, RI (2004) 53–61.

[46] D. M. Bennett, *An extension of Rellich's inequality*, Proc. Amer. Math. Soc. **106** (1989) 987–993.

[47] E. Berchio, F. Gazzola, *Some remarks on biharmonic elliptic problems with positive, increasing and convex nonlinearities*, Electronic J. Diff. Equa. No. 34 (2005), 20 pp.

[48] H. Berestycki, A. Kiselev, A. Novikov, L. Ryzhik, *The explosion problem in a flow*, J. Anal. Math. 110 (2010) 31-65.

[49] H. Berestycki, L. Nirenberg, S.R.S. Varadhan, *The principal eigenvalue and maximum principle for second order elliptic operators in general domains*, Comm. Pure Appl. Math., 47 (1994) 47-92.

[50] A. Blanchet, M. Bonforte, J. Dolbeault, G. Grillo, J.L. Vasquez, *Hardy-Poincaré inequalities and applications to nonlinear diffusions*, C. R. Acad. Sci. Paris, Ser. I 344 (2007) 431-436.
[51] G. Bliss, *An Integral Inequality*, J. London Math. Soc., 5 (1930) 40-46.
[52] G. Blower, *The Gaussian isoperimetric inequality and the transportation*. Positivity 7 no. 3 (2003) 203-224.
[53] S. G. Bobkov, M. Ledoux, *From Brunn-Minkowski to Brascamp-Lieb and to Logarithmic Sobolev inequalities*. Geom. Funct. Anal. 10 (2000) 1028-1052.
[54] S. Bobkov, F. Götze, *Exponential integrability and transportation cost related to logarithmic Sobolev inequalities*. J. Funct. Anal. 163, 1 (1999) 1-28.
[55] T. Boggio, *Sulle funzioni di Green d'ordine m*, Rend. Circ. Mat. Palermo (1905) 97-135.
[56] M. Bonforte, J. Dolbeault, G. Grillo, and J.-L. Vazquez, *Sharp rates of decay of solutions to the nonlinear fast diffusion equation via functional inequalities*, Arxiv preprint 0907.2986 (2009).
[57] R. Bosi, J. Dolbeault, and M. J. Esteban, *Estimates for the optimal constants in multipolar Hardy inequalities for Schrödinger and Dirac operators*, Commun. Pure Appl. Anal. 7 no. 3 (2008) 533-562.
[58] H. Brascamp, H. Lieb, *On extensions of the Brunn-Minkowski and Prékopa-Leindler theorems, including inequalities for log concave functions, and with an application to the diffusion equation*, J. Functional Analysis, 22 (4) (1976) 366-389.
[59] Y. Brenier, *Polar factorization and monotone rearrangement of vector-valued functions*. Comm. Pure Appl. Math. 44, 4 (1991) 375-417.
[60] H. Brezis, E. H. Lieb, *Sobolev inequalities with remainder terms*, J. Funct. Anal. 62 (1985) 73-86.
[61] H. Brezis, M. Marcus, *Hardy's inequality revisited*, Ann. Scuola. Norm. Sup. Pisa 25 (1997) 217-237.
[62] H. Brezis, M. Marcus, I. Shafrir, *Extremal functions for Hardy's inequality with weight*, J. Funct. Anal. 171 (2000) 177-191.
[63] H. Brezis, L. Nirenberg, *Positive Solutions of Nonlinear Elliptic Equations involving Critical Exponents*, Communications on Pure and Applied Mathematics, Vol 34 (1983) 437-477.
[64] H. Brezis, L.A. Peletier, Asymptotics for elliptic equations involving critical Sobolev exponent. In *Partial Differential equations and the calculus of variations*, eds. F.Colombini, A.Marino, L.Modica, and S.Spagnolo, Birkhaüser, Basel, 1989.
[65] H. Brezis, J. L. Vázquez, *Blowup solutions of some nonlinear elliptic problems*, Revista Mat. Univ. Complutense Madrid 10 (1997) 443-469.
[66] J. Byeon and Z.-Q. Wang, *Symmetry breaking of extremal functions for the Caffarelli-Kohn-Nirenberg inequalities*, Commun. Contemp. Math., 4 (2002) 457–465.
[67] R. Brown, *On a conjecture of Dirichlet*, Amer. Math. Soc., Providence, RI, 1993.
[68] X. Cabré, *Regularity of minimizers of semilinear elliptic problems up to dimension 4*, Comm. Pure Appl. Math. 63 no. 10 (2010) 1362-1380.
[69] X. Cabré, *Extremal solutions and instantaneous complete blow-up for elliptic and parabolic problems*, Contemporary Math., Amer. Math. Soc. in: Perspectives in Nonlinear Partial Differential Equations: In honor of Haim Brezis (2007)
[70] X. Cabré, A. Capella, *On the stability of radial solutions of semilinear elliptic equations in all of $\mathbb{R}^n$*, C. R. Math. Acad. Sci. Paris, 338 no. 10 (2004) 769–774.
[71] X. Cabré, A. Capella, *Regularity of radial minimizers and extremal solutions of semilinear elliptic equations*, J. Funct. Anal. **238** no. 2 (2006) 709–733.
[72] X. Cabré, Y. Martel, *Weak eigenfunctions for the linearization of extremal elliptic problems*, J. Funct. Anal. 156 (1998) 30-56.
[73] X. Cabré, Y. Martel, *Existence versus explosion instantanée pour des équations de la chaleur linéaires avec potential singulier*, C. R. Acad. Sci. Paris Sér. I 329 (1999) 973-978.
[74] L. Caffarelli, *Allocation maps with general cost function*, in Partial Differential Equations and Applications (P. Marcellini, G. Talenti and E. Vesintin, eds), Lecture notes in Pure and Appl. Math., 177. Decker, New-York (1996) 29-35.
[75] L. Caffarelli, B. Gidas, J. Spruck, *Asymptotic symmetry and local behavior of semilinear elliptic equations with critical Sobolev growth*, Comm. Pure Appl. Math., 42 (1989) 271-297.
[76] L. Caffarelli, R. Kohn, L. Nirenberg, *First order interpolation inequalities with weights*, Compositio Mathematica 53 (1984) 259-275.
[77] P. Caldiroli, R. Musina, *Rellich inequalities with weights*, Preprint (2011).

[78] P. Caldiroli, R. Musina, *Caffarelli-Kohn-Nirenberg type inequalities for the weighted biharmonic operator in cones*, Milan J. Math. (2011)

[79] J. Carrillo, R. McCann, C. Villani, *Kinetic equilibration rates for granular media and related equations: entropy dissipation and mass transportation estimates*, Revista Matematica Iberoamericana, Volume 19, Number 3 (2003) 971-1018.

[80] J.A. Carillo, A. Jüngel, P.A. Markowich, G. Toscani, A. Unterreiter, *Entropy dissipation methods for degenerate parabolic problems and generalized Sobolev inequalities*, Monatsh. Math. 133, **1** (2001) 1-82.

[81] E. Carlen and M. Loss, *Logarithmic Sobolev inequalities and spectral gaps*, in Recent advances in the theory and applications of mass transport, vol. 353 of Contemp. Math., Amer. Math. Soc., Providence, RI (2004) 53-60.

[82] L. Carleson, A. Chang, *On the existence of an extremal function for an inequality of J. Moser*, Bull. Sc. Math. 110 (1986) 113-127.

[83] D. Cassani, J. M. Do O, N. Ghoussoub, *On a fourth order elliptic problem with a singular nonlinearity*, Advances Nonlinear Studies, **9** (2009) 177-197.

[84] T. Cazenave, M. Escobedo, M. Assunta Pozio, *Some stability properties for minimal solutions of* $-\Delta u = \lambda g(u)$. Port. Math. (N.S.) 59 no. 4 (2002) 373-391.

[85] F. Catrina, Z.-Q. Wang, *On the Caffarelli-Kohn-Nirenberg Inequalities: Sharp constants, Existence (and Nonexistence), and symmetry of Extremal Functions*, Comm. Pure & App. Math., 54 (2001) 229-258.

[86] S.-Y.A. Chang, P. Yang, *Conformal deformations of metrics on* S^2, J. Diff. Geom., 27 (1988) 215-259.

[87] S.-Y.A. Chang, P. Yang, *Prescribing Gaussian curvature on* S^2, Acta Math., 159 (1987) 214-259.

[88] N. Chaudhuri, *Bounds for the best constant in an improved Hardy-Sobolev inequality*, Zeitschrift für Analysis und ihre Anwendungen, Vol. 22, No. 4 (2003) 757-765

[89] W. X. Chen and C. Li, *Classification of solutions of some nonlinear elliptic equations*, Duke Math. J., 63 (1991) 615–622.

[90] K. S. Chou and T. Y.-H. Wan, *Asymptotic radial symmetry for solutions of* $\Delta u + e^u = 0$ *in a punctured disc*, Pacific J. Math., 163 (1994) 269–276.

[91] K. S. Chou and T. Y. H. Wan, *Correction to: "Asymptotic radial symmetry for solutions of* $\Delta u + e^u = 0$ *in a punctured disc" [Pacific J. Math.* **163** *(1994), no. 2, 269–276]*, Pacific J. Math., 171 (1995) 589–590.

[92] K. S. Chou, C. W. Chu, *On the Best Constant for a Weighted Sobolev-Hardy Inequality*, J. London Math. Soc. (2) 48 (1993) 137-151.

[93] A. Cianchi and A. Ferone, *Hardy inequalities with non-standard remainder terms*, Ann. Inst. H. Poincaré Anal. Non Linéaire, 25 (2008) 889–906.

[94] O. Costin, V. Maz'ya, *Sharp Hardy-Leray inequality for axisymmetric divergence-free fields*, Calculus of variations and PDE, Vol. 32, no. 4 (2008) 523-532.

[95] C. Cowan, *Optimal Hardy inequalities for general elliptic operators with improvements*, Comm. Pure Appl. Anal. 9 no. 1 (2010) 109-140.

[96] C. Cowan, M. Fazly, *On stable entire solutions of semi-linear elliptic equations with weights*, Proc. Amer. Math. Soc. 140 (2012) 2003-2012.

[97] C. Cowan, M. Fazly, *Regularity of the extremal solutions associated to some elliptic systems*, Preprint (June 2012), 17 pp.

[98] C. Cowan, P. Esposito, N. Ghoussoub, *Regularity of extremal solutions in fourth order eigenvalue problems on general domains*, Discrete and Continuous Dynamical Systems - Series A, Vol 28, 3 (2010) 1033-1050

[99] C. Cowan, P. Esposito, N. Ghoussoub, A. Moradifam, *The critical dimension for a fourth order elliptic problem with singular nonlinearity*, Arch. Ration. Mech. Anal., 198 (2010) 763-787.

[100] C. Cowan, N. Ghoussoub, *Regularity of the extremal solution in a model for MEMS involving advection*, Methods and Applications of Analysis, Vol. 15, No 3 (2008) 355-362.

[101] C. Cowan, N. Ghoussoub, *Estimates on pull-in distances in MEMS models and other nonlinear eigenvalue problems*, SIAM Journal on Math. Analysis, Vol. 42, No. 5 (2010) 1949-1966.

[102] C. Cowan, N. Ghoussoub, *Regularity of semi-stable solutions to fourth order nonlinear eigenvalue problems on general domains*, Calculus of Variations and Partial Differential Equations (2012)

[103] D. Cordero-Erausquin, B. Nazaret, C. Villani, *A mass-transportation approach to sharp Sobolev and Gagliardo-Nirenberg inequalities*, Adv. Math. 182, 2 (2004) 307-332.

[104] D. Cordero-Erausquin, *Some applications of mass transport to Gaussian-type inequalities.* Arch. Rational Mech. Anal., 161 (2002) 257-269.

[105] D. Cordero-Erausquin, W. Gangbo, C. Houdré, *Inequalities for generalized entropy and optimal transportation.* Proceedings of the Workshop: Mass transportation Methods in Kinetic Theory and Hydrodynamics (2003).

[106] M. G. Crandall, P. H. Rabinowitz, *Some continuation and variational methods for positive solutions of nonlinear elliptic eigenvalue problems*, Arch. Ration. Mech. Anal., 58 (1975) 207-218.

[107] C. Cacazu, *Hardy Inequalities with Boundary Singularities*, arXiv:1009.0931v1, (2010).

[108] C. Cacazu, E. Zuazua, *Improved multipolar Hardy inequalities*, preprint (2012).

[109] E. B. Davies, *A review of Hardy inequalities*, Oper. Theory Adv. Appl. 110 (1999) 55-67.

[110] E. B. Davies, A.M. Hinz, *Explicit constants for Rellich inequalities in $L_p(\Omega)$*, Math Z. 227 (1998) 511-523.

[111] L. D'Ambrosio, *Hardy-type inequalities related to degenerate elliptic differential operators*, Ann. Sc. Norm. Super. Pisa Cl. Sci. (5) 4 no. 3 (2005) 451-486.

[112] J. Davila, L. Dupaigne, I. Guerra, M. Montenegro, *Stable solutions for the bilaplacian with exponential nonlinearity*, SIAM J. Math. Anal. 39 (2007) 565-592.

[113] M. Del Pino, J. Dolbeault, *The optimal Euclidean L^p-Sobolev logarithmic inequality*, 197 no. 1 (2003) 151-161.

[114] M. Del Pino, J. Dolbeault. S. Filippas, and A. Tertikas, *A logarithmic Hardy inequality*, Journal of Functional Analysis, Vol. 259, Issue 8 (2010) 2045-2072

[115] R. A. DeVore, *Approximation of functions*, Proc. Sympos. Appl. Math. vol. 36, Amer. Math. Soc., Providence, RI (1986) 34-56.

[116] E. DiBenedetto. C^1 *local regularity of weak solutions of degenerate elliptic equations*, Nonlin. Analysis 7 (1983) 827-850.

[117] J. Dolbeault, M. Esteban, G. Tarantello, and A. Tertikas, *Radial symmetry and symmetry breaking for some interpolation inequalities*, Calculus of Variations and Partial Differential Equations, 42 (2011) 461-485.

[118] J. Dolbeault, M. J. Esteban, M. Loss, and G. Tarantello, *On the symmetry of extremals for the Caffarelli-Kohn-Nirenberg inequalities*, Advanced Nonlinear Studies, 9 (2009) 713-727.

[119] J. Dolbeault, M. J. Esteban, and G. Tarantello, *The role of Onofri type inequalities in the symmetry properties of extremals for Caffarelli-Kohn-Nirenberg inequalities, in two space dimensions*, Ann. Sc. Norm. Super. Pisa Cl. Sci. (5), 7 (2008) 313–341.

[120] J. Dolbeault, B. Nazaret, and G. Savaré, *A new class of transport distances between measures*, Calc. Var. Partial Differential Equations, 34 (2009) 193–231.

[121] J.W. Dold, V.A. Galaktionov, A.A. Lacey, J.L. Vazquez, *Rate of approach to a singular steady state in quasilinear reaction-diffusion equations*, Ann. Sc. Norm. Super Pisa Cl. Sci. 26 (1998) 663-687.

[122] O. Druet, *The best constants problem in Sobolev inequalities.* Math. Ann., 314 (1999) 327-346.

[123] O. Druet, *Elliptic equations with critical Sobolev exponent in dimension 3*, Ann. I.H.P., Analyse non-linaire, 19, 2 (2002) 125-142.

[124] O. Druet, *From one bubble to several bubbles: the low-dimensional case*, Journal of Differential Geometry, 63 (2003) 399-473.

[125] O. Druet, E. Hebey, F. Robert, *Blow up theory for elliptic PDE's in Riemannian geometry*, Mathematical Notes, Princeton University Press, 45 (2003).

[126] O. Druet, F. Robert, *Asymptotic profile for the sub-extremals of the sharp Sobolev inequality on the sphere*, Communications in Partial Differential Equations, 26 (2001) 743-778.

[127] H. Egnell, *Positive solutions of semilinear equations in cones*, Tran. Amer. Math. Soc., 11 (1992) 191-201.

[128] I. Ekeland, N. Ghoussoub, *Selected new aspects of the calculus of variations in the large*, Bull. Amer. Math. Soc., 39 no. 2 (2002) 207-265.

[129] P. Esposito, N. Ghoussoub, *Uniqueness of solutions for an elliptic equation modeling MEMS*, Methods and Applications of Analysis, Vol. 15, No 3 (2008) 341-354.

[130] P. Esposito, N. Ghoussoub, Y. Guo, *Compactness along the branch of semi-stable and unstable solutions for an elliptic problem with a singular nonlinearity*, Comm. Pure Appl. Math. 60 (2007) 1731-1768.

[131] L. C. Evans, *Partial differential equations and Monge-Kantorovich mass transfer*, Proc. Sympos. Appl. Math. vol. 36, Amer. Math. Soc., Providence, RI, 1986 34-56.

[132] L. Evans, *Partial differential equations*, Graduate Studies in Math. **19** AMS, Providence, RI, 1998.

[133] M. M. Fall, R. Musina, *Hardy-Poincaré inequalities with boundary singularities*, Proc. Roy. Soc. Edinburgh A 142 (2012), 1-18.

[134] M. M. Fall, *On the Hardy-Poincaré inequality with boundary singularities*, Commun. Contemp. Math., Vol. 14, No. 3 (2012) 125009. DOI No: 10.1142/S0219199712500198.

[135] M. Fazly, *Private communication*, 2011.

[136] M. Fazly, N. Ghoussoub, *De Giorgi type results for elliptic systems*, Calc. Var. Partial Differential Equations, published online June 22 (2012) 15 pp.

[137] M. Fazly, N. Ghoussoub, *On the Hénon-Lane-Emden conjecture*, Discrete and Continuous Dynamical Systems-A, special issue on "Recent development in nonlinear partial differential equations" (2013) 23 pp.

[138] V. Felli, S. Terracini, *Elliptic equations with multi-singular inverse-square potentials and critical nonlinearity*, Comm. Partial Differential Equations, 31 (2006) 469-495.

[139] V. Felli and M. Schneider, *Perturbation results of critical elliptic equations of Caffarelli-Kohn-Nirenberg type*, J. Differential Equations, 191 (2003) 121–142.

[140] J. Feldman, R. Froese, N. Ghoussoub, C. Gui, *An improved Moser-Aubin-Onofri inequality for axially symmetric functions on S^2*, Calc. Var. Partial Differential Equations 6 no. 2 (1998) 95-104.

[141] S. Filippas, V. Mazỳa, and A. Tertikas, *On a question of Brezis and Marcus*, Calc. Var. Partial Differential Equations, 25 (2006) 491-501.

[142] S. Filippas, A. Tertikas, *Optimizing improved Hardy inequalities*, J. Funct. Anal. 192 no. 1 (2002) 186-233.

[143] ———, *Corrigendum to:* [**142**], J. Funct. Anal., 255 (2008), p. 2095.

[144] J. Fleckinger, E. M. Harrell II, F. Thelin, *Boundary behaviour and estimates for solutions of equations containing the p-Laplacian*, Electron. J. Differential Equations 38 (1999) 1-19.

[145] V. Felli, M. Schneider, *Perturbation results of critical elliptic equations of Caffarelli-Kohn-Nirenberg type*, J. Differential Equations, 191 (2003) 121-142.

[146] M. Flucher, *Extremal functions for Trudinger-Moser inequality in 2 dimensions*, Comment. Math. Helv. 67 (1992) 471-497.

[147] S. Gallot, D. Hulin, J. Lafontaine, *Riemannian geometry*, Springer-Verlag, 1987.

[148] W. Gangbo, *An elementary proof of the polar factorization of vector-valued functions*, Arch. Rational Mech. Anal. 128 no. 4 (1994) 381-399.

[149] W. Gangbo, R. McCann, *The geometry of optimal transportation*, Acta Math. 177 2 (1996) 113-161.

[150] J. P. Garcia Azorero, I. Peral Alonso, *Hardy Inequalities and Some Critical Elliptic and Parabolic Problems*, J. of Differential Equations 144 (1998) 441-476.

[151] F. Gazzola, H.-C. Grunau, E. Mitidieri, *Hardy inequalities with optimal constants and remainder terms*, Trans. Amer. Math. Soc. **356** (2003) 2149-2168.

[152] F. Gazzola, H.-Ch. Grunau, *Critical dimensions and and higher order Sobolev inequalities with reminder terms*, NoDEA **8** (2001) 35-44.

[153] I. Gentil, *The general optimal L^p-Euclidean logarithmic Sobolev inequality by Hamilton-Jacobi equations*, Journal of Functional Analysis 202 (2003) 591-599.

[154] A. Ghigi, *On the Moser-Onofri and Prékopa-Leindler inequalities*, Collect. Math. 56 (2005) 143-156.

[155] N. Ghoussoub, C. Yuan, *Multiple solutions for quasi-linear PDEs involving the critical Sobolev and Hardy exponents*, Trans. Amer. Math. Soc., 12 (2000) 5703-5743.

[156] N. Ghoussoub, *Duality and Perturbation Methods in Critical Point Theory*, Cambridge Tracts of math. Cambridge University Press (1993) 265 pp.

[157] N. Ghoussoub, X.S. Kang, *Hardy-Sobolev Critical Elliptic Equations with Boundary Singularities*, AIHP-Analyse non linéaire, Vol. 21 (2004) 767-793.

[158] N. Ghoussoub, F. Robert, *The effect of curvature on the best constant in the Hardy-Sobolev inequalities*, Geom. And Funct. Anal. Vol. 16, 6 (2006) 1201-1245.

[159] N. Ghoussoub, F. Robert, *Concentration estimates for Emden-Fowler Equations with boundary singularities and critical growth*, International Mathematics Research Papers (2006) 1-86.

[160] N. Ghoussoub, F. Robert, *Elliptic equations with critical growth and large sets of boundary singularities*, Trans. Amer. Math. Soc. 361 (2009) 4843-4870.

[161] N. Ghoussoub, Y. Guo, *On the partial differential equations of electrostatic MEMS devices: stationary case*, SIAM J. Math. Anal., 38 no. 5 (2006/2007) 1423-1449.

[162] N. Ghoussoub, A. Moradifam, *On the best possible remaining term in the improved Hardy inequality*, Proc. Nat. Acad. Sci., vol. 105, no. 37 (2008) 13746-13751.

[163] N. Ghoussoub, A. Moradifam, *Bessel potentials and improved Hardy-Rellich inequalites*, Math. Annalen 349 (2011) 1-57.

[164] N. Ghoussoub, A. Moradifam, *Remarks on improved Hardy inequalities with boundary singularity*, unpublished note (2010).

[165] N. Ghoussoub, C.S. Lin, *On the best constant in the Moser-Onofri-Aubin inequality on $\mathbb{S}2$*, Communications in Mathematical Physics, Vol. 298, No. 3 (2010) 869-878.

[166] M. Giaquinta, G. Modica, J. Soucek, *Cartesian Currents in the Calculus of Variations, I, Cartesian Currents 37*, Springer-Verlag, Berlin, 1998.

[167] B. Gidas, W.M. Ni, L. Nirenberg, *Symmetry and related properties via the maximum principle*, Comm. Math. Phys., 68 (1979) 209-243.

[168] B. Gidas, W.M. Ni, L. Nirenberg, *Symmetry of positive solutions of nonlinear elliptic equations in $\mathbb{R}^n$*, Mathematical Analysis and Applications, Vol. 7a, Advances in Mathematics. Supplementary Studies, Academic Press, New York, 1981.

[169] G. Gilbarg, N.S. Trudinger, *Elliptic partial differential equations of second order. Second edition*, Grundlehren der mathematischen Wissenschaften, 224, Springer, Berlin, 1983.

[170] M. Guedda, L. Veron, *Local and Global Properties of Solutions of Quasi-linear Elliptic Equations*, Journal of Differential Equations 76 (1988) 159-189.

[171] M. Guedda, L. Veron, *Quasi-linear Elliptic Equations Involving Critical Sobolev Exponents*, Nonlinear Anal TMA, 13 (8) (1989) 879-902.

[172] C. Gui, J. Wei, *On a sharp Moser-Aubin-Onofri inequality for functions on $\mathbb{S}^2$ with symmetry*, Pacific J. Math. 194 no. 2 (2000) 349-358.

[173] Y. Guo, Z. Pan, M.J. Ward, *Touchdown and pull-in voltage behavior of a mems device with varying dielectric properties*, SIAM J. Appl. Math 66 (2005) 309-338.

[174] Z. Guo, J. Wei, *On a fourth order nonlinear elliptic equation with negative exponent*, SIAM J. Math. Anal. **40** no. 5 (2008/09) 2034-2054.

[175] L. Gross, *Logarithmic Sobolev Inequalities*, Amer. J. Math., Vol 97, No. 4 (1975) 1061-1083.

[176] Z. C. Han, *Asymptotic approach to singular solutions for nonlinear elliptic equations involving critical Sobolev exponent*, Ann. Inst. H. Poincaré. Anal. Non Linéaire, 8 (1991) 159-174.

[177] G. H. Hardy, J. E. Littlewood, G. Polya, *Inequalities*, Cambridge University Press, 1952.

[178] P. Hartman, *Ordinary differential equations*, Wiley, New York, 1964.

[179] I. W. Herbst, *Spectral theory of the operator $(p^2 + m^2)^{1/2} - Ze^2/r$*, Comm. Math. Phys. 53 no. 3 (1977) 285-294.

[180] E. Hille, *Non-oscillation theorems*, Tran. Amer. Math. Soc. 64 (1948) 234-252.

[181] C. Huang, *Oscillation and Nonoscillation for second order linear differential equations*, J. Math. Anal. Appl. 210 (1997) 712-723.

[182] E. Hebey, *Sobolev spaces on Riemannian manifolds*, Lecture notes in mathematics 1635, Springer, 1996.

[183] E. Hebey, *Asymptotics for some quasilinear elliptic equations*, Differential Integral Equations, 9 no. 1 (1996) 71-88.

[184] E. Hebey, M. Vaugon, *The best constant problem in the Sobolev embedding theorem for complete Riemannian manifolds*, Duke Math. J., 79 (1995) 235-279.

[185] E. Hebey, M. Vaugon, *From best constants to critical functions*, Math. Z., 237 (1996) 737-767.

[186] M. Hesaaraki, A. Moradifam, *Oscillation criteria for second-order nonlinear self-adjoint differential equations of Euler type*, Methods Appl. Anal., 13 (2006) 373-385.

[187] C. Hong, *A best constant and the Gaussian curvature*, Proc. Amer. Math. Soc., 97 (1986) 737-747.

[188] T. Horiuchi, *Best constant in weighted Sobolev inequality with weights being powers of distance from the origin*, J. Inequal. Appl., 1 (1997) 275-292.

[189] J.-S. Hyun, S. D. Kim, *Hardy's inequality related to a Bernoulli equation*, Bull. Korean Math. Soc. 39 No. 1 (2002) 81-87.

[190] C.G.J. Jacobi, *Ueber die Reduction der Integration der partiellen Differential- gleichungen erster Ordnung zwischen irgend einer Zahl Variabeln auf die Integration eines einzigen Systems gew'ohnlicher Differential gleichungen*, Crelle, t.xxvii. pp. 97-162, and translated into French, Liouville, t. iii. pp. 60-96 and 161-201 (1837).

[191] J. Jang, N. Masmoudi, *Well-posedness of compressible Euler equations in a physical vacuum*, Preprint. http://arxiv.org/abs/1005.4441, 35 pages (May 24, 2010)

[192] R. Jordan, D. Kinderlehrer, F. Otto. *The variational formulation of the Fokker-Planck equation*, SIAM J. Math. Anal., Vol 29, No. 1 (1998) 1-17.

[193] B. Kawohl, *Rearrangements and Convexity of Level Sets in PDE*, Lecture Notes in Mathematics, Vol. 1150, Springer, Berlin, 1985.

[194] B. Klartag, *Marginals of geometric inequalities*, in Geometric aspects of functional analysis, Lecture Notes in Math. 1910, Springer, Berlin (2007) 133-166.

[195] A. Kufner, L-E Persson, *Weighted Inequalities of Hardy Type*, World Scientific, New Jersey-London-Singapore-Hong Kong (2003).

[196] A. Kufner, L. Maligranda, and L-E Persson, *The prehistory of the Hardy inequality*, American Mathematical Monthly 113(8) (2006) 715-732.

[197] I. Kombe, M. Ozaydin, *Improved Hardy and Rellich inequalities on Riemannian manifolds*, Tran. Amer. Math. Soc. 361 (2009) 6191-6203.

[198] B. Opic, A. Kufner, *Hardy Type Inequalities*, Pitman Research Notes in Mathematics, Vol. 219, Longman, New York, 1990.

[199] V. G. Maz'ya, *Sobolev Spaces*, Springer-Verlag, Berlin, 1985.

[200] M. Ledoux, *The concentration of measure phenomenon*, Mathematical Surveys and Monographs 89, American Mathematical Society (2001).

[201] J. Leray, *Sur le mouvement visqueux emplissant l'espace*, Acta Math. 63 (1934) 193–248.

[202] E. H. Lieb: *Sharp Constants in the Hardy-Littlewood-Sobolev and Related Inequalities*, Ann. of Math. 118 (1983) 349-374.

[203] E. H. Lieb, M. Loss: *Analysis*, Graduate Studies in Mathematics Vol. 14 (1997).

[204] C.S Lin, Interpolation inequalities with weights, *Comm. Part. Diff. Eq.* **11** (1986) 1515-1538.

[205] C.S Lin, L. Marcello, *One-dimensional symmetry of periodic minimizers for a mean field equation*, Ann. Sc. Norm. Super. Pisa Cl. Sci. (5) 6 no. 2 (2007) 269-290.

[206] C.S. Lin, *Uniqueness of solutions to the mean field equations for the spherical Onsager vortex*, Arch. Ration. Mech. Anal. 153 no.2 (2000) 153-176.

[207] C.S. Lin, M. Lucia, *One-dimensional symmetry of periodic minimizers for a mean field equation*, Ann. Sc. Norm. Super. Pisa Cl. Sci. (5) 6 no. 2 (2007) 269-290.

[208] *Symmetry of extremal functions for the Caffarelli-Kohn-Nirenberg inequalities*, Proc. Amer. Math. Soc., 132 (2004) 1685-1691 (electronic).

[209] ———, *Erratum to: "Symmetry of extremal functions for the Caffarelli-Kohn-Nirenberg inequalities" [Proc. Amer. Math. Soc. **132** (2004), no. 6, 1685-1691]*, Proc. Amer. Math. Soc., 132 (2004) 2183 (electronic).

[210] C-H Hsia, C. S. Lin, H. Wadade, *Revisiting an idea of Brézis and Nirenberg*, Journal of Functional Analysis, Vol. 259, Issue 7, 1 (2010) 1816-1849.

[211] V. Liskevich, S. Lyakhova, V. Moroz, *Positive solutions to nonlinear p-Laplace equations with Hardy potential in exterior domains*, Journal of Differential Equations, 232 (2007) 212-252.

[212] P. Li, J. Wang, *Weighted Poincare inequality and rigidity of complete manifolds*, Ann. Scuole Norm. Sup. (4) 39 no. 6 (2006) 921-982.

[213] F.H. Lin, Y.S. Yang, *Nonlinear non-local elliptic equation modelling electrostatic acutation*, Proc. R. Soc. London Ser. A 463 (2007) 1323-1337.

[214] M. Marcus, V.J. Mizel, Y. Pinchover, *On the best constant for Hardy's inequality in $\mathbb{R}^n$*, Transactions of the American Mathematical Society, Vol. 350, No. 8 (1998) 3237-3255.

[215] N. Masmoudi, *About the Hardy Inequality*, An Invitation to Mathematics: From Competitions to Research, edited by Dierk Schleicher, Malte Lackmann, Springer (2011) 165-180.

[216] R. McCann, *A convexity theory for interacting gases and equilibrium crystals*, Ph.D thesis, Princeton Univ., 1994.

[217] R. McCann, *A convexity principle for interacting gases*, Adv. Math 128, 1 (1997) 153-179.

[218] J. B. McLeod, L. A. Peletier, *Observations on Moser's inequality*, Arch. Ration. Mech. Anal., **106** 3 (1989) 261-285.

[219] F. Mignot, J. P. Puel, *Sur une classe de problèmes non linéaires avec non linéarité positive, croissante, convexe*, Comm. Partial Differential Equations, 5 no. 8 (1980) 791-836.

[220] V. D. Milman and G. Schechtman, *Asymptotic theory of finite-dimensional normed spaces*, Lecture Notes in Mathematics, vol. 1200, Springer-Verlag, Berlin, 1986, With an appendix by M. Gromov.

[221] A. Moradifam, *On the critical dimension of a fourth order elliptic problem with negative exponent*, Journal of Differential Equations, 248 (2010) 594-616.

[222] A. Moradifam, *The singular extremal solutions of the bilaplacian with exponential nonlinearity*, Proc. Amer. Math Soc. 138 (2010) 1287-1293.

[223] A. Moradifam, *Optimal weighted Hardy-Rellich inequalities on $H^2 \cap H_0^1$*, J. London. Math. Soc., 85(1): 22-40 (2012).

[224] J. Moser, *A sharp form of an inequality by N. Trudinger*, Indiana U. Math. J. 20 (1971) 1077-1091.

[225] B. Muckenhoupt, *Hardy's inequality with weights*, Studia Math., 44 (1972) 31–38. Collection of articles honoring the completion by Antoni Zygmund of 50 years of scientific activity, I.

[226] B. Nazaret, *Best constant in Sobolev trace inequalities on the half-space*, Nonlinear Anal. 65, no. 10, (2006) 1977–1985.

[227] G. Nedev, *Regularity of the extremal solution of semilinear elliptic equations*, C. R. Acad. Sci. Paris Série. I Math., 330 (2000) 997-1002.

[228] Z. Nehari, *On the principal frequency of a membrane*, Pacific J. Math. Volume 8, Number 2 (1958) 285-293.

[229] E. Onofri, *On the positivity of the effective action in a theorem on random surfaces*, Comm. Math. Phy. 86, (1982) 321-326.

[230] B. Osgood, R. Phillips, P. Sarnak, *Extremals of determinants of Laplacians*, J. Func. Anal. 80 (1988) 148-211.

[231] B. Opic, A. Kufner, *Hardy type Inequalities*, Pitman Research Notes in Mathematics, Vol. 219, Longman, New York, 1990.

[232] F. Otto, *The geometry of dissipative evolution equation: the porous medium equation*, Comm. Partial Differential Equations, 26 (2001) 101-174.

[233] F. Otto, *Doubly degenerate diffusion equations as steepest descent*, Preprint. Univ. Bonn (1996).

[234] F. Otto, C. Villani, *Generalization of an inequality by Talagrand, and links with the logarithmic Sobolev inequality*, J. Funct. Anal. 173, 2 (2000) 361-400.

[235] J.A. Pelesko, *Mathematical modeling of electrostatic mems with tailored dielectric properties*, SIAM J. Appl. Math. 62 (2002) 888-908.

[236] J.A. Pelesko, A.A. Bernstein, *Modeling MEMS and NEMS*, Chapman Hall and CRC Press, 2002.

[237] I. Peral, *Multiplicity of Solutions for the p-Laplacian*, Lecture Notes at the Second School on Nonlinear Functional Analysis and Applications to Differential Equations at ICTP of Trieste (1997).

[238] I. Peral, J. L. Vázquez, *On the stability and instability of the semilinear heat equation with exponential reaction term*, Arch. Rat. Mech. Anal. 129 (1995) 201-224.

[239] J. H. Petersson, *Best constants for Gagliardo-Nirenberg inequalities on the real line*, Nonlinear Anal., 67 (2007) 587–600.

[240] G. Pisier, *The volume of convex bodies and Banach space geometry*, Cambridge Tracts in Mathematics, 94. Cambridge University Press, Cambridge (1989).

[241] P. Pucci, J. Serrin, *A general variational identity*, Indiana Univ. Math. J. 35, no. 3 (1986) 681-703.

[242] F. Rellich, *Halbbeschrenkte Differentialoperatoren heherer Ordnung*, in: J.C.H. Gerneretsen, et al. (Eds.), Proc. of the International Congress of Math., North-Holland, Amsterdam, 1954, 243-250.

[243] F. Robert, *Asymptotic behaviour of a nonlinear elliptic equation with critical Sobolev exponent, The radial case*, Advances in Differential Equations, 6 (2001) 821-846.

[244] F. Robert, *Critical functions and optimal Sobolev inequalities*, Math. Z., 249 (2005) 485-492.

[245] O. Rothaus, *Diffusion on compact Riemannian manifolds and logarithmic Sobolev inequalities*, J. Funct. Anal. 42 (1981) 102-109.

[246] J. Serrin, *A symmetry problem in potential theory*, Arch. Rational Mech. Anal. 43 (1971) 304-318.

[247] R. Schaaf, *Uniqueness for semilinear elliptic problems: supercritical growth and domain geometry*, Adv. Differential Equations 5, 10-12 (2000) 1201-1220.

[248] R. Schoen, *Variational theory for the total scalar curvature functional for Riemannian metrics and related topics*, Topics in calculus of variations (Montecatini Terme, 1987), Lecture Notes in Mathematics, 1365, Springer, Berlin (1989) 120-154.

[249] R. Seiringer, *Inequalities for Schrödinger Operators and Applications to the Stability of Matter Problem*, Lectures given at "Entropy and the Quantum", a school on analytic and functional inequalities with applications, Tucson, Arizona (2009).

[250] Y.-T. Shen, Z.-H. Chen, *Nonlinear degenerate elliptic equation with Hardy potential and critical parameter*, Nonlinear Analysis 69 (2008) 1462-1477.

[251] B. Simon, *Schrödinger semigroups*, Bull. Amer. Math. Soc. 7 (1982) 447-526.

[252] D. Smets, M. Willem, *Partial symmetry and asymptotic behavior for some elliptic variational problems*, Calc. Var. Partial Differential Equations, 18 (2003) 57-75.

[253] M. Struwe, *Variational methods*, Springer-Verlag, Berlin-Heidelberg-New York, 1990.

[254] J. Sugie, K. Kita, N. Yamaoka, *Oscillation constant of second-order non-linear self-adjoint differential equations*, Ann. Mat. Pura Appl. (4) 181 (2002) 309-337.

[255] T. Suzuki, *Global analysis for a two-dimensional elliptic eigenvalues problem with the exponential nonlinearity*, Ann. Inst. H. Poincaré-Analyse Non Linéaire, 9 (1992) 367-397.

[256] V. Sverak, *On Landau's solutions of the Navier-Stokes equations*, J. Mathematical Sciences, Volume 179, Number 1 (2011) 208-228.

[257] M. Talagrand, *Transportation cost for Gaussian and other product measures*, Geom. Funct. Anal. 6, 3 (1996) 587-600.

[258] G. Talenti, *Best Constant in Sobolev Inequalities*, Ann. Math. Pura Appl. 110 (1976).

[259] G. Tarantello, *Selfdual gauge field vortices: an analytical approach*, PNLDE 72, Birkhäuser ed. Boston MA, USA, 2007.

[260] A. Tertikas, *Critical phenomena in linear elliptic problems*, J. Funct. Anal., 154 (1998) 42-66.

[261] A. Tertikas and K. Tintarev, *On existence of minimizers for the Hardy-Sobolev-Maz'ya inequality*, Ann. Mat. Pura Appl. (2007).

[262] A. Tertikas, N.B. Zographopoulos, *Best constants in the Hardy-Rellich inequalities and related improvements*, Advances in Mathematics, 209 (2007) 407-459.

[263] P. Tolksdorf, *Regularity for a more general class of quasilinear elliptic equations*, J. Diff. Eqns, 51 (1984) 473-484.

[264] G. Toscani, *Sur l'inégalité logarithmique de Sobolev*, C. R. Acad. Sci. Paris Sér. I Math., 324 (1997) 689–694.

[265] N. S. Trudinger, *On embeddings into Orlicz spaces and some applications*, J. Math. Mech. 17 (1967) 473-483.

[266] J. L. Vázquez, *Domain of existence and blowup for the exponential reaction diffusion equation*, Indiana Univ. Math. J. 48 (1999) 677-709.

[267] J. L. Vázquez, E. Zuazua, *The Hardy inequality and the asymptotic behaviour of the heat equation with an inverse-square potential*, J. Funct. Anal. 173 (2000) 103-153.

[268] C. Villani, *Topics in optimal mass transportation*. Graduate Studies in Math. 58, AMS, 2003.

[269] C. Villani, *Optimal transport, Old and new*, Vol. 338, Grundlehren der Mathematischen Wissenschaften [Fundamental Principles of Mathematical Sciences], Springer-Verlag, Berlin, 2009.

[270] Z-Q. Wang, M. Willem, *Caffarelli-Kohn-Nirenberg inequalities with remainder terms*, J. Funct. Anal. 203 (2003) 550-568.

[271] F. Wang, *Functional Inequalities, Markov Processes, and Spectral Theory*, Science Press, Beijing, 2004.

[272] J. Wei, *Asymptotic behavior of a nonlinear fourth order eigenvalue problem*, Comm. Partial Differential Equations, 21 (1996) 1451-1467.

[273] F. B. Weissler, *Logarithmic Sobolev inequalities for the heat-diffusion semigroup*, Trans. Amer. Math. Soc., 237 (1978) 255–269.

[274] A. Wintner, *On the nonexistence of conjugate points*, Amer. J. Math. 73 (1951) 368-380.

[275] A. Wintner, *On the comparision theorem of Knese-Hille*, Math. Scand. 5 (1957) 255-260.

[276] J. S. W. Wong, *Oscillation and nonoscillation of solutions of second order linear differential equations with integrable coefficients*, Trans. Amer. Math. Soc. 144 (1969) 197-215.

[277] D. Yafaev, *Sharp constants in the Hardy-Rellich inequalities*, J. Funct. Anal. 168 no. 1 (1999) 121-144.

Selected Published Titles in This Series

187 **Nassif Ghoussoub and Amir Moradifam,** Functional Inequalities: New Perpectives and New Applications, 2013
186 **Gregory Berkolaiko and Peter Kuchment,** Introduction to Quantum Graphs, 2013
185 **Patrick Iglesias-Zemmour,** Diffeology, 2012
184 **Frederick W. Gehring and Kari Hag,** The Ubiquitous Quasidisk, 2012
183 **Gershon Kresin and Vladimir Maz'ya,** Maximum Principles and Sharp Constants for Solutions of Elliptic and Parabolic Systems, 2012
182 **Neil A. Watson,** Introduction to Heat Potential Theory, 2012
181 **Graham J. Leuschke and Roger Wiegand,** Cohen-Macaulay Representations, 2012
180 **Martin W. Liebeck and Gary M. Seitz,** Unipotent and Nilpotent Classes in Simple Algebraic Groups and Lie Algebras, 2012
179 **Stephen D. Smith,** Subgroup Complexes, 2011
178 **Helmut Brass and Knut Petras,** Quadrature Theory, 2011
177 **Alexei Myasnikov, Vladimir Shpilrain, and Alexander Ushakov,** Non-commutative Cryptography and Complexity of Group-theoretic Problems, 2011
176 **Peter E. Kloeden and Martin Rasmussen,** Nonautonomous Dynamical Systems, 2011
175 **Warwick de Launey and Dane Flannery,** Algebraic Design Theory, 2011
174 **Lawrence S. Levy and J. Chris Robson,** Hereditary Noetherian Prime Rings and Idealizers, 2011
173 **Sariel Har-Peled,** Geometric Approximation Algorithms, 2011
172 **Michael Aschbacher, Richard Lyons, Stephen D. Smith, and Ronald Solomon,** The Classification of Finite Simple Groups, 2011
171 **Leonid Pastur and Mariya Shcherbina,** Eigenvalue Distribution of Large Random Matrices, 2011
170 **Kevin Costello,** Renormalization and Effective Field Theory, 2011
169 **Robert R. Bruner and J. P. C. Greenlees,** Connective Real K-Theory of Finite Groups, 2010
168 **Michiel Hazewinkel, Nadiya Gubareni, and V. V. Kirichenko,** Algebras, Rings and Modules, 2010
167 **Michael Gekhtman, Michael Shapiro, and Alek Vainshtein,** Cluster Algebras and Poisson Geometry, 2010
166 **Kyung Bai Lee and Frank Raymond,** Seifert Fiberings, 2010
165 **Fuensanta Andreu-Vaillo, José M. Mazón, Julio D. Rossi, and J. Julián Toledo-Melero,** Nonlocal Diffusion Problems, 2010
164 **Vladimir I. Bogachev,** Differentiable Measures and the Malliavin Calculus, 2010
163 **Bennett Chow, Sun-Chin Chu, David Glickenstein, Christine Guenther, James Isenberg, Tom Ivey, Dan Knopf, Peng Lu, Feng Luo, and Lei Ni,** The Ricci Flow: Techniques and Applications: Part III: Geometric-Analytic Aspects, 2010
162 **Vladimir Maz'ya and Jürgen Rossmann,** Elliptic Equations in Polyhedral Domains, 2010
161 **Kanishka Perera, Ravi P. Agarwal, and Donal O'Regan,** Morse Theoretic Aspects of p-Laplacian Type Operators, 2010
160 **Alexander S. Kechris,** Global Aspects of Ergodic Group Actions, 2010
159 **Matthew Baker and Robert Rumely,** Potential Theory and Dynamics on the Berkovich Projective Line, 2010
158 **D. R. Yafaev,** Mathematical Scattering Theory, 2010
157 **Xia Chen,** Random Walk Intersections, 2010
156 **Jaime Angulo Pava,** Nonlinear Dispersive Equations, 2009
155 **Yiannis N. Moschovakis,** Descriptive Set Theory, Second Edition, 2009
154 **Andreas Čap and Jan Slovák,** Parabolic Geometries I, 2009
153 **Habib Ammari, Hyeonbae Kang, and Hyundae Lee,** Layer Potential Techniques in Spectral Analysis, 2009

SELECTED PUBLISHED TITLES IN THIS SERIES

152 **János Pach and Micha Sharir,** Combinatorial Geometry and Its Algorithmic Applications, 2009
151 **Ernst Binz and Sonja Pods,** The Geometry of Heisenberg Groups, 2008
150 **Bangming Deng, Jie Du, Brian Parshall, and Jianpan Wang,** Finite Dimensional Algebras and Quantum Groups, 2008
149 **Gerald B. Folland,** Quantum Field Theory, 2008
148 **Patrick Dehornoy, Ivan Dynnikov, Dale Rolfsen, and Bert Wiest,** Ordering Braids, 2008
147 **David J. Benson and Stephen D. Smith,** Classifying Spaces of Sporadic Groups, 2008
146 **Murray Marshall,** Positive Polynomials and Sums of Squares, 2008
145 **Tuna Altınel, Alexandre V. Borovik, and Gregory Cherlin,** Simple Groups of Finite Morley Rank, 2008
144 **Bennett Chow, Sun-Chin Chu, David Glickenstein, Christine Guenther, James Isenberg, Tom Ivey, Dan Knopf, Peng Lu, Feng Luo, and Lei Ni,** The Ricci Flow: Techniques and Applications: Part II: Analytic Aspects, 2008
143 **Alexander Molev,** Yangians and Classical Lie Algebras, 2007
142 **Joseph A. Wolf,** Harmonic Analysis on Commutative Spaces, 2007
141 **Vladimir Maz'ya and Gunther Schmidt,** Approximate Approximations, 2007
140 **Elisabetta Barletta, Sorin Dragomir, and Krishan L. Duggal,** Foliations in Cauchy-Riemann Geometry, 2007
139 **Michael Tsfasman, Serge Vlăduţ, and Dmitry Nogin,** Algebraic Geometric Codes: Basic Notions, 2007
138 **Kehe Zhu,** Operator Theory in Function Spaces, Second Edition, 2007
137 **Mikhail G. Katz,** Systolic Geometry and Topology, 2007
136 **Jean-Michel Coron,** Control and Nonlinearity, 2007
135 **Bennett Chow, Sun-Chin Chu, David Glickenstein, Christine Guenther, James Isenberg, Tom Ivey, Dan Knopf, Peng Lu, Feng Luo, and Lei Ni,** The Ricci Flow: Techniques and Applications: Part I: Geometric Aspects, 2007
134 **Dana P. Williams,** Crossed Products of C^*-Algebras, 2007
133 **Andrew Knightly and Charles Li,** Traces of Hecke Operators, 2006
132 **J. P. May and J. Sigurdsson,** Parametrized Homotopy Theory, 2006
131 **Jin Feng and Thomas G. Kurtz,** Large Deviations for Stochastic Processes, 2006
130 **Qing Han and Jia-Xing Hong,** Isometric Embedding of Riemannian Manifolds in Euclidean Spaces, 2006
129 **William M. Singer,** Steenrod Squares in Spectral Sequences, 2006
128 **Athanassios S. Fokas, Alexander R. Its, Andrei A. Kapaev, and Victor Yu. Novokshenov,** Painlevé Transcendents, 2006
127 **Nikolai Chernov and Roberto Markarian,** Chaotic Billiards, 2006
126 **Sen-Zhong Huang,** Gradient Inequalities, 2006
125 **Joseph A. Cima, Alec L. Matheson, and William T. Ross,** The Cauchy Transform, 2006
124 **Ido Efrat,** Valuations, Orderings, and Milnor K-Theory, 2006
123 **Barbara Fantechi, Lothar Göttsche, Luc Illusie, Steven L. Kleiman, Nitin Nitsure, and Angelo Vistoli,** Fundamental Algebraic Geometry, 2005
122 **Antonio Giambruno and Mikhail Zaicev,** Polynomial Identities and Asymptotic Methods, 2005
121 **Anton Zettl,** Sturm-Liouville Theory, 2005

For a complete list of titles in this series, visit the
AMS Bookstore at **www.ams.org/bookstore/survseries/**.